EVOLUTIONARY DEVELOPMENTAL BIOLOGY

But no, I am mistaken; from the beginning of all things the Creator knew, that one day the inquisitive children of men would grope about after analogies and homologies, and that Christian naturalists would busy themselves with thinking out his Creative ideas; at any rate, in order to facilitate the discernment by the former that the opercular peduncle of the *Serpulae* is homologous with a branchial filament, He allowed it to make a *détour* in its development, and pass through the form of a barbate branchial filament.

Fritz Müller, *Facts and Arguments for Darwin* (1869), p. 114

I see no reason for doubting that the embryo in the earliest periods of development is as subject to the laws of natural selection as is the animal at any other period. Indeed, there appear to me grounds for the thinking that it is more so.

Francis Maitland Balfour (1874) A preliminary account of the development of the elasmobranch fishes. *Quart. J. Microsc. Sci.*, **14**, p. 343

TL

EVOLUTIONARY DEVELOPMENTAL BIOLOGY

Second edition

BRIAN K. HALL

Killam Professor of Biology
Dalhousie University
Halifax, Canada

KLUWER ACADEMIC PUBLISHERS
DORDRECHT / BOSTON / LONDON

A C.I.P. Catalogue record for this book is available from the Library of Congress.

ISBN 0-412-78590-0

Published by Kluwer Academic Publishers,
P.O. Box 17, 3300 AA Dordrecht, The Netherlands.

Sold and distributed in the North, Central and South America
by Kluwer Academic Publishers,
101 Philip Drive, Norwell, MA 02061, U.S.A.

In all other countries, sold and distributed
by Kluwer Academic Publishers,
P.O. Box 322, 3300 AH Dordrecht, The Netherlands.

02-1199-500 ts

Printed on acid-free paper

CONTENTS

PREFACE

Links between development and evolution have been sought for so long that in introducing the first edition of *Evolutionary Developmental Biology*, I discussed the difficulty of choosing a title relating development to evolution that had not already been used. In my final choice, I sought to reflect the major themes of the book, which are how developmental processes effect evolutionary change and how development itself has evolved. I also sought to emphasize how tightly interwoven the two fields are and to draw attention to a distinct, emerging sub-discipline within biology. For evolutionary developmental biology (EDB or 'evo-devo') is not merely a fusion of the fields of developmental and evolutionary biology, the grafting of a developmental perspective onto evolutionary biology, or the incorporation of an evolutionary perspective into developmental biology. EDB strives to forge a unification of genomic, developmental, organismal, population and natural selection approaches to evolutionary change. It draws from development, evolution, palaeontology, molecular and systematic biology, but has its own set of questions, approaches and methods.

Evolutionary developmental biology is now firmly established, as several lines of evidence attest. This edition is more than twice the size of the first; its 2200 references (more than double the number in the first edition), reflect the enormous amount of activity in the field. Symposia, conferences and TV programs have been devoted entirely to EDB. Advertisements for academic positions target evolutionary developmental biologists. There are books about subsections of EDB such as the origin of body plans (Raff, 1996; Arthur, 1997). The journal *Science* devoted a six-page special news report to EDB in its issue of 4 July, 1997 (*Science*, 277, 34–9) and the 1996 *McGraw-Hill Yearbook of Science and Technology* (the annual supplement to the *McGraw-Hill Encyclopaedia of Science and Technology*), has an entry on the topic (Hall, 1996a), while the Marine Biological Laboratory at Woods Hole, in cooperation with the North American Space Agency (NASA) has targeted its 1998 summer research fellowships proposals in the area of EDB. Clearly, EDB has arrived.

It is my hope that this volume will serve as the first comprehensive textbook in the area. To that end, I have attempted an overview of all that is embraced by this new sub-discipline of biology. Because the search for connections between development and evolution (or ontogeny and phylogeny, as the two fields are sometimes called) has been such a long one, I treat both historical and recent studies. Because EDB represents a hierarchical approach to biological organization, I discuss approaches ranging from the molecular/genetic to the population level and show how embryonic development relates to life-history evolution, adaptation, and interactions with the biotic and abiotic environment.

There are eight parts to the book. Chapter 25 in Part Eight provides an overview of the principles and processes, scope and prospects of evolutionary developmental biology. It is a summary of the book. The reader anxious for the take-home messages may want to read it first.

Part One places EDB in the historical context of the search for relationships between development and evolution by examining how the two terms and fields came to be related. A major thrust of this book and of evolutionary developmental biology is that development and evolution are neither mutually exclusive, nor under independent control. How development (proximate causation) impinges on evolution (ultimate causation) to effect evolutionary change and how development itself has evolved are therefore important topics. As much of the impetus for linking development and evolution is embedded in a search for ancestors of, and relationships between, groups of organisms, I discuss how animals are grouped into phyla and super-phyla and how the animals found in the Burgess Shale have influenced our thinking about metazoan origins and relationships.

Part Two chronicles the tensions between approaches to animal relationships and between the individuals who have sought those relationships. Of the key events, ideas and discoveries, an early milestone was the debate between Geoffroy and Cuvier over whether form determines function or function form. The search for ordered embryonic and adult structure and for equivalence of structures and features (homology) through types or Archetypes, and the establishment of an evolutionary embryology is explored, as is the concept of body plans or *Baupläne*. Part Two ends with a chapter on types of inheritance systems available to organisms and introduces epigenetics and epigenetic regulation of development, establishing a basic conceptual premise of EDB: epigenetics is an integrating component linking development to evolution.

In Part Three, I explore embryos in development. I discuss conserved embryonic and life-history stages, the inevitability (and drawbacks) of using model organisms in developmental studies, and introduce the concept that early developmental stages are as subject to evolutionary change as is any stage of ontogeny or life history. How germ cells segregate from body (somatic) cells in those (few) phyla in which such a separation occurs, is the topic of Chapter 9, which also introduces the establishment of body plans in insects and the conserved genetic signalling between arthropods and vertebrates. I then treat the early development of vertebrate embryos and describe how organ systems are established. Once formed, organ systems become integrated, coupled or canalized. These topics, along with a discussion of situations in which developmental stability breaks down (such as fluctuating asymmetry), complete Part Three.

In considering embryos in evolution in Part Four, I devote two chapters (13 and 14) to the origins of multicellularity and the origins of the metazoans. Included in the latter is a discussion of the vexing problem of how we determine whether organisms have become more complex during their evolution, and a discussion of conserved stages typical of phyla (phylotypic stages) and of the Animal Kingdom (zootypes). I then discuss the developmental mechanisms and genes involved in the origin of the chordates; the origin and diversification of vertebrates into jawless and jawed; major transitions within animal evolution (fins → limbs; reptiles → birds); and the evolution of such structures as turtle shells, feathers, pharyngeal jaws and insect wings. I round off Part Four with a chapter devoted to integrated change during evolution, as organisms adapt to intensification of function or adapt to a new function. Both gain and loss of structures are treated.

Part Five moves the discussion to relationships between embryos, evolution and environment by documenting that inductive interactions are not limited to embryonic development but can be extended to interactions among individuals, between species, and between organisms and their environment. The importance of identifying inter- and intraspecific causal links between embryonic inductive interactions, ecological adaptation, and evolutionary change is stressed. Indeed, such inductive interactions are key elements in any linking of development, evolution and ecology, or of development and life-history theory. In Chapter 19, I discuss how hidden genetic variability can be brought forth in phenotypic change as organisms respond to environmental signals, while Chapter 20 provides a model system for morphological change in development and evolution, and applies it to the development and evolution of mammalian bone and butterfly wing patterns.

A theme introduced in Chapter 8 – that early embryonic development is subject to evolutionary change – is expanded in Part Six with discussions of two topics: the dilemma for homology of the fact that development evolves; and that larvae can be highly modified or eliminated from the life cycle of organisms (amphibians, frogs, ascidians) for whom larvae are regarded as diagnostic. The importance of understanding how embryos measure time and place both in development and evolutionarily through heterochrony and heterotopy is the topic of Part Seven. An important aspect of the latter discussion is that heterochrony and heterotopy link development to ecology in the generation of adaptive changes. Part Eight brings the discussion full circle and sets out the principles and processes that underlie EDB.

I have highlighted key concepts and terms in bold throughout. Other parts of the book that contain relevant information are cross-referenced by section number (e.g., s. 1.1, or ss. 1.1, 9.1). Boxes are used for items of interest which may be at a tangent from the main argument. To avoid interrupting the flow of the text I have placed most references and some supporting statements in Endnotes which serve as an annotated bibliography through which access to the literature may be readily obtained. The text is extensively illustrated and referenced, and there is a detailed index. Finally, there is a list of the most common abbreviations; to some degree, this list may also be used as a glossary of the terms covered by those abbreviations.

I am very grateful for having had the benefit of comments on one or more chapters from Bob Carroll, Benedikt Hallgrimsson, Peter Holland, Chris Klingenberg, Tom Miyake, Gerd Müller, Kevin Padian and Jim Valentine. I am especially indebted to three individuals: Wendy Olson read and provided insightful comments on a draft of the entire manuscript. June Hall, my wife, edited drafts of the manuscript both carefully and exhaustively for style, content and comprehensibility and provided a warm, supportive environment. It would have been a very different book without her input. A large number of the figures were adapted from the literature or transformed from my rough drafts into their final forms by Tom Miyake. His artistic abilities and scientific acumen are doubly appreciated. Many thanks Wendy, June and Tom.

Much of the first draft was written in the first half of 1997 at the University of California, Berkeley, during a sabbatical leave. I thank Marvalee and David Wake for their kindness, hospitality and fellowship; the Miller Institute for Basic Research in Science for a Visiting Miller Research Professorship; the students, postdoctoral fellows and faculty who participated in a wide-ranging graduate seminar on 'evo-devo'; and the members of the Department of Integrative Biology and the Museum of Vertebrate Zoology for the warm welcome extended during our stay. My own research is currently supported chiefly by the Natural Sciences and Engineering Research Council (NSERC) of Canada and the Killam Trust of Dalhousie University.

Brian K. Hall
Halifax

PART ONE

EVOLUTION AND DEVELOPMENT, PHYLA AND FOSSILS

1

EVOLUTION AND DEVELOPMENT: TERMS AND CONCEPTS

'Evolution. development (of organism, design, argument, etc.); Theory of E. (that the embryo is not created by fecundation, but developed from a pre-existing form); origination of species by development from earliest forms.'

Concise Oxford Dictionary, 5th edn., 1969

'The dead hand of the past still produces effects in the present through the conservatism of language.'

Woodger, 1945, p. 95

Given the changing definitions of evolution and development, and the long-standing and much debated relationships between the two terms and concepts, it is appropriate at the outset to define evolution and development, to prescribe the limits of the concepts embraced by the two terms, and to examine the history of the interactions and conflicts between these two fields of scientific, philosophical, metaphysical, ethical and often religious enquiry.

Evolution is descent with modification. Development is the production of an individual (s. 1.7).

Although past endeavours are regarded by many as irrelevant to modern biology, no apology is made for beginning with a historical overview. The topic at hand has a long and involved history, and changes in organisms over time are its very foundation. That said, one enters this arena with trepidation. Indeed, one of the best historians in the area commented recently that 'The history of evolutionary embryology . . . is a particularly difficult area and historians should exercise particular caution in dealing with it' (Ghiselin, 1996, p. 126).

A historical approach has a double meaning for the study of evolution and development: the history of the discipline(s) and the history of life:

'The role of historical contingency cannot be disregarded in biology, so all concepts based on ahistorical reasoning alone are fundamentally inappropriate. There has been but one history of life, and historical contingency is a primary factor in evolution.'

Wake and Roth, 1989, p. 373

We think of the attitude 'anything more than a few years old is irrelevant', as a product of the molecular biology of the 1980s and 1990s, but it is in reality a long-standing problem. Witness Russell's lament in the Preface to *Form and Function*, penned over 80 years ago:

'It is unfortunately true that modern biology, perhaps in consequence of the great advances it has made in certain directions, has to a considerable extent lost its historical consciousness, and if this book helps in any degree to counteract this tendency so far as animal morphology is concerned, it will have served its purpose.'

Russell, 1916, p. vi

The same hope is expressed for the present book so far as evolutionary developmental biology is concerned.

1.1 EVOLUTIONARY AND FUNCTIONAL CAUSATION: GERM PLASM AND SOMA

One major difficulty in looking for relationships between development and evolution is that we are dealing with what have been regarded as two self-contained biologies:

1. functional/physiological (proximate or internal) causation, which deals with all aspects of the activation and operation of genetic programmes within an individual lifetime; and
2. evolutionary (ultimate or external) causation, viewed traditionally as dealing with how genetic programmes are altered through evolutionary time so that new variants arise.[1]

The notion of internal versus external causes of evolutionary transformations has a long history. Richard Owen, in discussing how 'one form or grade of animal structure may be changed into another', saw the possible causes as 'external and impressive, or internal and genetic' (1861, p. 204). The dichotomy of internal differentiation and external factors guided such cell biologists as Charles Whitman and Edwin Conklin in their studies on cell lineages in early embryonic development. Such dichotomous ways of thinking view development and evolution as self-contained and mutually exclusive. A major thrust of this book and of evolutionary developmental biology is that development and evolution are neither mutually exclusive nor independently controlled.[2]

Understanding the relative roles of internal and external factors lies at the heart of any attempt to reunite development with evolution and to integrate how changes in individuals relate to changes in populations. As Stephen Stearns succinctly stated in 1986, we need to know 'what is the relative importance of the outside and inside of organisms' (p. 42). How development (proximate causation) impinges on evolution (ultimate causation) to effect evolutionary change and how development itself has evolved are topics covered in this book.

The physical embodiment of proximate versus ultimate causation, in some animals, is in the separation of their eggs into soma and germ plasm. In such animals the somatoplasm, from which body cells develop, provides the information for the individual organism during its lifetime. The germ plasm is inherited by germ cells that produce gametes (eggs or sperm). Genes within these germ cells, together with gene products inherited maternally, provide the heritable information for subsequent generations and the raw material for future evolutionary change.

Importantly, germ plasm is not segregated from the soma in most animals or in any plants or fungi. Somatic inheritance is a theoretical possibility in such organisms and modes of inheritance therefore an important topic. The two self-contained biologies are not so self-contained in these many organisms. The consequences for inheritance of segregation or failure of segregation of germ plasm from soma are discussed in section 9.1.1.

1.2 EVOLUTION AS PREFORMATION

Part of the connection between evolution and development is etymological. Evolution (L. *evolutio*, unrolling), was first applied to the unfolding of an individual during embryonic development in theories that minimized interaction with the external world and maximized constancy and destiny. The term can be traced to Albrecht von Haller (1708–77), a wealthy Swiss child prodigy, poet, lawyer, botanist, physiologist and polymath, whose lifetime writings totalled 650 individual pieces. He used the term evolution in 1774 to describe the development of the individual in the egg:

'But the theory of evolution proposed by Swammerdam and Malpighi prevails almost everywhere [*Sed evolutionem theoria fere ubique obtinet a Swammerdamio et Malphighio proposita*] . . . Most of these men teach that there is in fact included in the egg a germ or perfect little human machine . . . And not a few of them say that all human bodies were created fully formed and folded up in the ovary of Eve and that these bodies are gradually distended by alimentary humor until they grow to the form and size of animals.'

Haller, 1774, cited from Adelmann,
1966, pp. 893–4

Evolution therefore described the development of individual embryos. In fact, evolution was used in an even more restricted sense to describe a particular preformationist view of individual development. At its most extreme, preformation maintained that evolution consisted only of the unfolding through growth of an individual whose essential features pre-existed in the egg. If evolution was an unfolding, akin to unrolling a parchment, or to the unfolding of the wings of a newly emerged butterfly, then its usage for preformation was axiomatic.

From Aristotle onwards, natural historians and biologists have sought a causal explanation for development by advocating one or other of two major theories:

1. **preformation**, the unfolding and growth of structures already present in the germ. For preformationists, form exists in the egg.
2. **epigenesis**, the gradual and progressive appearance of new structures, each dependent on preceding structures and processes. For adherents to epigenesis, information exists in the egg and form appears only gradually.

Introduced below, these explanations for embryonic development are discussed in more detail in section 7.1.

1.2.1 Preformationists

Preformationists belonged to many camps. I say belonged and not belong, because few are left today (but see section 7.1 for important preformed elements of eggs and embryos). Of the various permutations on preformation, all share the common theme of structures unfolding during development. **Ovists**, who were

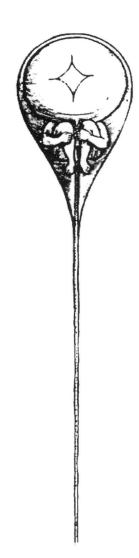

Fig. 1.1 A homunculus in a human sperm as 'seen' by Hartsoeker (1694).

Fig. 1.2 Richard Owen and Thomas Huxley examining a 'water baby' as depicted in Charles Kingsley's *The Water Babies* (1885) by the *Punch* political cartoonist (Edward) Linley Sambourne.

especially prevalent in the 17th and 18th centuries, maintained that organisms were preformed in eggs (ova). They thought that sperm, discovered in 1677, activated the preformed organism within the egg, were parasites of the seminal fluid, and/or played no role in development. **Spermists** (animalculists) maintained that organisms were preformed within sperm.[3]

It is not difficult to see how preformationist views could accommodate and even require the concept of individuals within individuals – the familiar homunculus 'seen' in sperm or drops of semen by Dalenpatius and D'Agoty, and Vallisnieri's notion of the whole human race and all human parasites nestled within the ovaries of Eve. Nicholaas Hartsoeker (1656–1725), a Dutch microscopist, lens grinder, instrument maker and self-taught naturalist, promulgated an extreme view of preformation, shared by many others, of adults fully formed in miniature within eggs or sperm. Figure 1.1, from a 1694 publication by Hartsoeker, shows the homunculus encapsulated in a human sperm. As illustrated in Figure 1.2, such imagery of embryonic development persisted for centuries.

1.2.2 Encapsulation

Use of evolution for individual development was propounded and extended by Charles Bonnet (1720–93), another Swiss, who like Haller was trained in the law. Bonnet's interest in natural history was coupled with a passionate religious zeal (his scientific writings are full of expositions on God, the angels, and the plight of the soul), a fervently anti-free-thinking stance, and a belief in recapitulation.

Bonnet used evolution as a label for his particular view of development through preformation. In his *emboîtment* (encapsulation) theory, all the members of all subsequent generations are present, preformed, within the egg. This was not a theory without observational foundation. Bonnet saw encapsulation in the cotyledons within plant seeds, in the insect imago visible through the skin of the pupa, and in the parthenogenetic female aphid, in whose body the rudiments of future offspring reside. Bonnet regarded preformation as one of the greatest triumphs of rational over sensual conviction. Charles Singer, the eminent historian of science, concluded that 'On such a shaky foundation have been built whole systems of biological philosophy. Nay, the very hope of salvation of men such as Bonnet was erected upon it!' (1959, p. 467)

Perhaps not surprisingly, Bonnet was an advocate of the Great Chain of Being originally propounded by Aristotle. The natural world, from 'atoms' at one extreme to cherubim at the other, was linked as a continuum. Although spontaneous generation was much in vogue during his lifetime, Bonnet strongly opposed it; it violated preformation and encapsulation, and represented a seemingly impossible break in the Great Chain of Being. In this pre-Darwinian view, the world and all life beneath, upon and above it, were created preformed, having only to unfold with the passage of time, whether the life span of an individual, the duration of a species, or the entire history of Earth itself.[4]

1.2.3 Regeneration

Any discussion of preformation in embryonic development must be interwoven with a discussion of the regeneration of lost parts. The only way a preformationist could explain regeneration of, say, the tail of a lizard or the limb of a salamander, was to postulate the existence of eggs containing preformed tails or limbs in the tails or limbs of adults. The new tail or limb would unfold from these eggs. Regeneration really was 're-generation' – the resumption of embryonic processes in adults. It did not matter where along the tail or limb amputation occurred; eggs lay everywhere, ready to unfold when needed. An important relationship therefore existed between preformation and regeneration in the 17th and 18th centuries.[5]

Not everyone in the early 18th century was a preformationist, however. For adherents of epigenesis, regenerating newt limbs, lizard tails, and bodies of *Hydra* did not exist preformed in the limb, tail or body remnant awaiting only the stimulus of amputation to spring forth. Montesquieu (Charles de Secondat, 1689–1755) – the logician, influential

French philosopher and astute observer of Nature – came to quite a different conclusion from the preformationists. Cogently, and with ruthless logic, he argued that:

'In order to be able to say that all trees to be produced from now to all eternity were contained in the first seed of each species that God created, it is first necessary to prove that all trees are produced from seed. If a green twig is planted in the earth, it grows roots and branches and becomes a perfect tree. It produces seeds which in their turn will grow into trees; and therefore if one wants to contend that a tree is nothing but an unfolded seed, it will be necessary to assert that a seed was secreted in this willow twig, which is more than I can admit.'

Cited in de Beer, 1962, p. 137

Not only are structures replaced with the correct structure – a tail for a tail and a limb for a limb – they are replaced with a structure of appropriate size and functional integration to be of maximum use to the organism. Eventually, knowledge that many animals could regenerate lost parts, coupled with increasing awareness of the amazing fidelity with which lost structures could be replaced, raised major difficulties for the preformationists.

1.2.4 Development

During the late 18th and first half of the 19th centuries, evolution was still equated with preformation in England as a hypothesis about development. If used at all the term development was applied to the views of Lamarck and others who believed in environmentally directed change. In Germany the word *Entwicklungsgeschichte* was used for development of the individual. We view this term as synonymous with embryonic development and with Wilhelm Roux (1850–1924), who founded the field of *Entwicklungsmechanik* and the journal *Wilhelm Roux Archives für Entwicklungsgeschichte*, which recently published its 200th volume.

In Germany this term also was used for the development of races.[6]

Evolution is often used for individual development even today. When used for change through time, evolution may still be defined in terms of individual development, as illustrated by the epigraphs. J. T. Bonner, who has contributed much to the integration of development with evolution, spoke of two kinds of development:

'that are occurring simultaneously: evolutionary development and life-cycle development. The life-cycle developments are in general connected one to another by a single-celled stage, a fertilized egg or some sort of asexual spore, but if one looks at a whole series of life cycles which change through the course of time, then one has an evolutionary development.'

Bonner, 1974, p. 5

My concern is the relationship between the two 'developments' or the two 'evolutions'.

1.3 EVOLUTION AS CHANGE

Ruse cites a question from a Cambridge University examination of 1851, the first year examinations were offered in the natural sciences tripos. The question indicates that the prevailing view associated past events with development and not with change.

'Reviewing the whole fossil evidence, shew that it does not lead to a theory of natural development through a natural transmutation of species.'

Ruse, 1979, p. xii

The transition of the use of evolution from the unfolding of embryonic development to progressive change through time from ancestor to descendant has been the subject of much discussion and review.[7]

Although Darwin proposed the theory of descent with modification or transmutation in *The Origin of Species*, he did not use the word evolution except as the last word:

'There is grandeur in this view of life, with its several powers, having been originally breathed by the Creator into a few forms or into one; and that, whilst this planet has gone cycling on according to the fixed law of gravity, from so simple a beginning endless forms most beautiful and most wonderful have been, and are being *evolved.*'

Geologists were among the first to use evolution in the modern, post-Darwinian sense of transformation of species and progressive change through time. Robert Grant (1793–1874), an Edinburgh-trained comparative anatomist, was for a long time Professor of Comparative Anatomy at the University of London. An expert on the anatomy of sponges (which he named the Porifera), Grant used evolution for the gradual origin of invertebrate groups in an 1826 article on the importance of geology. The geologist Charles Lyell (1797–1875) used the term with reference to gradual improvement and transformation of aquatic to land-dwelling organisms: 'the testacea of the ocean existed first, until some of them, by gradual evolution, were *improved* into those inhabiting the land' (1832, vol. 2, p. 11).

During the 120 years between Haller/ Bonnet and Herbert Spencer (1820–1903), evolution was progressively applied to a view of development that emphasized interactions between parts and with the external environment. Spencer enshrined modern usage. An engineer by training, he educated himself in the sciences and became an influential journalist, writer, social philosopher and proponent of education. A pioneer in sociology – he was fervently agnostic and devoted to the pursuit of truth, whatever the cost – Spencer did much to educate the general public to an evolutionary way of thinking. His two-volume treatise, *The Principles of Biology*, first published in 1864, influenced his contemporaries profoundly. Oxford University established the Herbert Spencer lectures to reinforce the influence on various disciplines of the theory of evolution

BOX 1.1
THOMAS HENRY HUXLEY AS POPULARIZER OF SCIENCE

Huxley was one of the greatest popularizers of science of his day, a man who throughout his life used a rare combination of scientific knowledge, acumen and literary skill to bring science to the people. Below is a sample from *On the Advisability of Improving Natural Knowledge*, a lay sermon delivered in St. Martin's Hall, London on January 7, 1866 and subsequently published in the *Fortnightly Review*. It is, in my view, the best justification for science ever written and one of the finest pieces of writing, scientific or otherwise.

I cannot but think that the foundations of all natural knowledge were laid when the reason of man first came face to face with the facts of Nature; when the savage first learned that the fingers of one hand are fewer than those of both; that it is shorter to cross a stream than to head it; that a stone stops where it is unless it be moved, and that it drops from the hand which lets it go; that light and heat come and go with the sun; that sticks burn away in a fire; that plants and animals grow and die; that if he struck his fellow savage a blow he would make him angry, and perhaps get a blow in return, while if he offered him a fruit he would please him, and perhaps receive a fish in exchange. When men had acquired this much knowledge, the outlines, rude though they were, of mathematics, of physics, of chemistry, of biology, of moral, economical, and political science, were sketched.

Cited from Huxley, 1910, pp. 48–9

proposed by Darwin. Spencer used evolution in its modern sense as early as 1852 in an essay entitled *The Developmental Hypothesis*, in which he emphasized progression toward greater complexity and interaction with 'outer forces', rather than the playing-out of internal programmes as emphasized in earlier preformationist usage.[8]

1.4 EVOLUTION AS PROGRESS

Lyell, Spencer, Huxley and Darwin all associated evolution with improvement or progress. Reformers like Huxley and Spencer used Darwin's theory and the evolutionary concept of progress to promote their views of the political and social changes that they felt should occur within society (see Box 1.1). Indeed, as Michael Ruse argues in *Monad to Man* (1996),

progress, consciously or unconsciously, influences the thinking even of present-day evolutionary biologists. This provocative book has evoked considerable debate.

Perceived parallels between the evolution of life on the planet and individual development arose from ideas of progress. In both there is, or appears to be, a progressive increase in complexity. This is why the parallel between embryonic development and the history of life has been so seductive for so long. Embryonic development proceeds from single-celled 'simplicity' to multicellular complexity. Evolution of life appears to have proceeded from single-celled simplicity to the manifold, multicellular complexity of later organisms. Organisms increase in size during development and increasing size is often perceived to indicate progress. I discuss complexity in section 14.1 in relation to increasing cell number, the origins of multicellularity and the origin of the Metazoa, and in Box 14.1 in relation to the evolutionary trends toward increasing size (Cope's rule, recognized in the 1870s).

Although biologists no longer view evolution as a progression from amoeba to man, notions of the parallel persist. Evolution for many organisms (prokaryotes and amphioxus, for example) is the continuing capacity to persist unchanged (or with limited change) over hundreds of millions of years. Even so, evolution is often associated in the minds of the non-scientist with progress and the ascent of family trees. The long-perceived parallel between development and evolution – which owes much to this belief – was summarized succinctly by Gavin de Beer in a BBC Science Survey broadcast aired on 19 December, 1950:

'The two series, developmental and evolutionary, have an important feature in common; in each there is a progressive increase of complexity of shape from the amoeba to man in the case of evolution, and from the egg to the adult in the case of development. The two series appear to run parallel with one another.'

De Beer, 1962, pp. 59–60

De Beer did much to promote a rational approach to this parallel and to integrate embryology and evolution, especially in several influential books: *Embryology and Evolution* (1930), *Embryos and Ancestors* (1958) and *Atlas of Evolution* (1964).

The two-fold parallel of evolutionary and individual history was extended to a threefold parallelism that included the history of the earth itself, by men such as Carl Friedrich Kielmeyer (1765–1844), an influential teacher of Georges Cuvier and Louis Agassiz. Notions of progressive change and directionality in evolution led to a search for the meaning of the assumed parallelism and to the doctrine of recapitulation, in which each organism climbs its family tree during its own lifetime. These

themes are dealt with in Chapters 4 and 5 in the context of types, archetypes, and the Geoffroy-Cuvier debates over how structural organization is maintained in different groups of animals.

1.5 EVOLUTION BEYOND BIOLOGY

Evolutionary thinking is pervasive. Indeed, it is inconceivable to consider the study of history, sociology, art, literature, music, law, science, technology or ethics, against any background other than evolution (de Beer, 1962). The titles of books and articles are sufficient to reveal the influence of evolutionary thinking. Steadman's *The Evolution of Designs: Biological Analogy in Architecture and the Applied Arts* (1979) typifies the biological and evolutionary approach to architecture and design. So too does the 1989 article by Gombrich (appropriately one of the Herbert Spencer lectures at Oxford University) which dealt with 'Evolution in the Arts: the Altar Painting, its Ancestry and Progeny'.

Examples of the generic use of evolution also abound in fields such as medicine and technology. For instance, articles with the following titles appeared in three consecutive issues of *Current Contents in the Life Sciences*:

'The *evolution* of anaesthesia as a specialty; The Omni-design-*evolution* of a valve; Histologic, morphometric and biochemical *evolution* of vein bypass grafts in a non-human primate model; A new step in the *evolution* of the journal; *Evolving* role of angiotensin-converting enzyme inhibitors in cardiovascular therapy; *Evolving* patterns in the surgical treatment of malignant ventricular tachyarrhythmia.'

Pervasive too is the use of evolution for change, especially for progressive change or change with improvement. An advertisement entitled 'In-Flight Evolution', placed by the de Havilland Division of Boeing Canada, captures all the essential elements of biological evolution. The Dash 8 is said to be:

BOX 1.2
THE HARDY–WEINBERG LAW

The history of this law is an interesting one. In the course of studies on gene frequencies, the Cambridge geneticist Reginald Punnett sought out his friend the mathematician and fellow cricket enthusiast, G. H. Hardy, for advice on the expected frequency of alleles of a gene in the population. Hardy quickly devised the now-classic formula, $pr = q^2$, with p, $2q$ and r as the proportions of the homozygous dominant, heterozygous and homozygous recessive alleles in the population. At Punnett's suggestion, this formula was named Hardy's Law. So it remained until 1943 when it was discovered that the German physician Wilhelm Weinberg had published the same formula six weeks before Hardy independently devised it.

'the next step in the evolution of flight . . . As our evolution continues, we create opportunities for individuals to explore their potential in an environment that thrives on the innovative and progressive contributions of a talented group of professionals.'

The Globe and Mail, 2 March, 1990, p. B23

Conrad Waddington (1975) used *The Evolution of an Evolutionist* as the title of his autobiography, with the explicit intention that the title have two meanings. No doubt Peel (1971) had a double meaning in mind when he subtitled his biography of Herbert Spencer *The Evolution of a Sociologist*. Spencer did not confine evolution to living things.

1.6 EVOLUTION AS GENETICS OR POPULATION GENETICS

Evolution as genetic change alone is especially emphasized, perhaps not surprisingly, by geneticists. Thus King and Stansfield define evolution as 'potentially reversible gene frequency changes within a population gene pool', but also link it to speciation, including as alternatives 'irreversible genetic changes within a genealogical lineage producing . . . speciation . . . splitting of one species into

another; production of novel adaptive forms worthy of recognition as new taxa' (1985, p.132). The *Glossary of Genetics and Cytogenetics* contains a comprehensive survey of views on evolution, including evolution as:

'a change in the genetic composition of a population the starting point of which is the formation of individuals with different genotypes. The unit process in evolution is gene substitution and the elementary quantity is the gene frequency which is the measure of genetic change in a population. Evolutionary changes are brought about by the primary evolutionary forces which produce and sort out genetic variations and operate in a field of space and time.'

Rieger, Michaelis and Green, 1976, p. 191

Population genetics, the branch of genetics that deals mathematically with frequencies and interactions of genes in populations, is especially concerned with how such factors as mutation, selection and migration alter gene frequency. Population genetics effectively began in 1908 with the independent elucidation by Hardy and Weinberg of the theorem of gene frequencies in populations under conditions of natural selection (Box 1.2).

The field was developed by J. B. S. Haldane, R. A. Fisher and Sewall Wright in the early 1930s, and by Theodosius Dobzhansky in the late 1930s. Fisher provided a genetic basis for natural selection in *Genetical Theory of Natural Selection* (1930). The 'modern' synthesis of evolution was forged by Theodosius Dobzhansky in *Genetics and the Origin of Species* (1937) and by Julian Huxley in *Evolution: The Modern Synthesis* (1942). Indeed, Huxley coined the term 'synthesis' for the forging of genetics and evolution. Three further books (all published in the Columbia University Biological Series) forged a synthesis with systematics (Mayr, 1942), palaeontology (Simpson, 1944) and extended it to plants (Stebbins, 1950). *Genetics, Paleontology and Evolution* (Jepsen *et al.*, 1949) and *Animal Species and Evolution* (Mayr, 1963) sealed the synthesis.[9]

Evolution as change of gene frequencies in populations was the most widely accepted definition after the early 1940s, reflecting the domination of evolution by population genetics. This domination is now over. Not that population genetics is an inappropriate means to study evolution – a quantitative genetics model of morphological changes in evolution is presented in Chapter 20. Rather it is neither sufficient nor inclusive. The search for the basis of the adaptiveness of small- and large-scale gene effects by Palopoli and Patel (1996) reveals that identification of genes does not necessarily reveal the developmental pathways underlying evolutionary change. Evolution is hierarchical.[10]

Although population genetics is not often used to shed light on macroevolution (see below), carefully designed studies show how such an approach can illuminate the evolutionary history of a group. The analysis by Futuyma, Keese and Funk (1995) of an ecologically important character – host affiliation – in the leaf beetle *Ophraella* is such a study in which limitation on genetic variation influences the evolution of host affiliation, i.e. macroevolution is limited by a genetic constraint (see s. 6.7.1).

1.7 EVOLUTION AS HIERARCHY

Evolution may be defined as:

1. change at the genetic level – the substitution of alleles;
2. change at the organismal level – the appearance of new characters (structures, functions and behaviours); and
3. change at the supraorganismal level – the origin, radiation and adaptation of species.

With the publication of *Tempo and Mode in Evolution* by G. G. Simpson in 1944 (one of the books that established the modern evolutionary synthesis), the three levels became known as:

1. **microevolution** (changes within species);
2. **macroevolution** (speciation); and
3. **megaevolution** (the origin of higher taxa).

Nowadays macroevolution is used for both speciation and the origin of higher taxa, the term megaevolution having fallen into disuse.[11]

Although all evolutionary change is based on changes in genes, evolutionary change can occur at the organismal level without being reflected at the supraorganismal level. As will be evident from the discussion of key innovations in sections 13.2 and 13.3, the appearance of new characters need not be linked to speciation. Nor need change at the supraorganismal level be qualitatively different from change within individuals. A specific microevolutionary explanation for an apparent macroevolutionary transformation – the origin of external from internal cheek pouches in mammals – is discussed in section 24.2.1.

Because evolution acts at all these levels, no single, universally applicable definition exists even today for change over time. Texts which deal explicitly with evolution, such as Simpson's *The Meaning of Evolution* (1950), Patterson's *Evolution* (1978), and Stanley's *The New Evolutionary Timetable* (1981), neither define the term nor, in Patterson's case, include it in the glossary. In *Evolution for Naturalists*, a folk tale is provided in place of a definition:

'Four myopic evolutionists looked at evolution. One said, 'It is survival of the fittest.' One said, 'It is differential reproduction.' One said, 'It is change in gene ratios.' And one said, 'It is a molecular process.' They all saw something real, but each magnified what he saw, and none saw evolution as a whole.'

Darlington, 1980, p. 1

There is much to be said for Darwin's 'descent with modification' as a definition of evolution. General enough to encompass all levels, it emphasizes the temporal, genealogical and variational aspects of evolution. Therefore, in this book, **evolution** is taken to be descent with modification, whether change is at the genetic, organismal or supraorganismal levels. Development is also hierarchical, but over the time-scale of individual ontogeny. I take **development** to be those complex hierarchical processes within an individual that produce and order cells, tissues and organs through the processes of differentiation, morphogenesis and growth.

1.8 ONTOGENY AND PHYLOGENY

I should briefly consider one further definitional issue, that of ontogeny and phylogeny. **Ontogeny** (G. *ontos*, birth or existence, and *genesis*, origin) is equated with development of the individual from fertilization to maturity. **Phylogeny** (G. *phylon*, a race) is equated with the evolution of species or lineages.

In *Archetypes and Ancestors*, Desmond gives a particularly succinct definition of phylogeny: 'the study of the evolutionary routes followed by particular organisms' (1982, p. 148). The use of phylogeny for the 'racial history' of a group originated with Haeckel (1834–1919), who also coined the term 'phylum'. Haeckel even proposed that Bismarck be awarded a Doctorate in Phylogeny

to mark Germany's position at the top of the European 'phylogenetic tree'!

Phylogeny reflects relationships: 'the relationships of groups of organisms as reflected by their evolutionary history' (King and Stansfield, 1985, p. 298). Patterson placed equal weight on relationship and descent with his definition of phylogeny as the 'study of the evolutionary history and relationships of species' (1978, p. 190). It is in this sense that phylogenetic trees, which emphasize relationships and not actual ancestors, are constructed. Phylogenetic systematics thus becomes a branch of evolutionary biology, although it is important to note that phylogenetic systematists intentionally reject any search for ancestors. In concentrating on relationships between sister groups they differ from those systematists who establish more traditional phylogenetic trees of life.[12]

1.9 ONTOGENY CREATES PHYLOGENY

The old notion that ontogeny recapitulates phylogeny (s. 5.5.2) has given way, albeit reluctantly, to the view that phylogenetic change in morphology reflects transformation in ontogeny, a major topic of this book. An advocate of this approach to ontogeny and phylogeny was C. O. Whitman, whose endorsement of ontogeny as the cause of phylogeny is captured in the following quotation:

'All that we call phylogeny is today, and ever has been, ontogeny itself. Ontogeny is, then, the primary, the secondary, the universal fact. It is ontogeny from which we depart and ontogeny to which we return. Phylogeny is but a name for the lineal sequences of ontogeny, viewed from the historical standpoint.'

Whitman, 1919, p. 178

That ontogeny causes phylogeny is epitomized in Garstang's phrase 'ontogeny does not recapitulate phylogeny, it creates it' (1922, pp. 21, 81). An explicit analysis by Northcutt

(1992) reaffirms Garstang's postulate in a phylogenetic framework in which the evolution of ontogeny can be assessed. David Wake encapsulated the view developed in this book that 'it is the whole ontogeny that evolves, and thus a true phylogeny will be based on evolutionary transformations of ontogeny' (1989, p. 371).

Not all evolutionary change is mediated by changes in development. Aspects of organismal evolution that are mediated by development include development itself, embryonic stages and processes, morphology (be it embryonic, larval or adult) and phenotypic plasticity.

Patterson (1983) provided a particularly useful discussion of the relationship of ontogeny to phylogeny, emphasizing that most evolutionary laws also describe ontogeny, pertaining as they do to growth, size and specialization of individuals. Thus, Dollo's law of the irreversibility of evolution concerns the inability of an organism to return to a state identical to an ancestral condition; Cope's law of the unspecialized states that evolutionary novelties in new taxa are more likely to arise from generalized than from specialized ancestral taxa. Cope's rule of the tendency of animals to increase in size during evolution carries connotations of progress, trends and increase in complexity (Box 14.1), while Williston's rule is that there is a decrease in the number of serially repeated parts as individual parts specialize during evolution.

Dollo's law is discussed in section 17.4.1 to illustrate how, at the level of developmental mechanisms, transformations in ontogeny produce phylogenetic change. Because such laws and rules deal with development as much as with evolution, Patterson argued that phylogeny hardly differs from ontogeny but saw ontogeny as economical and phylogeny as a 'random walk'. Atavisms, phylogenetic character reversals, and the reactivation of silenced genes, all indicate how knowledge of developmental processes reveals the ontogenetic basis of Dollo's law (ss. 17.5, 21.5). Olivier Rieppel, in an analysis of the role of ontogeny in systematics, encapsulated the dichotomy beautifully:

'Any change in the 'rules of construction', changing the developmental patterns, causes a phylogenetic change in the lineage represented by the developing organism(s): ontogeny creates phylogeny. Each phylogenetic change affecting an evolving lineage is reflected in the changing ontogeny of the organism(s) representing that lineage: phylogeny creates ontogeny.'

Rieppel, 1990, p. 178

Although ontogeny is individual development and phylogeny evolutionary relationships, once we begin to consider mechanisms of change through time in different organisms, the two are integrated as evolutionary developmental biology. The challenge taken up by evolutionary developmental biologists is to integrate development with genomic, organismal and population approaches to evolution. In Gabriel Dover's words: "to Dobzhansky's 'nothing in biology makes sense except in the light of evolution', I would add that nothing in evolution makes sense except in the light of processes, starting within genomes, that affect developmental operations and ultimately spread through a population" (1992, p. 283). Wed this to natural selection – 'the main unifying idea in biology is Darwin's theory of natural selection' (Maynard Smith, 1958, p. 11, in the opening sentence of the preface of *The Theory of Evolution*, perhaps the most reprinted and widely distributed book on evolution written this century) – and we have the hierarchical, yet integrative approach required.

Evolutionary developmental biology strives to forge a unification of genomic, developmental, organismal, population and natural selection approaches to evolutionary change. All these elements have been brought to bear on how phyla and the body plans that typify phyla arose. In Chapter 2, I discuss how we group animals into phyla and super-phyla, while in Chapter 3 I discuss how the animals of the Burgess Shale have influenced our thinking about the origin of, and relationships between, animals and phyla.

1.10 ENDNOTES

1. For discussions of proximate and ultimate causation in biology, see Mayr (1961, 1982), Jacob (1977), Taylor (1987), Rieppel (1986, 1988a), Alberch (1989) and Thomson (1991a).

2. For examples of the internal–external conflict in their work, see Whitman (1894) and Conklin (1896). For a recent analysis, see Maienschein (1978).

3. John Farley discusses in depth the shades of preformation and the various views held by ovists and spermists in *Gametes & Spores* (1982).

4. The definitive treatments of the Great Chain of Being and spontaneous generation are Lovejoy (1936) and Farley (1977) respectively.

5. Bodemer (1964) and several authors in *A History of Regeneration Research* (ed. Charles Dinsmore, 1991), discuss the relationship between preformation and regeneration in the 17th and 18th centuries. Regeneration in the contexts of selective gene activation/repression and homology is discussed in sections 6.7.2 and 21.3 respectively.

6. For the intersection of evolution and development in Germany, see Mayr (1959), Oppenheimer (1967) and Nyhart (1995).

7. For transfer of the term evolution from unfolding of development to descent from ancestors, see Oppenheimer (1967), Carneiro (1972), Bowler (1975, 1989a) and Gould (1977).

8. Freeman (1974), J. C. Greene (1981), Bowler (1988) and (especially) Godfrey (1985) discuss Herbert Spencer's efforts to educate the public to an evolutionary way of thinking.

9. See Smocovitis (1996) and Mayr (1997a) for recent accounts of the history of the modern synthesis. Three books by the Russian evolutionary biologist Ivan Schmalhausen, two in Russian (1938, 1942) and *Factors of Evolution* (1949) in English translation, are now seen as important contributions to evolutionary thought; see David Wake's Foreword to the 1986 reprinting of *Factors of Evolution*, a recent evaluation (D. B. Wake, 1986), and Chapter 12 and s. 19.5 for analyses of Schmalhausen's role.

10. For discussions of the domination of evolution by population genetics, see Lewontin (1974), Maynard Smith (1958, 1989a), Dobzhansky *et al.* (1977), Roughgarden (1979), Hedrick (1983), Provine (1986) and Hartl (1989).

11. The terms micro- and macroevolution were used with similar meanings by Dobzhansky (1937) and by Richard Goldschmidt in *The Material Basis of Evolution* (1940). According to Rieger, Michaelis and Green (1976), the terms were first used by J. Philipschenko in 1927 for intra- and interspecific evolution respectively. See De Braga and Carroll (1993) for a helpful recent discussion of macroevolution as illustrated by the origin of marine reptiles, and see Pettersson (1996) for an analysis of types of hierarchies.

12. See Panchen (1992), Rieppel (1993a) and Grande and Rieppel (1994) for discussions of phylogenetic systematics.

2

TYPES OF ANIMALS: KINGDOMS, PHYLA, RELATIONSHIPS

'Form is never trivial or indifferent; it is the magic of the world.'

Dalcq, 1968, p. 91

During their development and evolution organisms differ from one another most obviously in the physical appearances that reflect their anatomical plans. These differences appear anew in each generation and arose through time by evolution from forms with different anatomical plans. How types of organisms arose and became distinct from one another is a central quest of evolutionary developmental biology. Our investigation of the interrelationships of development and evolution continues with an examination of the origin of anatomical plans at the highest taxonomic level, the phylum.

2.1 KINGDOMS AND DOMAINS

For many centuries, life was recognized as falling within two kingdoms, animals and plants. In 1959, however, Whittaker recognized five kingdoms: Prokaryotae, Protoctista, Fungi, Plantae and Animalia, comprising 93 phyla (Table 2.1). The groupings of these kingdoms changed dramatically when Woese and his colleagues sequenced rRNA from Archaebacteria and proposed that they belonged to a separate domain or division of life, the Archaea (although most of us have taken a long time to accommodate to this change).[1]

In 1996 Bult and colleagues sequenced every gene in *Methanococcus jannaschii*, a methane-

producing member of the Archaea that lives (indeed thrives) at temperatures of 85°C, depths of 2600 metres, and pressures of over 200 atmospheres. Fifty-six per cent of the genes from *M. jannaschii* are unlike any found in bacteria or eukaryotes. This remarkable feat, the first sequencing of the entire genome of an organism, cemented the 20-year-old proposal that life could be organized into three domains – Bacteria, Archaea and Eukarya (Figure 2.1). The Eukarya domain includes animals, plants, fungi, ciliates, flagellates, slime molds, and other organisms with genomes enclosed in a nuclear membrane (Figure 2.2). The first complete genome of a species from the Eukarya domain – the budding yeast *Saccharomyces cerevisiae* – was sequenced by Clayton *et al.* in 1997.

2.2 PHYLA

A phylum (division in plants) is the taxonomic category immediately below the kingdom. Phyla are primary groupings of animals that share a common plan and evolutionary relationship. Within phyla of animals are one or more classes, orders, families, genera and species.

How many phyla are there in existence now and how many have there ever been? Experts differ. Opinions vary. Designations are not absolute even at this level in the taxonomic

Table 2.1   The five kingdoms of life

Kingdom	Derivation	Common names	Phyla or Divisions
Prokaryotae (Monera)	G. *pro*, before; *karyon*, nucleus, nut	Bacteria	17
Protoctista	G. *proteros*, very first; *ctistes*, to establish, build	Algae, protozoans, slime moulds	27
Fungi	L. *fungus*	Mushrooms, moulds, lichens	5
Plantae	L. *planta*, plant	Mosses, ferns, cone-bearing and flowering plants	10
Animalia	L. *anima*, breath, life	Animals with and without backbones	37

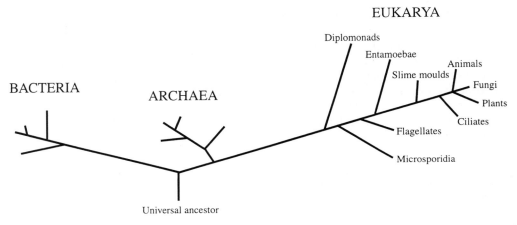

Fig. 2.1 The three divisions of life and some of the major groups within the kingdom Eukarya. Modified from Madigan and Marrs (1997).

hierarchy; witness the changing views of the number of kingdoms. (On the basis of 18S rRNA evidence, for example, Smothers *et al.* (1994) gave phylum status within the Metazoa to the Myxozoa, a group of obligate, multicellular parasites, traditionally placed within the protists. Other phylogenetic analyses place the myxozoans as a sister group to the nematodes or to all animals with bilateral symmetry.) A complete list of 37 extant phyla along with their diagnostic features is provided in Table 2.2. These fundamental features represent the type, body plan, or

Bauplan, concepts that are elaborated further in Chapters 4 and 6.

The 37 phyla comprise between 1.7 and 2 million named living species. Approximate species numbers for each phylum are listed in Table 2.2. Species diversity within phyla varies enormously, from the single species *Trichoplax adhaerens* and *Symbion pandora* of the Placozoa and Cycliophora, through six to ten species in each of the Vestimentifera, Loricifera, Phoronida and Priapulida, to the half million or more nematodes and more than a million arthropods. Total species numbers are con-

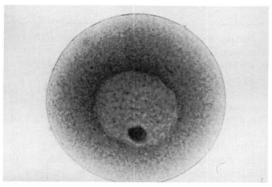

Fig. 2.2 Typical eukaryotic cells with nucleus enclosed in nuclear membrane and containing the rRNA-organizing centre, the nucleolus. Top as seen in bright-field microscopy, below as seen in an echinoderm egg viewed with Nomarski phase contrast microscopy. The minute cells to the right are sperm.

siderably higher, given the innumerable unidentified living species and that I have not included fossil species.

2.2.1 Phyla discovered this century

Given that phyla are so fundamental – 'probably the most satisfactory taxon after the species' (Willmer, 1990, p. 4) – it may come as a surprise to discover that at least six new phyla were erected this century, the most recent in 1995. The six are listed on page 23 in order of their discovery and with phylum numbers keyed to Table 2.2:

Table 2.2 A summary of 37 phyla within the Animal Kingdom, along with common names where available, etymology, approximate number of living species, organization into major superphyla on the basis of symmetry, type of coelom and fate of the blastopore, and the basic features that define the *Bauplan* of each phylum[1]

Kingdom Animalia (L. *anima*, breath, life)

Subkingdom Parazoa (G. *par*, near; *zoion*, animal)
Lack germ layers, tissues or organs; shape is indeterminate as is symmetry

1. **Placozoa** (G. *plakos*, flat; *zoion*, animal)
   Minute (0.3 mm), marine, soft-bodied, no obvious symmetry, variable shape, amoeboid, ciliated, asexual (fission) and sexual reproduction, no nerve or muscle cells, low DNA content (1000 daltons), 12 chromosomes, one species *Trichoplax adhaerens*
2. **Porifera** (sponges; L. *porus*, pore; *ferre*, to bear)
   Mostly marine (2 freshwater families), loosely aggregated cells, sessile adult, porous body, flagellated cells partially line the internal cavities, internal spicular skeleton of calcium carbonate, no nervous system, great regenerative ability, 10 000 species.

Subkingdom Eumetazoa (G. *eu*, true; *meta*, later; *zoion*, animals)
Tissues organized in organs; organ systems

Radial symmetry

3. **Cnidaria** (coelenterates; G. *knide*, nettle; *koilos*, hollow; *enteron*, intestine)
   Mostly marine, some freshwater, polymorphic with polyp and medusa, solitary or colonial, planula larva, radial symmetry, tentacles around mouth (generally), coelenteron as single body cavity, three-layered body wall, nematocysts as defence cells, nervous system as a net, hermaphrodite or dioecious, 10 000 species
4. **Ctenophora** (comb jellies; G. *kteis*, comb; *pherein*, to bear)
   Marine, free-swimming, cydippid larva, possess tentacles usually with colloblast (prey-capture) cells, transparent, gelatinous, luminescent, three-layered body wall, 8 radial ciliated comb rows, branched canal digestive system (coelenteron), ciliary locomotion, hermaphrodite, 100 species.

Table 2.2 Cont'd.

Bilateral symmetry
Acoelomata: lacking a body cavity or coelom

5. **Mesozoa** (dicyemids, heterocyemids, orthonectids; G. *mesos*, middle; *zoion*, animal) Minute parasites in invertebrate body cavities, bilaterally symmetrical, two-layered body lacking endoderm or mesogloea, sexual and asexual generations, 50 species

6. **Platyhelminthes** (flatworms; G. *platys*, flat; *helmis*, worm) Parasitic or free-living, bilaterally symmetrical, flattened, triploblastic, flame cells and ducts as excretory system, nervous system with brain, hermaphrodite, 25 000 species

7. **Nemertina** (ribbon worms; G. *Neremtes*, a sea nymph) Mostly marine, some terrestrial, a few parasitic, bilaterally symmetrical, unsegmented, worm-like, elongate, often flattened, proboscis, ciliated epithelium on outer and inner (gut) surfaces, nephridial excretory system (usually), blood-vascular system, parenchyma packs body, separate sexes, serially repeated gonads, mostly direct development but 3 larval forms known, 900 species

8. **Gnathostomulida** (jaw worms; G. *gnathos*, jaw; *stoma*, mouth) Microscopic, marine, bilaterally symmetrical, unsegmented, single-layered epidermis, each cell with only one cilium, muscular pharynx, paired jaws (usually), no anus, hermaphrodite, 100 species

9. **Gastrotricha** (G. *gaster*, stomach; *thrix*, hair) Microscopic, aquatic (marine and freshwater), free-living, unsegmented, bilaterally symmetrical, worm-like, ciliated bands, bristles scales and spines, one or more pairs of adhesive tubes, protonephridia (sometimes), 450 species

Pseudocoelomata: possessing a body cavity but not a true coelom

10. **Cycliophora** (G. *cyclion*, *phoros*, small wheel-bearing) Microscopic, epizoic, single species *Symbion pandora* confined to the mouth parts and appendages of the Norway lobster *Nephrops*. Three motile, non-feeding stages produced by budding. Bilaterally symmetrical sac-like body attached by an adhesive disc; cuticular, sessile, recurved gut. Sexual and asexual stages.

11. **Rotifera** (wheel animals; L. *rota*, wheel; *ferre*, to bear) Microscopic, aquatic, mostly free-living, unsegmented, bilaterally symmetrical, spherical/cylindrical body with bifurcate foot, anterior wheel organ; jaws (mastax) in pharynx, cuticle, protonephridia, parthenogenesis common, 1500 species

12. **Kinorhyncha** (G. *kinein*, to move; *rynchos*, snout) Microscopic, marine, free-living, bilaterally symmetrical, superficial segmentation, cuticle, retractile head bearing spines, cilia only in sense organs, one pair of protonephridial tubules, each with a flame cell, 100 species

13. **Loricifera** (L. *loricus*, girdle; *fero*, to carry) Minute, marine, meiobenthic in coarse sand and gravel or ectoparasitic, Higgins larva, several larval stages and moults; adult is bilaterally symmetrical with a cuticle, spiny head (introvert, with cross-striated retractor muscles), thorax and abdomen with a lorica (girdle) of 6 spiny plates; introvert and thorax retractable and covered with appendages (scalids); 8 or 9 oral stylets surround the mouth, extrusible buccal canal, paired salivary glands; large dorsal brain, 8 circumoral ganglia, ventral thoracic and caudal ganglia; separate sexes, 10 species

14. **Acanthocephala** (spiny-headed worms; G. *akantha*, thorn; *kephale*, head) Adults endoparasitic in vertebrates, unsegmented, bilaterally symmetrical, protrusible hooked proboscis, lack gut, three larval stages (acanthor, acanthella, custacanth), 1000 species

15. **Entoprocta** (G. *entos*, inside; *proktos*, anus) Marine except *Urnatella* which is freshwater, sessile, circlet of ciliated tentacles, U-shaped gut opening within tentacles, protonephridia, 150 species

16. **Nematoda** (roundworms; G. *nema*, thread) Aquatic, terrestrial and parasitic cylindrical unsegmented worms, triploblastic, cuticle without cilia, longitudinal muscle fibres, triradiate pharynx, gland cells/canals as excretory system, perhaps 500 000 species

17. **Nematomorpha** (gordian worms; G. *nema*, thread; *morphe*, form)

Table 2.2 Cont'd.

Freshwater (one marine genus), free-living adults, juveniles parasitic in arthropods, unsegmented, bilaterally symmetrical, filiform worms, gordioid larva, non-functional gut in adult, no excretory system, separate sexes, 250 species

Coelomata: possessing a true coelom

Protostoma: blastopore becomes mouth, spiral or modified radial cleavage, schizocoelic coelom, ventral nervous system, trochophore larva

18. **Ectoprocta** (Bryozoa; moss-animals; G. *ektos*, outside; *proktos*, anus)
Mostly marine but some freshwater and estuarine; sedentary, colonial, composed of zooids; retractable lophophore with ciliated tentacles; rigid or gelatious wall secreted by zooid, U-shaped gut, hermaphrodite, 400 species

19. **Phoronida** (G. *pherein*, to bear; L. *nidus*, nest)
Marine, planktonic actinotrocha larva, adults produce a chitinous tube or burrows into mollusc shells; elongate, cylindrical, bilaterally symmetrical, horseshoe-shaped lophophore with tentacles anteriorly, U-shaped gut, pair of metanephridia which also function as gonoducts, closed circulatory system, 10 species in 2 genera

20. **Brachiopoda** (lamp shells; L. *brachium*, arm; G. *pous*, foot)
Benthic, marine, most sedentary and attached, some burrow, free-swimming larva; body of adult in two shell valves to which adult attached by fleshy pedicle; lophophore as feeding organ (may have internal skeleton); no locomotory organs, no complex sense organs, 330 species

21. **Mollusca** (L. *molluscus*, soft)
Marine, freshwater and terrestrial, modified trochophore larva; body of head, foot and visceral hump covered by mantle which often secretes a calcified shell; complex alimentary canal of a muscular buccal mass, toothed radula, salivary glands, stomach; nervous system of circumoesophageal ring, pedal cords and visceral loops; gills, coelom may be reduced; blood system with propulsive heart, arterial and venous system and haemocoel, haemocyanin as respiratory pigment, 100 000 species

22. **Priapulida** (L. *priapulus*, little penis)
Marine, benthic, free-living, bilaterally symmetrical, unsegmented, worm-like, eversible anterior end, protonephridia with solenocytes, nervous system associated with epidermis, no ganglia, large body cavity, 9 species

23. **Sipuncula** (peanut worms; L. *siphunculus*, little pipe)
Marine, trochophore larva, unsegmented, bilaterally symmetrical, elongate, worm-like, no chaetae or prostomium, terminal mouth surrounded by tentacles, U-shaped gut, one or two metanephridia, 320 species

24. **Echiura** (spoon worms; G. *echis*, snake; L. *ura*, tailed)
Marine, mostly sublittoral, trochophore larva, unsegmented, bilaterally symmetrical, non-retractable proboscis, one pair chaetae, pair of anal vesicles, 135 species

25. **Annelida** (segmented worms; L. *anellus*, little ring)
Marine, freshwater and terrestrial, most free-living, trochophore larva, bilaterally symmetrical, segmented, hydrostatic skeleton, cuticle, chaetae, triploblastic, external circular and internal longitudinal muscles, preoral ganglia and pair of ganglionated ventral nerve cords, nephridia, coelomoducts, schizocoelic coelom, closed tubular circulatory system, 14 000 species

26. **Tardigrada** (water bears; L. *tardus*, slow; *gradus*, step)
Very small ($\geq$1 mm length), mostly freshwater, some marine, bilaterally symmetrical, some sign of segmentation, 4 pairs of stumpy legs ending in claws, sucking pharynx with stylets, cuticular exoskeleton, moults, 400 species

27. **Pentastoma** (tongue worms; G. *pente*, five; *stoma*, mouth)
Parasites of carnivorous vertebrates, bilaterally symmetrical, worm-like, 2 pairs of retractable claws surround mouth, cuticle is moulted, 100 species

28. **Onychophora** (velvet worms; G. *onyx*, claw; *pherein*, to bear)
Terrestrial, free-living, tropical or semi-temperate, flexible cuticle; circular, oblique and longitudinal muscle layers in body wall; paired, stumpy, unjointed appendages, cilia-lined coelomoducts, 70 species

29. **Arthropoda** (G. *arthron*, joint; *pous*, foot)
Marine, freshwater, terrestrial, triploblastic,

Table 2.2 Cont'd.

bilaterally symmetrical, metameric segmentation; paired, jointed appendages, at least one pair of jaws, chitinous exoskeleton, develop via moults; tubular gut, striated muscles, ventral nerve cord of segmented ganglia, cilia only in sense organs, reduced coelom, body cavity a haemocoel, heart, separate sexes, >1 000 000 species

Deuterostoma: blastopore becomes anus, radial or spiral cleavage, enterocoelic coelom, dorsal or superficial nervous system, dipleurula larva

30. **Pogonophora** (beard worms; G. *pogon*, beard; *pherein*, to bear)
Benthic marine, free-living, sedentary, tube-dwelling, no pelagic larva, tentacles present, polymeric segments, lack digestive system or alimentary canal, median nerve cord, closed blood-vascular system, sexes separate, 100 species

31. **Vestimentifera** (red tube worms; G. *vestibulum*, entrance)
Benthic, marine, may be 3 m long, associated with hydrothermal vents at depths of up to 2500 m, bright red because of a special haemoglobin that carries both O_2 and H_2S, no gut but a trophosome filled with symbiotic, sulphide-oxidizing, chemo-autotrophic bacteria, lateral body folds as an external tube or vestimentum, thousands of tentacles or branchial filaments, trochophore larva, 6 species

32. **Echinodermata** (sea lilies, sea cucumbers, sea urchins, sand dollars; G. *echinos*, sea urchin; *derma*, skin)
Marine, free-living; bilaterally symmetrical pluteus larva with complex metamorphosis; adult lacks head, pentamerous symmetry, coelom divided into three (axocoel, water vascular system, coelomic cavity), unique water vascular system of tubes and tube feet; internal, mesodermal skeleton of spicules of calcium carbonate; no excretory organs, separate sexes, 6000 species

33. **Chaetognatha** (arrow worms; G. *chaite*, hair; *gnathos*, jaw)
Marine, planktonic, free-living, bilaterally symmetrical body of head, trunk and tail, divided by vertical septa; paired eyes and grasping spines; lateral and caudal fins; longitudinal muscles in quadrants, no circular muscles, nervous system of cerebral and ventral ganglia connected by circumoesophageal connectives; no circulatory or excretory systems, 70 species

34. **Hemichordata** (acorn worms, *Balanoglossus* and its allies; G. *hemi*, half; L. *chorda*, cord)
Marine, benthic, free-living, solitary or colonial, tornaria larva, worm-like, elongate, bilaterally symmetrical, tripartite coelom (proboscis, collar, trunk), no postanal tail, gill slits, stomochord, superficial nervous system, 100 species

35. **Urochordata** (tunicates, ascidians, salps, larvacea; L. *uro*, tail; *chorda*, cord)
Marine, free-living or sessile, solitary or colonial; the adult lacks a coelom, is unsegmented and lacks cartilage or bone; a cellulose test (tunic) surrounds the body; notochord confined to the tail of the larva, not present in adult; nervous system degenerate in adult, 2000 species.

36. **Cephalochordata** (lancelets, amphioxus and its allies; L. *cephalo*, head; *chorda*, cord)
Marine and free-living, elongate, bilaterally flattened, tailed, unpigmented; anterior buccal cirri, wheel organ and velar tentacles; many gill slits, 23 species.

37. **Chordata** (chordates, craniates or vertebrates, L. *chorda*, cord)
Single dorsal nerve cord, notochord, gill slits in pharynx or throat, bilaterally symmetrical, tetrablastic, well-developed digestive tract, sense organs, sexual reproduction, separate sexes. The Agnatha and Gnathostomata have a brain, skull and paired appendages. Agnatha are jawless, Gnathostomata are jawed with a middle ear with three semicircular canals. All except agnathans have vertebrae and so are vertebrates, 45 000 species[2]

---

[1] Primarily based on data in Laverack and Dando (1987) and Margulis and Schwartz (1988). Organization into super-phyla is on the basis of three embryological criteria: symmetry, coelom and fate of the blastopore; see Table 2.3 for these and other criteria. Bilaterally symmetrical animals (phyla 5–37) are known as the Bilateria.

[2] The terms chordates and vertebrates are usually used as synonyms although they are not. All members of the phylum possess a notochord and so are all chordates. Hagfishes lack vertebrae and so are not vertebrates. All other chordates are vertebrates (craniates); see Fig. 15.2.

1. Pogonophora, 1937 (phylum 30);
2. Archaeocyatha, 1955 (an extinct phylum confined to the Cambrian era; see s. 2.2.2);
3. Gnathostomulida, 1969 (phylum 8);
4. Loricifera, 1983 (phylum 13);
5. Vestimentifera, 1985 (phylum 31); and
6. Cycliophora, 1995 (phylum 10).[2]

Debate over phylum status for some of the above is on-going.

The Vestimentifera – red, tubular worms that live in and near hydrothermal vents in the deep oceans – were initially classified as a class within the Pogonophora. Exciting descriptions of developing vestimentiferans from two genera (*Lamellibrachia* and *Escarpia*) collected in 1996 from cold deep sea vents in the Gulf of Mexico by a manned submersible, have given us our first glimpse of the embryology of these fascinating creatures. Cleavage of these embryos is absolutely typical of the spiralian cleavage seen in embryos of polychaete annelids. Craig Young and his colleagues concluded from the embryology that vestimentiferans are highly modified annelids and not a separate phylum, but debate continues.[3]

The Loricifera are minute (max. 300 μm length), marine organisms with no fossil record. Their unusual mix of characters includes features of mouth parts found in tardigrades, larval sense organs seen in rotifers, cross-striated muscle seen in kinorhynchs, a cuticle reminiscent of priapulans, and larval spines structured like those in nematomorphs; the invertebrate equivalent of the platypus.

Symbion pandora, the single species in the Cycliophora, is also minute (347 × 113 μm). Its sac-like body is attached by an adhesive disc to the lips of its host, the Norway lobster *Nephrops*. Its nearest affinities are to entoprocts and ectoprocts.

Although other unusual organisms were discovered this century, all can be assigned to classes or orders within phyla. One is *Neopilina*, a primitive, bilaterally symmetrical mollusc discovered in 1956. A second is a representative of the Cephalocarida, an ancient group of crustaceans discovered in 1955 (Mayr, 1982). A third is *Latimeria chalumnae*, the only living representative of the coelacanths; its discovery in 1938 was retold in 1991 by Keith Thomson in *Living Fossil*. The phylogenetic position of the coelacanth as a monophyletic group with lungfishes, each of which is equally closely related to land vertebrates, appears to have been settled finally with molecular analyses (s. 2.3.2).

2.2.2 Problematic extinct phyla and fauna

More controversial, because their morphology is even more subject to interpretation, is assignment of phylum status to extinct animals, including Archaeocyatha, the Tommotion fauna, Agmata, Hyolitha, Ediacaran fossils, and until recently, conodonts.[4]

(a) Archaeocyatha and Tommotion fauna

The Archaeocyatha are cone-shaped, double-walled, reef-dwelling organisms restricted to the Cambrian. Although their affinities are uncertain, Debrenne and Wood (1990) place them with the sponges. The Tommotion fauna, best known from Russia and China but world-wide in distribution, consists of unassignable calcified elements and archaeocyathids that became extinct by the end of the Cambrian.

(b) Agmata and Hyolitha

The Agmata consists of a single Early Cambrian species, *Salterella rugosa*, comprised of an 8–9 mm-long, radially symmetric, calcified cone, filled with inclined laminae surrounding a central tube.

The Hyolitha are bilaterally symmetrical, concave, worm-like (>150 mm long), marine fossils of the Lower Cambrian to Middle Permian. They have a calcareous skeleton, an operculum, and possibly lateral 'fins'.

(c) Ediacaran faunas

The 500–600 million-year-old Precambrian Ediacaran fauna, discovered in 1947, consists of flat, multicellular (some think them to be plasmodial), soft-bodied organisms, some up to a metre in length. Originally described from the Pound quartzite in the Ediacara Range of South Australia, this fauna is now known to be world-wide. Similar fossils are found in England, South Wales, Namibia, Russia, California, Alaska and, perhaps my favourite name for a geological formation, the Mistaken Point Formation in Newfoundland. The latest indications are that some Ediacaran organisms persisted into the Cambrian and so were contemporaneous with the Burgess Shale fauna (Chapter 3); Ediacaran fossils in Namibia are younger than 565 million years. (The base of the Cambrian is 543 mya.)

These enigmatic organisms have been classified in many ways:

1. as coelenterates, annelids and/or arthropods;
2. as polypoid or medusoid cnidarians;
3. as belonging to a separate, parallel, but unsuccessful sister group of the Eumetazoa, the phylum Vendobionta, with affinities to cnidarians; and
4. as belonging to a totally new kingdom (which may have been ancestral to the Metazoa) – the Vendobionta, named for the Vendian, the last period of the Precambrian era.[5]

Regarding these organisms as a sister group to the Eumetazoa, as proposed by Leo Buss and Adolf Seilacher, is the least parsimonious arrangement. It requires that the metazoan mouth had been acquired many times. A recently described Ediacaran biota in 600 (?) million-year-old rocks from Sonora, Mexico does little to resolve these problems. The Ediacaran fauna may neither be monophyletic nor consist of related organisms. That Ediacaran fossils represent various groups of organisms is reinforced by the reconstruction of *Kimberella quadrata* from the Ediacaran fauna of the White Sea in Russia. Previously described from the Ediacaran fauna of Australia as a jellyfish, this simple bilaterally symmetrical organism with a single shell may be a common Late Precambrian ancestor of molluscs and allied invertebrates.[6]

(d) Conodonts

Until this decade conodonts were among the most enigmatic of all fossils. Known only from minute (0.25–3 mm), tooth-like, phosphatic structures known as conodont elements (Figure 2.3), they represent a diverse assemblage of fossils with an enormously long geological history extending from the Lower Cambrian to the Upper Triassic. (Consequently, conodonts are of considerable utility in stratigraphic reconstructions.) Conodont elements could not be assigned to a phylum, in large part because no 'conodont animal' was known.

Three diverse sets of data now demonstrate that conodonts are chordates:

1. astonishing histological analysis demonstrating that conodont elements are composed of classic vertebrate skeletal and dental tissues (bone, dentine and enamel);
2. demonstration of microwear on conodont elements consistent with their use as cutting teeth in feeding, and reconstruction of the conodont apparatus (Figure 2.4); and
3. discovery and reconstruction of the 'conodont animal' as bilaterally symmetrical with paired eyes, tail fin, blocks of muscles and a notochord.[7]

Chordate affinities are shown in Figure 2.5. Cephalochordates are the closest sister group to the two extant taxa of jawless vertebrates, the hagfishes and the lampreys. Using the lines of evidence listed above, conodonts are a subphylum within the Chordata. Precisely

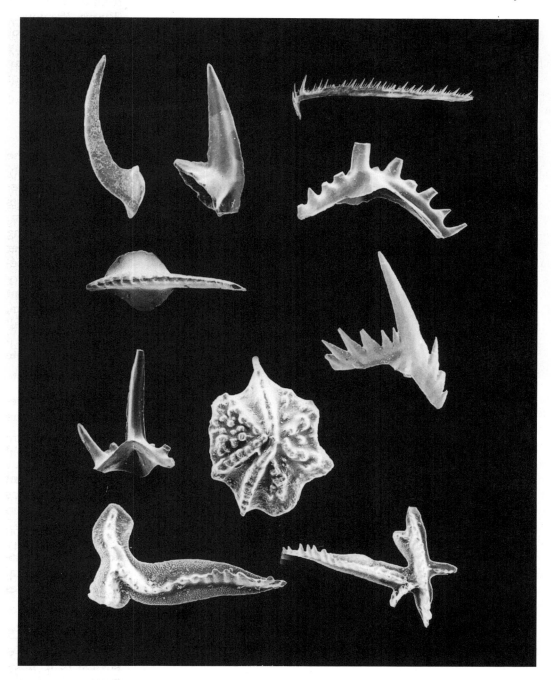

Fig. 2.3 A montage of conodont elements to illustrate their general structure and diversity. Adapted from Aldridge *et al.* (1994) from a figure provided by Richard Aldridge.

Fig. 2.4 A scale model of conodont elements forming a conodont apparatus as it would have functioned in feeding. Figure provided by Mark Purnell.

where conodonts fit within the chordates remains problematical; they could occupy any of the six positions shown in Figure 2.6. None is without its problems. Positions a–d in Figure 2.6 require independent evolution of characters of the muscular and skeletal systems in conodonts and vertebrates (independent loss in hagfishes would also fit situation d). Position e requires secondary loss of skeletal tissues in lampreys.

2.3 RELATIONSHIPS AMONG PHYLA

Relationships among phyla and how phyla arose – topics of abiding interest to naturalists for centuries – are again attracting much attention. The impetus is at least fourfold:

1. increased understanding of invertebrate and vertebrate structure and development (s. 2.3.1);
2. the availability of molecular phylogenies (ss. 2.3.2, 2.3.3);
3. the rediscovery of Ediacaran-age faunas and Cambrian faunas such as those of the Burgess Shale (Chapter 3); and
4. increased knowledge of the soft tissues of fossil species.

Knowledge has advanced so rapidly in these four disparate spheres that there are calls for discourse between them, as eloquently argued by Simon Conway Morris in his 1994 article entitled 'Why molecular biology needs palaeontology'. Evolutionary developmental biology is the arena in which that discourse is taking place.[8]

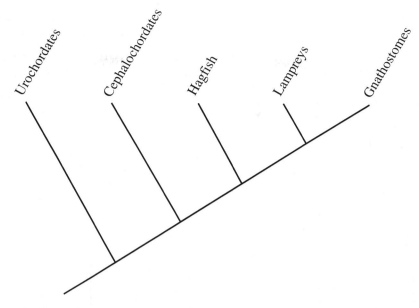

Fig. 2.5 Proposed relationships between living chordates. Cephalochordates are the closest sister group to the two groups of extant jawless vertebrates (hagfishes and lampreys). Adapted from Aldridge and Purnell (1996).

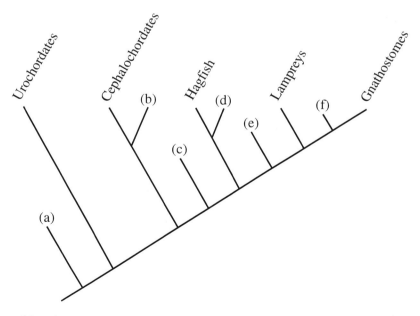

Fig. 2.6 Six possible relationships (a–e) of conodonts to urochordates, cephalochordates, jawless vertebrates (hagfish, lampreys) and gnathostomes. See text for details. Adapted from Aldridge and Purnell (1996).

2.3.1 Embryological evidence

Each phylum originated from a single founding species, is characterized by a unique body plan or *Bauplan* (Chapter 6), and had an independent evolutionary history for over 540 to 560 million years according to most current estimates. In traditional (Linnean) classification, phyla share common features and ancestors but are not necessarily monophyletic. In cladistic classifications, phyla are monophyletic.

Various schemes are used to group phyla into 'super-phyla'. Much evidence comes from fundamental features of embryological development, summarized in Table 2.3. The embryological criteria used and the grouping of phyla or super-phyla (coded to numbered phyla in Table 2.2) include:

1. radial or bilateral symmetry – Radiata (3, 4) and Bilateria (5–37);
2. two or three germ layers – Diploblastica (2–5) and Triploblastica (6–37);
3. spiral or radial cleavage – Spiralia (1–29) and Bilateria (30–37);
4. no coelom, pseudocoelom or true coelom – Acoelomata (5–9), Pseudocoelomata (10–17) and Coelomata (18–37);
5. original opening of the embryo forms the mouth or mouth opens secondarily – Protostomia (18–29) and Deuterostomia (30–37);
6. ventral, dorsal or variably positioned nervous system – Gastroneuralia (18–29), Notoneuralia (30–37) and Heteroneuralia (3–17); and
7. moulting (ecdysis) as a feature of the life cycle – Ecdysozoa (12, 16, 17, 22, 26, 28, 29).[9]

These schemes and groupings are not immutable. Even as such schemes were coming into existence over 120 years ago, the Russian embryologist Elias Metschnikoff was presenting evidence that sponges are triploblastic (Metschnikoff, 1876).[10] In Box 10.2, I summarize evidence detailed elsewhere (Hall, 1997a)

that the subphylum vertebrata is tetrablastic, not triploblastic, the neural crest forming a fourth germ layer. The simple dichotomy of spiral versus radial cleavage (3 above), which has resided in textbooks for a hundred years, may not withstand close analysis; James Valentine (1997) has re-evaluated cleavage patterns, evolutionary modifications in cleavage patterns, and their relationship to the classic divisions of protostomes with spiral cleavage and deuterostomes with radial cleavage. At least five major schemes for relationships between and origins of protostomes and deuterostomes have been proposed in the past (Figure 2.7).

The phylogeny produced by Valentine in 1992 is also informative, in that he incorporated levels of cell and tissue organization (diploblastic, triploblastic) and the status of the coelom, the sequence of appearance of phyla in the fossil record, extinct and living phyla, and morphological evidence for relationships among phyla. Valentine is sanguine about the fate of his scheme: 'It is expected that this tree will become obsolescent as have its predecessors' (Valentine, 1992a, p. 542). It has survived for at least five years; an updated version is produced as Figure 2.8.

Relationships among phyla reappear when I consider the origins of multicellularity and of the Metazoa in Chapters 13 and 14.

2.3.2 Molecular evidence

Molecular evidence for relationships among phyla comes primarily from amino acid sequences of such proteins as alpha crystallin and myoglobins, and from 18S and 28S ribosomal RNA. The highly conserved sequences of these RNA molecules and their relative independence from mode of development, life history or ecology, make them ideal candidates for tracking phylogenetic history.[11]

A telling example of the utility of molecular phylogeny is what may be a resolution of the relationships of the coelacanth, lungfishes

Table 2.3 Proposals for organizing phyla into 'super-phyla'

Phyla	Sub-kingdom		Symmetry			Coelom		Germ layers		Mouth opening		Nervous system			Cleavage	
	P	E	R	B	A	Ps	C	Di	Tr	Pr	D	H	G	N	Sp	Ra
Placozoa	P		lack		A							H			Sp	
Porifera	P		lack		A			Di				H			Sp	
Cnidaria		E	R		A			Di				H			Sp	
Ctenophora		E	R		A			Di				H			Sp	
Mesozoa		E		B	A			Di				H			Sp	
Platyhelminthes		E		B	A				Tr			H			Sp	
Nemertina		E		B	A				Tr			H			Sp	
Gnathostomulida		E		B	A				Tr			H			Sp	
Cycliophora		E		B	A				Tr			H			Sp	
Gastrotricha		E		B	A				Tr			H			Sp	
Rotifera		E		B		Ps			Tr			H				R
Kinorhyncha		E		B		Ps			Tr			H			Sp	
Lorificera		E		B		Ps			Tr			H			Sp	
Acanthocephala		E		B		Ps			Tr			H			Sp	
Entoprocta		E		B		Ps?			Tr			H			Sp	
Nematoda		E		B		Ps			Tr			H				M
Nematomorpha		E		B		Ps			Tr			H				M
Ectoprocta		E		B			C		Tr	Pr			G			M
Phoronida		E		B			C		Tr	Pr			G			R
Brachiopoda		E		B			C		Tr	Pr			G			R
Mollusca		E		B			C		Tr	Pr			G		Sp	
Priapulida		E		B			C		Tr	Pr			G			R
Sipuncula		E		B			C		Tr	Pr			G		Sp	
Echiura		E		B			C		Tr	Pr			G		Sp	
Annelida		E		B			C		Tr	Pr			G		Sp	
Tardigrada		E		B			C		Tr	Pr			G			M
Pentastoma		E		B			C		Tr	Pr			G			M
Onychophora		E		B			C		Tr	Pr			G			M
Arthropoda		E		B			C		Tr	Pr			G			M
Pogonophora		E		B			C		Tr		D			N	Sp	
Vestimentifera		E		B			C		Tr		D			N	Sp	
Echinodermata		E		B(1)			C		Tr		D			N		Ra
Chaetognatha		E		B			C		Tr		D			N		Ra
Hemichordata		E		B			C		Tr		D			N		Ra
Urochordata		E		B			C		Tr		D			N		Ra
Cephalochordata		E		B			C		Tr		D			N		Ra
Chordata		E		B			C		Tr(2)		D			N		Ra

Subkingdoms: Parazoa (P), Eumetazoa (E).

Symmetry: Radial (R), Bilateral (B), Modified radial (M); see Valentine (1997) for a re-analysis of cleavage patterns in the metazoans.

Coelom: Acoelomate (A), Pseudocoelomate (Ps), Coelomate (C)

Germ layers: Diploblastic (Di), Triploblastic (Tr)

Mouth opening: Protostome (Pr), Deuterostome (D). In protostomes, the mouth opening is primitive; the original opening of the embryo forms the mouth. In deuterostomes, the mouth opening is secondary and not derived from the original embryonic mouth.

Nervous system: Heteroneuralia (H), Gastroneuralia (G), Notoneurulia (N). The nervous system is organized on the ventral axis (Gastroneuralia), dorsally and associated with a notochord (Notoneuralia) or variable in position (Heteroneuralia).

Cleavage: Spiral (Sp), radial (Ra).

(1) Many echinoderms are built on the basis of pentameric symmetry.

(2) I argue that vertebrates are tetrablastic, the neural crest forming a fourth germ layer (see Box 10.2 and Hall, 1997a).

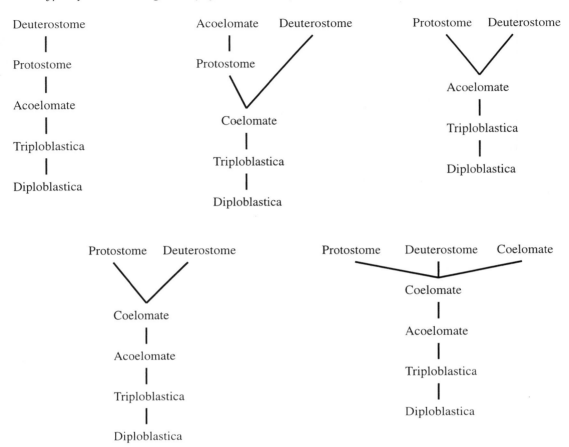

Fig. 2.7 Five scenarios for relationships between and origins of protostomes and deuterostomes. Based on data in Willmer (1990).

and tetrapods. Three possible relationships had their advocates in the past:

1. lungfish as the sister group of tetrapods (Figure 2.9 top);
2. coelacanths as the sister group of tetrapods (Figure 2.9 middle);
3. lungfishes and the coelacanth as equally closely related as sister groups of tetrapods (Figure 2.9 bottom).

Rafael Zardoya and Axel Meyer sequenced almost the entire gene (3500 base pairs) of the large nuclear 28S rRNA gene from three species of lungfish, the coelacanth, rainbow trout, an eel, and the short-nosed sturgeon. Such is the state of sequencing 28S rRNA that this study more than doubles the known sequences for all the vertebrates. No matter which of a variety of phylogenetic analyses was used, lungfishes and the coelacanth formed a monophyletic group, each equally closely related to tetrapods, i.e. the third option above.[12]

In two additional studies, Meyer's laboratory has sequenced the complete mitochon-

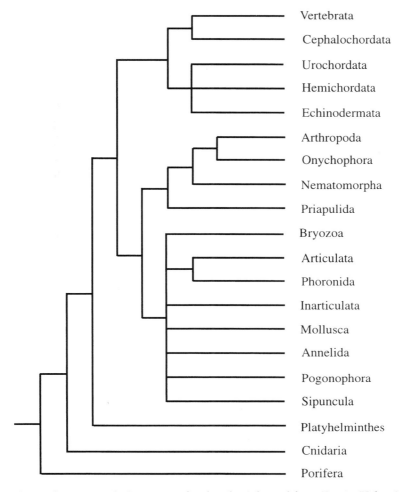

Fig. 2.8 Relationships of 20 major phyla; see text for details. Adapted from Erwin, Valentine and Jablonski (1997).

drial DNA sequences (16646 and 16624 base pairs respectively) from the African lungfish *Protopterus dolloi* and the bichir *Polypterus ornatipinnis*. Phylogenetic analyses using these sequence data allowed them to conclude that the African lungfish was intermediate between ray-finned fishes and tetrapods and that the bichir is the most basal living ray-finned fish and not a lobe-fin (sarcopterygiian) fish (Figure 2.10). One implication of this analysis is the independent evolution of the bichir fin from the fins of lungfishes, teleosts, sharks or *Amia*.[13]

Another approach to phylogenetic relationships is to use rates of divergence of sequences of selected molecules. Greg Wray, Jeffrey Levinton and Leo Shapiro (1996) used rates of sequence divergence of seven molecules to test whether phyla arose in an explosive burst at the beginning of the Cambrian, which is the

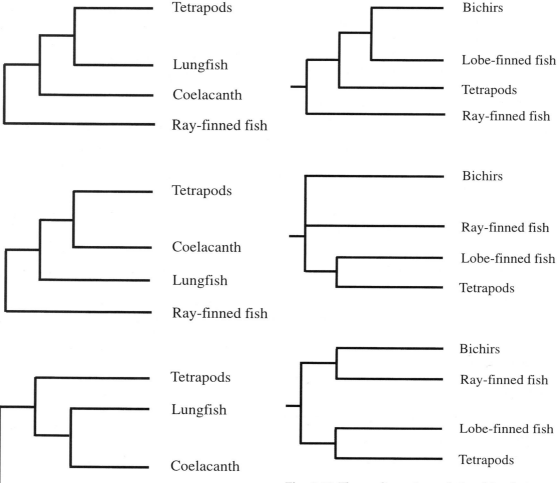

Fig. 2.9 Three alternative relationships of the coelacanth, lungfish, ray-finned fish and tetrapods; see text for details. Adapted from Zardoya and Meyer (1996a).

Fig. 2.10 Three alternative relationships between bichirs, ray-finned and lobe-finned fish. Top: As advocated by Huxley and Cope last century, bichirs are lobe-finned fish. Middle: The relationship of bichirs to ray-finned and lobe-finned fish is unresolved. Bottom: Bichirs are ray-finned fishes. Adapted from Noack, Zardoya and Meyer (1996).

conventional view (Figure 2.11). The seven molecules selected were 18S rRNA, ATPase6, cytochrome C, two cytochrome oxidases, alpha- and β-haemoglobins, and NADH1. Their conclusions include that chordates separated from invertebrates a billion years

ago and that divergence of phyla was spread over an extended time and not in a single explosion, Cambrian or otherwise (Figure 2.11).

These conclusions are controversial for two major reasons. A divergence over a billion years ago is neither in accord with morphological data nor with phylogenies generated using single conserved molecules such as 18S rRNA. Secondly, there is much debate over whether the rate of divergence of molecular sequences is constant, either over time, or between molecules. Wray and his colleagues are aware of this potential weakness in the approach; much of their paper addresses ways of countering such criticisms. They are confident that despite variations in rates of sequence divergence through time, across taxa, and between molecules, analysis of so many taxa over such a prolonged period of their history (114 invertebrate-vertebrate divergences involving 16 invertebrate phyla) allows rates to be averaged into a meaningful mean rate of divergence. Others are not so sanguine.

I share Simon Conway Morris's (1997) view that simple averaging over such a range of molecules and organisms does not increase the reliability of the data, especially when divergence times for pair-wise comparisons of groups cluster on the basis of the particular molecule used for the comparison. Divergence times for the same comparisons range as widely as 773 and 1621 million years for the separation of annelids from chordates and 788 and 1511 million years for the separation of molluscs from chordates. Resolution of molecular approaches is also bounded. Hervé, Chenuil and Adoutte (1994) argue from a study of 18S rRNA sequences from 69 members of 15 phyla, that phylogenetic events less than 40 million years apart cannot be resolved. Their conclusion was that phyla arose over some 20 million years between 540 and 520 mya.

While urging caution on the accuracy of molecular clock evidence, Richard Fortey and his colleagues see a new paradigm emerging in which the Cambrian 'explosion' of phyla was decoupled from the origin of the evolutionary innovations upon which those phyla are based. Adoption of this position is not contingent on origin of the innovations a billion years ago. Their caution – that first appearance of the fossil record need not coincide with time of origination – is an important principle to keep in mind in such studies (see s. 14.2.1).[14]

2.3.3 Molecules and morphology

Although congruence is often high, phylogenies based on morphology or molecular sequences do not always coincide. Lack of congruence should not surprise us, although it often does. Structures and features of organisms evolve at different rates, starting from different times in the history of the group. Why should the evolution of alpha crystallin track the evolution of paired appendages, for example? Indeed, as emphasized by Patterson, Williams and Humphries (1993) in their review of molecular and morphological phylogenies, incongruences between molecular trees are often as striking as those between morphological trees. Their claim is that no well-supported phylogeny generated using morphological data has been overthrown by molecular data. At the same time they do acknowledge that molecular sequences have contributed much, especially where morphological data are inconclusive.

Genealogical relationships between inbred strains of mice appear quite different depending on whether morphology, subsets of proteins, the immune system or viruses are used to construct the trees (Atchley and Fitch, 1991). The same situation holds for construction of phylogenies on the scale of the animal kingdom. Naylor and Brown (1997) reported that not all molecules produce trees that fit those produced using morphological criteria. Close analysis of nucleotides from mitochondrial genes showed that:

1. codons for genes involved in the three-dimensional structure of proteins produced trees with the best fit;

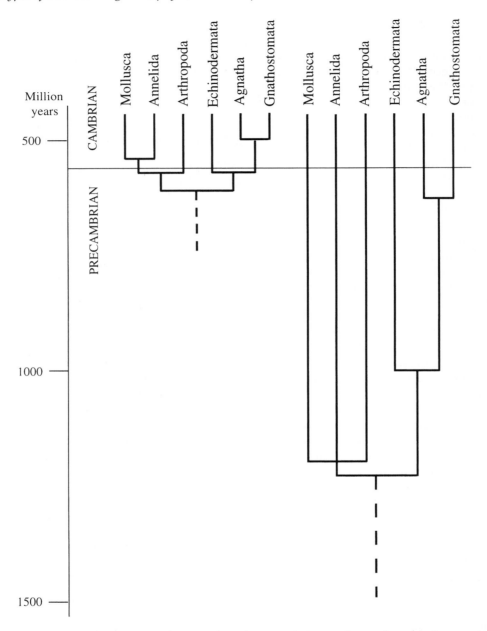

Fig. 2.11 Two views of the timing of the origins of major phyla. Analyses of morphology or 18S rRNA produce the scenario on the left, in which major invertebrate phyla originated in a 'Cambrian explosion' at the pre-Cambrian-Cambrian boundary while craniates (Agnatha, Gnathostomes) diverged in the Cambrian. On the right is the scenario resulting from analysis of rates of sequence divergence of seven molecules, giving much earlier origins for invertebrate phyla and divergence of craniates a billion years ago. Adapted from data in Wray, Levinton and Shapiro (1996) and Conway Morris (1997).

2. codons for hydrophilic or charged molecules the next best fit; while
3. codons for hydrophobic molecules produced what amounted to nonsense trees.

The phylogenetic information in molecules is not equal.[15]

The 1991 analysis by Bradley Shaffer and colleagues, although not at the level of phyla, is illustrative of difficulties that can arise between morphological and molecular phylogenies. Their phylogenetic analysis of North American salamanders of the family Ambystomatidae, based on variation in 26 allozyme loci, conflicted strongly with previous phylogenies based on morphology. On closer analysis it transpired that the major conflict involved five species in one subgenus for which independence of characters was an issue. Resolving that issue would resolve the conflict. Characters must both be selected with care and be independent from one another.

2.4 ORIGINS OF PHYLA

When we move beyond the relationships among phyla and ask questions such as 'What factors were responsible for the origin of phyla?', we are on even more intuitive and less empirical ground. The two poles are usually set as **internal** and **external** factors. Internal factors include:

1. an early-evolving genome;
2. many mutations of small scale or few of large;
3. the evolution of embryonic development;
4. origins and diversification of cell and tissues types; and
5. segregation of germ line from soma.

External factors include:

1. availability of ecological niches;
2. climatic and geographical factors;
3. competition or lack of competition; and
4. invasion of new adaptive zones.

In an approach that deviates from the norm by focusing on factors underlying radiations rather than the more typical morphological or molecular analyses, Erwin (1992) identified four patterns of evolutionary radiations:

1. novelty events that generate increased morphological complexity such as the new body plans associated with the origin of phyla;
2. broad diversification events involving many lineages driven by ecological factors, beautifully exemplified by Foote's 1996 analysis of diversification within crinoids over 150 million years;
3. economic radiations involving the exploitation of new ecological opportunities as would have occurred in the diversification of animals such as the vestimentiferans of the deep sea hydrothermal vents; and
4. adaptive radiations, a number of which are discussed in Chapters 16 and 17.

One particular fauna, the fossils of the Burgess Shale, brings origins, relationships and radiations into sharp relief. Although not unique, these fossils illustrate the issues of morphology, types of organisms, species, and the origin of body plans, and serve as a background for future chapters in which the relationship of development to these issues is discussed.

2.5 ENDNOTES

1. For discussion of the five kingdoms, see Margulis and Schwartz (1988). For the Archaea, see Woese (1981) and Woese, Kandler and Whellis (1990).
2. Basic references for these phyla are: Pogonophora (Southward, 1963; Nørrevang, 1970, 1975 and Ivanov, 1963, 1975); Archaeocyatha (Okulitch, 1955); Gnathostomulida (Riedl, 1969; Sterrer, 1972 and Boaden, 1975); Loricifera (Kristensen, 1983); Vestimentifera (M. L. Jones, 1985; Jones and Gardiner, 1989); Cycliophora (Funch and Kristensen, 1995; Conway Morris, 1995).
3. Formation of a polar lobe and consequent unequal first cleavage, and formation of a second

polar lobe before the second cleavage to give three small blastomeres and a large 'D' cells, typical of the spiralian cleavage seen in polychaete annelids, has been described for vestimentiferans by Young, Karatajute-Talimaa and Smith (1996), who regard these animals as highly modified annelids (see s. 21.7.2 for annelid development).

4. For literature on the Agmata, Hyolitha and conodonts, see Runnegar *et al.* (1975), Yochelson (1977), Runnegar (1980), M. M. Smith and Hall (1990, 1993), Aldridge *et al.* (1993, 1994) and Janvier (1995, 1996b). Additional literature on conodonts is cited in the text.

5. For literature on possible classifications and relationships of Ediacaran fossils, see Glaessner (1984), Seilacher (1989, 1992), Valentine (1992b), Conway Morris (1993b, 1994b), Buss and Seilacher (1994) and Knoll (1996). For difficulties associated with the relationships of the Cnidaria and whether they form a true phylum, see Moore and Willmer (1997).

6. See McMenamin (1996) for the Mexican Ediacaran fauna, and Fedonkin and Waggoner (1997) for the Late Precambrian *Kimberella*. The youngest Ediacaran fossils from Namibia, estimated to be 600 million years old, and thus to have extended to the base of the Cambrian, have been interpreted as supporting the Vendobionta (Narbonne *et al.*, 1997).

7. See Sansom *et al.* (1992, 1994) and Sansom (1996) for histology of conodont elements; Purnell and von Bitter (1992) and Purnell (1995) for functioning of conodont elements; Briggs, Clarkson and Aldridge (1983), Aldridge and Theron (1993), Aldridge *et al.* (1993) and Gabbott, Aldridge and Theron (1995), for reconstruction of the conodont animal as a chordate; and Aldridge *et al.* (1986, 1993, 1994), Janvier (1995, 1996a,b) and Aldridge and Purnell (1996) for chordate phylogenies incorporating conodonts.

8. Also see Hall (1998a) for a discussion of fields in which embryos and fossils meet. The 26 papers presented in the major symposia – 'Evolutionary Relationships of Metazoan Phyla' organized by D. McHugh and K.M. Halanych, and 'Development and Evolutionary Perspectives on Major Transformations in Body Organization' organized L. Olsson and B.K. Hall – held at the annual meeting of the Society for Integrative and Comparative Biology in Boston in January 1998, provide a rich and thoroughly up-to-date analysis of integrated use of developmental, geentic, palaenotologial and phylogenetic approaches to understanding relationships among, and origins of, phyla and of the body plans of metazoans. Both symposia will be published in *American Zoologist* in 1998.

9. Willmer (1990) and Nielsen (1995) contain comprehensive treatments of super-phyla in the context of invertebrate relationships, albeit approaching the topic from quite different perspectives. Moore and Willmer (1997) address topics such as segmentation, body cavities, symmetry and germ layers and conclude that there is evidence for substantial convergence (or that convergence cannot be ruled out on the available evidence). A recent analysis of 18S rDNA sequences places the nematodes close to arthropods in a new clade, the Ecdysozoa, or animals that moult (Aguinaldo *et al.*, 1997).

10. See Ghiselin and Groeben (1997) for an analysis of the embryological work of Metschnikoff.

11. See Lake (1989) and Willmer and Holland (1991) for molecular evidence and phylogenetic reconstructions, and Huelsenbeck and Rannala (1997) for the latest methods to test those relationships.

12. The seven species used by Zardoya and Meyer (1996a) were the South American, African and Australian lungfishes *Lepidosiren paradoxa*, *Protopterus aethiopicus* and *Neoceratodus forsteri*; the coelacanth *Latimeria chalumnae*; rainbow trout *Oncorhynchus mykiss*; eel *Anguilla rostrata*; and the short nosed sturgeon *Acipenser brevirostrum*. The issue may not be settled finally. An analysis of mitochondrial DNA by Axel Meyer and his colleagues (reported in the Sept. 5 , 1997 issue of *Science*), places the lungfish closer than the coelacanth to the tetrapods.

13. See Zardoya and Meyer (1996b) and Noack, Zardoya and Meyer (1996) for analyses of *Protopterus* and *Polypterus* mitochondrial DNA.

14. See Vermeij (1996), Conway Morris (1997) and Fortey, Briggs and Wills (1997) for critical evaluations and partial re-analysis of these data. Using 18 protein coding loci, Ayala *et al.* (1998) estimated divergence times of protostomes from deuterostomes and of echinoderms from chordates of 679 and 600 mya respectively, and estimated the origin of the phyla at 700 mya, a dating that is consistent with estimates derived from fossils. They raise

important points to counter the much older divergence times determined by Wray *et al.* (1996).

15. Two edited volumes (Patterson, 1987; Fernholm, Bremer and Jornvall, 1989) are devoted specifically to conflict or compromise between molecular and morphological evolution; see also A. B. Smith (1992) for echinoderms, Patterson (1990), Willmer and Holland (1991), Eernisse and Kluge (1993) and Valentine *et al.* (1991, 1996) for metazoan origins; and Lecointre (1994, 1996) for fish phylogeny. See Erwin, Valentine and Jablonski (1997) and Moore and Willmer (1997) for recent general accounts (the latter for a detailed analysis of convergent evolution within the invertebrates).

3

FOSSILS OF THE BURGESS SHALE

'It may not be generally appreciated by biologists that first occurrence in the fossil record is not necessarily the same as time of origination.'

Fortey, Briggs and Wills, 1997, p. 430

Until a decade or so ago, we contented ourselves with notions of a gradual progression of structural organization from few simple types in the Lower Cambrian to the abundance of organisms alive today. The reinterpretation of the Burgess Shale fossils initiated by Simon Conway Morris and Harry Whittington in 1985 changed all that. Representatives of every phylum except the Bryozoa are found from Lower or Middle Cambrian rocks. Organisms currently unassignable to any phylum also existed 530 mya, not as some isolated early evolutionary experiment, but as a world-wide metazoan radiation.

In the Lower to Middle Cambrian Period, in what is now part of Yoho National Park, British Columbia, existed a large limestone reef some 160 metres deep and more than 20 kilometres long. Deep water containing a profusion of animal and plant life lay adjacent to this reef. Normally only animals with such hard parts as mineralized skeletons or teeth are preserved as fossils, but at this site, exceptional conditions existed. Mud periodically broke away from the reef face, rapidly burying everything in its path under conditions of low oxygen and high hydrogen sulphide that delayed organic decay. Many plants and soft-bodied animals were thus buried beside the reef front, leaving an exquisite representation of Cambrian life. Some dozen fossil-containing localities are found along the reef, but the finest is the Phyllopod Bed of the

Burgess Shale, beautifully illustrated by Derek Briggs and his colleagues in *The Fossils of the Burgess Shale* (1994).[1]

The fossils were discovered in 1909 by Charles Doolittle Walcott (1850–1927), who at the time was Secretary of the Smithsonian Institution in Washington, DC. Walcott had previously discovered the fossil fauna of the Silurian Harding sandstone formation at Canon City, Colorado, opening the first window onto early vertebrate life to see such jawless vertebrates as *Astraspis* and *Eriptychius* (Walcott, 1892).[2]

Walcott spent four subsequent seasons excavating fossils from the Burgess Shale, describing his finds in a series of papers published between 1911 and 1926 in the *Smithsonian Miscellaneous Collections* and in a major posthumous addendum. Subsequent work on the site was done by a team from the Museum of Comparative Zoology of Harvard University in the early '30s, but the expeditions from the Geological Survey of Canada, organized by Harry Whittington in 1966 and 1967, revealed the unique features of these fossils and rekindled interest in the Burgess Shale.[3]

At least 140 species belonging to 124 genera have been identified. One hundred and seven genera can be assigned to phyla known to exist today. In decreasing order of abundance they consist of arthropods (44) > sponges (18) > brachiopods and lophophorates (8 each) > priapulans (7) > polychaete annelids (6) > echinoderms (5) > coelenterates (4) > molluscs and hemichordates (3 each) and one (?) cephalochordate. The remaining genera have not yet been assigned to any known phyla and may belong to new, undescribed phyla.

This simple statement, that some Burgess Shale fossils cannot be assigned to known phyla, raises important issues. Some of these – how phyla are identified; how organisms are grouped into higher taxa – were addressed in the last chapter. Others – whether there are phyla still to be discovered; how different morphologies, body plans, types of organisms and species are recognized and generated – are addressed in this chapter.

3.1 BURGESS SHALE FOSSILS ASSIGNABLE TO PHYLA

The most common Cambrian fossils in most assemblages are arthropods, specifically trilobites. The only fossilized embryos from this period are also thought to be trilobite – structures interpreted as cleavage-stage embryos as late as the 128-cell stage, recovered from Middle Cambrian deposits in southern China by Zhang and Pratt (1994). (Later stage embryos/larvae (possibly cnidarians, certainly early metazoans) have now been described from the Early Cambrian of China and Siberia by Bengtson and Zhao (1997)). Arthropods are the most abundant individuals, species and genera in the Burgess Shale, but here some 30 different types of soft-bodied arthropods predominate, representing all four major groups:

1. Trilobita, now extinct;
2. Crustacea (crabs, shrimp, lobsters and copepods);
3. Chelicerata (spiders, scorpions, horseshoe crabs, mites); and
4. Uniramia (centipedes, millipedes, insects).[4]

To provide some idea of this diversity, I describe two Burgess Shale arthropods (*Marrella splendens* and *Canadaspis perfecta*), an

onychophoran (*Aysheaia pedunculata*), and a lobopodan (*Hallucigenia sparsa*).

3.1.1 Arthropods

Marrella splendens, the most common species in the Burgess Shale, represents a third (15 000 specimens) of all fossils collected. Walcott called *Marrella* the lace crab in reference to the identical sets of limbs projecting from its many body segments (Figure 3.1). A basal, unspecialized arthropod, *Marrella* ranged in size from 2.5 mm to almost 20 mm.

Canadaspis perfecta, the next most common arthropod in the Burgess Shale, is the oldest and most completely known crustacean. The 10–50 mm-long carapace (formed from two valves connected by a hinge), biramous limbs, and characteristic head appendages such as antennae, maxillae and mandibles, reveal the crustacean affinities of *Canadaspis*.

3.1.2 Onychophorans and their allies

Aysheaia pedunculata, an arthropod-like organism, bears a striking resemblance to the terrestrial onychophoran *Peripatus* (Figures 3.2 and 3.3). The ten pairs of walking appendages (lobe-limbs) are distinctive, as is the shape of the 1–6 cm-long body (Figure 3.2). Initially thought to represent a common ancestor of insects and myriapods (millipedes and centipedes), *Aysheaia* is now recognized as an early member of the the phylum Onychophora. Onychophoran relationships are not,

Fig. 3.1 *Marrella splendens*, an arthropod and the most common fossil in the Burgess Shale. Top, as seen from the underside showing the curved spines, antennae (a), long appendages (b), and multiple pairs of limbs. Bottom as reconstructed. Top figure from Conway Morris and Whittington (1985), reproduced with the permission of the Minister of Supply and Services Canada, 1990. Bottom figure reproduced with permission from Whittington (1985).

(a)

(b)

10mm

Fig. 3.2 *Aysheaia*, an arthropod with considerable similarities to the onychophoran *Peripatus* (see Fig. 3.3). The thick legs end in claws. Reproduced from Conway Morris and Whittington (1985) with the permission of the Minister of Supply and Services Canada, 1990.

however, fully resolved. Ballard and colleagues (1992), using 12S mitochondrial ribosomal RNA in a cladistic analysis, identify the arthropods as monophyletic and as including onychophorans. Monge-Najera (1995), however, undertook an extensive cladistic analysis using morphological data that placed a separate onychophoran phylum between polychaetes and arthropods. Sister-group status as a separate phylum within an arthropod grade of organization currently seems most likely.[5]

Hallucigenia sparsa is perhaps the most bizarre of all the Burgess Shale fossils (Figure 3.4a). According to the original reconstructions seven pairs of unsegmented spines pro-

jected from and appeared to support a 0.5–3 cm-long elongated body (Figure 3.4b). Seven dorsal tentacles, each terminating in a bifurcated 'claw', were reconstructed. A further cluster of tentacles was interpreted as situated posteriorly on the forward-facing 'tail'. Details of the 'head' are unclear; indeed, palaeontologists were not even sure which end of *Hallucigenia* was which. Hence the generic name. Some thought these specimens might even be parts of a larger organism. Initially, therefore, *Hallucigenia* was not assigned to any phylum.

The reconstruction of *Hallucigenia* in Figure 3.4b is as just described: spines supporting the body ventrally with tentacles on the dorsal

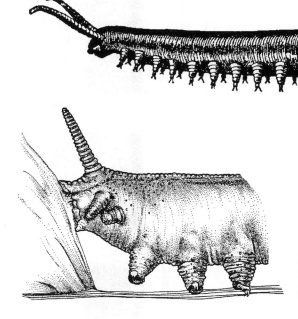

Fig. 3.3 *Peripatus capensis* (top) and *P. sedgwicki* (bottom) to show the resemblance to *Aysheaia*. Adapted from Parker and Haswell (1960) and Brusca and Brusca (1990).

surface. Further analysis and new specimens from another location have turned *Hallucigenia* upside down and back to front and given it a taxonomic home. With the benefit of Early Cambrian fossils from southern China dated at 520–530 mya, Ramsköld and Hou (1991) reinterpreted the 'tentacles' as legs and 'legs' as spines (Figure 3.4c) and concluded that *Hallucigenia* is a lobopodan (an animal with sac-like, fluid-filled appendages) with affinities to *Aysheaia* and the phylum Onychophora. Ramsköld (1992) and Hou and Bergström (1995) discuss the various interpretations and relationships of *Hallucigenia* and the Cambrian lobopodans as ancestors of the living onychophorans.

The first representatives of other groups are also found within the Burgess Shale fauna.

3.1.3 Ctenophores and coelenterates

Fasciculus vesanus has an 8 cm diameter body supporting many elongate, transversely striated bands. These are similar to, although greater in number than, the comb rows of ctenophores (comb jellies). Conway Morris and Collins (1996) interpret *Fasciculus* as the first ctenophore.

Mackenzia costalis had a ridged, sac-like body some 85 mm in diameter but as much as 16 cm in length. Its appearance and location on the seabed, or attached to sessile echinoderms, allies it with sea anemones.

3.1.4 A cephalochordate?

One putative cephalochordate, the 4 cm long *Pikaia gracilens*, is found in the Burgess Shale. The amphioxus-like (cephalochordate) appearance of *Pikaia* is striking (Figure 3.5a,b). Such is the preservation of these fossils that the outline of the notochord and the typical sigmoidally deflected (zig-zag) pattern of segmented chordate muscles (myomeres) are readily identifiable.

Notochord and myotomes are unequivocal evidence of chordate status, provided that they can be identified unequivocally.

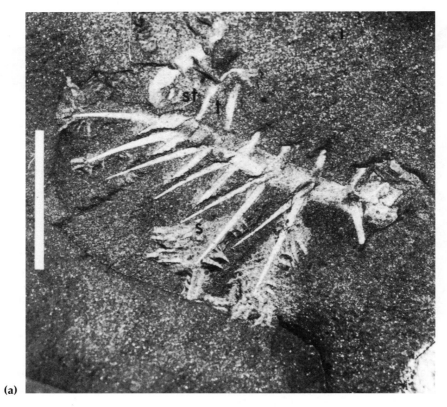

(a)

Fig. 3.4 *Hallucigenia sparsa.* (a) A specimen photographed in reflected light showing the recurved body, spines (s), tentacles (t) and short tentacles (st). Scale bar: 5 mm. Reproduced from Whittington (1985) from a photograph provided by S. Conway Morris. (b) The original reconstruction of an animal supported on its spines with tentacles in the air. Head and tail cannot be readily identified. Reproduced from Campbell (1987). (c) The reinterpretation which turns *Hallucigenia* upside down. Adapted from Ramsköld and Hou (1991).

(Myomeres in the conodont animal convinced Briggs, Fortey and Wills (1983) that conodonts should be assigned to the Chordata; see section 2.2.2d.) Heightened confidence in interpreting soft tissues comes from increased understanding of how soft tissues are preserved – the developing field of taphonomy, in which organisms such as amphioxus are allowed to decay under controlled conditions. Such experiments demonstrate that notochord sheath, cartilaginous rods and muscles are the most readily preserved chordate structures (Briggs and Kear, 1994).

One important word of caution.

Pikaia has not yet been fully described. The ongoing analysis of over 100 specimens by Conway Morris and Collins is awaited with anticipation, especially when (from his report at a recent conference) it is clear that Conway Morris is not yet convinced that *Pikaia* is a chordate. Butterfield (1990b), who examined some of the *Pikaia* specimens, believes their surface to be cuticular, excluding an affinity with cephalochordates.

The possibility that chordates were present in the Cambrian gains support from the description by Shu, Conway Morris and Zhang (1996) of a 22 mm long, slender,

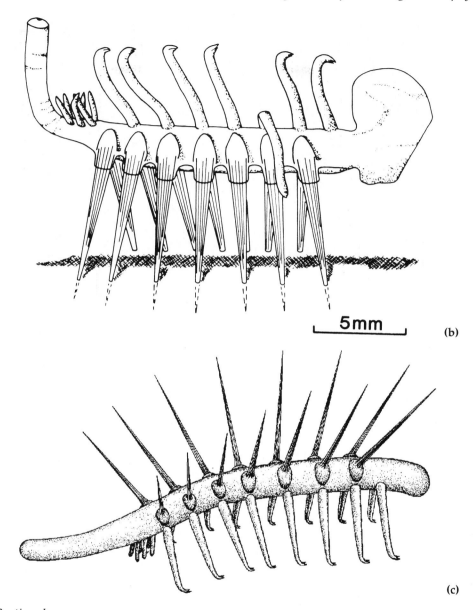

5mm

(b)

(c)

Fig. 3.4 *Continued.*

elongate specimen from the Lower Cambrian Chengjiang formation in Yunnan Province, China. Named *Cathaymyrus diadexus*, this species is rare indeed: 10 000 specimens of other species have been collected from this site, but only one specimen of *Cathaymyrus*.

The age of the Chengjiang formation places *Cathaymyrus* some ten million years older than *Pikaia*. Shu and colleagues identify a pharynx, gill slits, segmented elements (which they interpret as myomeres) and possibly a notochord. These features of cephalochordates

Fig. 3.5 (a) *Pikaia gracilens*, the single cephalochordate (?) species from the Burgess Shale. Myomeres are clearly evident. The dorsal band is presumed to be the notochord. Reproduced from Conway Morris and Whittington (1985) with the permission of the Minister of Supply and Services Canada, 1990. (b) Amphioxus, demonstrating the similarity to *Pikaia*. Adapted from Romer (1960) after Gregory.

and the absence of vertebrate features such as those derived from the neural crest, suggest that *Cathaymyrus* should be placed in the phylum Cephalochordata.

Given interest in the pivotal position of the neural crest in craniate origins and the origins of skeletal tissues in chordates (s. 10.4), it is unfortunate that no skeletal structures are evident in *Pikaia* or *Cathaymyrus*. The earliest vertebrate skeletal remains are conodont elements and the scales of jawless ostracoderm 'fishes' in the Upper Cambrian and Lower Ordovician. *Pikaia*, which is Early to Middle Cambrian in age or *Cathaymyrus*, which is Lower Cambrian, may represent an elusive stage in craniate skeletal evolution.[6]

Another Early Cambrian 'chordate', *Yunnanozoon lividum*, has been found in the Chengjiang fauna. Described as having a notochord, filter-feeding pharynx, endostyle, segmented muscles, branchial arches and gonads, *Yunnanozoon* was classified as a cephalochordate by J.-y. Chen *et al.* (1995b). Shu and colleagues have reinterpreted

Yunnanozoon as a hemichordate, the branchial arches being branchial tubes and the notochord the upper portion of the pharynx. Absence of a notochord would be decisive in removing *Yunnanozoon* from the cephalochordates, but in palaeontology, evidence of absence is not absence of evidence. Whether *Yunnanozoon* remains a hemichordate awaits further analysis. The presence of cephalochordates and/or hemichordates in the Lower Cambrian would speak to how early these phyla diversified and raise the real possibility that urochordates might be found also.

These fossils are assigned to phyla on the basis of structures and basic body plans. This is not the case for the remaining genera in the Burgess Shale.

3.2 BURGESS SHALE FOSSILS NOT CURRENTLY ASSIGNABLE TO PHYLA

Species in 17 genera (some of which are known only from the Burgess Shale), cannot

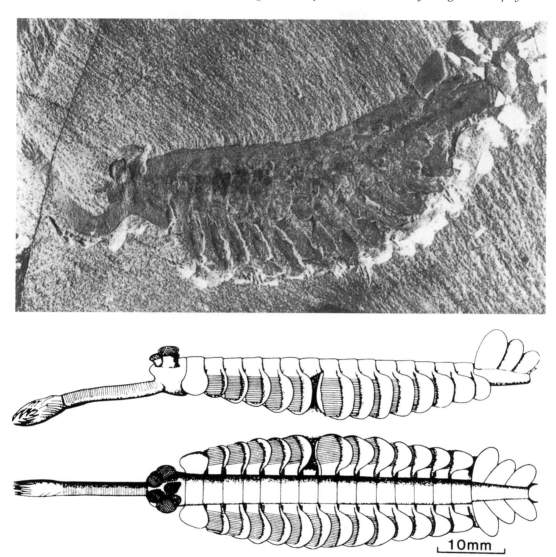

Fig. 3.6 *Opabinia regalis,* one of the unassignable Burgess Shale species. Note the segmentation, lateral extensions bearing gills and posterior locomotory fins. The limbs are unjointed. Three of the five median eyes can be seen. Top is a fossil reproduced from Conway Morris and Whittington (1985) with the permission of the Minister of Supply and Services Canada, 1990. Below are dorsal and lateral views of a reconstruction from Campbell (1987).

be assigned to any known phylum; their body plans do not 'fit' any fossil or extant forms. The 'unassignable' genera of the Burgess Shale are mostly coelomate, free-swimming invertebrates. *Opabinia, Anomalocaris, Wiwaxia* and *Dinomischus* are introduced here briefly. Other species, including *Amiskwia sagittiformis, Nectocaris pteryx* (a single specimen) and *Chancelloria eros,* are discussed and illustrated in Briggs *et al.* (1994).[7]

Fig. 3.7 *Wiwaxia corrugata* as seen from the dorsal surface. The body was covered with scales with spines at the edges. It has recently been shown to have affinities with the halkieriids and/or polychaete annelids (see text). The dark stain at the bottom of the fossil was shown by Butterfield (1990b) to be fused and decayed organic material. Reproduced from Conway Morris and Whittington (1985) with the permission of the Minister of Supply and Services Canada, 1990.

Anomalocaris canadensis was the largest Burgess Shale animal, reaching lengths of 0.5 m. Even larger anomalocaridids are found in China, Australia and Greenland. Anomalocarids were giant and active predators; those in the Chengjiang fauna of China reached two metres in length. At the front of the head, giant limbs surrounded a circular jaw with a formidable array of teeth (Chen, Ramsköld and Zhou, 1994, and see the reconstruction in Briggs, 1994).

Opabinia regalis, 4.3–7 cm long, was segmented with a flexible proboscis (Figure 3.6). Each of the 15 trunk segments carried a pair of lateral extensions on which a gill was located. With five compound eyes, grasping spines on the head and posterior fins, *Opabinia* was an active swimmer and effective forager. The limbs were not jointed and so *Opabinia* is not an arthropod. *Opabinia* may be related to a group ancestral to the annelids and arthropods or allied to anomalocarids within an

Fig. 3.8 *Dinomischus isolatus*. A stem supported a cup with terminal appendages (arms, tentacles?). Reproduced from Whittington (1985) from a photograph provided by S. Conway Morris. Scale bar: 0.5 cm.

Two rows of lateral spines were present, but the ventral surface was naked. The patterns and size of the sclerites suggest that wiwaxids moulted to add new sets of scales rather than increasing scale size through growth. Similar isolated scales are found in other Lower Cambrian locations.

Conway Morris and Peel (1990), who reported the first articulated specimens of a halkieriid, believe *Wiwaxia* to be a halkieriid (a group of organisms known only from isolated sclerites), but perhaps with affinities with molluscs. Butterfield (1990b) disputes interpretations that the sclerites provide evidence for moulting. After reexamining *Wiwaxia* and another Burgess Shale fossil (*Canadia spinosa*) using light and scanning electron microscopy, Butterfield concluded that the sclerites of *Wiwaxia* are polychaete paleae and that *Wiwaxia* is a polychaete annelid with close affinities to *Canadia*. In this context Lindberg and Ponder (1996) commented that the morphological differences between extant bivalve molluscs and chitons are as extreme as those between such Burgess Shale fossils, the inference being that morphological disparities between Burgess Shale forms may not be as extreme as oftentimes portrayed.

Dinomischus isolatus, known from only three individuals, represents less than 0.01% of Burgess Shale specimens (Whittington, 1985). A stem anchored by a holdfast supported a cup-shaped body or calyx bearing arms or tentacles (Figure 3.8). Only 1 cm tall (or long?) from the base of the stem to the tip of the calyx, *Dinomischus* had a U-shaped gut. A distant relationship to entoprocts has been suggested.

3.3 SETTING THE MORPHOLOGICAL BOUNDARIES FOR PHYLA

That these genera cannot be confidently assigned to phyla, the primary taxonomic category, indicates just how different their body plans are from all other known animals. Or does it? How do we determine whether these

arthropod super-phylum, as suggested by Chen *et al.* (1994) and by phylogenetic analyses.

Wiwaxia corrugata, which ranged in size from 3.5–55 mm, had a most bizarre morphology. The dorsal surface shown in Figure 3.7 was covered with flattened, scale-like sclerites.

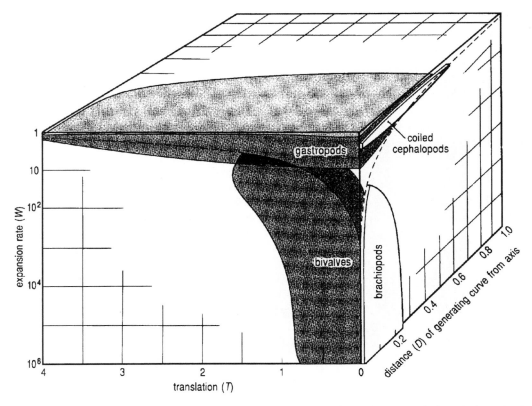

Fig. 3.9 Distribution in morphospace of gastropods, bivalves, brachiopods and coiled cephalopods. Only a small portion of theoretically available morphospace is occupied, indicating limits on organismal design. Adapted from Skelton (1994) after Raup (1966).

specimens possess unique body plans of otherwise unknown phyla, or whether they represent the extremes of body plans of existing phyla?

A phylogenetic analysis based on accurate identification of morphological characters is essential, but how do we set the bounds to the morphological diversity used in such analyses? The issue is to know whether the range of possible designs for an organism or group of organisms is limited. If it is, are the limiting factors internal to the organisms? Some designs may be inefficient or sub-optimal and therefore selected against (or not produced) because of limitations of the organism's development. Or are limits determined by the envi-

ronment or ecology of the organisms? David Raup (1966) pioneered a way to tackle this problem in a study of the coiled shells of gastropods, brachiopods, ammonoids and pelecypods.

If one could define the theoretical morphologies for the range of designs an organism could exhibit, then actual designs found in nature could be plotted three-dimensionally as what has been called a **morphospace**. Occupation of all the space would indicate no limitations on designs, occupation of only portion of the space that the number of possible designs was limited. Knowledge of the portion of morphospace occupied could provide clues to the factors limiting the range of forms displayed by the organism(s).

Raup constructed the morphospace for shelled invertebrates (gastropods, bivalves, brachiopods, coiled cephalopods) by plotting three axes defined by three parameters of shell growth: rate of shell expansion along the axis of coiling (W), translation rate of the coil down the axis of coiling (T), and distance from the coiling axis of the curve of the shell (D). By plotting these axes onto a cube of morphospace Raup demonstrated that the four shelled groups were clustered, and that much of morphospace was unoccupied by either extant or fossil species (Figure 3.9). Knowing this, it is possible to ask questions concerning the range of theoretical morphologies organisms can display. This approach has proved invaluable to paleontologists, for whom shapes are often the only characters of an organism available for analysis; see Raup and Stanley (1971) for further details. Yet although morphology can be bounded, the degree of confidence in the boundaries varies substantially from group to group. Consequently, analysis must be performed group by group.

Direction of change over evolutionary time can be determined when morphology is analysed in a cladistic or phylogenetic context. In a cladistic analysis of Cambrian and living arthropods, Briggs, Fortey and Wills (1992a) concluded that morphological disparity was no greater in the Cambrian than it is now, and provided compelling evidence that an emphasis on unusual features in Cambrian animals has led to an overestimation of the range of morphological designs in the Cambrian. For these workers, problematic taxa reflect an inadequate taxonomy and premature conclusions drawn in advance of detailed analysis of the forms. Inability to assign problematic taxa is a statement of our imperfect knowledge and not necessarily evidence of morphological disparity.

The fossil species problem also contributes.

3.4 TAXONOMY OF FOSSILS

An inability to categorize some Burgess Shale forms could be a consequence of their unique anatomy, or because of more general difficulties identifying fossil species. A discussion of fossil species and the taxonomy of fossils, especially Cambrian fossils, is therefore appropriate.

The limitations of morphology as the sole criterion for distinguishing fossil species, come into stark relief in Burgess Shale fossils. There is at least a two-fold problem:

1. how to identify morphological structures and interpret their functional roles; and
2. how to determine which phyla are represented by animals with these features.

Beginning with Simpson in 1944, many have discussed the problems associated with using different criteria and essentially different concepts to define species in different contexts. Fossil entities recognized as species on the basis of morphology (**palaeontological species**) are not comparable directly with species of extant forms recognized on the basis of reproductive isolation (**biological species**). To overcome these problems, Simpson (1961) developed a definition that would apply to living and fossil species. The **evolutionary species** is a series of populations connected as ancestors and descendants. While tidy, it begs the question of the criteria to be used in determining ancestor–descendant relationships.[8]

An analysis from breeding experiments with cheilostome bryozoans by Jackson and Cheetham (1990) provided evidence that morphological characters such as the skeletal characters found in fossils can be sufficient to discriminate extant biological species within the Bryozoa. These authors therefore argue that similar morphological structures could be used to identify fossil species as biological entities. However, the degree of morphological change associated with speciation varies tremendously between taxa. The cichlid fishes discussed in section 16.5 show much, amphibians show little, cephalochordates even less. It is easier, therefore, to set boundaries for some groups than for others. Major taxa have to be analysed case by case to delineate species. Positioning organisms in morphospace as described in the previous section deals in part with this problem.

These problems are even more confounded when the 'essential' features of a group can change over periods as short as a single generation, as Broadhead (1988) maintains is the case for the major traits of calyx and arms used in classifying Palaeozoic crinoids as members of the Class Crinoidea (sea lilies and feather stars) within the phylum Echinodermata. Further examples are discussed in section 6.2.

If Burgess Shale specimens cannot be placed into phyla and if major characters in higher taxonomic groups have the potential for rapid change, is it appropriate (or even possible) to distinguish between species or genera at early stages in metazoan evolution? It may not be realistic to apply a taxonomic hierarchy to animals of totally unknown affinities that existed at a time when groups were only beginning to appear and when many intermediates and/or a continuum may have been the order of the era. The latter possibility is supported by the cladistic analysis of Burgess Shale arthropods by Briggs, Fortey and Wills (1992a), discussed in the previous section, which supports the existence of many body plans forming a continuum, with subsequent loss of groups that now appear as intermediates in the fossil record, but that were, in reality, parts of the continuum.[9] Conway Morris raises similar concerns in the context of the delineation of phyla:

'Here (in the phylum) is the quintessence of biological essentialism, a concept that is almost inextricably linked with that of the body plan. For most of the Phanerozoic, the status of phyla (and classes) seems to be effectively immutable, the one fixed point in the endless process of taxonomic reassignments and reclassifications. However, for the Cambrian radiations, such precepts begin to fail and our schemes of classification lose relevance.'

Conway Morris, 1989, p. 345

In analysing the relevance of taxonomy to evolution, especially the level in the taxonomic hierarchy at which significant differences appear, Holman (1989) examined rate of origination, duration, and extinction of marine fossils at the ordinal, class and phylum levels and found that orders and classes had an 'evolutionary coherence' that phyla lacked. Analysis of diversity data from species abundance levels, coupled with the extensive elaboration of body plans at the class level in the Cambrian, prompted the conclusion that every 40th species represented a new class or phylum (Valentine, 1985, 1986). The clear impression is that novel body plans arose in Cambrian times only to disappear rapidly.

3.5 BURGESS SHALE FOSSILS ARE NOT UNIQUE

The animals of the Burgess Shale were neither an isolated evolutionary experiment nor unrepresentative of the general situation in the Lower to Middle Cambrian. At least 12 other sites contain animals with an equivalent range of body plans. These include Early Cambrian faunas in Pennsylvania, north Greenland, China, Spain, Poland and an additional fauna from northwestern Canada. Some of these can be assigned to the same genera as those in the Burgess Shale, indicative of the close relationships among at least some members of these faunas; I have already commented on anomalocarids and cephalochordates from the Chengjiang formation in China (s. 3.2).[10]

All animals in rocks younger than the Palaeozoic can be assigned to one of the known animal groups. Most of the Cambrian 'novel' body plans were short-lived in evolutionary and geological time. Although Burgess Shale species unassignable to phyla were geographically widespread, most had a short geological history. An exception, *Mimetaster* from the Hunsrück Shale of Germany, is related to the arthropod *Marrella splendens*, known from Lower Devonian times (Whittington, 1985). Some novel animals, however, persisted well into the Palaeozoic, in fact into the Pennsylvanian, which is some 300 million years after the deposition of the Burgess Shale. The paucity of novel animals could reflect the unlikelihood of soft-bodied animals fossilizing except under exceptional conditions as existed in the Burgess Shale, the existence of phyla containing only a few species or genera, and/or many intermediates between phyla recognized today but which themselves are not recognized as distinctive. The existence of these body plans raises the distinct possibility of polyphyletic and non-contemporaneous origins of the Metazoa both in time and space (s. 14.2).

3.6 SUMMARY

The 17 Burgess Shale genera that cannot be assigned to any of the 37 phyla raise substantial questions concerning radiations of body plans, relationships among phyla in the Cambrian, and the interplay of environmental, developmental and functional factors that winnowed out so many types, leaving the assortment that we have today. Despite, or perhaps because of this diversity of body plans in the Lower Cambrian, no new body plans have evolved since; in fact, a substantial number have been lost. Why? Filling of the ecological barrel, early constraints on development, and/or early relaxed selection are suggested answers.

Various levels of selection other than organismal have been proposed, including:

1. gametic and organelle (Lewontin, 1970);
2. between genes (Doolittle and Sapienza, 1980; Orgel and Crick, 1980); and/or
3. between families, populations or species (Eldredge and Gould, 1972; Williams, 1992; Michod, 1997).

Whether selection is hierarchical, an unresolved issue, is much discussed by paleontologists, evolutionary biologists, mathematicians, population geneticists, philosophers of biology, and more recently, developmental biologists. Distinguishing between levels of selection, units of selection and mechanisms of selection is critically important, as Lewontin so forcefully articulated almost 30 years ago.[11]

Anatomical diversity seems to have reached a maximum immediately after the initial diversification of multicellular animals. The situation of comparatively few species representing many body plans in the Burgess Shale has today been replaced by one with many more species but fewer body plans, body plans that were in place before the Burgess Shale fossils. In the over 500 million years since the Burgess Shale fauna, no new phyla or body plans have appeared.

Why did so few anatomical designs or structural plans develop and why have no new body plans evolved for over 500 million years? These questions are addressed in the next three chapters. Again, I take a historical approach in introducing some early explanations for anatomical design: unity of type, archetypes, form versus function, the Geoffroy–Cuvier debates, structural plans and relationships amongst organisms (Chapter 4); changing views of homology and the establishment of an evolutionary embryology (Chapter 5); and fundamental body plans or *Baupläne* (Chapter 6).

3.7 ENDNOTES

1. See Butterfield (1990a) for conditions under which Burgess Shale fossils were preserved.
2. M. M. Smith and Hall (1990), Forey and Janvier (1993, 1994), Briggs *et al.* (1994) and Forey (1995) provide recent evaluations of the fossils of the Harding sandstone.
3. Walcott (1911a,b) are two early papers on the Burgess Shale fossils; a posthumous addendum was published in Walcott's name in 1931. Conway Morris (1979) and Whittington (1985)

contain lists of Walcott's papers. For a sample of the now enormous literature on the fossils of the Burgess Shale, see Whittington (1985), Conway Morris (1979, 1989, 1993a,b), Conway Morris and Whittington (1985), Gould (1989) and Briggs *et al.* (1994).

4. The Burgess Shale fossils have kindled a resurgence of interest in arthropod relationships and origins. Samples from the vast and often conflicting literature are Gould (1991); Briggs, Fortey and Wills (1992a,b, 1993); Bowler (1994, 1996); J.-y. Chen *et al.* (1995); Fortey, Briggs and Wills (1996, 1997); Novacek (1996); and Waggoner (1996). Whether the Uniramia are a monophyletic group has been approached using a variety of techniques including phylogenetic analyses (J. W. O. Ballard *et al.*, 1992; Monge-Najera, 1995; Waggoner, 1996); recognition of intermediate and stem groups from the fossil record (Budd, 1997); recognition of common regional markers along the arthropod body which, when modified, can account for the diversity of organization of arthropod bodies (Schram and Emerson, 1991); and comparisons of mechanisms of limb development using molecular probes such as the *Distal-less* (*Dll*) gene (Panganiban *et al.*, 1995, 1997; Popadic *et al.*, 1996; Williams and Müller, 1996).

5. Recent demonstrations that onychophorans and arthropods (and therefore their common ancestor) share the same set of *Hox* genes provides a powerful tool for exploring arthropod relationships; see Grenier *et al.* (1997).

6. For introductions to the vast literature on neural crest, vertebrate origins and neural crest derivation of most cranial skeletal tissues, see Hörstadius (1950), Gans and Northcutt (1983), Northcutt and Gans (1983), Maisey (1986, 1988), Thomson (1987), Gans (1987, 1989, 1993), Hall and Hörstadius (1988) and M. M. Smith and Hall (1990, 1993). For discussions of the earliest vertebrate skeletal remains, see Smith and Hall (1990, 1993) and Janvier (1996b). A possible Later Cambrian phosphatic skeleton, interpreted as dermal armour of enamel without dentine and as therefore having vertebrate affinities, has been described by Young *et al.* (1996). The traditional view is that the first skeleton consisted of dermal denticles of dentine and bone overlain by enamel (Smith and Hall, 1990, 1993).

7. See Simonetta and Conway Morris (1991) for additional discussion of problematic Burgess Shala taxa.

8. See Simpson (1944, 1961), Stanley (1979) and Levinton (1988) for discussions of the biological species concept.

9. This analysis did not appear in print without comment; see Foote and Gould (1992), Lee (1992) and the rebuttal by Briggs, Fortey and Wills (1992b) in the issue of *Science* in which the original analysis was published. Also see Briggs (1991), Shear (1991), Eernisse, Albert and Anderson (1992), Fortey, Briggs and Wills (1996), Novacek (1996) and Waggoner (1996) for independent analyses and discussions on this subject. The Burgess Shale fossils have been the focus of some intensely argued debates over whether a cladistic or disparity (analysis of morphospace) approach will best resolve relationships; see McShea (1993), Ridley (1993) and Gould (1991, 1993).

10. For discussions of worldwide Early to Middle Cambrian metazoan faunas, see Yi (1977), Conway Morris *et al.* (1987), Hou and Sun (1988), Conway Morris (1989) and Butterfield (1994).

11. For discussions of whether selection is hierarchical, see Lewontin (1970), Vrba and Eldredge (1984), Buss (1987), Gilbert (1992), Goertzel (1992), Lloyd (1992), Williams (1992) and Bell (1997a,b).

PART TWO

FORM AND FUNCTION, EMBRYOS AND EVOLUTION, INHERITANCE SYSTEMS

4

TYPES AND THE GEOFFROY–CUVIER DEBATES: A CROSSROADS IN EVOLUTIONARY MORPHOLOGY

'What is the essence of life – organization or activity?'

Russell, 1916, p. v

Pre-Darwinian 18th- and 19th-century morphologists used several interrelated terms and concepts to explain morphological similarities among different organisms and how groups of organisms could be united on the basis of common anatomical design: the Type (*Typus*), the Archetype, unity of type and unity of plan. Embryological criteria became increasingly important in this quest, but embryology too was typological. The eventual acceptance that all animal life had a common origin and that animal development was based on similar embryological processes, produced the evolutionary morphology and evolutionary embryology of the late 19th century, and led to the evolutionary developmental biology of the 20th.

Sections 4.1 and 4.2 provide a brief overview of unity of plan and the type concept. I highlight the perceived dichotomy between form and function through an examination of Geoffroy's unity of form and Cuvier's four *embranchements* in sections 4.3–4.5. A series of debates between Geoffroy and Cuvier resounded through the halls of the *Académie des Sciences* in Paris in 1839 and throughout the morphological world thereafter. The enormously important debates – a crossroads in evolutionary morphology – and their ramifica-

tions, round out the chapter in sections 4.6 and 4.7.

4.1 UNITY OF PLAN: ARISTOTLE'S LEGACY

Among Aristotle's greatest accomplishments were his recognition of the unity of plan within major groups of organisms and development of a classification of animals. Each group was built upon a single structural plan of shared features and correlation between structures; for example, animals with tusks lacked horns and *vice versa*. Aristotle divided animals into 'man' (humans), viviparous quadrupeds, oviparous quadrupeds, birds, fishes, Cetacea, Cephalopoda, Malacostraca, Insecta, and Testacea (molluscs, echinoderms, ascidians). These he grouped as sanguineous and exsanguineous on the basis of the presence or absence of red blood.

This approach to morphology changed little in the millennia between Aristotle and the Renaissance. Pre-Renaissance scholars followed the authority of the early Greeks rather than undertaking independent investigation themselves. Aristotle's schemes were revered, while Galen's physiology and anatomy dominated medicine through the Middle Ages.

BOX 4.1
PERSISTENCE OF CLASSICAL EDUCATION

While contemporaries of Prince Albert, the future consort, learnt Latin and Greek and immersed themselves in the classics and antiquity at Oxford and Cambridge, Albert (through tutors and study at the University of Bonn) was immersed in philosophy, political economy, natural sciences, mathematics, even the newly emerging statistics.

Within a week of his election as Chancellor of the University of Cambridge in February 1847, Albert received a letter in which William Whewell and Charles Lyell set out their critical view of the status of the University. To them, as to many others, it was no more than 'a vast theological seminary'. Only classics and mathematics counted toward a Bachelor of Arts. Many college and university statutes had not been changed for hundreds of years. Science, natural history, chemistry, astronomy, geography, modern languages, art, history, political science, economics, law, psychology and modern languages were all lacking. Second year students were examined on St. Mark's Gospel (in Greek, of course), Cicero, Euripides, Old Testament history, and William Paley's *View of the Evidences of Christianity*, published 55 years earlier.[1]

'When an argument arose as to how many teeth the horse has, one looked it up in Aristotle rather than in the mouth of a horse. '
Mayr, 1982, p. 93

'The professor of medicine would recite Galen, while an assistant ("surgeon") dissected the corresponding parts of the body. This was poorly done, and the oratory and the disputations of the professors, all of them merely interpreting Galen, were considered to be far more important than the dissection.'
Mayr, 1982, pp. 94–5

Galen was not eclipsed until after William Harvey discovered the circulation of the blood in 1628. The eclipse was slow; the classical education of the Renaissance persisted in the universities of countries such as Great Britain well into the 19th century (Box 4.1)

4.2 IDEALISTIC MORPHOLOGY AND BUFFON'S UNITY OF TYPE

G. L. Buffon (1707–88), perhaps the most influential 18th century biologist, initiated the school of **idealistic** (transcendental) **morphology** or form as the expression of an idealized archetype. Transcendental morphology came to dominate the German school of *Naturphilosophie* through the work of Spix, Carus, Oken, Schilling and others.[2]

Buffon's concept was the **unity of the type** based on community of descent from a small number of types, perhaps only one. Georges Cuvier (1769–1832) incorporated unity of type in his principle of the correlation of parts, discussed in section 4.4. Goethe developed a parallel idealistic morphology of plants in which the leaf was the fundamental unit from which all parts of the plant (including flowers) arose in the primordial plant or *Urpflanze*.

Ernst Mayr (1982, pp. 452–9) distinguished at least five independent areas of investigation that have existed under the rubric of morphology, the term introduced by Goethe in 1807. These five areas, along with some more recent definitions by Marvalee Wake (1992b), are:

1. idealistic morphology – form as the expression of an archetype;

2. the morphology of growth – structure arises through differential growth. Often now termed developmental morphology and defined as 'the integration of data on early development and ontogeny of organisms in the analysis of the evolution of structure' (p. 314);
3. functional morphology – structure serves function, or 'the study of the way that form, or structure, causes, permits, and even constrains organisms to function or perform' (p. 283);
4. phylogenetic morphology – form in relation to ancestral form, or morphology as pattern;
5. evolutionary morphology – form as response to environment or selection, or morphology as process, defined by M. Wake as 'the integration of development, ecological, biomechanical, and phylogenetic analysis (as appropriate) to answer questions of the evolution of organismal complexity' (p. 324).

A sixth area should be added:

6. ecomorphology – 'the assessment of form and function as it correlates with the organism's environment' (M. Wake, 1992, p. 308).

With reference to the third area in the above list, some historians such as Ghiselin (1996) argue that from a historical perspective the term 'functional morphology' is an oxymoron. This is the view that Darwin was an evolutionary and functional morphologist (which he was) and that morphology has long been tied to function. The Geoffroy–Cuvier debates over whether form or function defined groupings of organisms certainly set form against function (s. 4.5).

4.3 UNITY OF FORM: GEOFFROY'S LEGACY

E. Geoffroy Saint-Hilaire (1772–1844) attempted to impose idealistic morphology and ideas of a single structural plan onto all vertebrate and invertebrate animals. Transformation of organisms would be limited to variations on this basic plan. Geoffroy sought neither the ideal type of Buffon, nor to relate all animal structure to man as the Archetype, as Aristotle had done. Rather, he used morphological connections to seek what he termed **analogous** but we now call **homologous** structures.

Geoffroy's thesis was central to the early-19th-century approach to morphological change that formed the foundation of the relationship between embryology and evolution developed later in the century. I therefore consider it in some detail. Geoffroy's approach to the analysis of animal structure is also intimately interwoven with the dual problems of equivalence of structure and homology. Initially, idealistic morphologists made no attempt to segregate what are now known to be homologous and analogous structures. Geoffroy, however, did.[3]

According to Geoffroy, structures could be related to one another if their relative positions or **connections** were the same in different organisms and if they were composed of the same 'elements'. Thus, he compared the bones of the forelimbs and shoulder girdle of higher vertebrates such as mammals to the bones of the pectoral fins and girdles of fishes as having equivalent (homologous) connections and as being composed of the same elements. Between 1818 and 1822, he produced a series of papers under the title *Philosophie Anatomique*, in which he developed his principle of connections and proposed his ideas of the unity of composition of all the vertebrates, claiming, for instance, that the bones of the operculum in fishes corresponded to mammalian auditory ossicles. Geoffroy's homologous structures therefore were based on topology; they extended beyond organs to organ systems and whole regions of the body.[4]

For Geoffroy there was only one plan into which all animals fitted and from which all animals could be derived. He attempted an environmental (and physiological) explanation for deviations from the unity of plan, which saw environmental agents directly altering embryonic development. He sought experimental evidence in what were some of the first studies in experimental embryology, although they are not usually recognized as such.

With Bonaparte on a military/scientific expedition to Egypt in 1799, Geoffroy requested 600 chicken eggs, an incubator in which to raise them, an enclosure to house the hatched chicks, 112 pairs of pigeons and two dovecotes, ostrich eggs and breeding pairs, instruments for measurements and salary for two assistants. Why all this paraphernalia? He needed it to investigate his theory that the sex of individuals was mechanically determined as a function of the shape of the egg. By removing and/or grafting portions of the shells between the long and short eggs of pigeons and chickens, Geoffroy modified egg size. Such experiments were a prelude to his attempts in the 1820s to alter development by manipulating the embryonic environment by covering eggs with wax, varnish or plaster; soaking eggs in water; removing eggs from the incubator for one to two days; or tying the oviducts of the hen to create a 'uterine' environment.[5]

Geoffroyism is therefore one of the **epigenic** theories introduced in section 1.2 and elaborated in Chapter 7. Development cannot be preformed or predestined if climatic or environmental changes evoke structural changes, or if the form of those teratologies varies with the developmental stage when embryos are exposed to the environmental influence. In a very real sense, Geoffroy founded experimental embryology and the scientific study of terata or teratology, the term coined by his son Isidore in 1832.

Geoffroy used development and his experimental studies on malformations and teratology as evidence for constancy of the Type in the face of environmental influences expected to modify it. Connections between the elements of homologous structures are maintained by developmental processes; basic form limits potential structural variation. Abnormalities arising from arrested development were equivalent to normal stages of 'lower' vertebrates, anticipating von Baer (s. 5.1).

Geoffroyism was neo-Lamarckian: environmentally induced changes were transmitted to future generations. Geoffroy was not alone in this position. Most Lamarckians of his day held the same views, although Lamarck himself did not subscribe to such a direct influence of environmental factors. Rather, he proposed an indirect environmental effect acting through altered behaviour and/or function. Mayr (1980) discusses such views as 'soft inheritance'.[6]

Geoffroy (1825a) used analyses of experimentally altered blood flow, the origin of mammalian middle ear ossicles, and the presence of a secondary palate in crocodiles and its absence from lizards, to show that increases in one element were compensated for by decreases in an associated element. This was his developmental mechanism to preserve the basic ground plan – *loi du balancement des organes* – or the law of the equivalence between organs. According to this thesis, environmental influences may alter structure in an adaptive (functional) way, but the basic plan is always conserved. Development limits structural variation by ensuring the maintenance of a common ground plan. Morphological gaps between organisms with different ground plans are therefore predicted and expected. The modern counterparts to this view are canalization, developmental constraint, stabilizing selection, and external versus internal factors in evolution (ss. 6.5, 12.2, 19.4 and 19.5). Geoffroy viewed morphological change as saltatory and as involving key innovations; birds arose by a sudden transformation of their lungs, an event which ushered in development of the other avian characters. Key innovations are discussed further in Part Two.

Lorenz Oken (1779–1881), a contemporary of Geoffroy's, fiercely supported idealistic

---

BOX 4.2
JARDIN DU ROI

Jardin du Roi (*Jardin des Plantes*) was the predecessor of the *Muséum d'Histoire Naturelle*, founded by decree as one of the first acts of the Revolutionary Convention on June 10, 1793. The *Jardin* itself was established in 1626 by Louis XIII as an alternative to the Paris *Faculté de Mèdicine* as a centre for medical education. Botany, anatomy and chemistry were taught and the *Jardin* was administered by the physician to Louis XIII. Buffon was appointed intendant (Director) in 1739 and over the next 50 years made the *Jardin* a world centre for research into natural science. With the transformation of the *Jardin* into the *Muséum d'Histoire Naturelle* in June, 1793, Geoffroy became one of France's first official professors of zoology. From 1793 until the mid-1830s the *Muséum* was a world centre for the study of comparative anatomy, zoology, morphology, palaeontology and the developing field of taxonomy or systematics.

---

morphology. He extended Geoffroy's approach of similar structures in different organisms to dissimilar structures in the same organism and to segments, and developed a vertebral (segmental) theory of the skull. Richard Owen (1804–92) developed the concept of serial homology for serially repeated structures (s. 5.3).

4.4 *EMBRANCHEMENTS*: CUVIER'S LEGACY

Geoffroy's and Oken's views were vehemently opposed by Georges Cuvier, one of the greatest comparative anatomists of his, or indeed, any day. Speculative theories were anathema to Cuvier and politically inadvisable if science was to maintain its prominent position. Cuvier argued from specimens, not from the armchair. Today, it is taken for granted that fossils are the remains of organisms that lived in the past; Cuvier (1812b) founded palaeontology with a study of living and fossil elephants that demonstrated that fossils were the remains of past animals and palaeontology the zoology of the past.[7]

Given their long feud, it is ironic, that Cuvier was given his start as an anatomist by Geoffroy; the two men lived and worked together in Paris in 1795. As Professor of Comparative Anatomy at the Jardin du Roi, a position to which he was appointed at the age of 21 in March 1793, Geoffroy appointed Cuvier as an assistant to what in June 1793 had become the Museum of Natural History (Box 4.2). By 1799, at age 20, Cuvier was Professor of Natural History in the Collège de France. Among many works, his 22-volume *Histoire des Poissons*, written with Achille Valenciennes between 1828 and 1849, cemented his reputation. By the end of his career he was Baron Cuvier, one of the greatest scientists of Europe.

Before Cuvier, each major vertebrate group – fish, birds, reptiles – had been regarded as equivalent to individual major invertebrate groups such as molluscs and insects. The animal kingdom was divided in two: animals with backbones and animals without backbones. Reinforced by Lamarck, who coined the term 'invertebrates' to unite all those animals without backbones, this dichotomy persisted until Cuvier (1812a) arranged the animal kingdom into four unrelated *embranchements* – Vertebrata, Articulata (Crustacea, Insecta), Mollusca and Radiata (coelenterates, echinoderms) – for which present-day phyla or groups of phyla are the closest approximation. *Embranchements* were erected on the basis of fundamental differences in basic structural body plan, especially organization of the nervous system:

1. Vertebrata, with a spinal nerve cord expanded into an anterior brain, an internal bony skeleton, five sense organs, red blood and a muscular heart.
2. Articulata, with two ventral nerve cords, segmentally organized ganglia and a small brain.
3. Mollusca, lacking an organized nerve cord, but with a well-organized brain, dispersed ganglia and nerves, a double circulatory system and a shell.
4. Radiata, with radial symmetry.

Each was subdivided into four classes, although Cuvier later modified these subdivisions. Differences met equivalent functional requirements within an *embranchement*:

'If there are resemblances between the organs of fishes and those of the other vertebrate classes, it is only insofar as there are resemblances between their functions.'

Cuvier and Valenciennes, 1828, p. 550

Cuvier's system won instant appeal; it provided a natural classification that precisely fitted the search of the 19th-century morphologist for perfection and the ideal type.

Cuvier used his 'principle of the correlation of parts' to delineate *embranchements*. These correlations represented suites of features that always accompanied one another, e.g. hollow horns, cud-chewing, a stomach consisting of several chambers, and cloven hoofs in oxen and cows. Functional relationships among organs ensured the 'harmony' of the organism. Organisms falling between any two *embranchements* could not exist; the correlation of organs required for such intermediate organisms would be incompatible with life. So overriding was the correlation of parts for Cuvier that he believed that, given the right part, he could identify and reconstruct an

entire organism and place it into the appropriate *embranchement*, much as an architect can construct a domestic interior and the sociology of its inhabitants from a single chair (Rybczynski, 1986). For Cuvier but not Geoffroy, any change within an *embranchement* only represented a deterioration or series of deteriorations of individual organs from the common plan.

4.5 DOES FUNCTION DETERMINE FORM OR FORM FUNCTION?

Geoffroy and Cuvier's explanations for the great variety of animals were totally at variance. For Geoffroy, variety reflected variation and transformation of form within single body plans. For Cuvier, variation reflected response or adaptation to function and/or environmental conditions. Where Cuvier sought functional correlation between parts – **function determines form** – Geoffroy sought structural connections between parts – **form determines function**. Russell encapsulated their positions:

'Geoffroy and Cuvier in pre-evolutionary times well typified the contrast between the formal and the functional standpoints. For Geoffroy form determined function, while for Cuvier function determined form. Geoffroy held that Nature formed nothing new, but adapted existing "materials of organisation" to meet new needs. Cuvier, on the other hand, was always ready to admit Nature's power to form entirely new organs in response to new functional requirements.'

Russell, 1916, p. 305

The opposition between Cuvier and Geoffroy culminated in a famous series of debates before the *Académie Royale des Sciences* in Paris on eight Mondays between February 8 and April 5, 1830.[8]

The debates centred on a dispute about comparative morphology, unity of the type, and the relative roles of structure and function in the determination of form. Does form determine function or function determine form?

> ## BOX 4.3
> ## FORM OR FUNCTION
>
> The dichotomy of form and function is most evident today in architecture, where design has oscillated between form and function, structure and utility, ground plan and adaptation. This struggle reflects the affirmation by Louis Sullivan (1856–1924) that in architecture, 'Form follows Function'. The father of modernism in American architecture, Sullivan designed one of the first skyscrapers, the Wainwright Building, built in St. Louis in 1890. Also, he designed the major buildings at the Chicago World's Fair of 1893. Perhaps more than any other modern architect, Sullivan transformed architecture from 'form follows tradition' to 'form follows function'. His fellow architects carved this on his tombstone in Gracelands Cemetery:
>
> By . . . his philosophy, where, in 'Form Follows Function,' he summed up all truth in art, Sullivan has earned his place as one of the greatest architectural forces in America.

This was no idle preoccupation. Morphology held philosophical domination over the natural sciences as thoroughly in the 18th and 19th centuries as it had in Aristotle's time. The quest in the 18th and 19th centuries was 'What is the essence of life – organisation or activity' (Russell, 1916, p. v). To Gould, Geoffroy '. . . had a vision – perhaps the boldest, the most noble, the most comprehensive idea ever promoted in biology' (1986, p. 206). Gould's enthusiastic affirmation of Geoffroy's vision was also the view of many influential medical, anatomical and zoological scholars of the 19th century. Some of these views are summarized in section 5.2 in the context of the British response to the great debate. Even today, discussions of form evoke expressions of transcendental delight, as exemplified in the following from a symposium on Form in Nature and Art.

'Form is both deeply material and highly spiritual. It cannot exist without a material support; it cannot be properly expressed without invoking some supra-material principle. Form poses a problem which appeals to the utmost resources of our intelligence, and it affords the means which charm our sensibility and even entice us to the verge of frenzy. Form is never trivial or indifferent; it is the magic of the world.'

Dalcq, 1968, p. 91

Form versus function lives on in architecture (Box 4.3).

Because the debates focused the two poles of 19th century morphology and set the agenda for the remainder of that century (an agenda still incomplete today) they represent a crossroads in evolutionary morphology. An overview of the debates and issues follows.

4.6 THE GREAT *ACADÉMIE* DEBATE

The feud had been building for two decades. Geoffroy fired the first salvo in March 1828 with the announcement that he was preparing a new synthesis of animal variation based on unity of composition and plan, embryonic variation, and the sequences of fossil animals. This synthesis appeared in the preface to a series of lectures published in 1830. Geoffroy challenged Cuvier's views and authority by documenting how Cuvier had negated some of his own views in his zeal to hinder the propagation of Geoffroy's. Geoffroy used

A

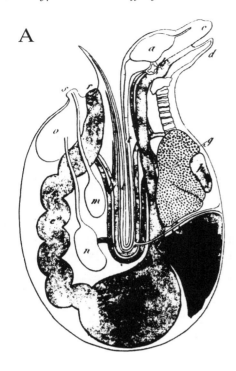

B

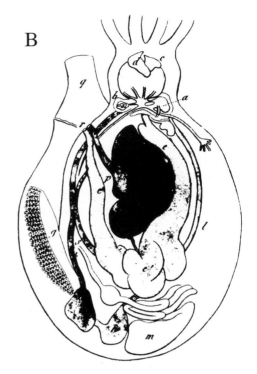

Fig. 4.1 A vertebrate bent back onto itself (A) for comparison with the cuttlefish, *Sepia* (B). Shading, letters, as used by Cuvier to refute Meyranx. Reproduced from Appel (1987) *The Geoffroy–Cuvier Debate*, by permission of the Oxford University Press.

embryological evidence against the primacy of function over form. For him, phases of embryonic development and rudimentary or vestigial organs that no longer served any function were retained as variations on a common plan. Function explained neither their presence nor their variation.

In October 1829 a paper was submitted to the *Académie des Sciences* proposing unity of the body plan of the cuttlefish *Sepia* with the vertebrates. The authors were two previously unknown naturalists, Pierre-Stanislas Meyranx and Laurencet, the latter so unknown that his first name(s) is (are) not known. Meyranx and Laurencet used Geoffroy's principle of connections to claim analogies (what we would call homologies) between cuttlefish and vertebrates – analogies of the vertebrate diaphragm, hyoid bone and

pelvis. Impatient with the slow response from the *Académie*, Meyranx and Laurencet requested it be examined by a commission. Geoffroy was one of two commissioners appointed.

The basic tenet of the paper was not new – Aristotle had posited a similar theory. The cuttlefish *Sepia* was portrayed as a vertebrate bent back onto itself with the middle of the abdomen as the fulcrum, the legs attached to the head, and anus and mouth in alignment, thus turning the vertebrate upside down and inside out to 'make it' a cephalopod (Figure 4.1).

Geoffroy sided with Meyranx and Laurencet's thesis – it fitted his philosophy of a universal, idealistic, animal morphology. In the first paper he read to the *Académie* – 'On Insects Reduced to the *Embranchement* of Vertebrates' – Geoffroy had sought to unite insects, crustaceans and vertebrates by positing that each separate invertebrate skeletal element equalled a vertebra; *'tout animal habite en dehors ou en dedans de sa colonne vertébrale'* (1820, p. 241). Two years later Geoffroy was using figures of lobsters on their backs or with legs widely separated to emphasize similarities to the vertebrae and ribs of vertebrates. Geoffroy saw Meyranx and Laurencet's work as bridging the unbridgeable gap between two of Cuvier's *embranchements,* the Mollusca and the Vertebrata, and presented their study to the *Académie* in that context in February 1830.

Cuvier opposed any such association; cephalopods and vertebrates occupied two separate *embranchements.* Geoffroy had cited from an 1817 memoir of Cuvier's on the anatomy of cephalopods to support his case. Cuvier objected angrily, especially to Geoffroy's use of his own work to bridge two *embranchements.* Cuvier's objection was so vehement and so persistent that Geoffroy agreed to remove reference to Cuvier's work from the printed version of the meeting. The great debate had begun!

A week after Geoffroy's presentation to the *Académie,* Cuvier came forth with his rebuttal of detailed evidence for the lack of 'analogy' between molluscs and vertebrates (subsequently published in 1830). The 1st of March saw Geoffroy's response and an attempt to enlarge the debate by introducing other examples (analogy of the hyoid bone throughout the vertebrates), and by an emphasis on philosophical differences in the two approaches. Cuvier resumed the attack three weeks later with his own paper on the hyoid bone, to which Geoffroy responded with a paper on the determination of homologous structures in fishes. The territory had been widened considerably by both protagonists.

Tactics became farcical, although they will be familiar to those who sit through departmental, faculty and senate meetings. [The meeting of March 29 began with an argument over who should speak first. Was Geoffroy's paper on fishes of March 22 a response to Cuvier, in which case Geoffroy should speak first, or was it a presentation of a new topic, in which case Cuvier should speak first? Geoffroy prevailed.] The debates continued back and forth until April 5, when Cuvier presented his last response, emphasizing the 'correct' (Cuvierian) application of analogy – correlation of parts so that form reflected function. Geoffroy proposed not to respond but rather to publish a treatise detailing his views; within two weeks his book, *Principes de Philosophie Zoologique,* was at the printers. By May it was on sale. This volume contained the report originally submitted to the *Académie* by Meyranx and Laurencet in October 1829, a record of the papers presented by Geoffroy and Cuvier to the *Académie,* and 'appropriate' introductions and commentary.[9]

It looked as if the debate was over, but, in fact, it had only just begun; the ramifications would be wide, varied and unanticipated. They continue to this day. As Gillian Beer commented with respect to the role of homology in ethology:

'. . . comparative morphology is still alive and well, even though now living on the fringes of the domain over which it used to rule. It survives by continuing the *Cuvier* and *Goodrich* tradition of attending to the detail of particular cases, adjusting the lines of inference according to the varieties of available evidence, and resisting the lure to grand transcendental Spencerian connections that loose analogy can present.'

Beer, 1984, pp. 301–2

In fact, even in 1984, the importance of comparative morphology for phylogenetic reconstructions was evident in the literature. In this era of molecular biology, however, comparative anatomy is often relegated to medical studies, and even then, often taught only in enough depth that a physician stands a reasonable chance of distinguishing muscle from blood vessel when performing a routine injection.

4.7 RAMIFICATIONS OF THE GREAT *ACADÉMIE* DEBATE

Although ostensibly about relationships between form, function, and natural classifications, the debate had a wide impact outside comparative anatomy. The issues raised fermented in the *Académie*, in journals, and in newspapers. Goethe fueled the flames in two articles in 1831 and 1832.

Two major studies of natural philosophy, produced by Geoffroy in 1835 and 1838, cemented his position. His view of the capability of man to understand the universe in a unified way through comparative anatomy appealed to the French intellectual community, and he corresponded extensively with novelists and social philosophers such as Honoré de Balzac and George Sand. Balzac included a Dr. Meyraux in his novels *Louis Lambert* (1835) and *Un Grande Homme de Province à Paris* (1839). Meyraux clearly represented Meyranx, co-author of the report that set the debates in motion. Balzac dedicated another novel, *Père Goriot* (1842), 'to the great and illustrious Geoffroy Saint-Hilaire as a tribute of admiration for his labors and his genius'. Geoffroy went blind in 1841 and died on 19 June 1844. Such was his influence and popularity that 2000 attended his funeral, among them Victor Hugo and leaders of the French scientific, medical, literary and artistic communities. Cuvier had died of cholera on 13 March 1832 while preparing a further reply to Geoffroy having devoted his life to the preservation of the status quo in French science and society.

The concern of the *Académie* with unity of the type persisted for decades. To settle a conflict between Henri Milne-Edwards (1800–85) and Étienne Serres (1787–1868), the *Académie* set as the 1849–54 topic for one of its prizes 'the positive determination of the resemblances and differences in the comparative development of Vertebrates and Invertebrates' (Russell, 1916, p. 206). In winning the prize, August Lereboullet vindicated Milne-Edwards' view of development of divergent types over the unity of the type, in what was round two in the Great *Académie* Debate.

Milne-Edwards saw embryos of organisms within a given *embranchement* as initially sharing a similar pattern of development that diverged later, but differed from the early development of embryos in other *embranchements*. Much of his life's work was inspired by the Geoffroy-Cuvier debates. He provided embryological evidence for Cuvier's four *embranchements* and, at the same time, saw a branching process of development that was equivalent to the features of the class or species to which the organism belonged. Milne-Edwards' insistence on the study of live animals produced a physiological zoology or anatomy that found favour quickly. By the mid-1800s he was the leader of French zoology. His influential *Introduction à la Zoologie générale* was published in 1853.[10]

Geddes and Mitchell captured the essence of the debates in their entry on Morphology in the 11th edition of the *Encyclopaedia Britannica*.

'On the point of fact (unity of structure in cephalopods and vertebrates) he (Geoffroy) was of course utterly defeated; the type theory was thenceforward fully accepted and the *Naturphilosophie* received its death blow. Such was the popular view; only a few, like the aged Goethe, whose last literary effort was a masterly critique of the controversy, discerned that the very reverse interpretation was the deeper and essential one, that a veritable "scientific revolution" was in progress, and that the supremacy of homological and synthetic over descriptive and analytic studies was thenceforward assured. The irreconcilable feud between the two leaders really involved a reconciliation for their followers; theories of homological anatomy had thenceforward to be strictly subjected to anatomical and embryological verification, while anatomy and embryology acquired a homological aim.'

Geddes and Mitchell, 1910–11, pp. 864–5

Thus, homology, the grandest of all comparisons, became the new metric for morphology.

This metric urged the search for relationships among organisms in their embryology, a search that has been 'rediscovered' today in evolutionary developmental biology.

Reports of the debates spread rapidly. Von Baer in Germany, and Richard Owen and other naturalists in Britain, devoted major efforts to reconcile the positions espoused by Cuvier and Geoffroy. In the next chapter I explore ramifications of the Geoffroy–Cuvier debates in the development and use of the concepts of the Archetype and homology as the attention of the morphological world focused on an evolutionary morphology, an embryological archetype, and an embryological basis for homology.

4.8 ENDNOTES

1. See Phillips (1981) and Cannadine (1990) for analyses of Prince Albert and the Victorian age, and James (1984, pp. 177–8) for the Cambridge curriculum in the mid-19th century. William Paley (1743–1805) was a leading exponent of argument from design. His book *Natural Theology: Or, Evidences of the Existence and Attributes of the Deity, Collected from the Appearances of Nature* (1802) had an enormous readership; see Thomson (1997) for an analysis of Paley's influence on Charles Darwin.

2. For recent analyses of transcendental morphology in Germany and Britain, see Sloan (1992), Rupke (1994) and Nyhart (1995). A recent scientific biography of Buffon is that by Roger (1997).

3. Geoffroy, the unity of type and/or various aspects of the structuralist (internalist) approach have been much discussed. The following provide overviews on the following aspects: morphology and development (Russell, 1916; Gould, 1977; Desmond, 1989; Mayr, 1982; Van der Hammen, 1988); Geoffroy

and Cuvier (Cahn, 1962; Bourdier, 1969; Appel, 1987); types (Farber, 1976; Ospovat, 1978); structuralism (Piaget, 1968, 1974; Webster and Goodwin, 1982; Lambert and Hughes, 1988; Rieppel, 1988a; Alberch, 1989; Goodwin, Sibatani and Webster, 1989; Goodwin, 1994); palaeontology (Desmond, 1982; Rudwick, 1985; Padian, 1995a); Lamarck (Corsi, 1988); and Darwin and Darwinism (Bowler, 1977, Ospovat, 1981; Mayr, 1982 and Asma, 1996).

4. Geoffroy set out his principle of connections in four papers (1807 a–d). For more detailed discussions and evaluations, see Appel (1987, pp. 86ff, 98ff and 140ff) and Corsi (1988, pp. 232ff).

5. For Geoffroy's reports of his embryological experiments, see Geoffroy (1802, 1818–22, 1825a,b). Appel (1987) discusses these experimental studies in some depth.

6. For discussions of Geoffroy's views on development, see Russell (1916, pp. 68–70, 75), Mayr and Provine (1980, pp. 5, 15), Rudwick (1985, pp. 150–2) and Rieppel (1988a, pp. 75–7, 115–16, 146–7).

7. Of the many biographies of Cuvier, I have found Coleman (1964) and Outram (1984) the most helpful.

8. The most thorough analyses of the feud between Geoffroy and Cuvier are the report by Geoffroy's son Isidore Geoffroy (1847), Bourdier (1969), *The Cuvier–Geoffroy Debate* by Appel (1987), and *The Age of Lamarck* by Corsi (1988).

9. The reading to the *Académie* on July 19 of an abstract of a paper on the dodo, delivered by Cuvier the week before, led to further confrontation over procedures. For the preceding century and a half, only titles, not abstracts of previous papers had been read. Cuvier and Geoffroy clashed again in October over a paper on crocodile anatomy and phylogeny read by Geoffroy. Cuvier claimed this as his territory.

10. See Patterson (1983) for a discussion of Milne-Edwards' views, including a reproduction of his maps of vertebrate relationships.

5

EMBRYOLOGICAL ARCHETYPES AND HOMOLOGY: ESTABLISHING EVOLUTIONARY EMBRYOLOGY

'Homologue . . . The same organ in different animals under every variety of form and function. Analogue . . . A part or organ in one animal which has the same function as another part or organ in a different animal.'

Owen, 1843, pp. 374, 379

Embryology increasingly came to be viewed as the way to understand relationships among different groups of organisms. As discussed in the first section of this chapter, Von Baer played a pivotal role in this development. His ideas were rapidly assimilated into Britain as scientists there responded to the implications of the Geoffroy–Cuvier debate (s. 5.2). Richard Owen made a lasting impact when he separated homology from analogy (s. 5.3), and Darwin took account of these developments in formulating his theory (s. 5.4). The search for embryological archetypes and establishment of an evolutionary embryology was carried furthest by Ernst Haeckel in Germany and Francis Balfour in England. By the late 19th century a solution to the generation of organismic form appeared to be at hand in homologous germ layers and conserved stages of embryonic development. This evolutionary embryology was applied to relationships among organisms and in a search for the ancestors of the vertebrates (s. 5.5).

5.1 LAWS OF DEVELOPMENT: VON BAER AND MECKEL

Karl Ernst von Baer (1792–1876), who discovered the mammalian ovum, built a series of laws of developmental transformation on the 'law of parallelism' proposed for the human embryo in Germany by J. F. Meckel (1781–1833) and independently in France by E. R. A. Serres, Geoffroy's chief disciple and the man who furnished philosophical anatomy with an embryological perspective.

Meckel (1811, 1821) saw a parallel between the embryonic stages of higher animals and the permanent stages of lower animals. Human embryos passed through a hierarchy of animal forms as they developed from fish → reptile → mammal and ultimately → human. Meckel's permanent stages were not necessarily adults. Rather they were stages with the archetypal characteristics of forms lower on the Great Chain of Being. He conceived of a single 'developmental' force, controlling the development of individuals and 'evolution', by which Meckel meant ontogenetic unfolding and not phylogenetic transformation. Serres (1830, 1860) also saw a clear parallel between human embryonic development and the stages of other vertebrates. That the order fish → reptile → mammal was the order in which animals were being discovered to appear in the fossil record provided a parallel between embryonic development and the progression of life on earth.[1]

Against this background von Baer proposed four laws of development (or a four-part law of development), which may be summarized as:

1. the more general characters of the group to which an embryo belongs appear in development before the more specialized characters;
2. less general structures form in development after more general structures until finally the most specialized structures appear;
3. during development embryos progressively diverge from embryos of other groups; and
4. embryos of higher animals resemble embryos and not adults of other animals (embryos do not recapitulate adults).

Like Aristotle, von Baer studied the development of many species at firsthand. In exhaustive and brilliantly executed studies, he refuted preformation, and in so doing outlined a progressive three-stage basis for embryonic development:

1. primary differentiation, in which the germ layers are formed;
2. histological differentiation, when cell types develop within the germ layers; and
3. morphological differentiation, the stage of initial organ formation.[2]

Von Baer dealt with comparative embryonic development throughout the animal kingdom, particularly the issue of whether 'higher' animals repeated in their development adult stages of 'lower' animals. He found structures in the former that were not present in the latter (the yolk sac in birds for example). Such observations proved to him that higher forms did not recapitulate lower ones. Nor were structures of lower animals found in embryos of higher forms; fins and tails characteristic of fishes were not found in embryonic birds. Therefore, although organs might be recapitulated, organisms or types were not. One of the nicest encapsulations of this progression is

Oken's rendering of von Baer in the second edition of his *Lehrbuch der Naturphilosophie*:

'The embryo successively adds the organs that characterize the animal classes in the ascending scale. When the human embryo, for instance, is but a simple vesicle, it is an infusorian; when it has gained a liver, it is a mussel; with the appearance of the osseous system, it enters the class of fishes; and so forth, until it becomes a mammal and then a human being.'

Oken 1931, p. 387, cited from Ospovat, 1976, pp. 4–5

Von Baer equated the type with the structural plan of organisms within particular groups; the embryo, not the 'idea', became the Archetype. Therefore, for von Baer, as for Milne-Edwards (s. 4.7), each of Cuvier's *embranchements* had a different embryological archetype. Von Baer proposed the peripheral for the Radiata, longitudinal for the Articulata, massive for the Mollusca, and a somewhat heterogeneous vertebrate type for the Vertebrata. [A parallel modern classification of phylotypic stages based on *Hox* genes is presented in section 14.2.3.] While von Baer's defence of the Archetype allied him with Geoffroy, his views on relationships among organisms were closer to those of Cuvier. Identification of the Archetype with the common ancestor, or the replacement of the Archetype by the ancestral form, was but a short step away.

The development of individual organs within each type was specific to the grade of organization and only varied within the limits set by the type; the embryo equalled the type. Embryos of one type repeated or recapitulated neither the embryonic development nor adult organization of organisms from other types. But, within a group, the earliest stages of embryonic development were more alike than were the later stages. Development proceeded from the general to the specific to parallel the hierarchical classification of that species. Characteristics of the type or phylum appeared first, followed by those of the class, order, etc., until species characteristics were determined. Von Baer therefore provided embryological criteria and an embryological rationale for taxonomic organization. For him there was no phylogeny apart from ontogeny, and the only way to understand phylogeny was through ontogeny.

The consequences of von Baer's postulates for morphology were enormous. Russell captured this beautifully:

'If the embryo develops from the general to the special, then the state in which each organ or organ-system first appears must represent the general or typical state of that organ within the group. Embryology will therefore be of great assistance to comparative anatomy, whose chief aim it is to discover the generalised type, the common plan of structure, upon which the animals of each big group are built. And the surest way to determine the true homologies of parts will be to study their early development . . . Parts therefore, which develop from the same "fundamental organ," and in the last resort from the same germ-layer, have a certain kinship, which may even reach the degree of exact homology.'

Russell, 1916, p. 126

Under von Baer, homology became a concept applied to types on the basis of development, rather than a concept applied between types on the basis of adult structure and connections. Thus, von Baer focused the attention of morphologists on embryos. The goal of morphology became to seek the Archetype in the embryo and to seek homology in development. The 'embryological criterion' of homology had arrived and the foundations of evolutionary developmental biology were laid.

5.2 THE BRITISH RESPONSE TO VON BAER AND THE GREAT DEBATE

British anatomists digested news of the great debate and took up Geoffroy's views along with Von Baer's ideas of the embryological archetype and the embryological criterion of homology. Geoffroy's influence in England radicalized the medical curriculum and polarized British anatomy. Influential anatomists such as R. E. Grant and R. D. Grainger in London, sought to base the teaching of anatomy in British medical schools on Geoffroy's unity of plan, 'the great law of the organic world, the one grand principle that reigned over zoological science' (Desmond, 1989, p. 198). According to Grant (1835), no anatomist was professionally competent unless schooled in the philosophical anatomy of Geoffroy. Almost 40 years later, J. A. Symonds, of the Bristol Medical School, saw Geoffroy's approach to anatomy as one of the achievements of the age (Symonds, 1871).[3]

In addition to the radicalization occurring in the medical schools, Barry, Carpenter, Spencer, Chambers and Owen brought the issues of the Geoffroy-Cuvier debate and von Baer's views to the teaching of zoology.

Martin Barry (1802–55) was a Scottish physician who won a Royal Society medal for his studies in mammalian embryology. In the mid 1830s, he used von Baer's embryological laws of the parallelism between ontogeny and phylogenetic history to argue for embryological development as the basis for the unity of nature and the classification of living things (Barry, 1836–7a,b). The physiologist William Benjamin Carpenter (1813–85) was a staunch supporter of Huxley and, after initial admiration, a critic of Richard Owen. Carpenter and Owen incorporated Barry's (and therefore von Baer's) ideas into their treatments of the *Académie* debates.

Carpenter and Owen were among the first in Britain to apply embryological concepts of progressive divergence during development to the fossil record. Agassiz (1857) championed a similar view in America in his threefold parallelism between ontogeny, fossils and morphology. Just as individuals diverged from the Archetype during development, so as one proceeded through the fossil record, fossils diverged progressively from an ancient archetype. (As noted in section 1.4, the original idea of continuity between the developmental history of the earth, life upon it, and embryonic development, goes back at least to Kielmeyer (1793). Kielmeyer, who was one of Cuvier's teachers, explained transformations between organisms as transformed developmental stages which became fixed in the new organism). In Agassiz' case, the parallelism was interpreted in the context of recapitulation; Agassiz was an anti-Darwinian. True to Cuvier's *embranchements* to the end, his last publication of 1874 was on the permanence of the type.[4]

To Carpenter, the parallel between development and the fossil record was powerful. Following the German tradition, in which *Entwicklung* was used for ontogeny and for evolution, he used evolution for embryonic development and for change through time: 'the evolution of structure . . . both in the ascending scale of creation, and in the growth of embryos' (1839, p. 170). Carpenter applied the theories of von Baer and developed his own in a paper that appeared in 1837 and in an influential book, *Principles of General and Comparative Physiology* (1839). In the third edition, Carpenter discusses von Baer's embryology in detail as a unifying concept. Echoing Geoffroy, he said:

'We arrive at the important truth that, where any new function, or great modification of function, is to be performed, no entirely new structure is evolved for the purpose;- the end being always attained by a corresponding modification in some structure already present.'

Carpenter, 1851, p. 190.[5]

In what he saw as a graded transition between molluscs and fishes, Carpenter used Geoffroy's principles to argue for analogy (homology) between the wings of flying insects and the gills of aquatic insects, and between squid cranial cartilages and the vertebrate chondrocranium. (A modern version of wings

from gills is discussed in section 16.6.2.) Carpenter also used examples like progressive reduction in horses' toes and changes in the morphology of foraminiferans over time to support his thesis.

Herbert Spencer, who strove tirelessly to promote the idea of evolution in Britain (s. 1.3), used the theories of progressive embryological divergence of von Baer and Carpenter in developing 'Social Darwinism' – the inevitability of change in industry and society. Spencer described von Baer's embryology as 'the law of all progress' (1857, p. 148).

A book entitled *Vestiges of the Natural History of Creation* was published anonymously in 1844. Widely read, it went to a tenth, considerably extended and amended edition by 1853, and had sold 28 000 copies in 12 editions by 1884. The subject of scathing attacks by theologians and natural scientists (for different reasons), *Vestiges* played no small role in conditioning the British public to the appearance of *The Origin*. A major aim of the anonymous author was to demonstrate that the development (evolution) of life on earth was progressive and paralleled the growth of the human embryo, in what was called 'the principle of progressive development'.

Authorship of *Vestiges* remained unknown throughout the lifetime of its author, Scottish publisher Robert Chambers (1802–71). Robert and his brother William founded the successful and influential *Chambers' Edinburgh Journal*. Robert, the respected editor of *Chambers' Encyclopaedia*, kept his authorship of *Vestiges* a secret as much to protect the family business as to avoid controversy. He believed in environmental causes of major changes in embryonic structure, spontaneous generation, evolution by large jumps, and a divine, progressive plan for life on earth.[6]

Chambers is included here for two reasons. First, his was an influential attempt to provide a developmental link to evolution; as summarized by Bowler (1988), the view that ontogeny recapitulates phylogeny is much more characteristic of the non-Darwinian approach to evolution than of the Darwinian. Rather than treating *Vestiges* as a paltry anticipation of Darwinian evolution we should regard it as the clearest expression of the developmental view of nature gaining ground in Germany, a view which Nyhart has analysed in depth in *Biology Takes Form: Animal Morphology and the German Universities, 1800–1900* (1995).

Second, Chambers was among those who introduced the ideas of von Baer and the Meckel-Serres law of parallelism to the British scientific community. Although Chambers relied on Carpenter's interpretation of von Baer's ideas, he advocated the reappearance of ancestral embryonic stages:

'. . . the new being passes through stages in which it is successively fish-like and reptile-like. But the resemblance is not to the adult fish or the adult reptile, but to the fish and reptile at a certain point in their foetal progress.'

Chambers, 1844, p. 212

Chambers drew upon fossil evidence of progression within animal groups from simple to complex, evidence from comparative morphology, unity of type within animal groups, von Baer's embryological law that embryos exhibited stages resembling ancestral stages, and a belief in divine providence and progression. From these diverse elements he fashioned a principle of progressive development that he believed revealed the universal gestation of nature, reconciled science and religion and revealed how the Deity had fashioned life on earth through time. Darwin was singularly unimpressed by *Vestiges* and by arguments not founded on a solid factual foundation as seen in this passage from a January 1845 letter to Joseph Hooker:

'I have also read the "Vestiges", but have been somewhat less amused at it than you appear to have been: the writing and arrangement are certainly admirable, but his geology strikes me as bad, and his zoology far worse.'

Darwin, F., 1887, I, p. 333

5.3 OWEN: HOMOLOGY AND ARCHETYPES

Sir Richard Owen (1804–92), a complex, enigmatic figure, was the first Superintendent (Director) of the British Museum (Natural History), a post he held from 1856 to 1883. Owen brought order and an evolutionary perspective into the morphological debate by distinguishing **homology** from **analogy** and recognizing **serial homology**.[7]

In 1827 Owen was hired to catalogue John Hunter's vast anatomical collection, which formed the basis of the Hunterian Museum of the Royal College of Surgeons. Owen married the daughter of the Conservator and took lodgings above the Museum to maximize the time available for this work. His capacity for work was enormous; some thought it his greatest asset! Cataloguing the Hunterian collection was no small task, for Hunter left no notes (or destroyed them) and Owen had to redescribe all the specimens. So ably did he meet this challenge that in 1834 the then President of the Royal College of Surgeons, G. J. Guthrie, wrote that the Hunterian Museum was of more practical value than the *Jardin des Plantes* in Paris (Box 4.2).[8] Owen, by then the acknowledged expert on the newly discovered marsupial and monotreme faunas of Australia, was appointed Hunterian Professor at the Royal College of Surgeons in 1836.

The Geoffroy–Cuvier debates made a great impression on Owen: 'I was guided in all my work with the hope or endeavour to gain inductive ground for conclusions on these great questions' (Owen, 1868, pp. 786–7). He saw five issues as central:

1. whether unity of plan or final purpose was a governing condition of development;
2. whether the series of species in the fossil record were uninterrupted or broken by intervals;
3. whether extinction was cataclysmic or regulated;
4. whether development occurred by epigenesis or by 'evolution' (preformation);

5. whether the origin of life was by miracle (created) or a result of the operation of natural laws.

Owen was the last of the idealistic morphologists. At age 42, under a commission from the British Association for the Advancement of Science, he produced a monographic work using the vertebrate skull as the basis for a synthesis of archetypes, idealistic morphology and homology. An extension of this position, based on a lecture to the Royal Institution in London, was published in 1849 as *Discourse on the Nature of Limbs*.

The term 'Archetype' was introduced by Owen in 1846 simultaneously with its introduction by Joseph Maclise, a London anatomist who saw the recognition of the Archetype as the aim of comparative science. Owen was searching for a vertebrate archetype while avoiding any notions of the unity of types across groups. For him the Archetype was an idealized form, the single, ideal type that embodied the potential for all the fundamental structures of the body and limited the variation possible within members of a single group of animals. He figured a vertebrate Archetype in his 1848 monograph which is still often reproduced (Figure 5.1). Investigations by Rathke and Huxley on the development of the skull of cyclostomes (lampreys) provided an embryonic archetype for the vertebrate skull.[9]

Although with the Archetype Owen projected segmentation and serial repetition to what are now seen as unreasonable limits, he did morphology and subsequent evolutionary studies an invaluable service by clearly and explicitly separating analogous or adaptive from homologous or essential structures or characters and by establishing and distinguishing among special, serial and general homology. He was quite prepared to take the best that Cuvier, Geoffroy and others had to offer. Owen used von Baer's concepts as developed by Martin Barry in taking ideas from embryology to explain the unity of type. Each

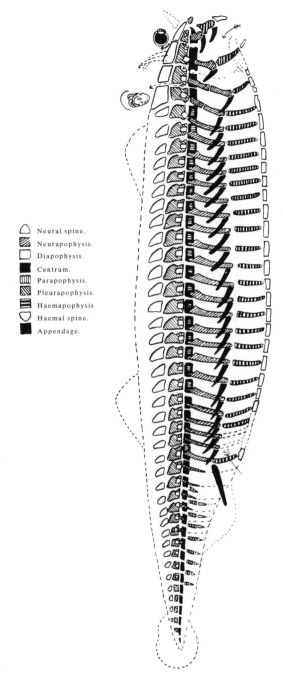

Neural spine.
Neurapophysis.
Diapophysis.
Centrum.
Parapophysis.
Pleurapophysis.
Haemapophysis
Haemal spine.
Appendage.

Fig. 5.1 The vertebrate archetype proposed by Owen (1848) on the basis of the segmental organization of the axial (vertebral) skeleton. Reproduced from Russell (1916).

embranchement had a common developmental plan that produced the morphology typical of that group and/or from which modifications of plan were derived. Using von Baer's concepts, Owen tried to limit Geoffroy's unity of type to transformations within single *embranchements*.[10]

Thus in Owen the fullest resolution of the dichotomy posed by the Great Debate is manifest. Types, archetypes, connections, *embranchements*, homology, early embryonic development and recapitulation are integrated into a synthesis of the central role of embryological changes in the production of new and in the maintenance of old morphologies. All this Owen did by his usage of homology and analogy, adhering to the principle of connections established by Geoffroy:

'Homologue . . . The same organ in different animals under every variety of form and function.'

Owen, 1843, p. 379

'Analogue . . . A part or organ in one animal which has the same function as another part or organ in a different animal.'

Owen, 1843, p. 374

The distinction between homology and analogy was at last clear.

Although Owen clarified these distinctions, recent interest in homology, including historical aspects, has revealed that he was not the first to do so and that he developed his own ideas earlier than 1843. Alex Panchen has demonstrated that the Englishman William Sharp Macleay distinguished homology from analogy (or affinity) and parallelism in relation to taxa, if not structure, as early as 1821. Owen had these distinctions between homology and analogy in mind as early as 1836. He used the concepts in his article on cephalopods in Todd's *Cyclopaedia of Anatomy and Physiology* to examine correspondence between the eyes of cephalopods and vertebrates, much as Geoffroy had done. Homology had been in use as an anatomical term from early in the 1800s, as Huxley (1894) demonstrated in an evaluation of Owen's work in the standard 'life' edited by Owen's grandson and namesake.[11]

5.4 DARWIN

Publication of *The Origin of Species* generated enormous interest in tracing phylogenetic histories. With his theory of descent from a common ancestor, Charles Darwin provided the evolutionary insight into why groups of organisms share a common plan or morphological type: common ancestors, not common archetypes, 'explain' common ground plans.

Explanations for homologous structures changed fundamentally after 1859 from a pre-Darwinian typological view of homology based on equivalent connections and elements to the post-Darwinian evolutionary view of homology based on possession of shared ancestors. Descent from a common ancestor provides the explanation for unity of the type and for homologous structures. Darwin's use of heredity and adaptation to explain change in morphology through time involved a further, parallel compromise between Geoffroy and Cuvier. Less appreciated are two major methodologies made popular by Darwin's approach to understanding the living world: the use of inductive reasoning based on the collection and analysis of a vast quantity of information, and the formulation of grand comprehensive theories.

Darwin's use of morphological evidence for the theory of descent was greatly influenced by the three prevailing concepts of unity of type, Geoffroy's law of connections, and homology. Two quotations from *The Origin* typify his view. The first also indicates that Darwin embraced the inheritance of acquired characters.

'It is generally acknowledged that all organic beings have been formed on two great laws – Unity of Type, and the Conditions of

Existence. By unity of type is meant the fundamental agreement in structure which we see in organic beings of the same class, and which is quite independent of their habits of life. On my theory, unity of type is explained by unity of descent. The expression of conditions of existence, so often insisted on by the illustrious Cuvier, is fully embraced by the principle of natural selection. For natural selection acts by either now adapting the varying parts of each being to its organic and inorganic conditions of life; or by having adapted them during past periods of time: the adaptation being aided in many cases by the increased use or disuse of parts, being affected by the direct action of the external conditions of life, and subjected in all cases to the several laws of growth and variation. Hence, in fact, the law of the Conditions of Existence is the highest law: as it includes, through the inheritance of former variations and adaptations, that of Unity of Type.'

Darwin, 1910, p. 156

'"What can be more curious," he asks, "than that the hand of man, formed for grasping, that of a mole for digging, the leg of a horse, the paddle of the porpoise, and the wing of the bat, should all be constructed on the same pattern, and should include similar bones, in the same relative positions?" . . . Geoffroy St Hilaire has strongly insisted on the high importance of relative position or connection in homologous parts; they may differ to almost any extent in form and size, and yet remain connected together in the same invariable order. We never find, for instance, the bones of the arm and fore-arm, or of the thigh and leg, transposed. Hence, the same names can be given to the homologous bones in widely different animals. We see the same great law in the construction of the mouths of insects: what can be more different than the immensely long spiral proboscis of a sphinx-moth, the curious folded one of a bee or bug, and the great jaws of a beetle? – yet all these organs, serving for such widely different purposes, are formed by infinitely numerous modifications of an upper lip, mandibles, and two pairs of maxillae. The same law governs the construction of the mouths and limbs of crustaceans. So it is with the flowers of plants.'

Darwin, 1910, pp. 358–9

Initially, Darwin went further than Owen (for whom the Archetype was an abstract model), seeing the Archetype as the ancestral form. Thus:

'I look at Owen's "Archetypes" as more than ideal, as a real representation as far as the most consummate skill & loftiest generalization can represent the parent form of the Vertebrate.'

Ospovat, 1981, p. 146, citing Darwin's notations in the back of his copy of Owen's *On the Nature of Limbs*

Subsequently, his views of the type changed. In a letter to Thomas Huxley of 23 April 1853, Darwin signified the importance he ascribed to homology and to an embryological view of the type:

'The discovery of the type or "idea" (in *your* sense, for I detest the word as used by Owen, Agassiz & Co) of each great class, I cannot doubt is one of the very highest ends of Natural History . . . I sh[d][should] have thought that the archetype in imagination was always in some degree embryonic, & therefore capable of *generally undergoing* further development.'

Darwin and Seward, 1903, Vol. 1, p. 73

Agassiz's sense of type was of recapitulation: 'the oldest representatives of every class may then be considered as embryonic types of their respective orders or families among the living' (Agassiz, 1857, p. 174).

Huxley and Darwin by this time shared a materialistic view of the unity of type that differed from the idealized type of Geoffroy and Owen. Huxley's attitude toward Owen, 'sense' of type, and some sense of the feud between the two men, are captured in the following quotations from his 1853 paper on the morphology of the cephalopods, and from a letter to the marine biologist Edward Forbes (Box 5.1):

BOX 5.1
EDWARD FORBES JR. (1815–54)

Edward Forbes abandoned art (insufficient talent) and medicine (insufficient interest) for the study of marine animals, especially their distribution, and subsequently produced one of the first general studies of oceanography. He was equally versed in biogeography, invertebrate zoology and invertebrate palaeontology. Living on a £150 annual allowance from his father, a wealthy businessman on the Isle of Man, on 1 April 1841 Forbes joined a survey ship charting the coastal waters of Greece and Turkey. This was one of the most extensive, systematic, and scientifically profitable dredging operations ever undertaken.

With three and a half years of exploration under his belt Forbes became Professor of Botany at Kings College, London, on less than £100 a year, an appointment taken reluctantly, and only because reversals in his father's fortunes brought his allowance to an end. He added the positions of curator for the Geological Society (£150) and Palaeontologist for the Geological Survey to augment his stipend. In April 1854 he was appointed to replace Jameson as Regius Professor of Natural History at Edinburgh, but died seven months later at age 39. Huxley replaced him in the Natural History position at the Royal School of Mines in 1854, refusing the position of paleontologist because he 'did not care for fossils'.

'From all that has been stated, I think that it is now possible to form a notion of the archetype of the Cephalous Mollusca, and I beg it to be understood that in using this term, I make no reference to any real or imaginary "ideas" upon which animal forms are modelled. All that I mean is the conception of a form embodying the most general propositions that can be affirmed respecting the Cephalous Mollusca, standing in the same relation to them as the diagram of a geometrical theorem, and like it, at once imaginary and true.'

Huxley, 1853, p. 50

'He [Owen] is not referable to any "Archetype" of the human mind with which I am acquainted.'

Huxley, 27 November, 1852, cited in Desmond, 1982, p. 22

Darwin's explanation for the unity of type was common descent: 'On my theory, unity of type is explained by unity of descent' (Darwin, 1910, p. 156). With common descent came a new and evolutionary definition of homology which can be summarized as: a feature is homologous in two or more taxa if it can be traced back to the same feature in the nearest common ancestor (Mayr, 1982).

For Darwin, commonality of embryological structure provided powerful evidence for common descent, for an embryological view of the Archetype, and a basis for homology. Darwin was influenced strongly, therefore, by von Baer and by the importance of embryology, although his use of comparative embryonic development was tempered with caution:

'Thus, community in embryonic structure reveals community of descent; but dissimilarity in embryonic development does not prove discommunity of descent, for in one of two groups the developmental stages may have been suppressed, or may have been so greatly modified through adaptation to new habits of life, as to be no longer recognizable.'

Darwin, 1910, pp. 371–2[12]

This still represents a major drawback to the use of common embryonic development or developmental processes as a (the) criterion for homology. As elaborated with examples in section 21.7 and in Hall (1995a), homologous adult structures do arise by non-homologous developmental processes.

Darwin expressed no similar caution with respect to the importance of embryology as evidence for evolutionary change in morphology. In letters to J. D. Hooker in 1859 immediately after the publication of *The Origin* and to Asa Gray in 1860 with reference to the reviews that had begun to appear, Darwin wrote:

'Embryology is my pet bit in my book, and, confound my friends, not one has noticed this to me.'

(letter to J. D. Hooker, cited in F. Darwin, 1887, Vol. 2, p. 39)

'Embryology is to me by far the strongest single class of facts in favour of change of form . . . [but] not one, I think, of my reviewers has alluded to this.'

letter to A. Gray, from Darwin's *Autobiography*, cited in F. Darwin, 1887, Vol. 1, p. 72

Debate over homology continued for many decades, as indeed it continues today, as evidenced in the variety of positions expounded by the contributors to *Homology: The Hierarchical Basis of Comparative Biology* (Hall, 1994a).

Edwin Ray Lankester (1847–1919), a former student of Robert Grant's, whom he succeeded as Professor of Zoology at University College, London, wrote a paper in 1870 on the use of homology in 'modern zoology'. A devoted follower of Huxley – his 'father in science' as he was Huxley's 'scientific son' – Lankester's intention was to remove any hint of ideal form or archetypes from homology, a position which brought him into direct confrontation with Owen and the idealistic morphologists. Lankester proposed abandoning the term; in

its place he introduced **homogeny** (homogenous) for similarity resulting from shared ancestry, and **homoplasy** (homoplastic) for similarities shared by unrelated organisms. His contributions to germ-layer theory are discussed in section 5.5.3.[13]

5.5 AFTER DARWIN: EVOLUTIONARY EMBRYOLOGY

Darwin's writing revolutionized evolutionary embryology. I can only skim this vast topic and discuss only some of the many players. In the sections that follow I concentrate on the first test of Darwin's theory using embryological evidence, Ernst Haeckel's grand synthesis of embryos and evolution, the influential role played by germ layers and germ-layer theory, and the integration of germ layers and homology in the comparative and evolutionary embryology established by Francis Balfour in England.

5.5.1 Embryological evidence put to the test

Immediately following publication of *The Origin*, Darwin's affirmation of the importance of embryological evidence was put to the test by Johann Friedrich Theodor (Fritz) Müller (1822–97) in the first application of the theory of descent with modification to a single major group of organisms, the Crustacea. A German-born and -educated naturalist, Müller's free-thinking religious views and depression over the failure of the 1848 revolution forced him at age 30 to flee Germany and seek employment in Brazil, where he remained until his death.

Although priority is usually given to Haeckel, Müller was the first to use recapitulation to reconstruct the original form of a major group. So varied are crustacean life history strategies that, inspired by Darwin's theory, Müller was able to use the details and varieties of life-history stages to construct a scheme of crustacean relationships in which embryology was the handmaiden of classification. Using notions of recapitulation that were taken to excess by Haeckel (see below), Müller identified the nauplius larva as the crustacean ancestor. Müller found a nauplius stage in such groups as shrimps, where it had not been observed previously. In support of two criteria expected of an embryonic stage that recapitulated an ancestor, all crustaceans possessed a nauplius stage early in their ontogeny. Ancestors of 'higher' forms of crustaceans such as the Malacostraca (crabs, shrimps, prawns, lobsters, crayfish) would be expected to appear later in ontogeny and only in embryos of those higher forms. The zoea stage fitted these expectations precisely, being present in crabs and shrimp (Figure 5.2). So Müller saw, both in ontogeny and in phylogeny sequences of nauplius → lower crustaceans and nauplius → zoea → higher crustaceans.

With his emphasis that evolutionary adaptations could occur at any stage during development, Müller was ahead of his time. He summarized his research in a book so devoted to Darwin's theory that he called it simply *Für Darwin* (1864). In turn, Darwin arranged to have the book translated into English in 1869 under the title *Facts and Arguments for Darwin*. The historian of biology Peter Bowler has taken the story of Müller's contributions further than anyone, especially in his 1994 paper on whether arthropods form a natural group, and in *Life's Splendid Drama* (1996), which includes discussion of *Für Darwin* and arthropod origins in the context of the relationship of embryology to morphology in the mid-1800s.

5.5.2 Haeckel's Gastræa theory

The embryological approach to morphology, evolution, and the search for archetypes in the last half of the 19th century is most closely identified with the **Gastræa** theory of Ernst Haeckel (1834–1919). Foreshadowed in his two-volume *Generelle Morphologie der Organismen* (1866), it was developed in an 1872 monograph on calcareous sponges, elaborated in three subsequent papers (1874, 1875, 1876a) and a popular book (1868), which was trans-

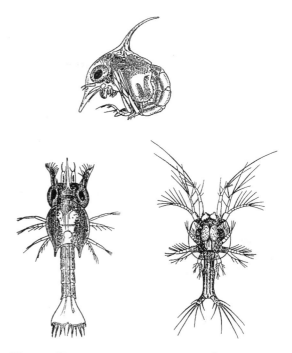

Fig. 5.2 Three zoea-stage crustacean larvae as depicted by Fritz Müller. Top, the Marsh crab; left, the hermit crab; right, the shrimp *Peneus*. Adapted from Müller (1869).

lated into English under Lankester's guidance as *The History of Creation* (Haeckel, 1876b).

Haeckel trained in idealistic morphology under the botanist Alexander Braun and the zoologist Johannes Müller (1801–58) at Jena, where he was introduced to the marine organisms on which he prepared his dissertation and which he studied for much of his life. Johannes Müller (not to be confused with Fritz Müller of *Für Darwin*) and Braun emphasized the importance of development and of the cell as the fundamental embryological unit. Rinard (1981) provides an insightful analysis of how these ideas formed the basis of Haeckel's contributions to ontogeny, phylogeny, the nature of the individual as each lower stage became part of the next highest stage, and the essentially repetitive or 'regenerative' nature of the relationship between ontogeny and phylogeny.

Table 5.1   Haeckel's five stages of metazoan embryonic development and the equivalent hypothetical primitive ancestral metazoans

Embryonic stage	Features	Ancestral equivalent
Monerula	Fertilized ovum after loss of germinal vesicle	Anucleate Monera, ancestral to all Metazoa
Cytula	Ovum with reformed (zygote) nucleus	Amoeba
Morula	Ball of cells	Synamoeba, the first multicellular stage
Blastula	Hollow, ciliated sphere, free-swimming	Planæa (Blastæa)
Gastrula	Two-layered sac, equivalent to a simple sponge	Gastræa, common metazoan ancestor

Haeckel's attempt was neither more nor less than to document that all multicellular organisms arose phylogenetically from an organism structurally equivalent to the early gastrula, an embryonic stage found early in the development of all multicellular animals (s. 8.3.3). This hypothetical, universal Archetype and ancestor was the Gastræa. Haeckel developed his theory and formulated his phylogenetic trees in the context of recapitulation: *'die Ontogenie ist eine Rekapitulation der Phylogenie'* (1891, p. 7). One of his earliest enunciations of recapitulation is in volume 2 of his 1866 treatises on *General Morphology*:

'Die Ontogenesis ist die kurze und schnelle Rekapitulation der Phylogenesis, bedingt durch die physiologischen Funktionen der Vererbung (Fortpflanzung) und Anpassung (Ernährung). (Ontogeny is a brief and rapid recapitulation of Phylogeny, dependent on the physiological functions of Heredity (reproduction) and Adaptation (nutrition).'

Haeckel, 1866, vol. 2, pp. 300[14]

Haeckel's theory required that new evolutionary stages had been added at the end of the embryonic development of ancestral forms. This meant that adult ancestral stages were to be found in the embryos of descendants. Whether Haeckel's Gastræa theory was right or wrong, it represented a brilliant synthesis of recapitulation, Darwin's theory of evolution, comparative morphology, homology, and comparative embryology which had revealed common embryological construction from the same germ layers.

Haeckel believed that organisms pass through a two-germ-layer stage equivalent to that identified by Huxley in the medusae of coelenterates (s. 5.5.3). He saw these stages as equivalent on the basis of the two-layered structure and because of the mechanics of their production, during which the inner (endodermal) layer formed from the outer (ectodermal) layer by inpushing (invagination) within the blastula to produce the structure shown as a gastrula in Figure 5.3. Haeckel therefore used dynamic developmental processes, not just static embryological stages, in formulating his theory.[15]

Haeckel identified five primordial stages in early metazoan development, each equivalent to an ancestral form: the Monerula, Cytula, Morula, Blastula and Gastrula (Table 5.1), which are shown in Figure 5.4 alongside the grades of animal life they represent. Subsequently, he added additional stages at the end of the sequence: the Coelomula (coelomic pouch) stage, Chordula (neural tube and notochord) stage, and Spondula (segmented mesoderm) stage. For Haeckel, embryos were archetypal ancestors.

A major phylogenetic consequence of the embryological criterion of the archetype and the Gastræa theory was that metazoans were monophyletic and embryos more informative than fossils. Haeckel was able to erect a phylogenetic tree to reflect this monophyly

Diploblastic

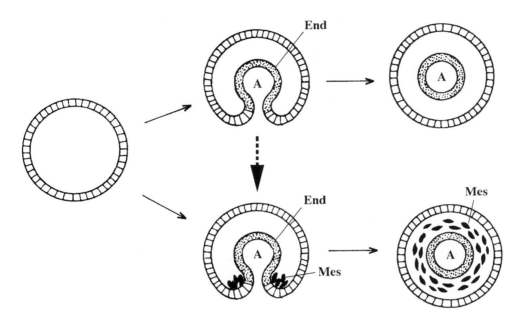

Triploblastic

Fig. 5.3 This figure serves a number of purposes. Solid arrows show ontogenetic transformations from the blastula (left) to the gastrula (centre and right). Ontogenetically it displays the infolding of the endoderm (stippled, End) in simple diploblastic and triploblastic embryos to form a hollow gut cavity surrounding the archenteron (A). Phylogenetically, the dashed arrow shows the origin of the triploblastic from the diploblastic condition, with mesoderm represented as mesenchyme (Mes) which forms a third germ layer in triploblastic embryos (bottom right). There is a progressive transformation of a two-layered to a three-layered condition during development of triploblastic embryos (bottom figures).

and the relationships between grades of animals as interpreted from progressive embryonic development. This tree and the roots that supported it stood totally at variance with the four *embranchements* of Cuvier and the type concept, but supported descent from a common ancestor.

5.5.3 Germ layers and germ-layer theory

Christian Pander first recognized germ layers in chick embryos in 1817 when he carried out an exhaustive study using 2000 embryos. By

examining embryos at 15-minute intervals over the first five days of incubation, Pander produced a most detailed analysis. Von Baer, who had actively encouraged Pander, extended his studies in 1828 by demonstrating that all vertebrate embryos are constructed on the same three-layered plan.[16]

Huxley placed germ layers into an evolutionary context in 1849 with his affirmation that the inner layers of ectoderm and endoderm of embryonic vertebrates could be equated with, indeed were homologous with, the two layers of adult coelenterates. In 1873

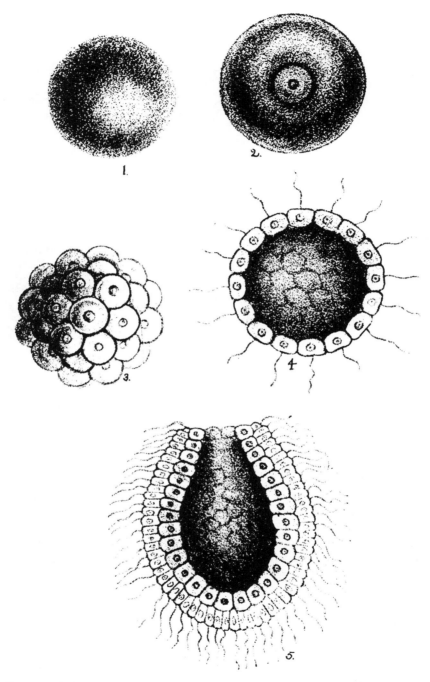

Fig. 5.4 The five primordial stages in metazoan development identified by Haeckel. 1. Monerula. 2. Cytula. 3. Morula. 4. Blastula. 5. Gastrula. Additional information in Table 5.1. Reproduced from Russell (1916).

(more fully in 1877), Lankester extended germ layers into phylogeny by dividing the animal kingdom into three grades on the basis of the number of germ layers present:

1. Homoblastica (protozoa);
2. Diploblastica (coelenterates); and
3. Triploblastica (the remaining phyla).

Lankester was the first to use the terms ectoderm, mesoderm and endoderm for the three germ layers, terms that Haeckel (1874) quickly adopted. Indeed, origination from common germ layers was the basis of homology for Haeckel in his scheme in which recapitulation, germ layers and homology were inextricably linked.

'True homology can only exist between two parts which have arisen from the same primitive "Anlage" and have deviated from one another by differentiation only after the lapse of time.'

Cited by Tait, 1928

Edouard Van Beneden (1846–1910), Belgian embryologist and pioneer student of the development of ascidians and the parasitic nematode *Ascaris*, was one of the first to demonstrate cell lineages in embryogenesis and the union of male and female nuclei at fertilization. In the 1870s, he used studies on germ-cell origins in coelenterates to demonstrate that while sperm arose from ectoderm, eggs arose from endoderm. Such a dramatic and opposing tendency of ectoderm and endoderm reinforced the importance of germ-layer origins and established close links between germ-layer theory, sexual reproduction and heredity. These links were reinforced by several workers, especially the Hertwig brothers (Oscar and Richard) and August Weismann.

Through such studies, recapitulation, germ layers, homology, reproduction and heredity became inextricably linked.[17]

Germ layers and the germ-layer theory remained central to the study of inheritance and evolution for 70 or 80 years. Times have changed. Germ layers are still important, but the germ-layer theory is not. I have argued that germ-layer theory be discarded but the number of germ layers be expanded from three (ectoderm, mesoderm, endoderm), to four (ecto-, meso-, endoderm and neural crest). The reason for this new division, 104 years after Lankester established the fundamental three-fold division of the animal kingdom, is that vertebrates are tetrablastic, the neural crest forming a fourth, exclusively vertebrate, germ layer. Ectoderm and endoderm are primary germ layers, whereas mesoderm and neural crest are secondary germ layers; see sections 10.1 and 10.4.1, Box 10.2 and Hall (1997a) for the evidence and arguments.[18]

5.5.4 Balfour: homology and germ layers

Evolutionary embryology was taken up by Francis (Frank) Maitland Balfour (1851–82), director of a Morphology Laboratory at Cambridge at the remarkably early age of 22 and one of the most promising embryologists of his age. Balfour is an unsung, little-studied and insufficiently appreciated pioneer of evolutionary embryology.[19]

Balfour established evolutionary embryology at Cambridge, attracting such students as William Bateson, D'Arcy Thompson, Henry Fairfield Osborn and Adam Sedgwick. Cambridge University was so concerned that Balfour would be lured away to Oxford or Edinburgh whose Universities were making overtures, that in 1882 it created a Professorship in Animal Morphology for him. His death just seven weeks later in an Alpine climbing accident, near Courmayeur in Switzerland, robbed evolutionary embryology of its most promising advocate and Cambridge of its enviable position in this new field. Cambridge established the Balfour studentship in his name. William Caldwell, the first Balfour Scholar and co-inventor of a microtome for cutting serial sections, settled the 80-year-old question of whether monotremes lay their eggs or give birth to their young. The work of

William Bateson, the second Balfour Scholar, is discussed in section 5.5.5.

Balfour's preoccupation with comparative and evolutionary embryology, recapitulation, and the origin of the vertebrates had their basis in homology of the germ layers and in von Baer's discovery of common developmental processes in embryos of different organisms. Balfour described chick embryo germ layers from first-hand observation and was the first to discuss their homology with germ layers in amphibian embryos; he established homology of the amphibian blastopore with Hensen's node of chick embryos in studies undertaken while still an undergraduate! (Balfour, 1873a,b). He placed his work with vertebrate embryos into distinct phylogenetic contexts in comparing vertebrate with invertebrate development. In distinguishing secondary embryonic adaptations from primary, ancestral patterns, Balfour avoided Haeckel's excesses (see further discussion in Hall, 1997a).

Balfour's description of the embryology of the dogfish in 1874 was an account of research undertaken in 1872 as the first British scientist to occupy a table at Anton Dohrn's new *Stazione Zoologica Napoli*. Balfour had much to say about homology in this paper. At age 27 he produced a monographic treatment on the embryology of elasmobranchs, a group regarded as primitive and therefore pivotal in the search for vertebrate ancestry (s. 5.5.5). This monograph was as illuminating and comprehensive for shark embryology as von Baer's studies had been for birds. Lankester reviewed the monograph in the pages of *Nature*. Balfour's identification of homologous structures throughout the vertebrates and homologous germ layers throughout the animal kingdom, are mentioned especially by Lankester as among important additions to fundamental knowledge. The chief significance of the monograph for Lankester, however, was in phylogenetic information revealed by the embryos of elasmobranch fishes:

'. . . we have in these fishes the nearest living representatives of the common ancestors of the great group of Gnathostomous Craniate Vertebrata, and it was to be expected that a full knowledge of their ontogeny or individual development would carry us yet further back in the line of primitive vertebrata, and yield a mass of explanatory evidence, exhibiting the development of complex and heterogeneous structures from simpler and more homogeneous forms, likely to serve as a satisfactory starting point for all Vertebrate morphology.'

Lankester, 1878, p. 113

Balfour's two-volume textbook of comparative embryology published in 1880–81 was the first attempt to compile this subject into a single treatise.[20] Balfour saw the future task of embryology as:

'To test how far Comparative Embryology brings to light ancestral forms common to the whole of the Metazoa . . . How far . . . larval forms may be interpreted as the ancestral type . . .'

Balfour, 1880–1 vol. 1, p. 4

These views are discussed further in section 14.2.4 in the context of larval types and the origin of the Metazoa.

5.5.5 Evolutionary embryology and the search for vertebrate ancestors

The 1870s and early 1880s were an active and exciting time for those involved in the classification and ordering of animals, in comparative embryology and in forging an evolutionary embryology. The grand schemes of animal clasification that emerged then are, by and large, with us today. Much energy went into searching for ancestors.

Theories about possible ancestors of the vertebrates fell into several major camps: insects (Geoffroy, 1818–22), ascidian tadpoles (Kovalevsky, 1867; Garstang, 1928), and annelids (Dohrn, 1875). The two major rival theories, however, pointed to an ascidian or

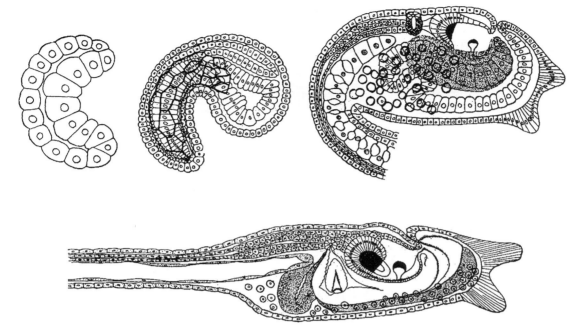

Fig. 5.5 Four stages in the development of the ascidian tadpole from blastula (top left) to tadpole with notochord. Adapted from Russell (1916) after Kovalevsky.

annelid ancestry. In the former (advocated by Haeckel and Kovalevsky), the ascidian tadpole was regarded as the closest living representation of the common vertebrate ancestor. Invertebrates did not come into play at all; vertebrates and invertebrates were separate. In the annelid theory, advocated by Anton Dohrn and Carl Semper, vertebrates and arthropods arose from a common segmented annelid or proto-annelid. This theory linked the two great animal groups, vertebrates and invertebrates. Haeckel and Kovalevsky based their theory on embryological evidence; Dohrn and Semper based theirs on adult anatomy.[21]

In 1866, Kovalevsky awakened interest in amphioxus as a primitive vertebrate with a brief note on its development. Because the paper was in Russian it attracted less attention than two substantial papers Kovalevsky wrote in English (one in 1867, the other ten years later), in which he demonstrated that embry-

onic development in amphioxus is typically vertebrate, with a notochord, multiple gill slits, an archenteron formed by invagination, and medullary folds that fuse to form a neural canal. Not only a typical vertebrate, amphioxus demonstrated the simplest pattern of vertebrate development and thus held the key to understanding vertebrate embryology. The key to embryology would unlock the door to vertebrate ancestry.

At the same time, Kovalevsky (1886b, 1871) demonstrated that the ascidian tadpole (not previously thought to be a vertebrate) had embryonic development that was similar to what he had seen in amphioxus. Ascidians had been grouped with the molluscs, but Kovalevsky's description of a notochord, gill slits and neural folds that developed from folds of the ectoderm – i.e. an amphioxus-like and simple vertebrate pattern (Figure 5.5) – dramatically demonstrated an affinity between ascidian tadpoles and vertebrates and

ANNELID

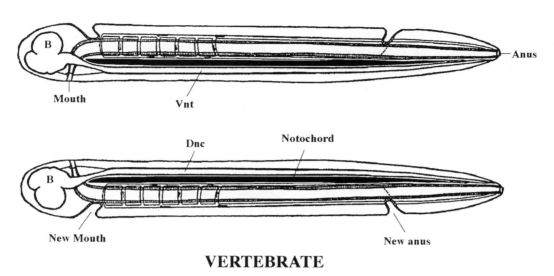

VERTEBRATE

Fig. 5.6 Representation of how an annelid could transform into a vertebrate. The annelid has an anterior brain (B), ventral nerve cord (Vnt), terminal posterior anus and a ventral gut that passes through the brain. The vertebrate has an anterior brain (B), dorsal nerve cord (Dnc) and notochord, sub-terminal anus, new mouth, and gut which does not pierce the brain. Although the notochord (black) is shown in the gastroneuralian annelid, it arose in the notoneuralian vertebrate. Complications associated with evolving a notochord, closing the annelid mouth and anus, and building new ones, did not deter adherents to the annelid theory (see text for details). Modified from Romer (1960).

suggested ascidians as possible vertebrate ancestors.[22]

Kovalevsky's findings and conclusions were quickly accepted by leading workers in evolution such as Darwin, Haeckel and Gegenbaur. That vertebrates arose from an ascidian tadpole-like ancestor became the favoured theory. In 1894, Arthur Willey (who had taken up the study of amphioxus while supported by a Balfour studentship from Cambridge), synthesized all the information on the embryology of amphioxus and ascidians into an influential book, *Amphioxus and the Ancestry of the Vertebrates.*

In 1875, however, just a few years after Kovalevsky produced his work on amphioxus and ascidian tadpoles, Dohrn and Semper independently developed an annelid theory of vertebrate origins using evidence from adult anatomy. They linked vertebrates and invertebrates as Geoffroy had done earlier in the century. The essence of Dohrn's and Semper's arguments was segmentation. Both annelids and vertebrates were segmented. Segmentation was an ancestral, primitive feature. Therefore annelids and vertebrates must have shared a common segmented ancestor. It seems extreme to compare worms with their ventral nervous system and mouth above the brain with vertebrates with their dorsal nervous system and mouth below the brain. Figure 5.6 shows just such a comparison. (Indeed, with modern molecular analyses, this comparison, indeed this same figure, is raising its head again; see section 9.4.) Which mouth position was primitive? Did the first vertebrate acquire a new mouth or a new brain as it arose from its annelid or pro-annelid ancestor?

Dohrn opted for a new mouth, whose origin was probably by coalescence of gill slits in the mid-line. Change in function (as occurs in the transformation of locomotory appendages to jaws in arthropods), a central element of Dohrn's theory, was laid out in detail in his 1875 book.[23]

The common ancestor of annelids and vertebrates would have had a notochord and segmentally arranged organs from which gill slits arose. Dorsal rather than ventral fusion of the nerve cord and the development of a new mouth gave this ancestor basic vertebrate characters. But what about amphioxus and ascidian tadpoles, which had no annelid characters? As far as Dohrn was concerned amphioxus and ascidians were degenerate, not primitive. There was precedence for degeneracy. Cyclostomes were regarded as degenerate vertebrates. Dohrn saw amphioxus as a very degenerate cyclostome and the ascidian tadpole as even more degenerate than amphioxus. Lankester gave much credence to degeneration as an evolutionary mechanism (s. 12.3.2), partly from his contacts with Dohrn, but also because of his bias as perhaps the most vigorous supporter of recapitulation in Britain. Recall that Lankester edited the translation of Haeckel's *History of Creation*. As far as Lankester was concerned, recapitulation was a given unless some disturbing factor obscured ancestral history.[24]

Semper, like Dohrn, followed Geoffroy's lead of a unity of plan between invertebrates and vertebrates. Semper concentrated on *Balanoglossus*, the acorn worm. Like Dohrn, Semper did not recognize amphioxus or ascidians as vertebrates – they lacked the fundamental segmentation required of vertebrates. Semper and Balfour's independent discovery that the excretory tubules of the shark kidney were segmented and resembled the arrangement of the nephridia of the annelids, gave additional support to the annelid theory, which Balfour adopted for a time. Segmentation continued to be considered a primitive vertebrate characteristic until an-

other holder of the Balfour studentship, William Bateson, demonstrated that segmentation was not primitive; see below.

Kovalevsky had discovered gill slits in *Balanoglossus* in 1866. Three years later Metschnikoff saw the resemblance of the free-swimming tornaria larva to *Balanoglossus* and echinoderm larvae. In the early 1880s *Balanoglossus* was grouped with the echinoderms on the basis of larval resemblances, because the mesoderm formed by outgrowth from the endodermally lined gut of the gastrula and because it lacked segmentation. These features induced Walter Garstang to propose, near the end of the century, the revolutionary theory that vertebrates arose from some 'proto-echinoderm' through a process of neoteny, the tornarian larva becoming, in effect, an adult chordate.[25]

Because of its segmentation and numerous gill slits, Semper, Kovalevsky, Haeckel and most other anatomists thought *Balanoglossus* was the nearest living relative of the common vertebrate-annelid ancestor. An ascidian tadpole-like organism, rather than an annelid, was seen as the most likely vertebrate ancestor. Weak comparisons with annelids, reliance on segmentation, and difficulties with the evolution of a new mouth, were too many obstacles to overcome in proposing an annelid ancestry.

In 1870, Karl Gegenbaur designated a new taxonomic group, the Enteropneusta, for the balanoglossid worms. Affinities recognized between vertebrates, amphioxus, and the ascidian larva of the tunicates, rendered existing classifications obsolete. In particular, a new group and name was needed to encompass vertebrates, amphioxus and ascidians. That group was the phylum Chordata, named by Balfour in 1880 to embrace the three classes created by Lankester in 1877: the urochordates (tunicates), cephalochordates (amphioxus) and vertebrates.

Little was known of the embryology of *Balanoglossus*. William Bateson, in the best Balfourian tradition, expected that embryology would sort out its systematic position and therefore its relationship to the great groups of animals (s. 5.6). He showed that lack of segmentation was not an impediment to the inclusion of *Balanoglossus* within the chordates. Bateson made a complete break with the theories of Dohrn and Semper on annelid ancestry and the nature of segmentation. Bateson argued that segmentation was not primary and ancestral but secondary and a consequence of patterns of growth. After all, he said, the notochord, the primary chordate character, is not segmented. He then made what appeared to be the pivotal discovery: *Balanoglossus* had a notochord, albeit a partial one, confined to the proboscis. With this affirmation of chordate affinities, Bateson (1884a,b), placed *Balanoglossus* and its allies in a new class of chordates, the Hemichordata, adding a fourth class to Lankester's Urochordata, Cephalochordata, and Vertebrata.

Bateson's research typifies precisely the approach advocated by Frank Balfour: use embryology to unravel phylogeny and to organize animals into natural groups.

Bateson, however, 'got it wrong'. One hundred and five years of debate over homology of the 'notochord' of *Balanoglossus* and the notochord of chordates has changed the picture. Hemichordates are now accorded the status of a separate pre-chordate phylum allied to the echinoderms (phylum 34 in Table 2.2, and see Figure 2.6). The notochord is actually a stomochord – a buccal pouch or diverticulum projecting from the mouth cavity. *Balanoglossus* has gone full circle from echinoderm to a fourth class of chordates and back to an association with the echinoderms, acquiring status in the phylum Hemichordata on the way.

Bateson produced four papers on *Balanoglossus* and a fifth more general paper ('The Ancestry of the Chordata') that contains the designation of the new chordate class, the Hemichordata. His last paper, which was highly regarded for quite some time, contains the first intimation that he felt that embryol-

ogy might not provide a reliable path, either to understand evolutionary history, or to shed light on relationships between and among organisms. He wrote the paper anyway, prefacing it with:

'Of late the attempts to arrange genealogical trees involving hypothetical groups has come to be the subject of some ridicule, perhaps deserved. But since this is what modern morphological criticism in great measure aims at doing, it cannot be altogether profitless to follow this method to its logical conclusions. That the results of such criticism must be highly speculative, and often liable to grave error, is evident.'

Bateson, 1885c, p.1;
pagination from 1928 reprint

This nice example of having your phylogenetic cake and eating it too, is symptomatic of the demise of evolutionary embryology after the mid-1880s.[26]

5.6 CONCLUSIONS

In conclusion, a transition in the way of thinking concerning types and development of the concept of the 'embryological archetype' followed von Baer's enunciation of the parallelism between individual ontogeny and phylogenetic history, discovery of germ layers in embryos, Haeckel's Gastraea theory, and demonstrations that embryological development provided the best evidence for determining homologies and establishing ancestry.

With understanding of the universality of the developmental stages blastula, gastrula and neurula, and with Haeckel's enunciation that ontogeny recapitulates phylogeny, embryologists felt that they finally had, in evolution, the 'cause' of individual development. Embryological stages represented an evolutionary solution to the generation of organismal form. Embryology was evolution, and embryology (and embryos) existed because of evolution. Consequently, the triad of morphology, embryology and evolution came to dominate late-19th-century zoology. William Bateson, who began in evolutionary embryology with Francis Balfour before switching to the study of variation and heredity, captured the enthusiasm of the 1870s in an evening lecture delivered to the American Association for the Advancement of Science 50 years later (December 28, 1921):

'Morphology was studied because it was the material believed to be the most favorable for the elucidation of the problems of evolution, and we all thought that in embryology the quintessence of morphological truth was most palpably presented. Therefore every aspiring zoologist was an embryologist, and the one topic of professional conversation was evolution.'

Bateson, 1922, p. 56[27]

In the next chapter, I pursue into the 20th century relationships between archetypes, conserved embryological stages and homology by examining the concept of common ground plans (*Baupläne*) and constraint in ontogeny and phylogeny. The developmental basis for the generation of conserved embryonic stages as reflected in basic body plans flows from common networks of developmental interactions and the hierarchical organization of embryonic development that follow from those processes. These are the topics of the chapters in Part Three.

5.7 ENDNOTES

1. For discussions of Meckel's parallels between embryonic stages, see Russell (1916), Oppenheimer (1967) and Gould (1977).
2. See von Baer (1828, 1835) for his laws of development. For discussions of his embryology, see Oppenheimer (1951) and Ospovat (1981).
3. Desmond (1989) provides a thorough treatment of Geoffroy's influence in Britain and Scotland. His quotation is from the *British and Foreign Medical Review* of 1839. In addition, Oppenheimer (1959, pp. 242–8) and Appel (1987, pp. 222–30) contain discussions of British reaction to the debate.

4. See Lurie (1960) for a life of Agassiz, and Benson (1981) and Maienschein (1981) for analyses of embryology in America at this time. See Coleman (1973) for an analysis of Kielmeyer and his views.
5. See Oppenheimer (1967, pp. 240–3) for an analysis of Carpenter's use of von Baer's laws.
6. For details on Robert Chambers, *Vestiges* and its influence, see Millhauser (1959), Hodge (1972), Bowler (1988) and Schwartz (1990).
7. Rupke's *Richard Owen: Victorian Naturalist* (1994) is the most extensive evaluation of Owen and his work since Owen's grandson Richard wrote the standard 'life' 100 years earlier. That volume contains an evaluation of Owen by Thomas Huxley, his adversary on so many scientific issues (Huxley, 1894). The other major studies on Owen are those by Desmond (1982, 1985, 1989), E. Richards (1987), R. J. Richards (1992), Sloan (1992) and the volume edited by Gruber and Thackray (1992). Padian (1995b,c, 1997) has also examined aspects of Owen's typological thinking and discovered and annotated a missing Hunterian lecture of Owen's.
8. See Desmond (1989, p. 246) for Guthrie's evaluation of Owen and Appel (1987) for details of *Jardin du Roi*. The Royal College was pressing Owen to publish the catalogue of the museum specimens.
9. For his theory of the archetype, see Owen (1846, 1848). For the embryonic studies, see Rathke (1839) and Huxley (1864).
10. See Ospovat (1981) for a discussion of common developmental plans within an *embranchement*. Desmond (1989, pp. 338–9, 346) argues that Owen 'cannibalized' Barry's article, using large pieces verbatim in his 1837 lecture. Large sections from J. Müller's *Elements of Physiology* were also apparently 'utilized' by Owen.
11. See Panchen (1994), Padian (1995b) and Wood (1995, 1997) for Macleay, Owen and the origination of the modern distinction between homology and analogy. Macleay spent much of his career in Australia as a Trustee of the new Australian Museum, encouraged Darwin to publish his findings, and promoted the career of the young Thomas Huxley and Huxley's own thinking on homology and analogy (see Lyons, 1995 for an analysis of Macleay). How a structure as perfectly adapted as the eye of a human and cephalopod could have evolved troubled Charles Darwin greatly. Indeed, the twofold problems of the long time required to

evolve an eye and how a partially evolved eye could serve a useful function continue to be raised. A theoretical paper by Nilsson and Pelger (1994) demonstrates how it is possible to derive vertebrate and cephalopod eyes through a comparatively small number of steps and stages, without invoking the large scale or quantitative changes often thought necessary. For the genes involved in eye development, see section 21.7.3.

12. For the most recent and thorough examination of Darwin's use of embryology, see Richards (1992).

13. See Lankester (1870, pp. 36, 41) for his definitions of homogeny and homoplasy; Lester and Bowler (1995) for a monographic treatment of Lankester and his work; and the chapters in Hall (1994a) and Sanderson and Hufford (1996) for the latest monographic treatments of homology and homoplasy.

14. For some of the more extensive discussions of Haeckelian recapitulation, see Russell (1916), Oppenheimer (1951), Gould (1977) and Mayr (1994). For a recent analysis in relation to patterns of ontogenetic transformation and altered timing and sequences in development, see Alberch and Blanco (1996). See Brusca and Brusca (1990) for a discussion of organisms, such as the placozoan *Trichoplax*, that have been regarded as conservative metazoan descendants similar to the ancestral metazoan type proposed by Haeckel.

15. Lankester (1873) developed an alternative Planula Theory in which the ancestral metazoan was postulated to have been a two-layered sac formed by delamination rather than by invagination; see Ghiselin and Groeben (1997) for a discussion.

16. Pander and von Baer were influenced strongly by an earlier but now little-known thesis on the embryology of the chick published by Louis Sébastien Tredern in Jena in 1808. Beetschen (1995) brought this thesis to light and placed it in its important historical context.

17. For these studies on germ layers, see Van Beneden (1874, 1875, 1876), Hertwig and Hertwig (1879) and Weismann (1883a); see Farley (1982) for analysis.

18. For general reviews or syntheses of germ-layer theory, see Jenkinson (1906), Russell (1916), Oppenheimer (1940), Berrill and Liu (1948), Churchill (1986), Hall and Hörstadius (1988) and Hall (1997a).

19. Ridley (1986) provides one of the few evaluations of Balfour's work. I am doing my best to acquaint the scientific community with Balfour's life, contributions to the forging of an evolutionary embryology and his position as England's foremost 19th-century evolutionary embryologist (Hall, 1997a, 1998b). My intention is to produce a full length scientific biography.

20. Although comparative embryologies of selected groups have been produced in the almost 120 years since Balfour's textbook appeared, the only other treatise for the entire animal kingdom is that edited by Gilbert and Raunio and published in 1997.

21. For recent historical accounts of proposed ancestors of the vertebrates and the anatomical inversions required to turn one into the other, see Ghiselin (1994), Maienschein (1994), Nübler-Jung and Arendt (1994), Bowler (1996), Gee (1996) and Arendt and Nübler-Jung (1997).

22. The most thorough analyses of Kovalevsky's work on vertebrate ancestry are the unpublished Ph.D. thesis by Roberta Beeson (1978) and the recent book by Peter Bowler (1996).

23. Dohrn (1875); Semper (1875, 1876–7). Dohrn followed up his initial study with a long series of papers between 1882 and 1907 in the journal of the Naples Zoological Station.

24. See Lankester's entry on Zoology in the 9th edition of the *Encyclopaedia Britannica* (1888) for his views on recapitulation.

25. For the discovery of gill slits in *Balanoglossus*, see Kovalevsky (1866a, 1867). For echinoderm-vertebrate relationships, see Metschnikoff (1869) and Garstang (1894, 1896).

26. Bateson (1884a,b, 1885a–c). For discussions of the unfolding story of *Balanoglossus*, see Barrington (1979), Margulis and Schwartz (1988) and Brusca and Brusca (1990), esp. pp. 850–8. For current studies on the phylogeny of the hemichordates, see P. W. H. Holland, Hacker and Williams (1991) and Holland and Graham (1995). For a recent evaluation of segmentation in relation to chordate origins, see Presley, Horder and Slipka (1996).

27. Zoology continued to attract the brightest and best well into this century. Witness Sir Peter Medawar's comment on why he chose zoology as the subject to study as an undergraduate:

'"Zoology, the best-fitted of all the science subjects to provide its students with a liberal education, partly because of the intrinsic interest –

even grandeur – of the concepts that inform it, such as evolution, heredity, and epigenesis; partly because of the qualitative exactitude of the formal study of one of its principle disciplines, comparative anatomy; and partly because zoology overlaps with and irrupts into anthropology, demography & ecology. In later years I came to take the view that a person who was really good at zoology in the broad sense of the above description would be qualified to turn his hand to most things".'

Medawar, 1986, pp. 54–5

6

BAUPLÄNE, CONSTRAINTS AND BASIC PHASES OF DEVELOPMENT

'It should be added, of course, that the eggs of vertebrates, no matter how great their anatomical diversity may be, must certainly have a fundamental ground-plan in common, and the discovery of this ground-plan is one of the principal agenda of embryological research.'

Medawar, 1954, pp. 172–3

The notion of common plans persists today in the term *Bauplan*, from the German meaning ground plan or an architect's sketch or plan (*Bau*, design, type of construction, structure, form; *plan*, plan, design, intention), which was introduced into morphology in 1945 by Joseph Henry Woodger (1894–1981). Although more or less assimilated into biology, the term has not been incorporated into general use outside of science, unlike terms such as 'evolution' and 'species', or phrases such as 'survival of the fittest'. This lack of general usage is exemplified by the fact that I could find only one English dictionary or encyclopaedia that defined or discussed *Baupläne*: 'generalized, idealized, archetypal body plan of a particular group of animals' (Lawrence, 1990, p. 56).

6.1 WOODGER AND THE *BAUPLAN* CONCEPT

Woodger was trained as an embryologist by J. P. Hill at University College, London, but is known as a philosopher of biology. His special interest was language – its philosophy, its use as a theoretical tool in biology, and the development of a theory-neutral language for biology. Woodger played an important role in translating the philosophical ideas of Karl Popper, A. N. Whitehead and Bertrand Russell to biologists of his generation. He was one of the first theoretical biologists and an influential member of the Theoretical Biology Club, which met in Cambridge and Oxford before the Second World War and in London thereafter. [The Theoretical Biology club included such influential biologists and philosophers as Joseph and Dorothy Needham, Conrad Waddington, Karl Popper, Peter Medawar, J. Z. Young, Francis Huxley, Avrion Mitchison, N. J. Berrill and Hans Motz.]

Woodger's career was spent teaching medical students at the Middlesex Hospital Medical School. In 1926 he took a sabbatical term to travel to Vienna to study transplantation in worms. Collected for him before his arrival, the worms turned out to be the wrong species. As it was winter and the ground was frozen, a more suitable species could not be collected, so Woodger spent his time in discussions with members of the Vienna Circle of scientists, philosophers and humanists. The unsuitability of the worms, and that his leave was taken in winter rather than spring, set Woodger on his lifelong philosophical course; a scant three years later he had produced his first influential book on the philosophy of biology, *Biological Principles: A Critical Study* (1929). Woodger followed this book with *The Axiomatic Method in Biology* (1937) and *Biology as Language* (1952). In 1933 he translated into English the first text in theoretical biology by von Bertalanffy.[1]

Woodger opposed mechanistic biology as practised by such embryologists and cell biologists as E. B. Wilson. He rejected the chromosome theory of inheritance (which he saw as a modern version of preformation) and the material basis of the gene. Given these views, it is not surprising that his philosophical approach had little appeal for embryologists or biologists. Woodger's views on biological transformations, however, are of particular interest and importance in the context of body plans, *Baupläne* and homology.[2]

In 1945 Woodger wrote a chapter entitled 'On Biological Transformations', part of the *Festschrift Essays on Growth and Form* published to mark D'Arcy Thompson's 60th year as Professor of Natural History at St. Andrews, Scotland. Woodger outlined a concept of correspondence between parts, illustrating it with the generalized pentadactyl limb skeleton. The pattern of proximo-distal elements in the pentadactyl limb is highly conserved throughout the tetrapods: humerus > radius/ulna > carpals/metacarpals > phalanges in the forelimb; femur > tibia/fibula > tarsals/metatarsals > phalanges in the hindlimb (Figure 6.1 and see s. 11.4). Woodger found three relationships – immediately distal to, postaxial to, and articulating with – sufficient to model morphological transformations of the limb skeleton among the vertebrates. In this context – comparing sets of parts – Woodger introduced *Bauplläne*:

'Now consider any particular set of parts of some life and the set consisting of it and all the sets in other lives which are isomorphic with it, then we shall call such a complete set of sets of parts, which are all isomorphic with one another, a *Bauplan* . . . This word is used in preference to "structural plan" because of its brevity, but also and chiefly because a *technical*

term is needed having just the significance given to *Bauplan* in the above definition, and this would not be so easily secured if a phrase already in common use were employed.'

Woodger, 1945, p. 104

Bauplläne replaced the type or archetype, but without the connotations given to those terms by *Naturphilosophen* and adherents of idealistic morphology such as Richard Owen. Woodger saw evolution as hierarchical and therefore *Bauplläne* as nested, overlapping and defining taxonomic groups.

'For example, there is a *Bauplan* which determines the Gnathostomata, every gnathostome exhibits it, and everything which exhibits it is a gnathostome. But it does not determine the Pisces although it characterizes them, for although every member of the Pisces exhibits this *Bauplan*, not everything which exhibits it also exhibits the determining *Bauplan* of the Pisces . . . By a *taxonomic group* (in a restricted sense) we shall mean any set of lives which is determined by a *Bauplan*.'

Woodger, 1945, pp. 104–5

He discussed morphological correspondences in relation to homology with the expectation that they were synonymous, as indeed they were for Geoffroy and Owen. Aristotle, Geoffroy and Cuvier all stressed connections and correspondence among parts in their respective morphologies, as discussed in Chapter 4. Noting that Darwin had effectively shown that shared *Bauplläne* reflect community of descent and gradual modification of structure within the *Bauplan*, Woodger developed an evolutionary postulate encompassing morphological connections, *Bauplläne*, homology, taxonomy and descent – no easy task!

'If two lives exhibit the same *Bauplan*, then there is an ancestor common to both which exhibits a *Bauplan* which is exhibited by all three lives. We shall refer to this as the evolutionary postulate.'

Woodger, 1945, p. 109

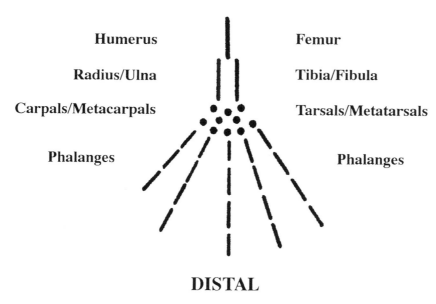

PROXIMAL

Humerus	Femur
Radius/Ulna	Tibia/Fibula
Carpals/Metacarpals	Tarsals/Metatarsals
Phalanges	Phalanges

DISTAL

Fig. 6.1 The tetrapod pentadactyl limb as seen in forelimbs (left) and hindlimbs (right), showing the conserved proximo-distal skeletal elements and the pentadactyl digits. For issues relating to homology and naming digits, see Box 16.1. Adapted in part from Woodger (1945).

Woodger argued explicitly the position advocated by von Baer; the most basic structures develop early in embryonic life. Deviations early in development would thus have much more drastic consequences for morphology than deviations later in development. He related this notion of features of *Baupläne* appearing early in development to the evolution of *Baupläne*, using terms such as 'highly evolved' and 'primitive' that are shunned today:

'If now we remember that parts which are low-level parts in highly evolved lines would be in morphological correspondence with high-level parts in primitive forms which are ancestral to them, we can see how what would in the former be drastic changes in the Bauplan would be comparatively small changes in the latter, but would in fact be laying the foundation for large differences between the Bauplans of later forms.'

Woodger, 1945, p. 111

6.2 *BAUPLÄNE* AND LEVELS OF ORGANIZATION

For most recent authors the *Bauplan* represents the basic organizational plan common to higher taxa at the level of the phylum, order or class, or the common, basic structural plan within a monophyletic taxon (Eldredge, 1989). Valentine distinguishes *Baupläne* as assemblages of homologous architectural and structural features among phyla and classes (1986, p. 209). Gould speaks of phyla representing the 'fundamental ground plans of anatomy', and of a 'vertebrate body plan', 'angiosperm body plan', 'molluscan body plan', and so forth (1989, pp. 99, 218). Morton (1958) identifies *Baupläne* for each of the six extant classes of gastropods and a single *Bauplan*, the *Urmollusk*, as the Archetype of the phylum. As with Owen and the vertebrate archetype discussed in section 5.3, Morton saw *Urmollusk* as an actual ancestor – a primitive, untorted gastropod.[3]

Table 2.2 summarizes the features of *Baupläne* for each of the 37 phyla. This approach to *Baupläne* is amply illustrated in Brusca and Brusca's beautifully produced textbook, *Invertebrates* (1990). They use *Baupläne* as one of three fundamental themes to organize invertebrate structure, relationships and evolutionary history; the other two are developmental patterns/life history strategies and evolutionary/phylogenetic relationships. Although the validity and utility of the *Bauplan* concept is superbly demonstrated in this volume, individual species living under unusual conditions or which display extreme paedomorphosis associated with miniaturization can stretch the limits of the *Bauplan* of a phylum. Palaeozoic crinoids were introduced as one example in section 3.4. Three further examples, each of which deviates from a fundamental character for the group, are:

1. a newly discovered motile stage in the life cycle of the parasitic barnacle *Loxothylacus panopaei* (barnacles are normally sessile);
2. the carnivorous sponge *Asbestopluma*, which lacks the choanocytes otherwise diagnostic for the phylum Porifera; and
3. *Girardia biapertura*, a freshwater planarian from Brazil that has separate openings for the vasa deferentia at the tip of the penis (a single opening is normal).

Other instances (and there are numerous examples) of alteration of the *Bauplan* of invertebrates and vertebrates are associated with parasitism or extreme sexual dimorphism in body size.[4]

For other workers such as McGhee (1980) and Ebbesson (1984), *Bauplan* is applied to a single morphological structure within a

group, as in the morphology and growth of brachiopod valves, or the circuitry of the nervous system. Application of the *Bauplan* concept to individual characters is, however, neither appropriate nor consistent with the essence of the concept, which is the suite of characters that unites members at higher taxonomic levels. Woodger discussed this issue and clearly came down against *Baupläne* as individual elements.

'But the view taken in this essay is that there is no such thing as the pentadactyle limb, or the crustacean appendage, or the chordate notochord *apart from* the determining Bauplans of the Vertebrata, the Crustacea, and the Chordata respectively.'

Woodger, 1945, p. 112

I agree.

In a 1995 analysis of mammalian species richness in relation to latitude gradients, Kaufman extended this notion of *Baupläne* as suites of characters to *Baupläne* as functional types with shared morphological, physiological, and biotic and abiotic ecological components. Verraes (1981) explicitly recognized several interrelated aspects of the *Bauplan* concept. His definition of the *Bauplan* is 'all morphological features from conception to death'. Additionally he recognized:

- the *Bauplan* stage as the *Bauplan* at a single time of its development;
- species (specific) and supra-specific *Baupläne* as features that characterize groups of individuals but are not seen collectively in a single individual;
- a part of a *Bauplan* as a morphological feature; and
- the element of a *Bauplan* as the supracellular components that make up a part and for which one can recognize shape, structure and size of the element.

These subdivisions highlight the organizational bases upon which *Baupläne* are built, while retaining the unity of the concept as suites of characters uniting individuals. Important unresolved issues remain. Those discussed in subsequent sections are:

1. nested *Baupläne* (s. 6.3);
2. the developmental processes that initiate and maintain *Baupläne* (ss. 6.4, 6.5);
3. whether *Baupläne* constrain development and/or evolution (ss. 6.5–6.7);
4. how *Baupläne* arose in evolution (s. 6.8);
5. how rapidly *Baupläne* were assembled (s. 9.2, Chapter 15);
6. why there are so few *Baupläne* (Chapters 2 and 3 and s. 25.3);
7. why no new *Baupläne* have arisen in the last half billion years (s. 25.3); and
8. whether the existence of *Baupläne* necessitate macroevolutionary processes.

Each of these themes represents a major question for evolutionary developmental biology.

6.3 NESTED *BAUPLÄNE* (*UNTERBAUPLÄNE*)

Any *Bauplan* beyond the level of the individual (species?) represents a nest of *Baupläne*. Thus, snakes possess a *Bauplan* that differs from each of the *Baupläne* of lizards, turtles and crocodiles, but all share the reptilian *Bauplan* (Figure 6.2). Similarly, reptiles, birds and mammals each have individual *Baupläne* but share the vertebrate *Bauplan*. The vertebrate *Bauplan* arose before the reptilian, which in turn arose before those of snakes, lizards and crocodiles. Each arose through the gradual emergence of the group; for origins of snake and turtle *Baupläne*, see sections 16.4 and 17.2.2. The characteristics of mammals are traced back through the mammal-like reptiles to the reptiles; they did not arise *de novo* with the mammals. Emerging *Baupläne*, as they appeared through time, gradually defined groups of animals as we now know them.

In the Lower Cambrian, as discussed in Chapter 3, there were *Baupläne* that persisted and those that did not. Both may be regarded as early experiments in body plan organization, some of which were successful, and some of which were not. Some may be intermediate

Fig. 6.2 A 7-mm-long embryo of the lizard *Lacerta muralis* with features of the lizard *Bauplan* clearly visible. Adapted from Balfour (1880–1), Vol. 2.

between known phyla and/or part of an early continuum of body plans.

Such a hierarchical or nested view of *Baupläne* and *Unterbaupläne* (the term used by Valentine (1986) for taxa below the phylum) seems appropriate, unless the concept is restricted to basic organization or ground plans at the highest taxonomic levels. To do so, however, would be to lose the opportunity:

1. of asking how similar structures arise at any taxonomic level (a topic discussed in section 13.3 in relation to novelty and speciation);
2. of searching for mechanisms of homology (Chapter 21);
3. of unravelling the relative roles of key innovations and/or constraints in the origin of *Baupläne* (the topics of ss. 9.2–9.5 and Chapter 15); and

4. of determining whether characters arise sequentially or in a coordinated manner (s. 13.2).

A hierarchical view of *Baupläne* is not only appropriate, it is essential for any investigation into the origin of body plans and the systematic organization and evolution of structures and organisms. Definitions of *Baupläne* are synonymous with different taxonomic levels (see Table 2.2) and are fundamental requirements for any comparative developmental and evolutionary analysis.

Palaeontology provides major evidence for long-term stability of *Baupläne*, indicating how difficult it is to modify them. The reverse statement also could be made. Given that palaeontology normally has only the morphological forms of fossils to study, palaeontological species are morphological types, and *Baupläne* provide the life-blood and *raison d'être* of the palaeontologist, despite the problems of setting boundaries for fossil species discussed in section 3.4.[5]

Fig. 6.3 Generalized amphibian tadpoles (top) and tadpoles of highly derived species such as *Xenopus* (below) can nevertheless readily be identified as the tadpole stage of the life cycle. Adapted from Balfour (1880–1), Vol. 2.

6.4 *BAUPLÄNE* AND MECHANISMS

Levinton (1988), in discussing whether *Baupläne* necessarily arose as integrated units or through a single key innovation, argued that neither need be so. *Baupläne* could have arisen by a gradual accumulation of features that initially were not correlated with one another but which became progressively correlated over time. He argued for the role of historical accident. Even if a suite of features is tightly correlated in, or even synapomorphic for, present-day members of a group, we do not know whether those characters arose in a contemporaneous or correlated way, although, in fact, they do so consistently. Langille and Hall (1989) provided a similar view in their analysis of the sequential origin of what we now know to be fundamental features of the vertebrate *Bauplan*: dorso-ventral polarity, the notochord, dorsal nerve cord and neural crest (s. 15.1). Mayr (1982) took the

view that much of evolution was restricted to key characters and to characters correlated with them. These issues are taken up in Part Two in the context of key innovations.

Because it has typological connotations, the *Bauplan* concept is not universally used by morphologists today. Some seek mechanisms for the organization and evolution of structure without reference to the *Bauplan* concept, regarding it as metaphysical rather than mechanistic. It is not that biologists disagree over whether animals possess basic body plans, for such plans exist, but the developmental and evolutionary significance of basic body plans that is in contention. Some individuals emphasize generative processes of development as a primary factor in morphological evolution, or (like Richard Hinchliffe) have placed the search for the common developmental processes that underlie structural homology within the need for reconstructing the Archetype, a topic discussed further in section 14.2.[6]

While concept is not mechanism, the *Bauplan* is no more metaphysical than are the designations blastula, neurula, tadpole and larva. Even when highly specialized or adapted, amphibian tadpoles can be readily identified as tadpoles (Figure 6.3). To search for the mechanism of metamorphosis is not to deny the existence of tadpoles and larvae. Rather, it is to use those well-recognized ground plans as the starting point in a search for the mechanisms that produce them and the adults that form from them. To do otherwise is to place form above function, concept above mechanism and description above causality. In short, it would be to return to the very problems that informed the Cuvier–Geoffroy debates almost 170 years ago. The need is not to regard the *Bauplan* as the idealized, unchangeable abstraction of Geoffroy but to treat it as fundamental, structural, phylogenetic organization that is constantly being maintained and preserved because of how ontogeny is structured.

I continue to examine structural organization and how ontogeny is structured with a discussion of developmental constraints.

6.5 CONSTRAINTS AND THE EVOLUTION OF *BAUPLÄNE*

'Conservation and constraint may be a strategy for diversification and evolvability.'
                              Gerhart, 1995, p. 336

Baupläne inevitably consist of a mix of ancestral and derived characters. To understand their origin the concept should not be limited to adults, but must incorporate larval stages, indeed, whole ontogenies. Dullemeijer (1974) enumerated structure, position, composition, shape and size as five essential components of the *Bauplan*. The origin of these components has to be sought in ontogeny. Hence, the need for life history and developmental perspectives.

The existence of *Baupläne* leads us to seek the basis for the permanence and invariance of common patterns of design of organisms. In part, that basis lies in constraints imposed on organization by development, for genetic, developmental and structural constraints channel ontogeny. Only within such limits is variation of each element in the *Bauplan* possible.[7]

Jones and Price (1996) provide a fine example of analysis of variation within a *Bauplan* in their study of the plant genus *Pelargonium*. By analysing variation in features that define the *Bauplan* in a phylogenetic analysis of 13 species, they identified two major clades and the major changes involved in the evolution and constraint of the *Bauplan* for this genus. With its emphasis on suites of characters and rules underlying the development of these characters, this is a model for future studies of multicellular organisms.

Although the neo-Darwinian view of evolution ascribes randomness to variants that are sorted by natural selection, there are, in fact, a variety of restrictions on the consequences of natural selection. Not all structures are possible and even when all appear theoretically possible, not all appear, as Raup's analysis of shell form among invertebrates confirms (Figure 3.9). Constraints channel the range of possible responses of an organism to selection. In the short term, constraints may prevent a population from reaching the nearest adaptive peak (Stearns, 1982). This is not the same as saying that constraints limit evolution or produce stasis. It may even be too strong to state that constraints are any 'component of the phylogenetic history of a lineage that prevents an anticipated course of evolution in that lineage' (McKitrick, 1993, p. 307).

One finds suggestions for the view that not all morphologies are possible in the writings of numerous evolutionary biologists. Whitman (1919, p. 11), in perhaps the first statement of the role of developmental constraints, concluded: 'but if organization and the laws of development exclude some lines of variation and favor others, there is certainly nothing supernatural in this, and nothing which is incompatible with natural selection'. Maynard Smith (1983) spoke of historical and contingent constraints arising from the existing pattern of development. In the context of taxonomic characters Ernst Mayr expressed the view that 'some components of the phenotype are built far more tightly into the genotype than others', and that 'certain components of the phenotype may remain unchanged during phyletic divergence' (1982, p. 213). With reference to unity of type and *Baupläne*, Mayr further asked:

'Why is the chordate type so conservative that the chorda still is formed in the embryology of the tetrapods and gill arches still in that of mammals and birds? Why are the relations of structures so persistent that they can form the basis of Geoffroy's principles of connections? Clearly this is a problem for developmental physiology and genetics, indicated by such terms as the cohesion of the genotype or the homeostasis of the developmental system, terms which at this time merely conceal our profound ignorance.'

1982, pp. 468–9

The somatic programmes discussed by Mayr in 1994, especially the inductive interactions that characterize the development of so many organisms, are elaborated in the chapters that follow.

A common thread runs through these statements:

- variation is channeled or bounded by genetic, developmental, and/or functional constraints;
- genetic control of some structures is protected from change, conservative, or inert; and therefore,
- some structures persist or are maintained unchanged throughout the history of a group.

This is not to imply that all evolutionary biologists place equal emphasis on the importance of constraints. Two analyses, one calling for a moratorium on the use of the term constraint in evolutionary biology, the other (in rebuttal) advocating the centrality of both term and concept, provide valuable if opposing viewpoints.[8]

Kurt Schwenk (1994/95) provides an in-formative analysis of constraint, drawing attention to two fundamentally different classes: class I constraints (evolutionary channeling) and class II (developmental and genetic) constraints. Within class I, functional and structural constraints channel the patterns sorted by natural selection. Evolutionary channeling occurs because of a number of underlying constraints; Schwenk considers two (developmental and genetic) under class II. He sees internal and external causation mirrored in these two classes, but emphasizes that the critical distinction between the two is their relationship to natural selection.

A clear hierarchy exists in these two categories. Class II constraints provide proximate mechanisms and processes independent of natural selection that are further constrained by functional and structural constraints and the actions of natural selection. Schwenk's analysis of constraint and of the distinction between phenotypic patterns and the processes that produce them is an important clarification of this topic and is recommended reading for any interested in fathoming constraints.

I have followed Schwenk's lead but grouped constraints under **pattern** and **process constraints**. Of course, this is not an absolute distinction; the convenient dichotomy of pattern and process draws too sharp a boundary, but it does focus our attention on separating constraints on what is produced, from constraints on what is ultimately sorted and retained. I have recognized cellular and metabolic to give four process constraints that result in the non-random production of structures, features and behaviours.[9] They are:

1. pattern or universal constraints (s. 6.6);
   1a. structural constraints (mechanical-architectural constraints; s. 6.6.1); and
   1b. functional constraints (s. 6.6.2).
2. process constraints (s. 6.7);
   2a. genetic constraints (s. 6.7.1);
   2b. developmental constraints (s. 6.7.2);

2c. cellular constraints (s. 6.7.3); and
2d. metabolic constraints (s. 6.7.4).

Although considered as six types in two cat-egories, interactions between levels of con-straint provide another level of control, and indeed of constraint. It is perhaps not unrea-sonable to generalize so far as to say that ho-mology can arise from constraint(s) on a single character, while *Baupläne* represent suites of constrained characters.

Understanding constraint, especially sepa-rating the various types of constraint, will only come from utilizing the population and quantitative genetics approaches that allow measurement of patterns of current and past selection, genetic variation and covariation. Such a union of phyletic, developmental and genetic realms, long since called for and ar-ticulated in models, remains an important task for evolutionary developmental biology.[10]

6.6 PATTERN OR UNIVERSAL CONSTRAINTS

6.6.1 Structural constraints

Structural constraints are examples of what Maynard Smith and colleagues in 1985 called 'universal constraints' in that all cells or organisms are subjected to them. Even though universal, their consequences for single-celled and minute organisms may be much greater than their effect on multicellular organisms.

Structural constraints could be called size constraints, for size can only be reduced so far before its consequences begin to affect organ-isms; evolutionary novelties often appear in organisms of small size. In this category are such mechanical constraints as effective sur-face-to-volume ratios that limit the size of individual cells and minute organisms. The breakaway from this constraint for unicellular organisms is colony formation and the evolu-tion of multicellularity (s. 13.5.1).[11]

Structural constraints imposed by the limi-tations of patterns in multipart organ systems such as limbs are the topic of section 12.2.1.

Such **constructional constraints** are mediated by developmental processes and so bridge structural and developmental constraints. Part of that constraint is the mechanical load his-tory of the developing skeletal elements that channels possible form transformations. But morphological constraints go beyond me-chanical load history, as Lessa and Stein (1992) elegantly demonstrated in an analysis of con-straints in the two modes of digging in pocket gophers: those which dig using claws and limbs, and those which dig using teeth and jaws. Structural variation is found only in the forelimb musculature; muscles of the jaws, neck and hyoid arches are constrained and presumably do not have the potential to display the variability seen in forelimb muscles.[12]

The literature of functional morphology, biomechanics and bioengineering also intro-duces the notion of **excess structural capacity**. This is the finding that many organisms are built to withstand forces considerably in excess of those they encounter in their daily lives or, indeed, in their lifetimes. To a degree, excess structural capacity can be set against structural constraint.[13]

6.6.2 Functional constraints

As embryos develop, and especially once functions such as feeding and respiration begin, organ systems become functionally connected. Well-studied examples are the integration of the skeleton of the jaws or branchial basket and the muscular systems in feeding and respiration, and dependence of the muscular system on the nervous system.[14]

The more coupled or interconnected fun-ctions become, the more difficult it is to uncouple one component from another or to change one component without concomitant change in another. Such coordination and functional interdependence imposes directionality and boundaries on further change. Coupling and interdependence influ-ences our approach to key innovations and integrated change, the topics of Chapters 12–17. Oyama (1993) stressed interdependence when she drew attention to the environment constraining development and entering as a constituent into developmental and behavioural processes. Environment as an element of functional constraint is taken up again in Chapter 18.

A potential counter to functional constraints is the ability of reptiles, birds and mammals to change (reversibly) the size of individual organs as an adaptive, often seasonal, response to such altered life history situations as reproductive status or ingestion of a large meal. It is well-known, for instance, that fat reserves are gained and shed seasonally and that gonad size can increase by as much as 100-fold leading up to and during the breeding season. Less well-known is the fact that other organs, such as adrenal glands, also cycle with breeding or migration for breeding, as in the eastern rosella, *Platycercus eximius* (Hall, 1968), or the ability of Burmese pythons (*Python molurus*) to increase intestinal mass two-fold, kidney and liver mass by almost 50% and metabolic rate seven-fold within a day of ingesting a large prey item (Piersma and Lindström, 1997).

6.7 PROCESS CONSTRAINTS

6.7.1 Genetic constraints

The size of the genome may constrain such cellular parameters as size, rates of division, number, metabolism, and differentiation. This is especially true in some salamanders and lungfish, where the haploid genome may contain upwards of 80 pg of DNA when the norm for animals is a tenth of that. (At the other extreme, *Trichoplax adhaerens,* the single species within the phylum Placozoa, contains only 1010 daltons of DNA.) Why DNA content should vary so dramatically across phyla is sufficiently puzzling to be known as the **C-value paradox**. DNA content and the accompanying increase in cell size are invoked as

explanations of paedomorphosis in amphibians. In such cases, constraint is an explanation for what would otherwise be interpreted as adaptive features.[15]

The length of the mitotic cycle was shown by Rothe and colleagues (1992) to impose a physiological barrier to the maximal size of transcripts that can be produced. Their studies with *Drosophila* illustrating the requirement for coordination of cell-cycle length and gene size, led them to conclude that this requirement constrained evolution of rapid development.

Maximum rates of mutation and recombination of individual alleles constrain the evolution of single traits. The lack of additive genetic variance that arises from mutations also constitutes a genetic constraint over the long term. A recent example (which also illustrates the utility of comparing closely related species in a phylogenetic context) is that of Zardoya, Abouheif and Meyer (1996a) on *sonic hedgehog* (*shh*) and *Hoxd-10* genes in 18 species of cyprinid fish related to the zebrafish *Danio rerio*. Exon 1 of *sonic hedgehog* is almost invariant among the 18 species with only 9.5% variable sites. Previous studies implicated exon 2 as the main functional domain of *sonic hedgehog*. Exon 1 is more constrained even though exon 2 is the main functional domain.

The evolution of particularly successful and phylogenetically highly conserved genes, such as the segmentation and homeotic genes that determine the basic body plan of animals (ss. 9.2, 15.2), have been considered as constituting a genetic constraint. Two arguments are advanced:

1. the classes of **homeodomain genes** arose early, and the features they determine are fundamental to the establishment of body plans;
2. **homeotic mutations** reflect constraints that are intrinsic to the systems that generate individual body regions, and go beyond conservation of individual gene function within cascades (s. 9.4.1).[16]

Heritable patterns of gene expression such as chromosomal imprinting also constitute a genetic constraint (s. 7.5.3). Suites of characters can be constrained because of genetic correlation and covariance among traits, although Björklund (1996) provides examples where genetic covariance promotes evolutionary change at least over ecological, if not evolutionary, time scales. Pleiotropy (the actions of a gene in multiple tissues) and epigenetic associations (s. 20.1) constitute genetic constraints, often revealed as developmental constraints.[17]

Genetic constraint on the evolution of ecologically important characters in wild populations has been little studied; the few studies are on inbred populations. Futuyma, Keese and Funk (1995), however, document minimal genetic variation (genetic constraint) in the evolution of host associations in *Ophraella*, a genus of leaf beetles. The authors took advantage of a phylogenetic reconstruction of the history of host associations, and an experimental approach that allowed larval survival, oviposition, and/or feeding responses on plants normally grazed by congeneric species, to be measured. No discernible genetic variation was detected in 46% of tests of feeding responses to congeners' host plants, and in an amazing 87% of tests of larval survival. Genetic constraint 'explains' conservatism of the diet in these species. Their study is a model for future analyses of genetic constraint in ecologically meaningful situations.

6.7.2 Developmental constraints

Developmental constraints are dealt with explicitly or implicitly throughout the remainder of this book.

The interdependent nature of developmental processes (the inductive interactions and cascades discussed in Chapters 9–11) constrains an organism's ability to produce all possible variations in morphology. These constraints are expressed as invariant patterns that persist throughout long periods of evolutionary time. Maynard Smith and colleagues (1985), in the most cited and insightful overview of the topic, defined developmental constraints as biases on the production of phenotypic variability because of the nature of developmental systems. Levinton defined a developmental constraint as 'a factor in development, such as an obligatory tissue interaction, that might prevent or channel an evolutionary change' (1988, p. 607).

How developmental constraints provide an interface between environmental influences and morphology is the subject, implicitly or explicitly, of much of the literature in vertebrate form and function, especially the rich literature on adaptation of fishes to shifts in diet discussed in sections 16.5 and 18.3.3. Here developmental constraints shape behaviour, as they do in a water strider (*Gerris odontogaster*), in which the length of abdominal ventral processes (a secondary secondary character in males) is positively correlated with length of the last moult. This correlation persists despite increased cannibalism of males with longer processes. Length of the processes is a trade-off between a developmental constraint (length of the moult) and selection.[18]

On the basis of empirical studies on regeneration of pectoral fin hooks in the blenny *Salario pavo*, Wagner and Misof (1993) proposed that developmental constraints arise because of interactions late in ontogeny. They refer to these as **maintenance constraints** that stabilize patterns generated earlier in ontogeny, and contrast them with **generative constraints** that bound the production of the structures. One difficulty with separating generative from maintenance constraints as extrapolated from a regenerative system is that the mechanisms responsible for regeneration are not those that produce the initial structure during ontogeny. Regeneration occurs from already differentiated cells that dedifferentiate before redifferentiating during regeneration. Such cells may well carry morphogenetic information with them; they certainly carry information pertinent to the types of cells they can become in the regenerate.

Furthermore, much of the evidence for the operation of developmental constraints is in the generation of morphology, although this may merely reflect the more concentrated research effort into generation than maintenance. Stabilization (canalization) of processes in the later stages of their formation or to maintain them, seems more likely to play a minor role, just as anomalies that arise later in ontogeny are of smaller effect than those that arise early (s. 12.1.2). The limited compensation later in ontogeny to perturbations that arise early in ontogeny, indicates that the ability to compensate is not great. Specific examples for the musculoskeletal system may be found in Hall (1978, 1985a).

Miriam Zelditch and her colleagues re-examined the ontogenetic complexity of developmental constraints by examining sets of local features of morphology at particular developmental stages. Using this approach in the analysis of skull development in the cotton rat, *Sigmodon fulviventer*, Zelditch, Bookstein and Lundrigan (1993) uncovered three types of developmental constraint characterized (using terms introduced by Waddington) as:

1. canalization – constraint that is relatively constant throughout ontogeny giving rise to pathways of development in individual embryos;
2. chreods – reduction of variance of canalization features ensuring close similarity from individual to individual; and
3. opposition – reduction of age-specific variance along the canalized path.[19]

This approach is important for at least two reasons. One is the recognition of **opposition** as a constraint counteracting the other two. The second is the emphasis on understanding how variance of the features measured changes throughout ontogeny. An understanding of the existence and basis for variability during ontogeny of individual structures, or indeed, of entire embryos, is critical if we are to understand levels of control operating throughout ontogeny and establish a mechanism to bridge individual ontogeny with population level evolution.

6.7.3 Cellular constraints

Because developmental constraints are expressed through cellular activity, the distinction between cellular and developmental constraints is not great. Andrews (1995) drew interesting parallels between Fungi and Metazoa with respect to mechanisms constraining the evolution of growth forms. Each is based on few basic cell types; each exhibits phylogenetic and developmental constraints; and each is based on a modular body plan design.

Limits to the rate of cell division, secretion of cell products, cell migration, and/or metabolic efficiency, are just four examples of constraints operating at the cellular level. In her 1992 analysis of tissue strategies as developmental constraints, Ellen Larsen identified six cell behaviours as sufficient to derive all morphologies. These six strategies therefore constitute cellular constraints. They are:

1. cell division;
2. cell growth;
3. cell death;
4. changes in cell shape;
5. secretion of extracellular matrices; and
6. cell migration.

Larsen reminds us of the two tissue strategies by which morphological diversity is generated – **autonomous** (essentially self differentiation), and **conditional** (interactive or inductive) – and discusses how dependence on tissue interactions could constitute a constraint. An example of generation of the same structures (germ cells) in different members of the same group (amphibians) using autonomous and conditional mechanisms, is discussed in section 9.1.3.

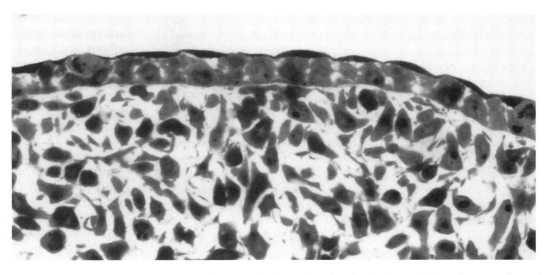

Fig. 6.4 A histological section through the mandibular arch of a developing chick embryo to show the difference in structure between the thin, sheet-like mandibular epithelium and the meshwork of mandibular mesenchymal cells below.

I would add that there are also only two ways in which cells are organized in multicellular animals: as sheets of connected cells known as epithelia or as loose mesh-works of cells embedded in extracellular matrix and known as mesenchyme (Figure 6.4). This fundamental dichotomy of cellular organization is a constraint (no other cellular organization is available) and an opportunity (epithelial-mesenchymal interactions have evolved to direct cellular differentiation and organogenesis). See section 10.5 for the origin of mesenchyme in the tail buds of vertebrate embryos and sections 11.1 and 11.2 for fuller discussions of epithelia, mesenchyme and interactions between them.

6.7.4 Metabolic constraints

Metabolic constraints affect single-celled organisms and set limits in multicellular organisms. Despite the associated metabolic penalties, large size may be advantageous ecologically in exploiting temporary resources or interacting with other organisms, and so be favoured by selection.[20]

A tissue that is dependent on a rich vascular supply because of high levels of oxidative metabolism, is more resistant to change than is a tissue with anaerobic or relatively anaerobic metabolism. This may be one of the reasons why cartilage, a tissue with relatively anaerobic metabolism, develops as the most common supporting tissue in vertebrates, as a surprisingly common supporting tissue in invertebrates, and appears so frequently ectopically. Highly vascularized tissues with aerobic metabolism, such as bone, are more constrained in their distribution.[21]

Maternal constraints are a special case of metabolic constraint. Especially seen in mammals, maternal constraints may be relevant in animals such as birds and fish which brood their young and in viviparous fish and reptiles. Here the notion is that embryos do not attain their genetic potential because of limitations of the uterine environment, interactions *in utero,* or maternal behaviour. Such maternal constraints may be genetic and/or environmental, pre- and/or postnatal.[22]

6.8 BASIC PHASES OF DEVELOPMENT AND SUITES OF CHARACTERS

The remainder of this chapter relates constraints to phases of embryonic development and then to two suites of characters in ontogeny, one of which represents the features of *Bauplände*. I use phases to refer to discrete periods of development. A phase may include several of the stages (gastrula, neurula and so forth) discussed in the next chapter. A suite of characters is a set of features that appear either independently or in concert but that are related by belonging to one of two classes of characters.

Von Baer identified three stages in embryonic development: primary, histological and morphological differentiation (s. 5.1). Wilhem Roux, whose experimental programme laid the groundwork for experimental embryology, distinguished two periods. One was an early period of self-differentiation (the 'embryonic' period or period of laying down of organ rudiments) during which organs were formed before any functional stimuli acted upon them. The second was a later period of 'functional form development', when further organ differentiation and growth depended upon function (Roux, 1905). The timing of these two periods was not synchronized throughout the embryo; each organ developed according to its own timetable.[23]

Roux's explanation for these two periods was an evolutionary one. Development during the first period, when germ layers, organ rudiments and segmentation took place, was inherited through predetermination in the germ plasm. Development during the second period was because of acquired inheritance. Features of development which form in response to functional demand (features of the function-dependent stage of development), form slowly, over many generations, and are assimilated as first-period characters through their inheritance as acquired characters. While we now decry such a mechanism, the designation by Roux of two periods of development subject to differing modes of inheritance is a theme I wish to pursue. [The mechanism of genetic assimilation proposed by Waddington (1942, 1953a,b, 1956, 1957a,b, 1961) and Schmalhausen (1949) provides a means for the assimilation of characters that first arise in response to environmental signals, a topic discussed further in Chapter 19. Such characters appear to represent acquired inheritance, but this is not so.]

Two periods of development are encapsulated below by the generation of two suites of characters (those associated with *Bauplände* and those that are more plastic and subject to variation), that follow from:

- the existence of *Bauplände*;
- the persistence of *Bauplände* unchanged throughout long periods of evolutionary history; and
- the fact that characters not associated with *Bauplände* are more constrained than others.

It is essential to emphasize that the same types of developmental processes produce both types of characters. What differs is the relative roles of similar processes.

6.8.1 Generation of conserved characters of *Bauplände*

The maintenance of *Bauplände* (and presumably their phylogenetic origins) and their resistance to change are fundamentally a result of internal factors acting as developmental constraints. This is Jacob's (1977) notion that history plays a greater part as complexity and constraint increases.

Genetic, epigenetic and cellular constraints ensure persistence of individuals that conform to the *Bauplan*. Embryos that cannot form the features of the *Bauplan* (because of mutations in key controlling genes, for example), or that produce defective features, either fail to form or are eliminated as a result of spontaneous abortion, failure to implant, or embryonic or larval mortality.

Variation associated with production of the *Bauplan* is minimal because of stabilizing selection; individuals that deviate from the *Bauplan* are selected against. Selection is for developmental uniformity. Phenotypically, this is expressed as constraint during ontogeny (s. 6.7) and conserved stages during phylogeny. A possible test of this thesis during ontogeny, correlating differential mortality at specific developmental stages with aspects of the body plan that arise at those stages, is outlined in Hall (1996b, pp. 241–2). Although such characters tend to appear by specific stages of development in particular phyla (phylotypic stages), they may develop sequentially, reflecting embryonic and larval adaptations (s.14.2).

6.8.2 Plastic characters, form and function

The second suite of characters are those that can be more readily modified than those of *Baupläne*. I call these plastic characters. Evolution of these structures and functions, their phylogenetic origin, and their variation in taxa, result from natural selection and chance mutations that ensure change and maintain variation in the population. Such characters tend to arise after those that define *Baupläne*. Wider limits are set on constraints of such characters. Selection is for functionality rather than conformity. Phenotypic variation in such structures or functions is greater but still bounded. Examples are discussed in Chapter 12.

6.8.3 Phylogenetic implications of basic phases of development and conserved and plastic characters

The existence of two suites of characters represents, on the one hand, a fundamentally hierarchical and interactive view of the organization of development and the consequences of that organization for evolutionary change, and on the other, a reconciliation of the Geoffroy–Cuvier dichotomy of whether form determines function or function form. During generation of the *Bauplan*, function follows form and form can be said to determine function. For more plastic characters, which tend to arise later in development, function determines form against the backdrop of past form.

It is generally true that differences between higher taxa, such as classes and orders, are manifest earlier in embryonic development than are differences between species and genera. The two phases above do not, however, equate with a dichotomy between microevolution and macroevolution.[24]

I do not side with Løvtrup (1986) in regarding innovation at the stage of 'form creation' as the source of the origin of new phyla, classes and orders, with changes in later 'growth processes' (more or less equivalent to when plastic characters develop) producing 'taxa of lower rank'. Løvtrup's view is essentially one of von Baerian recapitulation: '. . . during their ontogenesis, the members of sister taxa follow the same course up to the stage corresponding to their divergence into sister taxa' (Løvtrup, 1986, p. 76). Severtzov pioneered this idea in relation to intensification or change in function, caenogenesis (embryonic adaptation) and the generation of evolutionary novelties. His two phases were an initial, short phase of morphogenesis, followed by an extended period of growth. But development cannot be divided neatly into two phases on the basis of termination of morphogenesis and subsequent initiation of growth.[25]

Nor do the two phases or two suites of characters simplistically equate with the dichotomies of internal versus external factors, constraint versus selection, or genetic versus epigenetic control of development. I do, however, endorse the conclusion that the existence of stable ontogenies and the sharing of common ontogenies by closely related taxa 'have their explanation in a combination of internalist, externalist and historical factors' (Wake and Roth, 1989, p. 373). Eldredge expressed a similar view in discussing anatomical novelty in relation to the origin of higher taxa: 'the appropriate level of such theory (restructuring *Bauplän*) and investigation is in the developmental biology of organisms, not at the supposed level of taxa of higher categorical rank' (1985, p. 162).

In discussing characters that are highly conserved, Cain noted that:

'If it could be shown, however, that any class of characters is acted on only or primarily by selection conserving it within groups but keeping a distinction between groups, this would be more likely than any other to be a reliable guide to ancestors. Three classes of characters need to be considered.'

Cain, 1982, pp. 16–17

The three considered by Cain were characters of great physiological importance (such as intracellular mechanisms of respiration), secondary sexual characters, and conservation of the genetic code. In the past the latter would have been regarded as out of contention, conservation of the genetic code being taken for granted. The specific role played by segmentation and homeotic/*Hox* genes (helix-turn-helix transcription factors) in establishing body plans and functioning as region-specific selector genes, speaks to a genetic basis for conserved characters and to the unusually high conservation of such genes across the animal and plant kingdoms (ss. 9.4.2, 15.2–15.5).

Developmental mechanisms underlying constrained characters are conserved phylogenetically. Only comparative analyses of developmental processes can reveal which characters qualify as constrained. Examples of such comparative analyses are discussed in section 9.4 and in Chapter 15. The developmental issue in terms of the relationships among organisms and the origin of morphological diversity has been encapsulated neatly:

'In fact, the question of where in the course of evolution ontogenetic changes in growth and time become hierarchic and indicative of relationship subsumes the problem of when and in what way morphological traits become irreversible.'

Kluge and Strauss, 1985, p. 264

6.9 ENDNOTES

1. Woodger's *Biological Principles: a Critical Study* (1929), appeared in a second impression in 1948 and was reissued in 1967 with a new introduction by the author. Gregg and Harris (1964) is perhaps the best source for information on Woodger's career and publications.

2. For further discussion of Woodger's opposition to mechanistic biology, see Ruse (1975) and Roll-Hansen (1984). For Woodger's views on homology, see Withers (1964).

3. For additional discussions of *Bauplän* as the basic organization plans of phyla and higher taxa, see Valentine (1992a), Levinton *et al.* (1986), Arthur (1988), Rieppel (1988a,b), Simms (1988), Conway Morris (1989), Willmer (1990), Bock (1991), Atkinson (1992) and Presley, Horder and Slipka (1996).

4. See Glenner and Høeg (1995), Vacelet and Boury-Esnault (1995) and Sluys, Hauser and Wirth (1997) for the barnacle, sponge and planarian examples. See Ponder and Lindberg (1997) for a recent and comprehensive phylogeny of gastropods, encompassing 40 taxa, and for examples of the origin of new ground plans in gastropod molluscs in association with small size. For other examples, see Matsuda (1987) and Hanken and Wake (1993).

5. See Schindewolf (1969), Rudwick (1985) and R. L. Carroll (1987, 1997) for conservation of *Bauplän* in the fossil record.

6. For discussions of homology in the quest to reconstruct the archetype, see Webster (1984), Wake and Larson (1987), Rieppel (1988a), Goodwin *et al.* (1989), Hinchliffe (1989) and Shubin (1991, 1994a,b, 1995).

7. For discussions of variability within *Bauplän*, see Verraes (1981, 1989), Valentine (1986), Hall (1983a, 1984a,b, 1990a,b, 1996a,b), Thomson (1988), Sachs (1988) and D. B. Wake (1991).

8. See Antonovics and van Tiederen (1991) and Perrin and Travis (1992) for diametrically opposing views on the utility of constraints. For analyses of constraint in the context of adaptationism and response to selection, see

Amundson (1994), Scharloo (1990), Arnold (1992) and Seger and Stubblefield (1996).

9. These two categories correspond to class I (evolutionary chanelling) and class II constraints of Schwenk (1994/95). Moore and Willmer (1997) use phylogenetic, fabricational and functional as three categories of constraints.

10. For homoplasy as evidence of constraint, see D. B. Wake (1991, 1996b) and the papers in Sanderson and Hufford (1996). For population genetics approaches and their potential to integrate development with evolution or reveal the nature of constraints, see Gould and Lewontin (1979), Atchley and Hall (1991), Leroi, Rose and Lauder (1994), Björklund (1996) and Chapter 20.

11. See Wainwright *et al.* (1976) and Wainwright (1988) for mechanical constraints, and Bonner (1988) for colony formation as a breakaway from such constraints.

12. See Hanken (1983), Hanken and Wake (1993) and Emerson (1988a) for pattern constraints. Mechanical load history of the skeleton as a constraint is discussed, explicitly or implicitly, by Hall (1983a, 1985a), Oster *et al.* (1988), Wong and Carter (1990), Bertram and Swartz (1991), Carter, Wong and Orr (1991), Carter, Van der Meuleun and Beaupré (1996) and Carter and Orr (1992). Thomas and Reif (1991) discuss the resulting limited design elements.

13. For the concept of excess structural capacity and relevant literature, see Wainwright *et al.* (1976), Hilderbrand *et al.* (1985) and Galis and Drucker (1996). For the analagous concept of double assurance in embryonic induction, see section 11.2.2.

14. See Roth and Wake (1985), Miyake *et al.* (1992a) and Hunt von Herbing *et al.* (1996a,b) for the interdependence of muscular and skeletal systems.

15. For genome size in salamanders and lungfish and the C-value paradox, see Cavalier-Smith (1985), D. B. Wake (1991) and Hinegardner (1976). For effects on paedomorphosis, see Sessions and Larson (1987), V. L. Roth (1992) and references therein. For determinants of cell size, including the effects of temperature of rearing on cell size, see Atkinson and Sibly (1997).

16. See Kappen, Schughart and Ruddle (1993) and Duboule (1994b) for the evolution of conserved genes.

17. See Gaunt and Singh (1990) and Monk and Surani (1990) for chromosomal imprinting, and Cheverud (1984, 1996), Clark (1987) and Loeschcke (1987) for covariance.

18. For developmental constraints, see Alberch (1980), Goodwin *et al.* (1983), Maynard Smith *et al.* (1985), Stearns (1982, 1986) and Levinton (1988). For the shaping of behaviour by developmental constraints, see Galis (1993a). For water striders, see Arnqvist (1994).

19. The terms canalization and chreod are adopted from Waddington. Canalization is discussed further in s. 12.2; the chreod concept by Hall (1992).

20. For discussion of metabolic consequences of size, see Calow (1978), Bell and Koufopanou (1991) or any good physiology text.

21. For the frequent appearance of cartilage in invertebrates and vertebrates, see Hall (1978, 1983b, 1997b) and Person (1983).

22. For maternal constraints in viviparous animals, see Shine (1985), M. H. Wake (1989, 1990, 1992a), Blackburn (1995), Qualls and Shine (1996), Dulvy and Reynolds (1997) and Qualls, Andrews and Mathies (1997). For multiple genetic and/or environmental origins of maternal constraints, see Cohen (1979), Gluckman and Liggins (1984) and Cowley *et al.* (1989).

23. See Roux (1905) for the two periods of development. Although he was the founder of experimental embryology, Wilhelm Roux initiated little if any experimental work after his dramatic beginning in the early 1880s; see Hamburger (1997) for an analysis.

24. See Arthur (1988, pp. 75–81) for an interesting discussion of this point in relation to the insects and myriapods; see section 14.2.2 for a discussion of phylum-specific or phylotypic stages, and see Chapter 22 for a discussion of the evolution of embryonic development.

25. See Severtzov (1927, 1931) for caenogenesis; Maclean and Hall (1987) for ways of subdividing embryogenesis, and Hennig (1966), Nelson (1978) and Kluge and Strauss (1985) for the approach of phylogeneticists that earlier appearing stages in ontogeny are plesiomorphic.

7

INHERITANCE SYSTEMS: ZYGOTES, MATERNAL, EPIGENETIC

'Thus, there is a second inheritance system – an epigenetic inheritance system – in addition to the system based on DNA sequence that links sexual generations.'

Maynard Smith, 1989b, p. 11

Given that Aristotle began the study of development over two millennia ago, we might expect the mechanisms of development to be well understood, but alas, they are not. The problems must be intractable to have persisted unanswered for so long, and indeed, they are. In evaluating the state of knowledge of embryology some 80 years ago, Russell noted that development is one of the major and most difficult problems in biology. The situation has not changed. In 1993, John Moore devoted 159 pages and 20 concepts to just one aspect of development, differentiation. Moore, however, took a more optimistic view than Russell; he saw developmental biology as about to come into its own.[1]

Indeed, some of the most intractable, long-standing and central problems in developmental biology are beginning to reveal their secrets. They include the molecular and genetic bases of primary embryonic organization, axis formation, embryonic induction, control of the cell cycle, morphogenesis, and pattern formation. Differential expression of the embryonic genome is fundamental to all these events. **Epigenetics** is the umbrella term for processes regulating differential gene expression. Paradoxically, preformation also plays a role.

Heritable information is the raw material from which organisms are fashioned. All are familiar with zygotic nuclear genes, which are the genomes of embryos themselves. Less familiar are other classes of heritable information available to early embryos. Before beginning a discussion of early embryonic stages and processes in Chapters 8–11, I discuss these other classes of heritable information, which include products of the maternal genotype active at early stages of development and heritable patterns of gene expression. The latter has been called the **phenotype of the gene** or an **epigenetic inheritance system**. The existence of such heritable information outside the zygotic genome raises again the old questions of preformation or epigenesis, and nuclear or cytoplasmic control of development. Epigenetic inheritance systems are discussed in these contexts in this chapter in two contexts. One is historical; the roots of epigenetics are embedded in epigenesis, the major alternative to preformation as an explanation for how embryos develop (ss. 7.1–7.2; and see s. 1.2). The other is to emphasize that epigenetic control is not non-genetic control but rather an important component of selective control of gene activity (s. 7.3). Early phases of development are regulated by a combination of inheritance of preformed maternal cytoplasmic factors and heritable patterns of gene expression that result from methylation of DNA, chromatin structure and genomic imprinting (ss. 7.4, 7.5).

These inheritance systems, which are expressed epigenetically, are in addition to the zygotic genome inherited at fertilization.

7.1 PREFORMATION AND EPIGENESIS

How do so many different structures arise from so seemingly simple a beginning as the egg? Attempting to answer that question motivates the developmental biologist of the 1990s as much as it motivated Aristotle 2300 years ago.

Aristotle posed the major questions to which answers are still being sought today. In seeking to answer those questions, Aristotle observed as many embryos from as many different kinds of animals as he could to establish general principles, and provided the first detailed description of the embryology of any animal in his study of the developing chick. He saw fundamental similarities in the development of birds, fish and mammals, recognized the essential similarities between development and regeneration, and argued for progressive development (epigenesis) over preformation. Eighteen hundred years elapsed before Volcher Coiter (1534–76), a Dutch pupil of Fallopio, took up the Aristotelean tradition of careful observation and opened chicken eggs day by day to produce a description of chick development (published in 1572–3) that added substantially to Aristotle's.

Coiter was an exception, for preformationist views dominated comparative anatomy and natural history for almost two millennia. It was not until the 1651 and 1759 treatises of William Harvey (1578–1657) and Caspar Friedrich Wolff (1733–94) respectively, in which development of the hen's egg was redescribed, that Aristotelian epigenesis took hold again. Wolff saw that parts of embryos develop from elements that had no equivalence in adults or in earlier embryonic stages; for example, the appearance of blood vessels where none had existed previously, and the progressive development of the tubular gut

from a flat plate. For Wolff this meant epigenesis.

'When the formation of the intestine in this manner has been duly weighed, almost no doubt can remain, I believe, of the truth of epigenesis.'

    Wolff, 1767, cited by Gilbert, 1994a, p. 497

Pander's discovery of the germ layers in 1817, and von Baer's elaboration of that discovery in 1828 (s. 5.5.3), provided substance to Wolff's description of embryonic structures arising from material of a different type – germ layers that appear in embryos and give rise to tissues and organs of different types. In addition to parts arising anew, Wolff recognized that 'each part is first of all an effect of the preceding part, and itself becomes the cause of the following part' (Wolff, 1764, p. 211). This is a modern hierarchical and interactive cast to the interpretation of embryonic development, as will become evident as this chapter unfolds.

Geoffroy's experiments, in which malformed embryos were produced under experimental conditions (s. 4.3), suggested that if malformed embryos were not preformed, normal embryos must not be either. Despite these findings, preformation persisted as the fundamental explanation of development for a long time. Preformation was satisfying; in preexisting structure it provided a rational basis for development. But how could generation upon generation exist within the egg, sperm or embryo of a single individual?

Epigenesis was equally satisfying for its proponents. It accorded with the clear evidence provided by Aristotle, Harvey, Wolff, Malpighi and others, that structures arose from previously 'nonexistent' structures. But the nature of these 'nonexistent' structures provided a major stumbling block to the acceptance of epigenesis – the age-old problem of accepting a theory for which there is no observable explanation or mechanism. Epigenesis also had to account for how specificity of organism type was passed from generation to generation, for such 'information' was clearly preformed; hens arose from hen's eggs, frogs from frog eggs, and so forth. Such was the ebb and flow of opinion within and between preformation and epigenesis that Hartsoeker, who is identified with the extreme preformationism of the homunculus in sperm (Figure 1.1), later advocated an epigenic interpretation of crustacean limb regeneration and hence, an epigenic view of embryonic development.

E.B. Wilson, one of the most prominent and important cell biologists of the first quarter of this century, and a major proponent of cell lineage and determination in early development, placed epigenesis squarely within the cytoplasm, but only applied it in a limited way and to the development of few structures.

'Fundamentally, however, we reach the conclusion that in respect to a great number of characters *heredity is effected by the transmission of a nuclear preformation which in the course of development finds expression in a process of cytoplasmic epigenesis.*'

Wilson, 1925, p. 1112; his emphasis

Elsewhere in his monumental treatise on the cell, he makes it clear that, for him, epigenesis only applied to the external features of embryos:

'Biologists therefore gradually returned to the views of the fathers of embryology, and in the end universally accepted the fact that development, in its external aspects at least, is not a process of "evolution" or unfolding but one of progressive new-formation, or epigenesis. We know that the germ-cell contains no pre-delineated embryo: that the parts of the embryo are gradually formed in a typical order by a process which, externally at least, is one of epigenesis as understood by Aristotle, Harvey and Wolff.'

Wilson, 1925, pp. 6–7, 1036

For Wilson, the preformation-epigenesis dichotomy was a nuclear-cytoplasmic dichotomy, a logical position for one who studied organisms with fixed lineages of cells in which fate was fixed by the inheritance of pre-determined, cytoplasmic constituents. The issue of nuclear versus cytoplasmic control dominated interpretations of many of the early experiments in embryology.[2]

So what should the answer be when we ask, as we often do with dichotomies such as epigenesis and preformation, whether one has 'won out' over the other?

In one sense, epigenesis has triumphed. Embryonic structures are not all preformed in the egg. In another sense, preformation 'explains' some aspects of development that epigenesis cannot. The genetic basis for development lies preformed in the DNA of the egg and subsequently in the zygote. Basic raw material for protein synthesis is preformed in the nucleolus, ribosomes and endoplasmic reticula of the egg (Figure 2.1). Raw material for growth resides in the egg cytoplasm, ready to be partitioned among the newly formed blastomeres during cleavage (s. 8.3.2). Katz and Goffman (1981) produced a 'modern' preformationist model based on patterning of the cytoplasm in the egg and its subsequent epigenetic activation at fertilization. Others, such as Davidson (1968) and Ho (1984) have taken the view that only the genome, and then only the nuclear and not mitochondrial or chloroplast genomes, is preformed. But, given that the egg inherits its organelles and cytoplasm as structural components, and could not initiate development without them, it seems not unreasonable to include them as preformed structures and information available to embryos.

7.2 FROM EPIGENESIS TO EPIGENETICS

Epigenesis is development of embryonic structure from different earlier material. What is meant today when we speak of **epigenetics** or of development being under epigenetic control? Epigenetics adds a genetic component to epigenesis. Although definitions of epigenetics vary, most share the basic notion of development being organized through the control of gene expression by the environments and microenvironments encountered by embryos or parts of embryos – organs, tissues and cells. Differing traditions, fields of specialization, prevailing paradigms and biases have lent different emphases to one or other of these aspects.

In a series of studies beginning in 1940, Conrad Waddington (1900–75) coined epigenetics as an amalgam of epigenesis and genetics. Epigenetics, he said, is the causal analysis of development, in particular, the mechanisms by which genes express their phenotypic effects, a definition followed by Rieger and colleagues in their *Glossary of Genetics* (1976), and by King and Stansfield in their *Dictionary of Genetics* (1985).

Medawar and Medawar (1983) took a broad-brush approach in their definition when they stated:

'In the modern usage "epigenesis" stands for all the processes that go into implementation of the genetic instructions contained within the fertilized egg. "Genetics proposes; epigenetics disposes".'

Medawar and Medawar, 1983, p. 114

Epigenetic interactions are the basis of embryonic inductions, competence, the modulation of neoplastic cells to a normal state, and cell determination (other than determination that is based on cell lineages and the inheritance of preformed cytoplasmic constituents, as in embryos of ascidians and of *Caenorhabditis elegans*). Morphogenesis and growth also are highly epigenetic. In *Cell Commitment and Differentiation* (1987), Norman Maclean and I defined epigenetics as encompassing increasing hierarchical complexity and the influences of the environment on phenotypic expression through control of gene expression. The genotype is the starting point and the phenotype the endpoint of epigenetic control. A formal definition is:

'Epigenetics or epigenetic control is the sum of the genetic and non-genetic factors acting upon cells to control selectively the gene expression that produces increasing phenotypic complexity during development.'

7.3 EPIGENETICS IS GENETIC

It is a mistake to speak of epigenetics as non-genetic or of genetic versus epigenetic factors as if one is always in the ascendancy or operating to the exclusion of the other. As emphasized in the definition above, epigenetic control is control of gene expression. It is equally artificial to separate out initial gene action as somehow apart from epigenetics, although this is precisely what is done in *Henderson's Dictionary of Biology Terms,* where epigenetic is defined as 'the chain of processes linking genotype and phenotype, other than the initial gene action' (Lawrence, 1990, p. 166).

In an evaluation of concepts introduced by Waddington, Réne Thom provided the essential features distinguishing epigenetics from genetics and relating epigenetics to genetics:

'If you were to follow Aristotle's theory of causality (four types of causes: material, efficient, formal, final) you would say that from the point of view of material causality in embryology, every thing is genetic – as any protein is synthesised from reading a genomic molecular pattern. From the point of view of efficient causality, everything is also "epigenetic", as even the local triggering of a gene's activity requires – in general – an extra-genomal factor.'

Thom, 1989, p. 3

The latter point is crucial. Gene expression is regulated by a multitude of classes of factors including transcription factors, mRNA, growth factors, hormones, inductive cell and tissue interactions, and physical factors such as temperature, pH, and mechanical and electrical forces. Epigenetics or epigenetic control refers to the multiple genetic and nongenetic factors that influence or regulate gene activity during development. We see the result as constrained development.

An especially nice example of a nongenetic factor as regulator is the finding that relative timing and degree of expression of the programme of gene expression for myogenesis (and consequently muscle phenotype) in Clyde herring (*Clupea harengus*) is regulated by temperature. Timing of muscle fibre formation, onset of embryonic isoforms of myosin light chain 2, troponin I and troponin T, and time of muscle innervation are all temperature regulated. Temperature also influences morphogenesis and pattern formation, as documented for pigment stripes in embryonic alligators, where the length of the embryo at pattern initiation (which is temperature-dependent) influences the number of stripes that develop. Other effects of temperature on developmental programmes include induced phenotypic change and subsequent genetic assimilation of that change in *Drosophila* (s. 19.1), phenotypic (seasonal) polymorphism in butterfly wing patterns (s. 20.5), and temperature-dependent induction of the lens in amphibians (s. 24.1.5).[3]

Waddington spelled out the distinction between genetic and epigenetic aspects of development in his concept of **canalization** (ss. 12.2, 19.4, 19.5). Developmental programmes are so integrated (canalized) that individual genes involved in a programme can change without disrupting the running of the programme. Sewell Wright's concept of **universal pleiotropy** is similar, emphasizing as it does the importance of genetic correlations and the polygenic nature of developmental processes; see section 20.1 for further discussion.[4]

7.4 EARLY DEVELOPMENT AS A MIX OF PREFORMED AND EPIGENETIC INFORMATION

Organisms do not develop entirely epigenetically or entirely by preformation. Even in vertebrates where epigenetic control is predominant, some development is preformed. Three examples of fundamental developmental processes that are controlled by epigenetic expression of preformed information are:

1. maternal cytoplasmic control of early development (s. 7.4.1);
2. determination of germ cells in frogs by inheritance of polar granules and the gene-activating protein contained within them (s. 9.1.3); and
3. determination of the primary body axis by maternally inherited gene products (s. 9.2).

As discussed in section 22.2.2 in relation to direct-developing sea urchins, the distinction between cell lineage and cell fate is important. Only with specification through inheritance of maternal, or activation of zygotic, cytoplasmic factors is cell fate equated with cell lineage. In such situations, lineage, fate, specification and commitment are synonymous. Other generations of cells, which can be traced as cell lineages (using techniques such as vital dye injection), contribute to more than one cell type. Cell fate is not restricted until these lineages interact inductively with other and different cells; i.e. lineage reflects genealogy not specification, fate, or commitment. Maclean and Hall (1987) discuss these issues in some depth.

In organisms such as nematodes and tunicates, where much of development is preformed in cell lineages, epigenetic regulation nevertheless occurs. Thus in the tunicate *Styela partita*, most cells differentiate because of the inheritance of specific regions of cytoplasm containing specific cytoplasmic factors. This system of inheritance is so precise that a cell lineage can be determined and each cell identified by its specific position in that lineage. Individual cells follow their predetermined fate when isolated from neighbouring cells. Some individual cells, however, differentiate as nerve or muscle cells only in the presence of cells from another lineage. The fate of these cells is not predetermined, but arises through interaction with adjacent cells from different lineages. These two modes of developmental control, originally named mosaic and regulative development, are now often termed autonomous and conditional specification (ss. 6.7.3, 14.2.5 and 21.7.2).[5]

Preformed information is especially evident in maternal cytoplasmic control of early development and in cortical inheritance.

7.4.1 Maternal cytoplasmic control

Early development is under maternal cytoplasmic control. By this is meant that early development is not primarily controlled by the zygotic nucleus but rather by maternal factors, the products of maternal genes deposited into the egg during oogenesis. Both zygotic and maternal products therefore direct embryonic development.

Maternal cytoplasmic factors may consist of stored, long-lived mRNA, proteins, or precursors of structural proteins. Non-cytoplasmic maternal effects also influence embryonic development in mammals (Cohen, 1979). These effects – which include interactions between the maternal and zygotic genomes, uterine environmental effects, and uterine environment by genotype interactions – are considered further in sections 20.4.3–20.5. Such effects are not short-lived. The adult phenotype can be affected by epigenetic events involving interaction between maternal cytoplasmic and zygotic genes, as demonstrated by Reik *et al.* (1993) in nucleocytoplasmic hybrids in mice generated by nuclear transfer and in which growth was affected.

Development during this phase of maternal cytoplasmic control can occur in the absence of the sperm nucleus, as evidenced by the initiation of development of artificially activated or parthenogenetic frog eggs (s. 8.3.1). In some species, early development also occurs in the absence of the ovum nucleus, i.e., in the absence of the entire zygote nucleus. In these species, there is no essential role for epigenetic interactions involving the zygote nucleus

Table 7.1 The duration of maternal cytoplasmic control of early development as determined by the latest cleavage stage attained by parthenogenetically activated eggs maintained *in vitro*

Organism	Stage attained (number of cells)	Number of cleavage divisions
Mouse	2–4	1–2
Pig	4–8	2–3
Sheep, cattle	8–16	3–4
Xenopus	4096	12

during the phase of maternal cytoplasmic control.[6]

The amount of early development under maternal cytoplasmic control varies from as little as the first one or two cleavage divisions in the mouse to as much as all of cleavage, encompassing 12 divisions and 3,000–4,000 cells, in frogs such as *Xenopus*. In mammals the duration of maternal cytoplasmic control is dramatically shorter, terminating at the 2-cell stage in mice, 4–8-cell stages in pigs, and 8–16-cell stages in sheep and cattle (Table 7.1). In the mouse, cytoplasmic activators and repressors are absent from the 1-cell stage but present at the 2-cell stage (Henery *et al.*, 1995). The transition from maternal to zygotic gene control of development at the 2-cell stage is regulated by activation of repressors of promoters and by activators of enhancer activity.

Maternal cytoplasmic control can be used to argue that a preformed stage of development can (and perhaps should) be defined on the basis of preformed structures plus the duration of the phase of development that can take place in the absence of the zygote nucleus, i.e. that is under maternal cytoplasmic control.

A dramatic illustration of the role of maternal control over early development is **specification of the primary body axis** by maternal gene products deposited within the egg (s. 9.2). Despite our notions that early development is highly conserved and that something as basic as the major axis of the egg would be determined similarly across all phyla, an enormous diversity of mechanisms and times of axis specification exists. Mechanisms can vary enormously, even within members of the same phylum. As with other aspects of early development, mechanisms for axis specification have evolved (Goldstein and Freeman, 1997). In cnidarians and ctenophores respectively, the primary axis is specified as early as oogenesis and as late as the first cleavage. The secondary axis is specified as early as oogenesis and as late as gastrulation.

7.4.2 Cortical inheritance

A phenomenon that combines elements of preformed information with an epigenetic mode of inheritance is **cortical** (cytoplasmic) **inheritance,** seen in ciliated protozoa such as *Paramecium.* Row upon row of cilia all oriented in the same direction and all beating in coordinated waves cover the surfaces of these protozoa. The bases of the cilia are embedded in cortical cytoplasm, the most superficial cytoplasm nearest the cell membrane. If a piece of cortical cytoplasm is transplanted in reverse orientation between individuals of *P. aurelia,* the cilia in the graft beat in the opposite direction to those of the host. Astoundingly, subsequent generations inherit this reversed pattern of ciliary locomotion (Beisson and Sonneborn, 1965). Directionality of ciliary beat passes from generation to generation through inheritance of subcellular organization of this cortical cytoplasm, despite the fact that nuclear DNA is unaltered, and neither cilia nor their basal bodies contain organelle DNA. Inheritance of these organelles is based on somatic cytoplasmic organization.

7.5 HERITABLE PATTERNS OF GENE EXPRESSION: THE PHENOTYPE OF THE GENE

The notion that the 'phenotype' of the gene as well as the genotype should be considered

in modelling control of development is appearing in quantitative genetic models of development that attempt to incorporate epigenetics into morphological change during development and evolution; see s. 20.1 for discussions of current thinking on the genotype–phenotype dichotomy and the role of epigenetics. One practical instance where neglect of epigenetics confounds analysis of genotypic and phenotypic effects will be noted. The fundamental underlying rationale of mutagenesis experiments is that the only organisms that display a phenotypic change are those with the mutant genotype. Given that genes exert their phenotypic effects through epigenetic control and/or that mutated genes are compensated for by other genes, the rationale assumed when interpreting mutagenesis experiments is undermined. As aptly summarized by Arnold and colleagues:

'. . . the developmental biologist may ask whether the distinction between genotype and phenotype advances genetics by leaving out development. Does evolutionary genetics provide a sufficient theory of morphological evolution?'

Arnold *et al.,* 1989, p. 406 [7]

Several examples of the phenotype of the gene or transmission of epigenetic states of inheritance are discussed below. Essentially these represent patterns of parental genomic activity inherited by the progeny and active in early stages of embryogenesis. For additional examples of epigenetic control (inheritance of a particular pattern of cell-cell or tissue-tissue interactions), see sections 9.2 and 9.4.

Epigenetic variation and/or stable states of gene activity are transfered from generation to generation through:

1. patterns of methylation of DNA;
2. the structural conformation of chromatin; and
3. genetic imprinting.

7.5.1 Methylation of DNA

Methylation, the addition of a methyl group to some of the cytosine residues of DNA to form 5-methylcytosine, does not alter the message coded in the DNA but does influence its transcription, highly methylated DNA being less transcriptionally active than less methylated or unmethylated DNA. Although methylation does not create a new function for DNA, it stabilizes or modifies existing function. Methylation is therefore a heritable state influencing gene expression. It is also an ancient mechanism used in prokaryotes and eukaryotes, where it is passed from organism to organism through cell division as transmissible states of gene activity associated with differentiation. Here is a clear epigenetic mechanism that builds a second level of control onto the information contained in the sequences of DNA base pairs.

Not all genes are equally methylated. Housekeeping and inactive tissue-specific genes are often unmethylated; active tissue-specific genes are methylated. Patterns of methylation of specific genes can be followed through embryonic development. Patterns expressed in mouse sperm and eggs and inherited by the zygote are lost at the blastocyst stage, when all DNA is demethylated. New patterns are established later in development. Indeed, the parental gametic pattern is re-established in the germ line, having been disassembled during development, a remarkable example of inheritance, loss and regaining of epigenetic control. Although inheritance of patterns of methylation is high, patterns do change. Holliday (1987) named such random changes 'epimutations' to emphasize the parallel with mutations of base sequences, and that genetic and epigenetic patterns are heritable.[8]

The demethylation of mouse embryo DNA that occurs between fertilization and implantation is reversed, and 5-methylcytosine brought to adult levels between implantation and gastrulation. Maternal DNA methyl-transferase is found in mouse embryos until 9.5 days of gestation, when its concentration becomes limiting. These findings, and those demonstrating that embryos homozygous for a loss-of-function mutation (*Dnmt*[n]) in the DNA methyltransferase gene fail to develop beyond 25 somites, amply illustrate the developmental roles of methylation and the importance of maternal cytoplasmic control and inheritance of patterns of gene expression (Trasler *et al.*, 1996).

7.5.2 Chromatin structure

Evidence for the influence of chromatin structure on gene function is less compelling, although polytene chromosomes in flies provide an important class of evidence. Jablonka and Lamb (1989, 1995) introduced the term 'the gene's phenotype' for situations where the DNA sequence of an individual gene is constant, but chromatin structure variable.

The dynamic nature of chromatin structure, illustrated in several aspects of development, speaks to an important level of control directing development. Some examples are:

1. the genetic restructuring that occurs during gametogenesis, a restructuring that means that paternal and maternal chromosomes now differ in their chromatin structure;

2. genomic imprinting, as in the non-random inactivation of the X chromosome in female mammals and in male marsupials (s. 7.5.3);

3. the loss or heterochromatinization of paternal chromosomes during gametogenesis in male mammals and some insects;

4. inheritance of a specific order of genes along the chromosome and a corresponding order of expression as seen in homeobox-containing genes in *Drosophila* and mice (ss. 9.4, 15.2);

5. chromatin diminution (chromosome elimination) in many nematodes, dipteran insects, copepods and some hagfishes; and

6. the chromatin-associated proteins that maintain the state of activity of homeobox genes through cell divisions.[9]

Candidate genes maintaining heritable states of transcriptional activity through modification of chromatin structure are the *Polycomb* (*Pc-G*) and *trithorax* (*Trx-G*) response elements of the regulatory regions of such homeotic genes as *Sex combs reduced* (*Scr*). These elements open chromatin and maintain individual loci in conformational states that repress (*Polycomb*) or sustain (*trithorax*) transcription. How these response elements, which are expressed globally but act locally (e.g. at boundaries of expression of *Hox* genes), fit precisely into the regulatory cascade remains an open issue.[10]

7.5.3 Genomic imprinting

In genomic imprinting, two copies of a gene (either maternal versus paternal, or one allele) do not function equivalently during development. Genomic imprinting occurs at the single-gene level. Genomic inactivation is the integrated elimination of whole chromosomes, as in chromosome diminution. Methylation of DNA may be involved, both in imprinting and in inactivation. Perhaps the best-studied examples are the non-random inactivation of the X chromosome in female mammals and in male marsupials.

Autosomal genes can be imprinted, as evidenced when regions of chromosomes, whole chromosomes, or entire genomes are derived from the same-sex parent. Curiously, the effects of imprinting maternal and paternal genomes are not the same. Gynogenotic mice, which have two sets of maternal and no paternal chromosomes, show normal initial embryonic development but extraembryonic tissues fail to develop normally. In contrast, androgenotic mice, which have two sets of paternal and no maternal chromosomes, show normal development of extraembryonic but not embryonic tissues. Similarly, in transgenic mice, parental origin of the transferred gene can determine whether that gene is expressed. So too does the genetic background of the mouse strain. Such phenotypic patterns of

genes are heritable (exactly how remains obscure) and provide mechanisms for epigenetic control of development. In a real sense, cellular memory regulates gene expression.[11]

7.6 ENDNOTES

1. The histories of developmental biology most relevant to the topics covered in this book are those by Russell (1930), A. W. Meyer (1939), Needham (1959), Haraway (1976), Roe (1979, 1981), Gould (1977), Horder *et al.* (1986), Gilbert (1991a), Moore (1993) and Bowler (1996). The volume edited by Horder and colleagues is an invaluable source book as it contains a 28-page guide to reference works, dictionaries, encyclopaedias, bibliographies, catalogues, journals, histories, biographies and autobiographies.

2. I have discussed Wilson's views on nuclear preformation and cytoplasmic epigenesis elsewhere (Hall, 1983a). See Bowler (1989b) for a discussion of nuclear preformation.

3. See Johnston *et al.* (1997) for the influence of temperature on myogenesis in the Clyde herring, and Murray, Deeming and Ferguson (1990) for its influence on pigment patterning in the alligator *Alligator mississippiensis*.

4. For his pivotal analyses of epigenetics, see Waddington (1940, 1942, 1957a). For the development of the theory of canalization, see Waddington (1940, 1942). For universal pleiotropy, see Wright (1967, 1968). For further and more detailed discussions of Waddington's epigenetic legacy, see Polikoff (1981), Yoxen (1986), Gilbert (1991b, 1994b) and Hall (1992).

5. For nematodes and tunicates, see Kenyon (1985), Schierenberg (1987) and Nishida and Satoh (1983). For discussion of autonomous and conditional specification of cell fate, see Slack (1985), Maclean and Hall (1987), Davidson (1990), Gilbert (1994a) and s. 14.1.2.

6. For activation of parthenogenetic eggs, see Balinsky (1975) and Gilbert (1994a).

7. For the concept of the phenotype of the gene, see Jablonka and Lamb (1989, 1995), Atchley and Hall (1991), Lewontin (1992), K. C. Smith (1992a,b) and Wagner and Altenberg (1996).

8. For heritability and stability of patterns of methylation, see Haigh *et al.* (1982), Holliday (1987, 1990, 1991), Maclean and Hall (1987) and

Kafri *et al.* (1992). For further discussion of epimutations, see Holliday (1994) and the symposium to which this paper is the introduction.

9. For literature and discussions on chromatin structure, see Gaunt and Singh (1990), Monk and Surani (1990), Nakai, Kubota and Kohno (1991), Goday and Pimpinelli (1993), Lyon (1993) and Paro (1993).

10. See Paro (1990) and Paro and Hognes (1991) for possible genetic bases of maintenance of chromatin structure and states of transcription. See Grindhart and Kaufman (1995), Orlando and Paro (1995), Simon (1995), A. Gould (1997) and Schumacher and Magnuson (1997) for *Polycomb* and *trithorax* response elements and transcriptional regulation, including nine genes identified in mice.

11. For methylation in genomic imprinting and chromosome inactivation, see Lyon (1993). For studies relating to sex of origin of the chromosomes, see Surani *et al.* (1990) and Reik, Howlett and Surani (1990).

PART THREE
EMBRYOS IN DEVELOPMENT

8

MODEL ORGANISMS, CONSERVED STAGES AND PROCESSES

'A more intensive comparative study of development will be needed before we can know which aspects of ontogeny are historically constrained and which are free to vary, and to what extent patterns of constraint are similar in even closely related groups.'

Kluge and Strauss, 1985, p. 264

To begin a comparative analysis of developmental processes, I now discuss the fundamental stages and processes of embryonic development. I begin the chapter with a discussion of the (inevitable) use of model organisms and the desirability of a comparative approach, before proceeding to a discussion of conserved life-history stages, embryonic stages and processes of early vertebrate development. Chapters 9–11 are then devoted to how these processes build embryos and organ systems.

8.1 MODEL ORGANISMS

Only a small fraction of organismal diversity has been named, yet alone investigated developmentally. Stages of life history and ontogeny are typically studied in selected, 'model' organisms. I too take the model organism approach. But first, a major word of caution concerning the animals we choose to study.

A major difficulty encountered in uniting development and evolution is the dramatically different approaches of developmental and evolutionary biologists to the organisms they study. In part this is a difference in focus. Individuals develop, populations evolve. Consequently, developmental biolo-

gists study individual organisms, while populations are the focus of most evolutionary studies. This places a major obstacle in the way of integrating the two approaches. Furthermore, most developmental biologists assume that all individual embryos of a species at a given stage of development are identical.

When I remove two dozen fertile fowl eggs from the incubator after they have been developing for 48 hours I expect that the majority are at the same morphological stage. A few will be developmentally younger, having developed more slowly. Some will be older, having developed more rapidly. But once I 'stage' these embryos and remove those that are younger and older, I expect the remaining embryos of the same stage to be identical. Indeed, it is on this basis that we plan our experiments in the confident expectation that all embryos of the same stage respond in the same way to manipulation or perturbation. Variability is not an issue, although I recall being told by Dame Honor Fell, then Director of the Strangeways Laboratories in Cambridge, that if nine of her embryos did one thing and the tenth something different, she would follow up the exception. Variability, of course, is the stuff of discovery as it is of evolution.

BOX 8.1

MODES OF OSSIFICATION AND HOMOLOGY OF SKELETAL ELEMENTS

Model modes of development can be as misleading as model organisms when used as a basis for evolutionary constructs. Our deep-seated ideas of homology, which require that structures develop in identical (homologous) ways (ss. 21.7, 21.8), is undermined by a number of examples of homologous bones that develop by intramembranous ossification in one taxon but endochondrally in others: the caecilian orbitosphenoid; the median ethmoid in Atherinomorph fishes and elements of the skull in *Xenopus*.[2]

A second difficulty is evident in any attempt to describe the development of a group on the basis of a single species. This is essentially the approach taken in all developmental biology textbooks and, perforce, the approach that I take here, begging insufficient space to discuss all the patterns in all the embryos of all the species that have been studied – and only a small fraction of species have been studied. The species whose development is most well-known is the one presented as typical in the textbooks. Hence blocks to polyspermy are described as if all eggs produced a fertilization membrane as echinoderms do. Early cleavage is described as a uniform pattern typified by the South African claw-toed frog *Xenopus laevis*, relegating differences in other species to not much more than the effect of yolk. As we move into gastrulation and neurulation the focus becomes more species-specific, if only because the ways of making a gastrula or neurula seem so much more diverse than the ways of cleaving an egg to make a blastula.

When the species that is most convenient for analysis is a frog like *Xenopus*, we make the unspoken assumption that *Xenopus* is a typical amphibian and that its development, including genetic control over development, can be extrapolated to the many tens of thousands of other vertebrate species. But *Xenopus* is a most bizarre amphibian.

- *Xenopus* is not a basal amphibian. *Xenopus* is in the family Pipidae. Once thought to be among the most primitive frogs, pipids have now been shown through rigorous phylogenetic analyses to be among the most highly derived frogs.
- *Xenopus* is pseudotetraploid and speciation is based on polyploidy, which is unique among frogs.
- *Xenopus* displays many novel features associated with its aquatic mode of life and evolutionary origins from a terrestrial saltatorial ancestor (Figure 6.3).
- *Xenopus* lacks a tongue and so has a feeding mechanism (ram feeding) that is highly unusual among frogs.
- *Xenopus* has highly specialized keratinous claws.
- *Xenopus* has a skull that develops precociously in relation to other frogs and in which many of the bones surrounding the eye develop by intramembranous rather than endochondral ossification (Box 8.1).[1]

We should not expect development of a species such as *Xenopus* to be unmodified and indeed it is not; mesoderm develops from a two-layered, deep zone of cells, atypical of other amphibians, in which mesoderm arises from a single, superficial cell layer.

So why do developmental biologists study *Xenopus*?

- Not because it is primitive.
- Not because development of *Xenopus* is typical of amphibians.

- Not because *Xenopus* is the most likely species to provide generalizable onto-genetic and phylogenetic information.

No, *Xenopus* is chosen because it is easy to house in the laboratory, because breeding can be induced hormonally in any month of the year, because the eggs are large and relatively easy to manipulate, and because it has rapid development.

Other biases may enter with our choice of model organisms for developmental analysis. Much has been learned from mice, fowl, the fruit fly *Drosophila*, the nematode *Caenorhabditis elegans*, the zebrafish *Danio rerio*, and various species of echinoderms. All can be readily housed and bred. All have rapid development and short generation times, tremendous advantages when experimental, genetic or breeding studies are being undertaken. But as Bolker (1995) emphasized in her analysis of model systems, rapid development often means canalized development. If the species are inbred, as is true for many mouse, fowl and *Drosophila* strains, their responses differ between strains and may be quite different from those of animals captured in the wild. Mouse and chick embryos are such entrenched model species that we often don't list their species names – *Mus musculus, Gallus domesticus* – and rarely specify the strain being used.

8.1.1 Out of context

Model species are also studied totally independent of their natural history. Analysis of how development is regulated by ecology or how organisms adapt their development to environmental signals, predators, climate shifts or conspecifics is impossible with model, laboratory-raised animals.

Maintenance in the laboratory alters development, as documented in the divergent skeletal patterns that arise following regeneration in laboratory-reared amphibians in comparison with those obtained from the wild (s.

12.2.2). A combined analysis of laboratory strains and individuals from the wild, however, can be productive, as demonstrated in the studies of regeneration. But there are few such studies. Notable are studies by Brown (1980) on *Xenopus*; by Sage, Atchley and Capanna (1993) of house mice as models of hybrid zone biology, chromosomal evolution and speciation; by Flick (1992) on intermediate characters in hybrids between *Ambystoma tigrinum* and *A. texanum*; and by Scriven and Bauchau (1992) in a study of rapid (18 months) morphological change in hybrid mice following introduction of house mice to the Isle of May, in the Firth of Forth, Scotland.

8.1.2 The comparative method

Integration of studies from laboratory-reared and wild-caught species can be by the comparative method with as few as two species, not necessarily related. Or it can be in an analysis of related organisms whose relationships are confirmed by a rigorous phylogenetic analysis, in which case a minimum of a three-species comparison is required. Without the third species neither the direction nor the sequence(s) of change of the character(s) can be determined. Such comparative approaches were used in studies of the zebrafish, *Danio rerio*, and its relatives by Axel Meyer and his colleagues; of sea urchins by Greg Wray; and for evolutionary changes in developmental mechanisms between insects and arthropods by Nipam Patel.[3]

Marvalee Wake's analysis of the 1991 study on the evolution of multicellularity in plants by Niklas and Kaplan raises an especially telling set of arguments for comparative analyses even within groups for which there is no phylogenetic hypothesis of relationships.[4] So does Meyer (1996) with his emphasis (shared by others) that the comparative method:

1. allows predictions of likely conditions in common ancestors;

2. permits reconstruction of intermediate stages;
3. allows determination of the sequence of historical events; and
4. reveals the evolution of developmental processes.

Three recent edited books contain comprehensive overviews of the utility of the comparative approach, its theoretical underpinnings, and examples:

- *Phylogeny, Ecology and Behaviour: a Research Program in Comparative Biology* (Brooks and McLennan, 1991);
- *The Comparative Method in Evolutionary Biology* (Harvey and Pagel, 1991); and
- *New Uses for New Phylogenies* (Harvey et al., 1996).

It is probably no coincidence that the first and second were published in 1991, for the importance of comparative studies resurfaced at the turn of this decade. This timing may explain why neither book advocates the use of comparative developmental analyses as a way of furthering understanding of organismal evolution.

Development can contribute to such evolutionary problems as convergence, parallelism, homoplasy and homology if studied in a comparative or phylogenetic context. Conversely, understanding the evolution of the genetic, cellular and developmental mechanisms underlying convergence, parallelism, homoplasy and/or homology is impossible unless one uses comparative and/or phylogenetic approaches. Nor can we determine likely common ancestors, characters states in those ancestors, possible intermediate states, the direction of change of characters, or the developmental processes underlying them, other than in a phylogenetic context.[5]

8.1.3 Model organisms and the study of evolution

Despite disclaimers of some of the difficulties with model organisms, Elizabeth Kellogg and Brad Shaffer concluded their introduction to a symposium on model organisms with a positive affirmation for their use in evolutionary studies:

'Model systems can provide the vehicle for such a synthesis (of genetics, development, and systematics) and allow development of model systems for evolutionary studies.'

Kellogg and Shaffer, 1993, p. 413[6]

This is not because 'we assume [that model systems] are archetypical for their respective taxa' (Davidson, 1991, p. 9). The model organisms studied in developmental biology are highly derived. Organisms that raise some of the most interesting evolutionary questions are among the most difficult or impossible to study developmentally. They include hagfishes and the origins of and relationships between jawless vertebrates, jawed vertebrates and conodonts; lungfishes and the origin of the tetrapods; and onychophorans and arthropod relationships. Live hagfish embryos have not been obtained since Bashford Dean studied embryos of the Californian hagfish *Bdellostoma* (*Eptatretus*) *stouti* a century ago (Dean, 1899).[7]

The sections that follow should be viewed against this backdrop. The intention is to provide sufficient information to illustrate major life history stages (s. 8.2) and the major stages typical of vertebrate development (s. 8.3). The intention is not to create the impression that vertebrates all conform to a single pattern of development. The power of model organisms is that they allow us to uncover mechanisms that can then be compared with mechanisms operating in related organisms, the more closely related the better. Set these model organisms and mechanisms against the backdrop of a rigorous phylogeny to enable the direction of evolutionary change to be inferred, or the most likely condition in common ancestors to be predicted, and we have a basis for comparative evolutionary biology and a rationale for evolutionary developmental biology.

8.2 LIFE-HISTORY STAGES

Ontogenies of individual embryos are divided into stages – what Griesemer (1996) has called (appropriately if not euphoniously) periodization, the marking of time into parts. Periodization is central to the study of embryos and their development as it is to any historical analysis. The most obvious stages of life cycles (embryos, larvae and adults) are introduced in this section. The stages of ontogeny familiar to all developmental biologists (blastula, gastrula, neurula) and the developmental processes that produce them (cleavage, gastrulation and neurulation) are introduced in the following section. Less familiar are stages that mark major transitions in developmental control mechanisms:

1. the stage at which maternal cytoplasmic control of development is superseded by zygotic genomic control (s. 7.4.1);
2. the stage at which the germ plasm separates from the soma, from which the body arises (s. 9.1.1);
3. the zootype as a stage of expression of homeobox genes that defines all animals (s. 14.2.3); and
4. phylotypic stages, during which the body plans for types of animals are set aside (s. 14.2.2).

Periodization is central to how we structure our approach to organismal development. Periodization is not typology, although there is the danger that subdividing life history into stages channels our thinking. To divide the history of France into pre- and post-Napoleonic periods is to make a statement about the importance of Napoleon. To talk about European history as pre- and post-industrial revolution makes a statement about the forces that shape a continent's development. But periods, marks, parts, stages and landmarks are an essential first step in making any sense out of the continuity that is development or evolution. They are especially important if we want to study altered timing and

place in development and evolution, as discussed in Part Seven.

8.2.1 Embryos

Development begins in embryonic life but is not limited to embryonic life. Embryos are unborn or unhatched offspring – that stage of the life cycle between egg and juvenile. Embryonic development therefore covers the development of rudimentary, immature or underdeveloped stages. But the processes of embryonic development (differentiation, morphogenesis and growth) do not stop at birth or hatching. They continue until an organism attains adulthood. Indeed, some processes continue throughout life: to a limited extent in wound repair (healing a cut finger), more extensively in regeneration of body parts such as human fingertips, lizards' tails and amphibian limbs, and most extensively in the ability of organisms such as the flatworm *Planaria* and the coelenterate *Hydra* to recreate entire organisms from small body parts.[8]

Organisms are born or hatch at quite different and more or less advanced stages of embryonic development. For example, precocial birds are able to fend for themselves when they hatch, but altricial birds require parental feeding and further maturation. Furthermore, the various parts of an organism do not all begin their development at the same time or stage, develop at the same rate, or complete development at the same time and stage; see Chapter 23. Familiar examples are the advanced stage, relative to other organs, of the mammalian brain at birth, and the rudimentary status of the hindlimbs (but not the forelimbs) in newborn kangaroos, even though the hindlimbs are so much larger than the forelimbs in adults. Some organs, such as the thymus, attain their maximal development in the embryo and are regressing at birth.

All organ systems are not set to the same developmental clock, nor need those clocks be set by the same developmental mechanisms.

In section 23.2 is a discussion, based on Hall and Miyake (1995a), of how embryos and embryonic organs tell time. Alteration in timing can have enormous implications for embryonic development. Change in developmental timing in a descendant relative to the timing of the same process in an ancestor is **heterochrony**, viewed by many as an important mechanism for evolutionary change in morphology (s. 24.1).

8.2.2 Larvae

Many organisms have life histories comprised of more stages than embryo and adult; juvenile and fledgling stages in birds and childhood are familiar examples. Other organisms have a larval stage that is morphologically distinct from the adult and/or inhabits a different environment. A species with a life cycle that includes a larval stage displays **indirect development**. Indeed, some species of parasitic insects have multiple larval stages in their life cycle. One may be specialized to find a host to parasitize, another to develop within the host, while others have other functions.

Larvae must metamorphose to transform into adults; familiar examples are metamorphosis of aquatic tadpoles into terrestrial frogs, and caterpillars into butterflies or moths. In such organisms embryonic development produces two sets of structures, one larval and one adult. Among amphibians, some larval structures are retained to be used by the adult. Thus, despite a total turnover of larval jaw muscles at metamorphosis in the frog *Rana pipiens*, new adult muscles are retrofitted onto larval nerves that persist. In extreme cases, in insects such as the fruit fly *Drosophila* for example, cells that produce larval structures are completely separate from those that produce adult structures. *Drosophila* larvae set aside separate sets of cells in imaginal discs to be activated at metamorphosis to produce adult structures.[9]

Paradoxically, the independence of larval and adult stages is nowhere better illustrated than in organisms that have lost their larval stages. In many frog species, the embryos do not hatch as tadpoles but as miniature adults, bypassing the larval stage altogether. Similarly, many sea urchins and sand dollars hatch as miniature adults, bypassing the specialized pluteus larval stage. Such **direct development** is discussed in Chapter 22 in the context of the evolution of early development and of ontogeny itself.

Direct developers afford researchers a golden opportunity to investigate how embryonic development is partitioned into larval and adult cell lines and larval and adult developmental programmes. Organisms that have lost their larval stages illustrate the plasticity of developmental processes, even early in ontogeny. They teach us how the balance between embryo, larva and adult is maintained in the life cycle, how development can evolve, and that an ability to modify developmental processes is basic to an organism's capability for evolutionary change.

Larval evolution may provide a key to understanding the evolution of development itself and so is revisited on a number of occasions. Genetic and developmental mechanisms that regulate metazoan development, and aid in understanding origins and interrelationships of the Metazoa, are discussed in Chapters 13 and 14.

Variation in larvae or larval types is correlated with patterns of dispersal, an issue that is discussed in greater detail in Chapter 22. The lancelet *Epigonichthys lucayanum* has two adult forms which are distinguished on the basis of numbers of muscle blocks (myotomes) and gill slits. The two develop from two distinct larval forms (amphioxides and amphioxus) whose features are adapted for dispersal or non-dispersal respectively. Natural selection acts separately on these larval and adult stages. Larvae and adults evolve on different schedules and with considerable independence from one another. Indeed, natural selection acts throughout ontogeny, as Charles Darwin knew and discussed. The role played by natural selection on different parts of the life cycle has thus attracted considerable interest.[10]

8.3 CONSERVED STAGES AND PROCESSES OF EARLY VERTEBRATE DEVELOPMENT

The hierarchical organization of vertebrate development comes into stark relief when the stages and processes in a typical ontogeny are examined, for the major stages through which all embryonic vertebrates pass are remarkably similar:

zygote → blastula → gastrula → neurula.

Identification of such common embryological stages across the vertebrates, indeed across the whole animal kingdom, was a triumph of 19th century comparative embryology (s. 5.1). The stages are often more highly conserved than the developmental processes that produce them. Those processes (which are treated in greater depth in Chapters 8–11) are:

fertilization → cleavage → gastrulation → neurulation.

Constancy of stages in the face of variable mechanisms producing them (s. 8.4) goes to the heart of comparative and evolutionary biology, whether we can recognize phylotypic stages (s. 14.2.2) and what we mean by homology (Chapter 21).

8.3.1 The zygote and fertilization

The egg is transformed into a **zygote** by the process of **fertilization**, which is the first inductive event in ontogeny. Initiation of development in unfertilized eggs (parthenogenesis) represents an 'apparent' exception, 'apparent' only because some other epigenetic signal must substitute for that normally provided by the sperm. That signal might be temperature, a pH shock, or the mechanical penetration of the egg membrane, as when frog eggs are 'fertilized' or **activated** by pricking with a needle. Indeed, there is a positive correlation between the amount of epigenetic control of murine oocytes and the extent of development that can occur in parthenogenetically activated oocytes (Kono *et al.*, 1996).

Fertilization is the most remarkable of all interactions between two totally different cells. There are no other natural circumstances in which normal cell function is facilitated by one cell fusing with a different cell from the same organism, let alone with one from a different organism. There are examples of fusion between like cells, as in the fusion of pre-osteoclasts to form multinucleated osteoclasts and the fusion of myoblasts to form myotubes. Virally transformed cells can be fused with other cells. Bacteria and pathogens invade the cells of other organisms. Mutagenized cells can fuse with normal cells, but such fusions result in abnormalities in chromosome numbers, uncontrolled cell division, onset of disease and neoplastic transformation (Maclean and Hall, 1987). They lead neither to normal development nor to normal function.

Production of small numbers of large eggs and large numbers of small sperm is accomplished by unequal cell division in oogenesis but equal cell division in spermatogenesis (Figure 8.1). Eggs of animals with external fertilization and development are large and filled with yolk, as a consequence of unequal distribution of cytoplasm during oogenesis; nutrient in the yolk is sufficient to carry embryos to active, feeding stages. Eggs of animals with internal fertilization and development are smaller with less yolk, nutrient being supplied maternally and indirectly, rather than directly via the yolk.[11]

In addition to depositing yolk, the mother supplies mRNAs and proteins, some of which are asymmetrically distributed in the egg and play important roles in such early developmental processes as establishment of the A-P and D-V body axes and formation of the mesoderm (Figure 8.2. and section 10.1). Mitochondria and therefore mitochondrial genes are also inherited maternally. The female parent therefore plays a disproportionate role in early development through

Spermatogenesis   Oogenesis

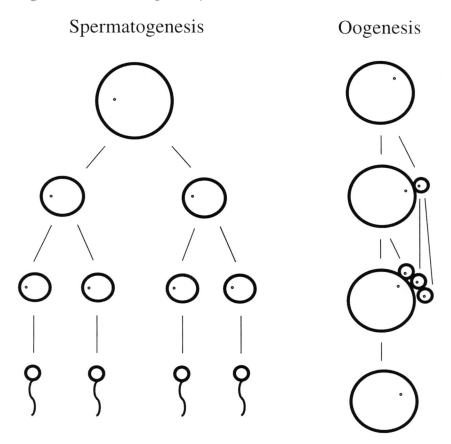

Fig. 8.1 Equal cell division during spermatogenesis produces numerous small secondary spermatocytes that mature into small sperm. During oogenesis, unequal distribution of cytoplasm to daughter cells produces a large future ovum and small polar bodies.

maternal cytoplasmic control, although such cytoplasmic constituents as centrioles are supplied to the egg by the sperm.[12]

In most species, oogenesis is not complete when fertilization takes place. Oocytes require the stimulus of sperm entry to complete one and sometimes both meiotic divisions. Only in cnidarians and sea urchins does fertilization occur after completion of meiosis, when the egg is a mature, haploid ovum (Table 8.1). A consequence of fertilization is thus completion of egg maturation and transformation of the primary or secondary oocyte, with its short life and limited potential, into the mature ovum, whose life is that of the individual and whose

potential includes providing germ plasm and gametes for future generations.

Irrespective of whether you regard embryonic development as starting with the development of the egg (oogenesis) and sperm (spermatogenesis), or with the zygote, fertilization is an epigenetic event that both completes oogenesis and activates the egg as a zygote. Other major consequences of activation are:

1. union of the male and female pronuclei to form the zygote nucleus, ensuring genetic recombination, the major 'advantage' of sexual reproduction;

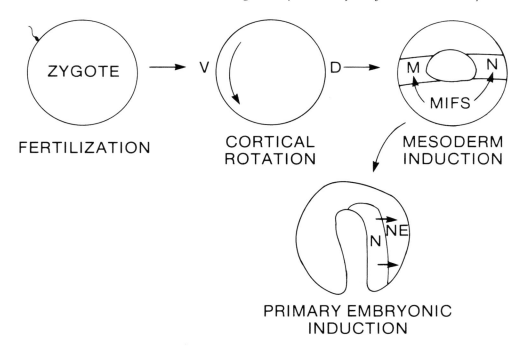

Fig. 8.2 Early epigenetic events in the development of a typical amphibian embryo. Fertilization elicits a 30° rotation of the cortical cytoplasm, establishing dorso-ventral (D-V) polarity. More dorsal endoderm induces notochord (N) and more ventral endoderm induces mesoderm (M) from marginal ectoderm at the equator. Mesoderm-inducing factors (MIFS) are peptide growth factors; see s. 10.1. Ectoderm 'above' the blastocoele at the animal cap is blocked from these inductions and so remains as ectoderm. During gastrulation the notochord (N) induces the adjacent ectoderm to become neural ectoderm (NE) during primary embryonic induction.

Table 8.1 Timing of fertilization in relation to phases of oogenesis[1]

Phase of oogenesis	Representative examples
Diploid, young primary oocyte	Worms, flatworms, *Peripatus*
Diploid, mature primary oocyte	Worms, sponges, dogs, foxes
Diploid, first metaphase stage of meiosis	Molluscs, many insects
Diploid, second metaphase stage of meiosis	Amphibians, amphioxus, most mammals
Haploid, mature ovum	Cnidarians, sea urchins

[1] See Gilbert (1994a) for further details.

2. formation of a fertilization membrane and the initiation of metabolic reactions, such as depolarization of the egg membrane, to ensure that excess sperm are prevented from entering the egg, thereby eliminating the genetic and developmental complications of polyploidy;

3. establishment of the primary and secondary (A-P and D-V) embryonic axes;

4. alteration in permeability of the egg membrane, allowing increased interaction between egg and environment;

5. dramatic increases in respiratory, metabolic and biochemical activities within the zygote. Respiration may increase by as

much as 500% immediately following fertilization;

6. rearrangement of cytoplasmic constituents. This may occur before or after the first cleavage, a timing difference that establishes the two classes of mosaic and regulative development; and

7. initiation of the first cleavage to begin the transformation of the zygote into a multicellular embryo.

Although activation is normally a consequence of fertilization, it need not depend on the presence of the sperm nucleus; eggs can be artificially activated using such techniques as pricking frogs eggs with a needle, or exposing sea urchin eggs to a temperature or pH shock. Such artificially activated eggs undergo the changes listed above, and in some species continue to develop until the end of cleavage (i.e. until they reach the blastula stage), when development stops. Stored maternal information within the egg cytoplasm and/or activity of the egg nucleus can carry embryos to the end of cleavage, but not beyond (s. 7.4.1). At the onset of gastrulation, the zygotic genome takes over from maternal cytoplasmic and ovum genomic controls that directed early development.

Without fertilization, the egg rapidly dies, its promise unevoked and its potential to produce another generation unfulfilled. Without activation of the zygotic nucleus at gastrulation, the blastula dies, its promise unevoked and its potential to produce another generation unfulfilled.

8.3.2 The blastula and cleavage

The zygote divides (cleaves) to produce a **blastula** of hundreds, thousands, or tens of thousands of cells (blastomeres), depending on the species (Figure 8.3). **Cleavage** is the process of cell division by which the zygote becomes multicellular.

A triad of developmental processes comes into play during the ontogeny of multicellular organisms:

1. **differentiation** – the process whereby cells specialize in space and/or over time;
2. **morphogenesis** – the generation of the shapes of cells, tissues and organs; and
3. **growth** – permanent increase in size.

Typically, no growth and morphogenesis occurs during cleavage. Typically, each cleavage division divides the egg into cells of unequal size (hence, unequal cleavage). Also typically, these divisions are synchronous during the early stages. Maternal cytoplasmic control may extend as far into development as the end of cleavage but never beyond (s. 7.4.1).

The blastula usually retains the shape of the zygote and is typically no bigger than it, reflecting minimal morphogenesis and growth at this stage (Figure 8.3). This is because there is no net increase of cytoplasm during cleavage, which therefore is an unusual form of cell division that does not lead to growth. During cleavage, one replication of DNA and division follows another without any intervening synthesis of cytoplasm. Later in development and in adults, the division of a cell into two is followed by synthesis of new cytoplasm and growth, preceding the next wave of DNA synthesis. In cleavage, therefore, a fixed amount of cytoplasm contained within the zygote is progressively allocated into smaller and smaller and more and more cells until cells the size of typical body cells are produced (Figure 8.3).

8.3.3 The gastrula and gastrulation

The zygote has maternally derived cytoplasm as a source of information, and maternal cytoplasmic control predominates during cleavage. The potential for cell-to-cell interactions exists within the blastula because individual groups (usually layers) of cells inherit different cytoplasmic components. Such an interaction specifies mesoderm as a secondary germ layer (s. 10.1). Further interactions require cell movements that transform the blastula into a **gastrula** by the process of **gastrulation**. This is the stage during ontogeny when differentia-

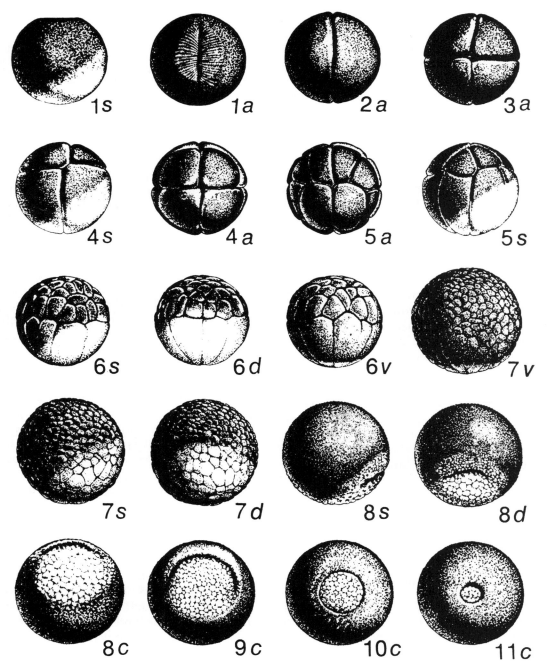

Fig. 8.3 Stages of early development (zygote to gastrula) of the frog *Rana pipiens* as seen in left lateral (s), animal pole (a), dorsal (d) and ventral (v) views. The stages shown are 1) fertilized egg (zygote); 2) 2-cell stage; 3) 4-cell stage; 4) 8-cell stage; 5) 16-cell stage; 6) early blastula; 7) late blastula; 8) early gastrula; 9) mid-gastrula; 10) yolk-plug stage; 11) late gastrula. Reproduced from Mathews (1986).

tion, morphogenesis and growth are initiated as the zygotic genome progressively asserts its control over development. Haeckel identified this stage with the universal metazoan ancestor (Figure 5.4).

Gastrulation is the first major stage when the parts of embryos are rearranged (Figures 5.3, 8.4). Rearrangement of layers establishes new cell associations and the physical relationships required for subsequent inductive interactions. Induction of the central nervous system begins during gastrulation, initiating a network of inductions that give rise to the different cells, tissues and organs of embryos and adults. A number of these are discussed in Chapters 8 and 9, especially those accounting for establishment of the central body axis and central nervous system.

Patterns of gastrulation vary considerably even within a single group. Collazo, Bolker and Keller (1994) gathered a considerable amount of the classical literature on gastrulation in teleost fishes in a phylogenetic analysis. They identified two key developmental transformations underlying teleost gastrulation: loss of the bottle cells so characteristic of amphibian gastrulation, and transformation from yolk contained in platelets to a continuous mass of yolk. The presence of a 'blastodisc' (typical of birds) rather than a spherical blastula (typical of amphibians) in direct-developing, egg-brooding hylid frogs of the genus *Gastrotheca*, an extreme modification of gastrulation, is discussed in section 22.2.1 and illustrated in Figure 22.1.

8.3.4 The neurula and neurulation

Because the central nervous system (brain and spinal cord) is the first organ system to develop, and because it is so evident along the 'back' of vertebrate embryos, the embryonic stage that follows the gastrula is known as the **neurula** and the process of formation of the neural tube as **neurulation**. Neural ectoderm is specified during gastrulation but elaborated during neurulation (Figure 8.5). There is,

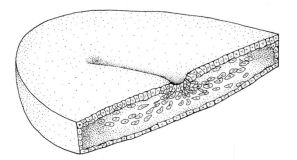

Fig. 8.4 Movement (ingression) of cells from the upper layer of the two-layered avian blastodisc through the primitive groove creates the third (mesodermal) layer.

however, more to the vertebrate nervous system than the central nervous system that arises from neural ectoderm: the peripheral nervous system arises from **neural crest cells** that form secondarily from neural ectoderm (Figure 8.5; ss. 10.2, 10.4) and from **placodes** that initially lie adjacent to neural ectoderm (or may arise from lateral neural ectoderm), but which move away as they differentiate or deposit their neuronal products. Other organs are also developing at this stage as mesoderm is segregated into axial and lateral plate, as the heart takes shape, as kidneys begin to form and so forth.

Hierarchical levels of control unfold progressively from fertilization → cleavage → gastrulation → neurulation:

1. The blastula possesses spatial heterogeneity arising from differential distribution of maternal cytoplasmic constituents.

2. The gastrula has spatial heterogeneity established through the morphogenetic movements of the three germ layers (Figure 5.3). These cell movements bring previously unassociated regions into association, setting up the next level in the hierarchy of interactions, the formation of organ systems.

3. The neurula possesses a diversity of differentiating cell types, spatial heterogeneity,

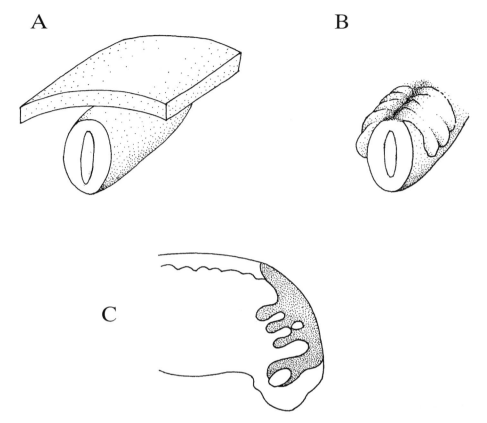

Fig. 8.5 A. The hollow neural tube and overlying ectoderm. B. Migration of streams of neural crest cells from the dorsal surface of the neural tube. C. A lateral view of an older embryo (anterior to the right) to show the distinct streams of neural crest cells migrating toward the visceral arches.

and temporal association/dissociation of parts as sources of epigenetic signalling.

8.4 MODIFICATIONS OF EARLY DEVELOPMENT

Von Baer's laws predict that early pre-gastrula embryos should bear the greatest resemblance to one another. Milne-Edwards developed a similar view of progressive divergence but within *embranchements* (s. 4.7). Despite von Baer's assertions of the uniformity of early embryos within or even between members of a phylum, early development does vary. Not that von Baer's law of the similarity of early embryonic stages is invalid. Rather, embryos do not necessarily display the greatest similarity at the outset of development. Several examples of the generality of this statement are:

1. the various forms of cleavage, blastula formation and gastrulation within invertebrate and vertebrate taxa: discoidal rather than spiral cleavage in cephalopods in comparison with all other molluscs; modified early cleavage patterns in leeches; single embryos that develop from two blastoderms in annular fishes of the genus *Cynolebias*; and the various germ-band stages in insects discussed below;

2. the presence of a 'blastodisc' rather than a spherical blastula in direct-developing, egg-brooding hylid frogs such as *Gastrotheca riobambae* and *G. plumbea* (s. 22.2.1);

3. the divergent patterns of egg size, cleavage and cell lineage seen in direct-developing echinoderms which have modified the feeding-larva or lost their larvae entirely (s. 22.2.2);

4. modified and asymmetrical gastrulation in the sea urchin *Heliocidaris erythrogramma*, which broods its larvae;

5. variable patterns of gastrulation in *Dynamena pumila* that are invariant patterns in more derived coelenterates;

6. the loss of primary mesenchyme in sea urchins such as *Eucidaris tribuloides*; and

7. the differences between oviparous and viviparous vertebrates.[13]

It is evident that patterns of cleavage vary widely across the animal kingdom. Variation includes reassignment of cells to different lineages in direct-developing sea urchins (3 above), and formation of single embryos from double blastomeres (1 above). The latter is a particularly interesting example. Embryos of two fishes (*Cynolebias whitei* and *C. nigripinnis*) develop from diblastodermic eggs in which two blastoderms are completely separate from the 1-cell to the late-blastula stages. Midway through epiboly, when the blastoderms are spreading over the surface of the yolk, they begin to fuse. By the end of epiboly the deep cells of each blastoderm have dispersed, intermingled and reaggregated to form a single population from which a single normal embryo will develop. Diblastodermic zebrafish eggs have been reported, but in this species the process is an aberration, resulting in the formation of embryos of different sizes.[14]

Mechanisms of gastrulation also vary greatly (2 above). However, even in direct-developing frogs that have lost the larval stage and possess a highly derived pattern of gastrulation from an embryonic disc, earlier and later stages are immediately recognizable as typical of species that retain larval stages (s. 22.2.1). As in many other areas, Olivier Rieppel has encapsulated the essentials of the argument:

'As a transformative process, gastrulation may indeed differ in closely related species, but the gastrula, a multilayered germ which minimally comprises the ectodermis and the gastrodermis, is, on topological grounds as much as by the test of congruence, a taxic homology of the Metazoa – no matter how it is arrived at, and even if preceded by a stage of blastomere anarchy as in triclad flatworms!'

Rieppel, 1992a, p. 709

Embryos of closely related species, with profoundly different patterns of cleavage and/or gastrulation, produce adults with similar morphologies. Modified patterns of development are adaptations for embryonic or larval stages of the life cycle; embryos, larvae and adults can evolve independently (s. 8.2.2). Divergence in germ-band stages in insects is an excellent example.[15]

8.4.1 Germ-band stages in insects

Three types of insect germ-band stages are recognized:

1. **long germ band** insects, such as *Drosophila*, with simultaneous specification of all segments by the end of the blastoderm stage. *Drosophila* displays extreme long germ band development;

2. **short germ band** insects, such as the grasshopper *Schistocerca americana*, in which body segments are generated progressively from anterior to posterior after the blastoderm stage;

3. **intermediate germ band** insects, such as the dermestid beetle *Dermestes*, in which some segments are specified before and some after the blastoderm stage.

In describing ontogenetic patterns that converge on the germ-band stage. Sander attrib-

Table 8.2 Distribution of embryo types among the major insect orders

Insect order	Embryo type
Diptera	Long germ band
Dragonflies, Hemiptera	Intermediate germ band
Orthoptera	Short and intermediate germ band
Lepidoptera	Long and intermediate germ band
Coleoptera and Hymenoptera	Long, short, and intermediate germ band

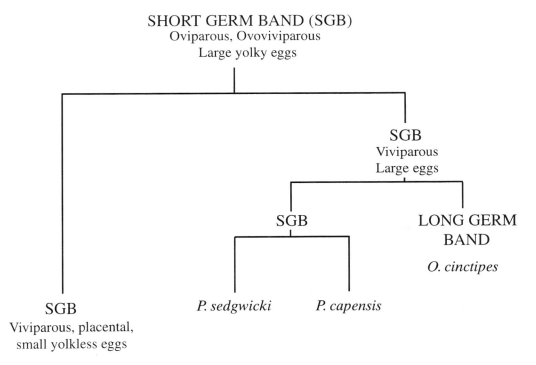

Fig. 8.6 Phylogenetic relationships of modes of embryonic development in onychophorans illustrating predominance of short germ band (SGB) development and the origin of long germ band development in *Opisthopatus cinctipes* from a viviparous ancestor with (SGB) development and large eggs. Adapted from Walker (1995).

uted the conservation and conservatism of the stage to the interactive, hierarchical, network character of embryonic development (a view shared by Raff from his studies on echinoderms). Cohen and Massey, however, attributed such conservation to maternal control (s. 7.4.1).[16]

Knowledge of the molecular bases of differences among short, intermediate and long germ band insects, reinforces the conserved nature of this stage (s. 9.3). Equivalent stages can also be identified even when the cells for that stage are recruited at different times during development and/or from different lineages. Thus the germ-band stage of the crustacean *Diastylis* is derived from ectoteloblasts (that produce this stage in other crustaceans), and from other cell lineages recruited at a later stage.[17]

These embryo types are not distributed so that monophyletic groups all possess a single germ-band type, although all flies examined so far are long germ band, and all dragonflies and hemipterans are intermediate. Table 8.2 displays the distribution of embryo types in the insect orders.

Nor need all members of a phylum share the same germ band type. Muriel Walker (1995) discovered that a South African onychophoran, *Opisthopatus cinctipes*, develops from a long germ-band embryo. All other onychophorans from either of the two families Peripatidae and Peripatopsidae develop from short germ band embryos. The phylogeny of *Opisthopatus* and its allies indicates that this shift arose within the last 30 million years from a species with short germ band development (Figure 8.6). Shortening of overall development time and formation of mesoderm earlier in ontogeny, are considered likely mechanisms for this shift.

As they initiate their own development and lay down basic body plans, embryos must also initiate mechanisms to ensure the production of gametes for the next generation; the functional and evolutionary causations introduced in section 1.1. How such disparate decisions are made is the subject of the next chapter.

8.5 ENDNOTES

1. Hanken (1986, 1993a), Trueb and Hanken (1992), Cannatella and De Sá (1993) and Trueb (1996) have documented many derived features of *Xenopus*. See Kobel and Du Pasquier (1986) for polyploidy. For a general, up-to-date overview of the biology of *Xenopus*, see the symposium organized by the Zoological Society of London and edited by Tinsley and Kobel (1996).
2. See Bellairs and Gans (1983), Tigano and Parenti (1988) and Trueb and Hanken (1992) for

the caecilian, fish and amphibian examples respectively. See Chapter 21 and Hall (1995a) for further discussions of homology and common development.

3. For discussion with examples of the need for a minimum of three species comparisons in phylogenetic analyses, see Huey (1987), Brooks and McLennan (1991), Garland and Adolph (1994) and Lauder *et al.* (1995). See Meyer *et al.* (1993, 1995), Meyer (1996), Zardoya *et al.* (1996a,b), Wray and Bely (1994) and Patel (1994) for comparative studies in fishes and sea urchins. For general analyses of the comparative approach, see Hanken (1993a), Kellogg and Shaffer (1993) and the six papers in the symposium to which the last paper is an introduction.
4. M. H. Wake (1992b); also see Niklas (1997) for multicellularity and other aspects of plant evolution.
5. The chapters in the volumes edited by Hall (1994a) and Sanderson and Hufford (1996) are the latest broad treatments of homology and homoplasy. The vexing problem of the amount of convergent evolution among phyla is dealt with in detail by Moore and Willmer (1997). If their analysis is even partially correct, then analyses of relationships are far more imprecise than we routinely take them to be.
6. See Shaffer (1993) and Shaffer and Voss (1996) for the phylogenetics of model organisms, as illustrated by the recently derived clade of 17 species of ambystomatid salamanders such as *Ambystoma mexicanum*.
7. Whether hagfishes and lampreys form a natural group and whether lampreys are closer to gnathostomes than either is to hagfishes continues to be debated ; see M. M. Smith and Hall (1990, 1993), Forey, (1995), Janvier (1996a,b) and Mallatt (1996). For me, the weight of evidence separates lampreys from hagfishes (Figure 2.4).
8. Goss (1969), Maclean and Hall (1987), Hodges and Rowlatt (1994), Stocum (1995), Tsonis (1996) and Ferretti and Géraudie (1998) discuss the resumption of embryonic developmental processes in adult organisms.
9. For up-to-date discussions of larval development and evolution throughout the animal kingdom, and the elusive problem of exactly how to define larvae or the larval stage (or metamorphosis), see the introduction and chapters in Hall and Wake (1998). For retrofitting larval neuromuscular circuits in *R. pipiens*,

see Alley (1989, 1990). Insects that display larval heteromorphosis (different larval stages within the life cycle) include blister beetles of the family Meloidae, and mantispids of the order Neuroptera (Wilson, 1971; Nijhout, 1998).

10. For divergent adult and larval forms of lancelets, see Willey (1894), Bone (1957) and Gibbs and Wickstead (1996). For the action of natural selection at different stages of the life cycle, see the books by Gould (1977), Calow (1983), Mayo (1983), Roff (1992), Stearns (1992) and Williams (1992) and the major review by Roff (1996).

11. The effect of yolk on cleavage, known as Balfour's law, was formulated to account for the effect of the uneven distribution of yolk: 'In eggs in which the distribution of food material is not uniform, segmentation does not take place with equal rapidity through all parts of the egg, but its rapidity is, roughly speaking, inversely proportional to the quantity of food material' (Balfour, 1875, p. 210). The embryologist R. Rappaport described this correlation between cleavage rate and yolk content as 'one of the oldest generalizations in developmental mechanics', and concluded that 'the correlation is attributed to an impending effect that the yolk exerts upon mitosis and cytokinesis. As yet, however, no experimental analysis of the impending effect has been made' (Rappaport, 1974, p. 79).

12. A fascinating exception to simple uniparental inheritance of mitochondria through the female parent has been discovered in the blue mussel *Mytilus* and its allies by Elefterios Zouros and his colleagues. Female mussels inherit mitochondrial DNA (mDNA) from their mothers and transmit it to both male and female offspring. Male mussels transfer only paternal mDNA to their sons (Zouros *et al.*, 1994).

13. For patterns of cleavage and gastrulation, see Boletzky (1989), Dohle (1989), Wray and Raff, (1991), De Loof (1992), Carter and Wourms (1993), Bolker (1994) and Collazo, Fraser and Mabee (1994). For the evolution of viviparity, see M. H. Wake (1989, 1990, 1992a), Blackburn (1995), Dulvy and Reynolds (1997) and Qualls *et al.* (1997). For morphological variation in gastrulation of the marine hydroid *Dynamena pumila*, see Cherdantsev and Krauss (1996).

14. See Carter and Wourms (1993) and Laale (1984) for diblastodermic eggs in *Cynolebias* and *Danio* respectively.

15. See del Pino and Elinson (1983), Dohle (1989) and Dohle and Scholtz (1988) for production of similar adult morphology despite variant cleavage patterns. See Hall and Wake (1998) for patterns of larval evolution.

16. See Sander (1983) and Raff (1987) for conservatism through interacting networks, and Cohen and Massey (1983) for maternal cytoplasmic control.

17. For molecular studies on different germ band insect embryos, see Akam (1987, 1995), Akam *et al.* (1988, 1991), Weisblat *et al.* (1994), Patel *et al.* (1992, 1994), Nagy and Carroll (1994) and Carroll (1995). See Dohle and Scholtz (1988) for the work on *Diastylis*.

9

WHERE GENERATIONS CONVERGE: GERM LINES AND BODY PLANS

'The issues of how major body plans arise, whether they require any special explanations over and above accumulated intraspecific divergences and speciations, and indeed whether there is any such thing as a body plan, are highly contentious.'

Arthur, 1988, p. 34

As embryos develop, a hierarchy of developmental events is expressed in the progressive differentiation of cell types, regionalization of embryos, and establishment of morphological differences between regions. Perhaps the most fundamental of such events are the segregation of body cells from germ cells and the laying down of the body plan.

9.1 DETERMINATION OF BODY CELLS AND GERM CELLS

A 'choice' made during the development of all vertebrates (very early during the development of some) is between future germ line (gametes) and future soma (body cells). Such a choice is not made during the development of many invertebrates. In some, separation of germ cells from soma is delayed until later in ontogeny; in others, no separation occurs. In the latter species, body cells can become gametes throughout life and somatic inheritance is theoretically possible.

9.1.1 A phylogenetic perspective

Timing of germ line specification during development determines the degree to which genetic variation arising during development will be heritable. Buss (1987) provided a novel discussion of the nine phyla in which separation of somatic from germ plasm does not occur. Development of gametes from somatic cells (**somatic embryogenesis**) occurs in all members of the Placozoa, Porifera, Cnidaria, Entoprocta, Ectoprocta (Bryozoa) and Phoronida, and some members of the Platyhelminthes, Annelida and Hemichordata. Somatic embryogenesis is universal within all members of the Protoctista, Fungi and Plantae, a total of 42 phyla or equivalent higher level taxa (Table 2.1).

'Early and late germ line determination' and 'somatically derived germ cells' are phrases suggested by Jablonka and Lamb to distinguish groups such as echinoderms (in which germ line segregation occurs late in embryogenesis) and nematodes (in which germ line segregation occurs early and is therefore more rigid) from sponges and coelenterates, in which germ cells are somatically derived. Experimental investigation of whether a germ line is set aside early in embryogenesis in a sea urchin (it is not) led Ransick, Cameron and Davidson (1996) to compare timing of germ cell determination across the Metazoa. They concluded that no species with indirect development (i.e. with a larva) sets aside its germ line early in development. Indeed, in many species, the germ line only arises after both embryonic and larval structures are established.

Buss (1988) identified interesting correlations between early germ line determination and low species number; late determination and greater species numbers; and somatic derivation with the highest species numbers. Whether these correlations are causal factors contributing to speciation is an important issue to address.

9.1.2 Early germ line determination and continuity of the germ plasm

Segregation of germ plasm from soma is a fundamental event and decision-making process during vertebrate embryonic development. To become somatic is to have your cell fate progressively restricted and only to be able to produce cells of the body. It is to be mortal and to disappear with the death of the individual. By becoming part of the germ line, cells gain (potential) immortality, totipotency and the ability to produce gametes for the next generation. Early segregation of body and germ cells in vertebrates means that it is highly unlikely (although not theoretically impossible) that acquired changes will be inherited.

In a series of papers between 1883 and 1889, August Weismann (1834–1914) proposed one of the truly revolutionary ideas in the history of biology, the theory of the **continuity of the germ plasm**. John Maynard Smith (1989b) has hailed Weismann as one of the greatest evolutionary biologists of the late 19th century. Viktor Hamburger (1988) calls him one of the leading Darwinists of his epoch. Weismann postulated that the protoplasm (germ plasm) that produces the germ cells is separate from the somatoplasm that produces the body or soma. Germ plasm is not 'used-up' in the production of the body. Rather it is conserved to produce the gametes for the next generation and maintained as a continuous living connection between generations. Bodies are temporary vehicles for transporting sperm and eggs to the next generation.

With this theory, Weismann pioneered much of our current thinking about the significance of sex, embryology, ageing and death. In sex, we see separation of germ plasm from soma. In embryology, we see embryos as cradles for germ plasm and germ cells. Ageing and death are byproducts of natural selection and the separation of germ plasm from soma. Germ plasm is the physical representation of past evolutionary history and future evolutionary potential, housed in the soma or bodies through which evolutionary change is transported.

Somatoplasm provides the raw material for individual development within a lifetime; germ plasm provides the raw material for future generations. Therefore, changes in germ plasm are heritable, changes in soma are not. The bookish son does not inherit the muscular arms of his blacksmith father. The genetic potential to possess muscular arms is inherited, but only expressed if the son's muscles are put to arduous use. Weismann's concept of separateness of germ plasm from soma and continuity of germ plasm sounded the death-knell of theories of inheritance of acquired characters. This is not to say that there are not somatic influences on germ cells, for there are. The maternal environment is one, requirement for interaction with somatic cells for germ cells to form another. Epigenetic control is played out in germ cells as it is in the soma.[1]

9.1.3 Mechanisms of germ cell determination

The separation of germ cells from body cells occurs either through early segregation of cytoplasmic factors, or later by embryonic induction. The mechanisms of determination of body cells versus germ cells in *Drosophila* and anuran amphibians on the one hand, and urodelean amphibians on the other, illustrate nicely the preformed and epigenetic inheritance in this event of such far-reaching developmental and evolutionary consequences for organisms and species.[2]

In insects such as *Drosophila*, germ cell determination is based upon inheritance of a pre-

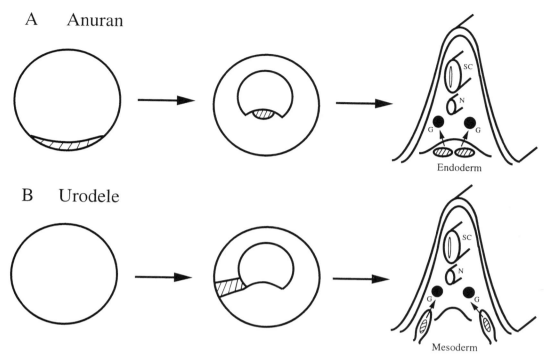

Fig. 9.1 A comparison of the origin and migration of germ cells in anuran and urodele amphibians. A. Anuran. Germ plasm is localized at the vegetal pole of the unfertilized egg (left). At gastrulation germ cells are localized in the endoderm (centre), from which they migrate to the gonads later in embryogenesis (right). B. Urodele. Germ plasm is not predetermined in the unfertilized egg. Germ cells arise in gastrula mesoderm (centre), from which they migrate to the gonads later in embryogenesis (right). G, gonad rudiments; N, notochord; SC, spinal cord.

formed cytoplasm, the germ plasm. Only cells that contain germ plasm during cleavage activate the genes that permit gametes to form. This is not because cells that lack germ plasm lack the genes for gamete production. In a dramatic demonstration of the power of preformed cytoplasmic information and a now classic experiment with *Drosophila*, Illmensee and Mahowald (1974) demonstrated that germ plasm transplanted into putative somatic cells 'turns them into' germ cells. Genetics proposes, epigenetics disposes, as Peter and Jean Medawar so aptly put it.

The germ plasm at the vegetal (south) pole of *Drosophila* eggs contains a 95 000 m.w. protein sequestered within granules, the polar granules. This protein was identified as the putative 'gamete-activating agent'. It is now known that at least eight genes (including *oskar*, *staufen*, *tudor* and *vasa*) are involved in polar granule formation. By controlling assembly of the products of the *vasa*, *tudor* and *nanos* genes into granules within the germ plasm, *oskar* effectively determines the number of posterior pole cells that become germ cells. Ectopic expression of *oskar* at the anterior pole of *Drosophila* eggs is sufficient to duplicate the results obtained by Illmensee and Mahowald, producing embryos with double posterior poles and inducing polar granules at the posteriorized anterior pole.[3]

In anuran amphibians, germ plasm is preformed and inherited maternally, as in

Drosophila. In urodele amphibians, however, germ cells arise epigenetically during induction of the mesoderm (Figure 9.1). This appears to be fundamentally different from germ cell specification in anurans, in which future germ cells inherit a cytoplasmic factor(s) in the primordial germ plasm – the classic difference between preformation and epigenesis. The difference may, however, be more apparent than real. Classes of molecules that function as determinants of cell fate (transcription factors, growth factors, and RNA-binding proteins) are localized within egg or early embryonic cytoplasm and can trigger later cell-to-cell interactions or be induced in response to such interactions. It may be that the determinants of germ cells are similar in anurans and urodeles, the difference being in where they are stored and how they are activated – by preformed localization in anurans, epigenetically in urodeles.[4]

I now turn to establishment of the major embryonic axes, the dorso-ventral (D-V) and antero-posterior (A-P). A model organism approach is unavoidable and inevitable. Because genetic control of axial specification is best known in *Drosophila*, I begin with a discussion of the process in that species and other insects (ss. 9.3, 9.4), before proceding to what is known of parallel processes in *Xenopus* (s. 9.5).

9.2 ORIGIN OF THE BODY PLAN IN *DROSOPHILA*

Insects are arguably the most ecologically successful group of organisms alive today. Holometabolous insects, with morphologically and ecologically distinct larval, pupal and adult stages, have radiated into the tremendous diversity of body plans represented by beetles, flies, bees, wasps, ants, fleas and butterflies, and into the diversity of environments they inhabit. In *Drosophila*, generation of the larval body plan involves activation of a series of maternal and zygotic genes belonging to several gene classes:

1. maternal-effect genes;
2. segmentation genes (segment polarity genes, gap genes and pair-rule genes); and
3. homeotic genes.

9.2.1 Maternal-effect genes

D-V and A-P polarities are established in the egg and embryo by the differential localization of protein or mRNA products of several groups of maternal-effect genes. At least 15 such genes are involved in establishing D-V polarity. *Snake, Toll, zerknullt, gurken, cactus,* and *sog* are key players. A further set of maternal effect genes (*bicoid, oskar, caudal* and *bicaudal*) establish A-P polarity. As expected of genes with major functions in establishing body patterns, mutations of maternal-effect genes result in embryos with missing or duplicated heads or tails, anterior or posterior body regions, and/or dorsal or ventral structures.[5]

9.2.2 Segmentation genes

Several sets of segmentation genes are activated in sequence following establishment of D-V and A-P polarity and axial organization:

1. **segment polarity genes** such as *wingless* and *engrailed* determine A-P compartments within individual segments. Mutations of these genes produce flies which have segments with double anterior or double posterior halves;
2. **gap genes** such as *Krüppel* and *hunchback*;
3. **pair-rule genes** such as *paired, even-skipped* and *fushi tarzu* establish groups of segments in pairs. Every second segment is missing in mutants of pair-rule genes; and
4. **homeotic genes**, which specify the identity of individual segments.

A summary of the sequence of expression runs as follows:

1. Maternal gene products of the genes *bicoid* and *nanos* originate from nurse cells at the

anterior end of the egg and localize to the anterior or posterior poles. The A-P embryonic axis is established through protein gradients of *bicoid* anteriorly and *nanos* posteriorly.

2. Anteriorly, where the concentration is highest, *bicoid* activates the zygotic gap-gene *hunchback*. Posteriorly, where nanos protein is highest, *nanos* prevents translation of *hunchback*. The gap-gene *orthodenticle* acts on *bicoid* in head specification.

3. The gradient of hunchback protein expression established by inhibition of *nanos* posteriorly, activates bands of expression of further zygotic gap genes such as *Krüppel* and *knirps*;

4. Diffusion of these gap genes from their sites of activation establishes overlapping gradients in combinations that vary along the A-P axis of what is still a syncytial embryo.

5. Specific combinations and concentrations of gap-gene proteins activate pair-rule genes such as *hairy* and *even-skipped* as seven stripes along the embryo. Development of cell membranes isolates embryonic regions and prevents diffusion of pair-rule proteins.

6. Pair-rule combinations activate segment-polarity genes such as *engrailed* and *wingless*, evident as stripes of expression specific to each segment.

7. And finally, homeotic genes (discussed below) establish the identify of each segment.[6]

In a comparative study of *Drosophila* and the house fly *Musca domestica*, Sommer and Tautz (1991) demonstrated that the cascade of genes that determines body plan in *Drosophila* is not peculiar to *Drosophila* among the flies. Primary expression domains of the major classes of genes are conserved in the house fly, although some divergence in secondary expression domains was found. Over 100 million years of evolution separates these two fly species, ample time for expression patterns or gene functions to have diverged. This does not guarantee that other dipteran species with modified primary expression domains will not be found. But for the moment the *Drosophila* pattern is extrapolated to all dipterans and used as the point of departure for considering mechanisms of specification in other groups (ss. 9.4, 9.5).

9.3 ORIGIN OF THE BODY PLAN IN OTHER INSECTS

Long and short germ band insects were introduced in section 8.4.1 in the context of conserved embryonic stages. Long germ band insects such as *Drosophila* have syncytial eggs with nuclei in a common cytoplasm for a large portion of their early development. Diffusion of maternal and zygotic gene products is therefore a sufficient mechanism to establish body axes, patterns of segmentation and segment specificity. Such a diffusion-based mechanism was not thought to operate in short germ band insects which have cellular embryos; the cell membranes would prevent diffusion of maternal gene products.

This view may turn out to be shortsighted.

Ralf Sommer and Diethard Tautz in 1993 investigated body plan formation in the short germ band embryos of the flour beetle *Tribolium castaneum* by isolating *Tribolium* homologues of two pair-rule genes, *hairy* and *Krüppel*, that define double-segmented regions in *Drosophila*. Both genes were expressed in several stripes at the cellular blastoderm stage in *Tribolium* in a manner similar to their expression in the syncytial *Drosophila* blastoderm. Sommer and Tautz hypothesize that the pre-patterning mechanism of the pair-rule genes that operates in *Drosophila* is involved in long and short germ band insects and may indeed be fundamental to insect development.

A year later, Lisa Nagy and Sean Carroll showed that a homologue of *wingless*, a gene that mediates cell-to-cell communication in *Drosophila*, is expressed in the cellular embryos of *Tribolium*. Expression was in the same

A

B

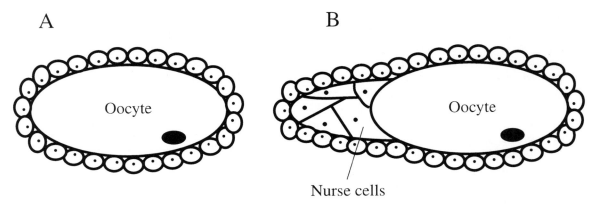

Nurse cells

Fig. 9.2 Types of oogenesis in insects. A. The panoistic oocyte, which is surrounded by follicle cells but lacks nurse cells, is typical of insects such as *Schistocerca*. B. A meroistic oocyte, also surrounded by follicle cells, except anteriorly, where nurse cells are present, is typical of dipterans such as *Drosophila* and beetles such as *Tribolium*.

pattern (although sequentially from anterior to posterior, rather than in simultaneous bands) as in syncytial *Drosophila* embryos. These results implicate a similar functional role for *wingless* in segmentation, despite the different cellular organization of long and short germ band embryos.

The gene *even-skipped*, which functions as a pair-rule gene in *Drosophila*, does not serve an early pair-rule function but does later in the development of the short germ band grasshopper *Schistocerca americana*. This difference reflects temporal differences in segmentation between short and long germ band insects (Patel, Ball and Goodman, 1992). These studies on *even-skipped* in single species of short and long germ band insects (*Schistocerca*, *Drosophila*) illustrate the difficulties in interpreting patterns for entire groups on the basis of studies of one or two species – the model organism problem raised in section 8.1.

As introduced in section 8.4.1, there are also intermediate germ band embryos, of which the dermestid beetle *Dermestes* is an example. At the onset of gastrulation, segmentation has progressed no further posteriorly than the head in the short germ band embryos of *Tribolium*, to the anterior abdominal region in intermediate germ band embryos of *Dermestes*, while in *Callosobruchus* (which has a long germ band embryo), all segments except the most posterior abdominal have formed. Patel, Condron and Zinn (1994) found pair-rule expression in all three germ band types of embryos, each expressing eight pair-rule stripes of *even-skipped*. The only differences between the three species were temporal and reflected in the relative number of stripes formed before, rather than after, onset of gastrulation. Patel and colleagues concluded that germ-band types do not correlate with mechanisms that establish segments but rather with temporal aspects of segmentation, and made the interesting observation that germ band types may correlate with one of the two ovary types known in insects (Figure 9.2):

1. panoistic ovaries or panoistic oogenesis, the ancestral mode found in all orthopterans in which the ovary lacks nurse cells. Pair-rule patterns have not been observed in beetle species with panoistic ovaries such as *Schistocerca*, although both short and long germ-band beetle embryos are known;

2. meroistic ovaries or oogenesis in which nurse cells are present but only at the anterior end of developing eggs. Diptera (*Drosophila*), representatives of the Hymenoptera and Lepidoptera, and beetles such as *Tribolium*, all have meroistic oogenesis and show pair-rule patterning, whether germ band embryos are long, short or intermediate.

Today there are more species of beetles than any other group of animals. With an evolutionary history spanning 300 million years, sufficient time to develop pair-rule patterning in insects with panoistic ovaries has been available. It may be that nurse cells and the ability to polarize deposition of maternal gene products into the egg are prerequisites for the evolution of pair-rule patterning.[7]

9.4 HOMEOTIC GENES

A series of homeotic genes regulates segmental structures in *Drosophila*. Homeobox gene products are transcription factors that regulate gene expression by binding to a 4-base core sequence of DNA of genes that encode secreted, structural and transcriptionally regulated proteins. As a single homeobox gene can control 100 or more genes, the regulatory potential of this class of genes is enormous. Such pivotal switching genes have been highly conserved structurally and functionally throughout evolution; stabilizing selection to maintain them is strong indeed.

Homeotic genes share a highly conserved nucleotide sequence (the homeobox) with only minor variation in the 180 nucleotides between different homeotic genes. There is 80–90% similarity between the six homeotic genes. Remarkable conservation of the homeobox domain also exists between widely separated organisms. For example, 59 of the 60 amino acid residues of homeotic gene product in frogs and *Drosophila* are identical, while similarity between *Drosophila* and mammalian homeobox genes is of the order of 70%.[8]

Homeotic genes in *Drosophila* are arranged into two major complexes in two regions of chromosome 3:

1. the Antennapedia (ANT-C) complex, which specifies anterior regions such as head, thoracic segments 1–3 and the limbs on the thorax; and
2. the Bithorax (BX-C) complex which specifies the identity of posterior regions of the body and suppresses limb development in the abdomen (Figure 9.3).

The Antennapedia complex contains three genes – *Deformed*, *Sex combs reduced*, and *Antennapedia* – expressed in that order along the A-P body axis. The Bithorax complex also contains three genes – *Ultrabithorax*, *abdominal A* and *Abdominal B* – which specify thoracic segment 3 and abdominal segments 1–9.

Bithorax and Antennapedia complexes in *Drosophila* appear to 'build the body' using thoracic segments 1 and 2 as the ground plan. Deletions of all Antennapedia and Bithorax complex genes produce flies in which all abdominal, all thoracic, and the two most posterior head segments become or remain thoracic segment 1, which represents the basic body plan segment. If only genes of the Bithorax complex are deleted, all abdominal segments and the third thoracic segment develop as thoracic segment 2, as elegantly demonstrated by Lewis (1978, 1985). Thus, these two thoracic segments represent basal undifferentiated states that are switched into particular and more specialized segments under the influence of homeotic genes.

9.4.1 Homeotic mutations in *Drosophila*

Mutations of homeotic genes produce **homeotic mutants**. William Bateson coined the term **homeosis** in his monumental compilation of variation in nature, *Materials for the Study of Variation* (1894). He used the term for those morphological variations that transform

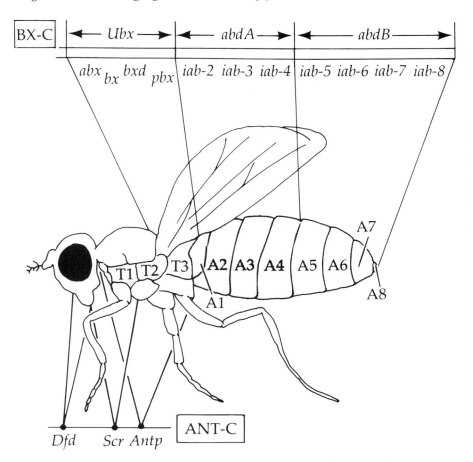

Fig. 9.3 The homeotic genes within the bithorax (BX-C) and *Antennapedia* (ANT-C) complexes in *Drosophila* to illustrate the anterior-posterior regionalization of the specification of body parts. The three genes of ANT-C [*Deformed* (*Dfd*), *Sex Combs reversed* (*Scr*) and *Antennapedia* (*Antp*)] specify head and thoracic segments T1–T3. Within the BX-C complex, *Ultrabithorax* (*Ubx*) specifies T3, while *abdominal A* and *Abdominal-B* (*abd-A, Abd-B*) specify abdominal segments A1–A4 and A5–A8 respectively. Reproduced from Gilbert (1988) *Developmental Biology* 2nd edn, Sinauer Inc., New York.

a structure into the likeness of another structure.

Homeotic mutants are best known from *Drosophila*, in which one region, such as the metathorax or antennae, is transformed into another 'related' region such as the mesothorax or legs. 'Relatedness' comes from shared developmental properties involved in making the body segment or appendage. Homeotic mutants are usually but not invari-

ably associated with mechanisms of segmentation, as in the body parts of insects (this section), and serial repetition, as in the parts of the nervous, muscular and skeletal systems in vertebrates (s. 15.5).

Homeotic mutations produce either:

1. **duplications**, in which one part is duplicated in the adjacent region of an animal. *Bithorax* is a classic example (Figure 9.4).

Fig. 9.4 A four-winged *Drosophila* in which the third thoracic segment, the metathorax, which normally has balancers and no wings, has been transformed into a second (mesothoracic) segment with wings. Reproduced from Gilbert (1988) *Developmental Biology* 2nd edn, Sinauer Inc., New York.

Even minor anatomical features such as bristles and veins are faithfully duplicated in the new site; or

2. **replacements**, when one region is replaced by another, as in wings for halteres and legs for antennae. *Antennapedia*, normally only expressed in the last thoracic and abdominal segments, is expressed in the head when the antenna is transformed into a leg.

Individual cells and tissues are not transformed by homeotic mutations. Rather, regions of the body are transformed. Waddington (1940) proposed that the genes controlling the production of homeotic mutations had all the properties of developmental switches activating new developmental programmes. This prediction has proven true to a

vastly greater extent than even Waddington, with his emphasis on integration of genetics, development and evolution, could have hoped.

9.4.2 Conserved homeotic genes in animals and plants

Homeotic genes, although not necessarily homeotic mutations, are found in sponges, planaria, echinoderms, annelids, molluscs, amphibians, birds and mammals but not in yeast, slime molds and tapeworms. Unsegmented metazoans such as planaria (favourite species for the study of whole-body regeneration), the nematode *Caenorhabditis elegans*, the two hydroids *Hydractinia symbiolongicarpus* and *Eleutheria dichotoma*, and the

---

BOX 9.1

ORTHOLOGY AND PARALOGY

Orthology, or an orthologous molecule or gene, is functional correspondence between molecules or genes from two or more different organisms. Paralogy, or a paralogous molecule or gene, is correspondence between molecules or genes within an organism. (Some would substitute homology or sequence similarity for correspondence.) Paralogous genes arise by gene duplication (Fitch, 1970).

---

hydrozoan *Sarsia*, contain homeobox genes, although sequence similarity of the planarian with homeobox sequences from other organisms is low, 30–50%. Balavoine and Telford (1995) identified ten *Antennapedia*-like genes in their analysis of three species of triclad planarians. Of five in *Polycelis nigra* from which the homeobox-coding exons could be amplified, two were orthologues of *labial*, four were orthologues of genes of the *Antp* class (see Box 9.1 for orthology and paralogy). With this stronger confirmation of the homology of these genes, the presence of homeobox genes in planaria and cnidarians speaks to the antiquity of this class of genes in the ancestor of triploblastic animals, and to a more general role in body organization than control of segmentation. Homeotic genes may generate cell diversity along or throughout the body in unsegmented metazoans. They certainly play a role in the A-P organization of the unsegmented intestinal tract in rodents, mediating epithelial–mesenchymal interactions. See Chapter 15 for homeotic genes in chordates and vertebrates.[9]

Homeotic genes and homeotic mutations have also been described from plants. *Bicalyx* is a naturally occurring mutation that replaces petals with sepal-like structures in a Californian population of *Clarkia concinna* (red ribbons), a member of the Onagraceae. Weigel and Meyerowitz (1994) proposed the ABC model of regulation of floral organs. Three classes of homeotic genes (designated A, B and C) act in combination to produce the four types of floral organs found in plants:

- A genes → sepals;
- A and B genes → petals;
- B and C genes → stamens; and
- C genes → carpels.

A further proposal was that carpels would transform into sepals if class A genes inhibited class C genes.

Members of the *Polycomb* group of response elements in the regulatory region of homeobox genes are important regulators of cell fate in *Drosophila*. *Curly leaf* in *Arabidopsis* is the homologue of *Enhancer of zeste*, a member of the *Drosophila Polycomb* group. *Curly leaf* represses floral homeotic genes of class C, providing strong evidence for conserved elements for determination of cell fates in plants and animals.[10]

9.5 CONSERVED GENETIC SIGNALLING IN DORSO-VENTRAL SPECIFICATIONS IN INSECTS AND VERTEBRATES

Many invertebrate body plans, such as those of insects, are built around a ventral axis and ventral nervous system (see Table 2.3). Vertebrates on the other hand are built around a dorsal axis and a dorsal nervous system and notochord. The only more divergent way of making an organism than turning its D-V axis upside down as Geoffroy wanted to do, is to turn it end to end, reversing anterior and posterior, as was done with the revised reconstruction of *Hallucigenia* (s. 3.1.2). Cuvier's reaction to radically different organization was to group organisms into totally separate groups and not to allow any crossover between groups. Geoffroy, on the other hand, sought to unite organisms of such diverse designs by turning one inside out to transform it into the other (Figure 4.1). Deriving vertebrates from a wide variety of invertebrate groups by just such transpositions, reversals, and inversions preccupied many anatomists and morphologists during the 19th century (s. 5.5.5). It now appears (at least at the level of genetic control), that such transpositions are not as far-fetched as they sound.[11]

Recent and totally unexpected information from studies into organization of the D-V and V-D axes in *Drosophila* and *Xenopus* has revealed a conserved genetic basis for how these two divergent organisms specify this body axis. The two conserved genes which encode related proteins are the zygotic genes *short gastrulation* (*sog*) in *Drosophila* and *chordin* in *Xenopus*.[12]

Sog, which is expressed ventrally in the early *Drosophila* embryo at the site of formation of the ventral nervous system (Figure 9.5), is involved in development of the ventral nervous system and of the muscle found in the ventral midline.

D-V polarity in amphibians is established through a 30° rotation of cortical cytoplasm immediately following fertilization (Figure 8.2). A subset of parallel microtubules is the putative cellular mechanism driving this rotation; UV irradiation prevents these microtubules from developing, prevents cortical cytoplasmic rotation, and prevents such embryos from forming D-V axes. Cortical cytoplasm in *Xenopus* anchors subclasses of RNA in animal (dorsal) and vegetal (ventral) cytoplasms. Entrapment and/or active transport of RNAs is emerging as a mechanism that establishes and maintains cytoplasmic specificity at these early developmental stages.[13]

In *Xenopus*, *Chordin* is expressed initially in the dorsal lip of the blastopore, then in dorsal mesoderm and notochord (Figure 9.5). Functionally, *Sog* in *Drosophila* is equivalent to *chordin* in *Xenopus*. Whereas *sog* is associated with the ventral nervous system, *chordin* is associated with the developing dorsal nervous system. That the two genes share sequence similarity suggests that essentially similar genes are specifying opposite axes in these single invertebrate and single vertebrate species. Such conserved sequence similarity is surprising. How can opposite poles be specified by the same genes?

This fascinating story would have ended here had Holley and colleagues (1995) not injected mRNA for *sog* and *chordin* into *Drosophila* eggs. Injected *sog* specified ventral development, but so did *chordin*. When mRNA was injected into *Xenopus* eggs, *chordin* specified dorsal structures, but so did *sog* (Figure 9.5). The gene responsible for specification of the D-V axis is conserved but the primary axis it determines is inverted.

Why should invertebrates and vertebrates use the same gene to specify the D-V axis? Vertebrates really may be upside down invertebrates as discussed in section 5.5.5 and illustrated in Figure 5.6.

But the story does not end there. Parallels between the control over *Drosophila* and *Xenopus* D-V axis determination go even further. In *Drosophila*, *sog* antagonizes the product of the *decapentaplegic* (*dpp*) gene, a member of the transforming growth factor-β family of growth factors. The vertebrate homologue of *dpp* is a bone morphogenetic protein, BMP-4. Six of the seven BMPs are members of the transforming growth factor-β superfamily. The seventh codes for a protease responsible for cleaving other BMP into active subunits. Low levels of *Dpp* in *Drosophila* are associated with ventral neural ectoderm, high levels with dorsal epidermis. *Sog* therefore promotes ventral development, in part, by preventing dorsalization by creating a functional gradient of *Dpp* across the embryos. *Dpp* is homologous

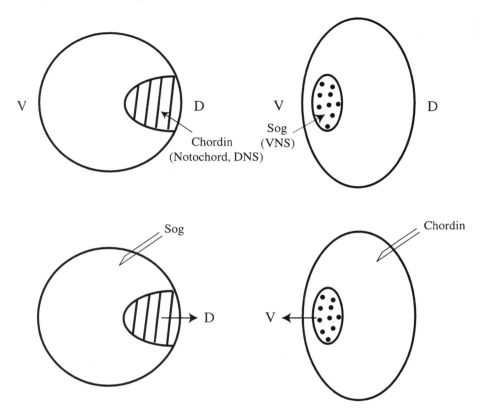

Fig. 9.5 The genes specifying the dorsal and ventral nervous systems (DNS, VNS) are conserved between *Drosophila* and *Xenopus* and, by implication, between arthropods and vertebrates. Chordin is expressed dorsally and specifies notochord and dorsal nervous system in *Xenopus* (top left), while sog is expressed ventrally and specifies ventral nervous system in *Drosophila* (top right). Sog mRNA injected into *Xenopus* specifies dorsal (bottom left), while chordin mRNA injected into *Drosophila* specifies ventral (bottom right).

to BMP-4, while BMP-4 in *Xenopus* promotes ventral development. When *dpp* is expressed in *Xenopus* it promotes the development of ventral structures and is antagonized by *sog*. Similarly, BMP-4 in *Drosophila* promotes dorsal structures.

Mutations in the genes *dino* and *swirl* disrupt D-V patterning in the zebrafish and in *Drosophila* (*swirl* dorsalizes zebrafish as does disruption of BMP-4; *dino* ventralizes as does overexpression of BMP-4), further linking the D-V patterning systems of flies and vertebrates. Human BMP rescues null *dpp* mutant embryos and confers normal D-V patterning.

Conserved genes and their antagonists function to specify the D-V axis in these diver-

gent organisms and do so by inverting the axis they specify, a remarkable illustration of how the function of the same gene is conserved yet modulated throughout evolution. Vertebrate embryonic development subsequent to axis specification is used in the next chapter to demonstrate the major developmental processes and genes involved in building embryos.[14]

9.6 ENDNOTES

1. See Weismann (1883a,b, 1885, 1889) for development of the theory of continuity of the germ plasm. For one aspect of somatic influences on germ cells, see the study by García-Castro *et al.*

(1997) on interactions between extracellular matrix glycoproteins and migrating primordial germ cells in mouse embryos.

2. For discussions of the mechanisms by which germ and body cells are generated and specified, see Illmensee and Mahowald (1974), Nieuwkoop and Sutasurya (1979), Maclean and Hall (1987), Buss (1988), and the issue of *Developmental Genetics* for which Eddy (1996) wrote the introduction.

3. See Gilbert (1994a) for a general discussion of germ plasm constituents. For the studies on transplantation and genetic control in *Drosophila*, see Illmensee and Mahowald (1974), Ephrussi and Lehmann (1992) and Lehmann and Ephrussi (1994). For the role of genes such as *oskar*, see Erdelyi *et al.* (1995).

4. See Nieuwkoop and Sutasurya (1979) and Hanken (1986) for comprehensive syntheses of germ cell determination in amphibians, and King (1996) for the molecular basis of germ cell determination.

5. Key papers establishing that maternal effect genes generate D-V and A-P axes in *Drosophila* are Nüsslein-Volhard and Wieschaus (1980), Anderson and Nüsslein-Volhard (1984), Drieuer and Nüsslein-Volhard, (1988a,b), MacDonald and Struhl (1986) and Holley *et al.* (1995).

6. See Baumgartner and Noll (1990), Hülskamp, Pfeifle and Tantz (1990), Cohen and Jürgens (1990) and Finkelstein and Perrimon (1990) for sequences of expression of the genes that specify the A-P axis in *Drosophila*.

7. See Sander (1994) for an evaluation of the correlation between oogenesis and pattern formation, and Brown, Parrish and Denell (1994) for genetic control of early stages of embryogenesis in *Tribolium castaneum*.

8. See M. M. Muller, Carrasco and de Robertis (1984) and Kappen, Schughart and Ruddle (1989a, 1993) for sequence similarity of homeotic genes.

9. For presence and sequences of homeobox genes in unsegmented metazoans, see Garcia-Fernàndez, Bagunà and Saló (1991), Kenyon and Wang (1991), Murtha, Leckman and Ruddle (1991), Degnan and Morse (1993), Salser and Kenyon (1994), Schierwater *et al.* (1991) and Balavoine and Telford (1995). See Duluc *et al.* (1997) for regionalization of the intestinal tract by *Hox* genes.

10. See P. W. H. Holland and Hogan (1986, 1988), Ford and Gottlieb (1992) and Vollbrecht *et al.* (1991) for studies on homeotic transformations; Paro (1990) and Grindhart and Kaufman (1995) for *Polycomb* as a regulator of transcription within imaginal discs; and Goodrich *et al.* (1997) for studies with *curly leaf* in *Arabidopsis*.

11. For analyses of the recent molecular and morphological work on formation of body plans in insects and vertebrates, see P. W. H. Holland and Graham (1995), Holley *et al.* (1995), Jones and Smith (1995), Arendt and Nübler-Jung (1996, 1997), De Robertis and Sasai (1996), Lacalli (1996a) and Holley and Ferguson (1997).

12. For conservation of *sog* and *chordin* in *Drosophila* and *Xenopus*, see Holley *et al.* (1995), De Robertis and Sasai (1996) and Holley and Ferguson (1997). For antagonism of *chordin* by BMP-4 and BMP-7, see Hawley *et al.* (1995).

13. Understanding of cortical rotation comes primarily from the laboratory of Rick Elinson in Toronto (Elinson and Rowning, 1988; Elinson and Pasceri, 1989; Elinson, King and Forristall, 1993; Holowacz and Elinson, 1993) and John Gerhart in Berkeley (Gerhart *et al.*, 1989, 1991). For an overview, see King (1996).

14. For levels of *Dpp* and ventralization or dorsalization, see Ferguson and Anderson (1992), Arendt and Nübler-Jung (1994) and Ferguson (1996). For *dino* and *swirl* mutants in *Danio rerio* and *Drosophila*, see Hammerschmidt, Serbedzija and McMahon (1996). For rescue of *Drosophila* embryos with human BMP, see Padgett, Wozney and Gelbart (1993). For an overview of BMPs, see Kingsley (1994a).

10

BUILDING VERTEBRATE EMBRYOS:
HEADS AND TAILS

'Development is therefore the immediate cause of introduction of variation at the level of the individual organism . . . Central to understanding the role of development in evolutionary mechanisms must be the study of the emergent and epigenetic properties of developing systems and their unique role in the processes by which variation is introduced among individual phenotypes.'

Thomson, 1988, pp. 16–17

Differentiation and morphogenesis of the soma of vertebrate embryos is epigenetic and hierarchical. Early events, as determined by experimental investigation of amphibian embryos, introduced in section 8.3 and discussed in this chapter, are:

1. establishment of the primary embryonic axis at the time of sperm entry into the egg;
2. dorsalization of the zygote and establishment of the D-V axis (ss. 9.4, 9.5);
3. induction and specification of the mesoderm (ss. 10.1–10.3);
4. differentiation of the chordamesoderm (the future notochord);
5. induction of neural ectoderm from presumptive ectoderm by the chordamesoderm;
6. establishment of the neural axis and morphogenesis of the neural tube (s. 10.2);
7. origin of the neural crest at the boundary between neural and epidermal ectoderm (s. 10.4); and
8. a process of secondary neurulation and tail-bud formation that is fundamentally different from how cranial and trunk regions form.[1]

The processes that bring about these developmental events are diverse, including inheritance of maternal gene products distributed in gradients across the egg, responsiveness to gravity, rotation of cortical cytoplasm, activation of mesoderm-inducing factors, morphogenetic movements of developing germ layers, and inductive interactions. These processes also set up the three-dimensional topography required for further secondary and tertiary inductions to initiate the differentiation of yet further cell types. Secondary interactions are first seen in the development of the caudal regions and tail buds of vertebrate embryos (s. 10.5). Consequently, these regions do not arise by primary differentiation from germ layers, but from a process of secondary neurulation and origin from epithelial and mesenchymal cells. Heads and tails do not arise in the same ways.

10.1 MESODERM INDUCTION

Although the gastrula contains the three germ layers – ectoderm, mesoderm, and endoderm – only future ectoderm and endoderm exist in the early blastula (see Figure 5.3). Mesoderm arises by sequential inductions within the early blastula.

Ectoderm forms those structures on the outsides of embryos and adults, such as the skin, as well as the brain and spinal cord of the central nervous system, which develop by a sinking in of neural ectoderm from the future dorsal surface. Endoderm forms the alimentary canal and structures such as the thyroid gland, lungs, liver and pancreas that bud from it. Many, but not all, of the remaining embryonic structures, essentially those in between the outer ectoderm and inner endoderm, form from mesoderm. These include the skeleton, muscles, blood vessels, blood cells, gonads, heart and kidneys. An important exception to this mesodermal origin occurs in the head, where many skeletal and connective tissues develop from an ectodermal derivative, the neural crest, whose cells initially lie in the folds of the developing nervous system (s. 10.4).

10.1.1 Sequential induction

The extensive studies by Slack, Cooke, and their colleagues, using injection of fluorescein-dextran-amine into individual blastomeres, have enabled us to visualize two inductive interactions pivotal to establishment and organization of mesoderm in *Xenopus*:

1. The vegetal half of the blastula induces equatorial cells to become mesoderm in a **mesodermal induction** (Figure 8.2).
2. During gastrulation and under the influence of the dorsal lip of the blastopore, mesoderm is regionalized into notochord, somites, pronephros and lateral plate mesoderm, each progressively more ventral and lateral to the dorsal embryonic axis. This is **dorsalization of the embryo**.[2]

Mesoderm induction thus occurs before the rearrangement of cells at gastrulation, illustrating how early in development epigenetic processes operate and how complex are the interactions involved in forming individual embryonic elements.

Ectoderm and endoderm are in contact around the equator of the zygote but not across the blastocoele (Figure 8.2). Endoderm signals to marginal ectoderm but not to apical ectoderm; interaction is blocked by the blastocoele. The induction involves both activation of muscle-specific genes and repression of epithelial genes, i.e., both positive and negative signals accompany induction. As discussed in more detail in section 11.6, induction is not the simple transfer of a single positive signal from inductor to responding cells.

The hope that a single mesodermal inducer would be found has turned out to be a faint hope indeed. Multiple inducers have to be considered, and various types of mesoderm are induced. The cells forming these regions of mesoderm are not naive at the outset of induction; specificity of induction of dorsal and ventral mesoderm, notochord, axial and head mesoderm rests as much with the responding cells as it does with the cells providing diffusible inductive signals.[3]

Mesoderm induction is due to the activity of growth factors in the TGF-β family and to bFGF, which are localized within the presumptive endoderm (s. 10.1.2). Basic FGF, TGF-β2 and -β3, activin-A (a long-range signalling member of the TGF-β superfamily) and BMP-4 (a short-range signalling member of the same superfamily) are all involved in mesoderm induction (Figure 10.1). These growth factors act as switches, activating such DNA-binding proteins as homeodomain proteins.[4]

FGF and FGF mRNA have been localized in blastulae, but FGF cannot normally act alone; exogenous FGF only induces vegetal-posterior mesoderm and not dorsal mesodermal structures such as notochord (Green *et al.*, 1990). FGF receptors are localized in the early embryo, including the apical ectoderm, from which they are lost unless apical ectoderm is exposed to FGF.

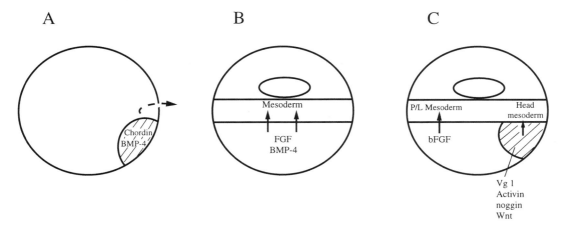

Fig. 10.1 Mesoderm formation as seen in *Xenopus*. A. Chordin and BMP-4, localized in the Nieuwkoop centre (hatched) specify the dorsal side of the embryo (dashed arrow). B. FGF and BMP-4 are involved in specification of marginal cells as mesoderm, with BMP-4 specifying ventral mesoderm. C. Molecules within the Nieuwkoop centre (hatched), including Vg-1, activin, noggin and Wnt, specify head mesoderm. Basic FGF (bFGF) is involved in specification of postero-lateral (P/L) mesoderm.

Sequentially, three inducing centres play a role in setting up the conditions required for induction of the various types of mesoderm:

1. Immediately after fertilization and in association with cortical rotation of the egg cytoplasm, a maternal determinant moves from the vegetal pole to one side of the egg. The reciprocal activities of *chordin* and BMP-4 (s. 9.5) specify the dorsal side of the embryo and the plane of bilateral symmetry, and define a vegetal centre known as the **Nieuwkoop centre** (Figure 10.1).
2. The Nieuwkoop centre of the early blastula signals to cells in the dorsal marginal zone. These cells form head mesoderm; the Nieuwkoop centre is thus a **head organizer**. Specification of blastomeres as marginal zone is controlled by bFGF, as is specification of posterior and lateral mesoderm, over-riding signalling from BMP-4, which otherwise specifies ventral mesoderm (Figure 10.1).
3. The head organizer signals cells immediately adjacent to it, specifying these cells as notochordal and as the centre for organiza-

tion of the body axis. Head and body axis organizers together make up the organizer discovered by Hilda Mangold and Hans Spemann that earned Spemann the 1935 Nobel prize for physiology and medicine.[5]

A continuum of body plans with varying proportions of D-V structures is produced when *Xenopus* cleavage-stage embryos are exposed to lithium. This procedure has therefore been used to understand mesoderm induction. Lithium evokes additional dorso-anterior mesoderm from what would otherwise have been ventral equatorial ectoderm. This additional or enhanced induction accompanies respecification of the D-V axis, enhanced expression of anterior neural markers, and diminished expression of such posterior neural markers as *XlHbox-6*. As with induction of mesoderm, there is positive and negative control of gene expression.[6]

10.1.2 Molecular basis and responses

There is a hierarchy of interactions in the induction of mesoderm. Multiple growth factors maintain synthesis of the appropriate receptors; bind to the receptor; and activate the synthesis of homeodomain proteins, which in turn bind selectively to DNA to activate downstream elements of the genetic programme for mesoderm formation.

(a) Head mesoderm

Maternal activin is localized in the egg, and blastomeres require activin to promote dorsal development (Figure 10.1). Embryonic amphibian ectoderm treated with activin forms a secondary embryonic axis. Activin-A induces mesodermal derivatives such as notochord and muscle with D-V polarity as a dose-dependent specification of cell fate and position. Activin-B appears earlier in development (at the blastula stage) than does activin A and at the site expected for an endogenous mesoderm inducer. Finally, activin B induces the primary body axis in *Xenopus* and in the chick.

Vg-1, another member of the TGF-β superfamily, appears to be a more potent mesoderm inducer than activin and is currently a more likely candidate as an endogenous inducer of dorsal mesoderm. Homologues of *Vg-1* have been found in the zebrafish, chick and mouse. When grafted into the periphery of the epiblast, the chick homologue, *cVg-1*, induces an ectopic primitive streak. Such results support conservation of signalling for axis formation across the vertebrates.[7]

The *wingless* (*wnt*) family of ligands certainly plays a role, as demonstrated by experiments in which injection into ventral blastomeres of mRNA for the *Xenopus wnt* genes *Xwnt-1* and *Xwnt-8*, induces a second body axis and head structures, and rescues ventralized embryos. These results mimic transplantation of blastomeres containing the Nieuwkoop centre. *Noggin* also rescues ventralized embryos and *noggin* mRNA is distributed in the blastula at the appropriate time (Figure 10.1). *Wnt* is, however, involved in

specification of portions of the neural axis, recent findings indicating that *Wnt* is inhibited by *Cerberus* and that that inhibition of *Wnt* and BMP are required for head induction (s. 10.2.2).[8]

One recently discovered gene regulating head development is *Cerberus*, which is localized within the anterior endomesoderm of the organizing centre of *Xenopus* embryos, where it suppresses formation of trunk or tail mesoderm. The important role of *Cerberus* in head formation was demonstrated when microinjection of *Cerberus* mRNA into *Xenopus* embryos elicited ectopic heads (and duplicated hearts and livers). Other genes such as the homeobox genes *Lim-1*, *Otx-1* and *goosecoid*, which are expressed in the organizer, also play important roles. Deletion of these genes in mice eliminates all head structures rostral to the posterior hindbrain without affecting caudal hindbrain and spinal cord.[9]

(b) Response to induction

The molecular basis of the response to these inductive signals (the ligands that bind them) is being established in ongoing studies as the early transcriptional activation and/or upregulation of gene expression is uncovered.

Goosecoid, a transcriptional factor and one of the earliest genes identified so far, is expressed at the appropriate time and place in normal embryos; is not expressed in embryos that lack a Nieuwkoop centre; and is activated in response to *Xwnt* and *activin*. As activation of *goosecoid* is an early response to induction, *goosecoid* may be involved in specification of the dorsal body axis. Slightly later in development *goosecoid* inactivates *Xwnt-8* expression, further supporting negative and positive elements to mesoderm induction. Removal of *Xwnt-8* may facilitate formation of ventral or posterior mesoderm (Christian and Moon, 1993).

Brachyury, a gene required for notochord induction, is activated by bFGF and activin and so is another candidate gene in the early response to induction of dorsal mesoderm. This is an ancient pathway; bFGF induces *Brachyury* and notochord formation in the ascidian *Halocynthia roretzi* (Nakatani *et al.*, 1996).

Subsequent inductions do not occur until gastrulation, when movements of embryonic regions produce the extensive areas of contact required for inductive interaction. The development of these extensive associations allows previously separated layers of the embryo to associate, interact and produce new cell types, which are themselves capable of generating further cell types through subsequent cascades of interactions. The most obvious of these cascades is induction of neural ectoderm, formation of neurons, specification of the central nervous system, and its organization along the primary body axis.

10.2 NEURAL INDUCTION

Morphogenetic movements during gastrulation bring the chordamesoderm (future notochord and mesoderm) and the overlying ectoderm into contact. Interaction between these two layers begins during gastrulation, following specification of ectoderm as neural ectoderm at the blastula stage (Figure 8.2). The interaction has at least three consequences for neural development:

1. induction of nerve cell differentiation during neurulation;
2. organization of those neurons into a neural tube, beginning during neurulation; and
3. morphogenesis and regionalization of that neural tube into a central nervous system consisting of brain and spinal cord and further subdivision of the brain into fore-, mid- and hindbrain.

This set of processes is called **primary embryonic induction**, not because it is the first induction, but because it establishes the primary embryonic axis centred on the nervous system and notochord.

Just as some ectoderm does not come in contact with endoderm during cleavage and so is not induced to become mesoderm, so some ectoderm does not come in contact with notochord during gastrulation and is therefore not induced to become neural ectoderm. Instead, it becomes epidermal ectoderm and forms the epidermis. This is not because epidermal ectoderm is incapable of undergoing neural differentiation; it does not have the opportunity to do so, just as animal cap ectoderm *in situ* does not have the opportunity to become mesoderm. If brought into association with notochord by experimental manipulation, or by some deviation in normal development, future epidermal ectoderm is quite capable of transforming into neural ectoderm, forming neurons and organizing those neurons into a central nervous system. This was the classic experiment performed by Spemann and Mangold (1924) in which the notochord was identified as the primary embryonic organizer.[10]

The textbook story is that epidermal differentiation is the default or ground-state condition, the tissue type that ectoderm becomes if not induced to form something else. This nice tidy tale is changing. Neural ectodermal fate has to be inhibited and epidermal ectodermal fate induced (by growth factors such as BMP-2, -4 and -7) in what may well be a rather complex series of interactions involving activation of *msx-1* (Suzuki *et al.*, 1997a,b).

Much is now known about molecular control of the formation of neural ectoderm and of primary embryonic induction. Just as positive and negative regulation is involved in specification of mesoderm, so positive and negative regulation is involved in specification of ectoderm as neural. One negative regulator is activin. Ectoderm takes on a neural fate if activin signalling is inhibited with a natural inhibitor such as follistatin or if activin is bound to its receptor. Another negative regulator, BMP-4, is, like activin, a member of the TGF-β superfamily of growth factors. BMP-4 is expressed in the ectoderm of *Xenopus* blastulae and is a ventralizing factor, inhibiting the action of activin in promoting dorsal mesoderm formation and inhibiting neuralization of ectoderm. Unlike activin, which diffuses over many cell diameters, BMP-4 signalling is short-range and involves *XVent-2*, a novel homeobox gene. Conservation of the dorsalizing role of BMP-2 and BMP-4 is supported by studies demonstrating expression of both molecules in zebrafish gastrulae and that overexpression of BMP-2 or BMP-4 eliminates dorsal mesoderm and notochord.[11]

It is unclear whether inhibition of BMP-4 is a sufficient signal for ectoderm to assume a neural fate, or whether a positive signal(s) also is required. Studies undertaken in the late 1980s and early '90s, in which ectodermal caps were dispersed and the cells maintained at low concentrations, demonstrated neuralization of these cells (Godsave and Slack, 1991). This is, of course, an artificial condition in which normal intercellular signalling is prevented. Nevertheless, it does speak to an autonomous neural differentiation within these cells and to much more intricate inductive mechanisms than suggested by the standard textbook story.

10.2.1 Properties of inductive interactions

An important attribute of these inductive interactions, implicit in the preceding discussion, is the specificity and non-overlapping nature of the differentiative end points. Reduction in the strength or duration of an induction does not result in induction of a tissue that is partly neural, but rather the induction of a smaller amount of neural tissue. Similarly, there is no grading-off of neural differentiation at the boundary between neural and epidermal ectoderm, with cells becoming less and less neural until they become epidermal. A sharp discontinuity exists at the boundary. Adjacent to the most lateral neural cell lies the most medial epidermal cell. The responding cells have two alternate cell states, neural or

epidermal. Embryonic inductions act as switching mechanisms, setters of thresholds, or establishers of boundary conditions. Some examples of how modifications of such interactions bring about evolutionary change in morphology are discussed in the chapters of Part Two.

The ability of future epidermal ectoderm to respond to notochordal induction (the embryological property of **competence**) is not retained through embryonic life, but is lost by the end of neurulation. Similarly, notochord is not always able to induce neural ectoderm; blastula ectoderm cannot respond to notochordal induction until notochord and ectoderm have reached a stage equivalent to early gastrulation. Nor can notochord from post-neurula-stage embryos induce. This temporal association of acquisition and loss of ability to induce (or respond to an induction), is typical of embryonic inductions.

Such data on the timing of primary embryonic induction reveal important general properties of embryonic inductions:

1. the ability to induce or to be induced is gained and lost during development;
2. only some embryonic regions act as specific inductors and only some regions are competent to respond (a spatial limitation); and
3. inductions only occur at particular times during development set by the combination of inductive capability and competence (a temporal limitation).[12]

Induction is therefore limited spatially and in its timing, ensuring that the right organ is formed in the right place at the right time during development – and probably also of the right size, for strength and/or duration of induction and the number of competent cells determine how much of a given cell type will form.[13]

The potential for modification of morphology and generation of new morphologies in evolution is realized when inductive interactions break these temporal and spatial bounds (Chapter 16). Spatial localization and temporal limitations of induction provide potential mechanisms for developmental constraint, heterochronic change, the multiplication of specific embryonic regions, increase or decrease in organ size, and development of novel tissues and organs during ontogeny and phylogeny, all of which are topics taken up later in the book.

10.2.2 Regionalization of the central nervous system

In response to neural induction, not only do neural cells differentiate from ectoderm, but neural ectoderm is organized into the major regions of the central nervous system (brain and spinal cord), and within the brain into fore, mid- and hindbrain.

Formation of neural cells represents **cell differentiation**. Organization of those cells into specific regions of the nervous system is an example of **regionalization**, one aspect of morphogenesis. When taken at any time during gastrulation notochord can evoke neural differentiation from competent ectoderm. Regionalization differs with the age of the gastrula providing the notochord. Invaginating notochord from early gastrulae specifies forebrain, while notochord from late gastrulae specifies spinal cord. Embryonic inductions therefore control differentiation and morphogenesis.[14]

Neural induction occurs in two steps. The first specifies neural cell differentiation, the second regionalization. This two-step induction is mediated by different cellular processes: regionalization requires direct cell-to-cell contact between notochord and neural ectoderm; neural differentiation does not (Nieuwkoop, Johnen and Albers, 1985). Three primary candidate secreted proteins – *noggin*, *chordin* and follistatin – are expressed in the organizer, bind to ventralizing signals such as BMP-2 and BMP-4, and induce anterior neural markers (ss. 9.5, 10.1.2).

Our understanding of how the neural axis is patterned into forebrain (which may or may not be segmented), midbrain, segmented hindbrain with its rhombomeres, and spinal cord (in which hints of segmentation are emerging), has increased immensely over the past decade with the identification of the *Hox* codes that pattern the vertebrate head. It is tempting to follow Gould (1986) and look to the homeotic genes as the genetic embodiment of serial homology. A series of findings on organization of the hindbrain in chick embryos takes us down this path:

1. the hindbrain is segmented into serially repeated units (**rhombomeres**);
2. rhombomere boundaries coincide with boundaries of expression of homeobox genes;
3. cells within individual rhombomeres behave as independent cellular units much as do the compartments in insects; and
4. similar patterns of expression of homeobox genes and rhombomeres are found in the mouse hindbrain.[15]

The source of the signals that pattern neural and mesodermal tissues is different in the hindbrain/branchial arch region and spinal cord/trunk. (Little is known about how the forebrain is patterned.) Hindbrain and branchial arches are patterned from signals that arise in the neural tube and are transmitted to the mesoderm, neural crest and epidermal ectoderm (s. 15.4). *Wnt*, the product of which is a secreted protein, along with the transcription factors *engrailed*, *Pax-2* and *Pax-5*, and the growth factor FGF-8, play important roles in establishing the boundary between mid- and hindbrain later in development. Indeed, in embryonic chicks and mice, this boundary acts as an organizing centre for that region of the brain.[16]

Homeotic genes are specifically activated by growth factors and by retinoic acid (RA), a molecule known to play a major role in embryonic patterning. As examples:

1. FGF plays a major role in mesoderm induction, including activation of a homeodomain protein (s. 10.1).

BOX 10.1
NOMENCLATURE OF HOX GENES

The nomenclature of *Drosophila* homeotic and vertebrate *Hox* genes is horrendous, especially for those making their first foray into the literature.

A committee convened at a 1992 international meeting attempted to rationalize the nomenclature (Scott, 1992). *Hox* is used exclusively for vertebrate genes that are related to the *Drosophila* Antennapedia and Bithorax homeotic genes. The four clusters of mouse *Hox* genes are known as *Hoxa*, *Hoxb*, *Hoxc* and *Hoxd*. Genes within clusters are numbered from 1 to 13, where 1 is 3′ and has the most anterior limit of expression along the A-P body axis, and 13 is 5′ and has the most posterior limit of expression. Lower case a–d is used for mouse genes, upper case A–D for human. Thus, *Hoxa-1* and *HoxA-1* are equivalent (orthologous, Box 9.1) mouse and human genes.

As our knowledge increases, *Hox* designations have been removed from some genes. *Hox-7.1* is now *Msx-1* and *Hox-8.1* is *Msx-2*, a new family of mouse homeobox genes that are not part of the *Hox* cluster. The name change reflects the fact that *Msx* genes are more closely related to *Drosophila msh* (muscle segment homeobox) than to Antennapedia and Bithorax genes. The nomenclature for such vertebrate genes is that the first two letters are taken from the homologous *Drosophila* gene (hence *Ms*) and an *x* added to signify a vertebrate gene (hence *Msx* in vertebrates for *Msh* in *Drosophila*). Italics are used for specific genes but not for classes of genes or for proteins produced by the genes.[19]

2. FGF selectively activates posterior-acting homeotic genes.
3. TGF-β selectively activates anterior-acting homeotic genes. RA 'induces' anterior neural tissue to become posterior.
4. In *Xenopus,* RA suppresses differentiation of anterior mesoderm in favour of posterior or ventral mesoderm and induces the expression of the posterior homeobox gene *XlHbox-6* but not *XlHbox-1*, which is expressed anteriorly.

These findings suggest involvement of RA in regionalization of the nervous system and mesoderm and in specification of the primary body axis; coupling of mesoderm and neural tissue is discussed further in the next section. Specification of the primary body axis is undoubtedly coupled to primary embryonic induction; anterior notochord preferentially elicits expression of anterior homeobox gene proteins such as *En-2* in *Xenopus.*[17]

In addition to these studies on vertebrate embryos, a role for RA in the induction of *Hox* genes has now been demonstrated in amphioxus and a tunicate (s. 15.2). When embryos in all three groups (vertebrates, cephalochordates and urochordates) are exposed to RA, *Hox-1* genes are expressed more anteriorly in the nervous system than their normal expression boundaries (see Box 10.1 for the nomenclature of *Hox* genes). Such a functional connection between RA and *Hox* genes suggests that, as discussed in Chapter 15, evolution of regulation of *Hox* genes by RA may have been a precondition for the evolution of the chordate dorsal neural tube from a flattened neural plate.[18]

The injection of homeotic gene DNA into mouse embryos to produce transgenic mice leads to ectopic expression of the gene and the formation of additional structures, such as an extra anterior vertebra, providing presumptive evidence for a role for homeotic genes in specification of the axial skeletal structures of the vertebrate body plan. Such studies, along with the anterior-posterior graded distributions of *XlHBox-1* in developing limb buds in *Xenopus*, of *Hox* genes in feather germs along the avian body axis and specification of the A-P organization of the mammalian intestinal tract, indicate that homeobox gene proteins play a role in anterior-posterior patterning (regionalization) that extends well beyond the establishment of the primary body axis.[20]

10.3 REGIONALIZATION OF MESODERM AND COUPLING OF MESODERMAL AND NEURAL TISSUES

The notochord is the primary embryonic inducer, and the induction of neural ectoderm lays down the primary embryonic axis. But further structures arise in relation to this central axis. This is especially evident in the development of paired elements such as the somites that presage the vertebrae, and paired organ rudiments such as left and right limb buds and primordia of the gonads, kidney, lung and heart. (Although the heart is a single, midline structure, it arises as left and right precardiac rudiments that subsequently migrate to the midline to form the left and right halves of the heart.)

The notochord, as primary organizer, couples neural induction and the formation of mesodermal organs. RA-mediation of *Hox* gene patterning is currently the most promising molecular mechanism for patterning the anterior regions of the nervous system and perhaps the posterior as well (s. 10.2.2). In the trunk, mesoderm imposes anterior-posterior polarity onto the spinal cord. A novel perspective on conserved patterns across the vertebrates emerged from the comparison made by Tam and Quinlan (1996) of fate maps of zebrafish, *Xenopus*, chick and mouse embryos, representing four of the five classes of vertebrates. Tam and Quinlan stress that there is a striking similarity (a striking homology, to use their phrase) between these fate maps when they are compared to the position of the primary organizer. Future dorsal mesoderm lies closest, and future ventral mesoderm furthest from the organizer in all species. Similarly future neural and epidermal ectoderms are closest and furthest from the organizer respectively.

A gradient of induction that spreads anteroposteriorly from the notochord (and which is responsible for induction of the spinal cord) spreads laterally away from the central axis and is responsible for specification of the major regions of the embryonic mesoderm. Mesoderm closest to the axis is specified as paraxial (somitic) and subsequently segments into pairs of somites. Paraxial mesoderm divides into three components (Figure 10.2):

1. sclerotome, from which the vertebrae arise;
2. myotome, from which body muscle arises; and
3. dermatome, from which connective tissue arises.

Sclerotome is determined under notochordal control. A key player is the product of the gene *sonic hedgehog*, which is expressed in the notochord and the ventral floor (floor plate) of the developing neural tube. *Shh* plays multiple patterning roles in the development of disparate organs. It ventralizes the neural tube, induces neuronal differentiation, and specifies axial mesoderm as sclerotome. Later in development, *Shh* is expressed at the posterior margin of the limb buds, where it polarizes limb mesenchyme and maintains the specialized apical ectoderm ridge (s. 11.4.3). *Shh* is also involved in development of some aspects of the eye; it is expressed in the ventral midline in association with the single median eye in zebrafish embryos carrying the *cyclops* mutation.[21]

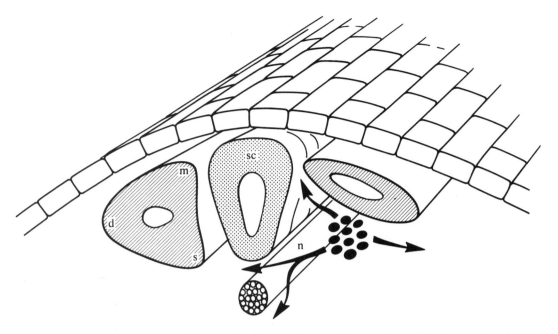

Fig. 10.2 A diagrammatic view of an early chick embryo with spinal cord (sc), notochord (n), and unsegmented somite on the left consisting of future dermatome (d), myotome (m) and sclerotome (s). Migrating sclerotome (arrows) on the right is shown as it appears in an older embryo.

As with the specification of mesoderm and neural ectoderm discussed earlier, lateral somitic mesoderm is regulated by antagonistic, diffusible signals. It appears from studies on early avian embryos that specification of the myotome and dermatome is directed by BMP-4 from lateral plate mesoderm and by several members of the *Wnt* gene family: *Wnt-1* and *Wnt-3a* from the dorsal neural tude and *Wnt-4* and *Wnt-6* from the surface ectoderm.

Mesoderm immediately lateral to the somitic mesoderm is specified as intermediate mesoderm from which the primordia of the kidneys form. More lateral mesoderm – lateral plate mesoderm – subsequently splits into two layers. One is associated with ectoderm and produces the dermis and superficial muscles of the body; the other is associated with the endoderm and forms the smooth muscle and connective tissue covering of the gut.[22]

10.4 THE NEURAL CREST AND NEURAL CREST CELLS

In 1868 the Swiss embryologist Wilhelm His discovered a band of cells situated between the neural tube and epidermal ectoderm in the embryonic chick, and named it *Zwischenstrang*. We know it as the neural crest and its cells as the neural crest cells that arise in, and subsequently migrate away from, the developing brain and spinal cord (Figure 10.3). The range of cell types derived from neural crest cells and the number of tissues and organs to which they contribute is enormous (Table 10.1).[23]

The neural crest was first studied for its role in the formation of pigment cells and its contribution to the nervous system in amphibians. Work in the 1960s by Jim Weston and Mac Johnston on migration of neural crest cells in chick embryos, and Nicole Le Douarin's dis-

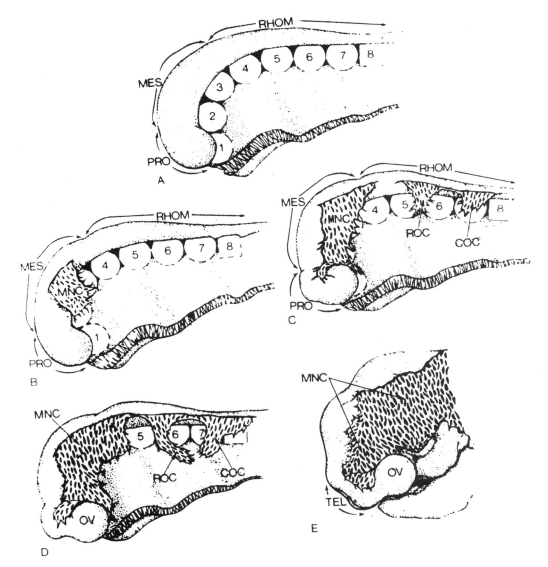

Fig. 10.3 Migration of cranial neural crest cells in the North American snapping turtle, *Chelydra serpentina*. (A) pre-migration showing the eight mesodermal somitomeres (1–8) and the major subdivisions of the brain (PRO, prosencephalon; MES, mesencephalon; RHOM, rhombencephalon). (B) Mesencephalic neural crest cells (MNC) are the first population to migrate from the neural tube. (C) Rostral and caudal rhombencephalic neural crest cells (ROC, COC) then begin their migration as increasing numbers of mesencephalic cells migrate more caudally. These three migrating populations gradually coalesce (D) to provide the bulk of the head mesenchyme (E). OV, optic vesicle. Reproduced from Meier and Packard (1984) with the permission of the Publisher.

Table 10.1 Cell types derived from the neural crest or tissues and organs that contain neural-crest-derived cells

Cell types	
Sensory neurons	Cholinergic neurons
Adrenergic neurons	Rohon-Béard cells
Satellite cells	Schwann cells
Glial cells	Chromaffin cells
Parafollicular cells	Calcitonin-producing (C) cells
Melanocytes	Chondroblasts, chondrocytes
Osteoblasts, osteocytes, fibroblasts	Odontoblasts
Striated myoblasts	Cardiac mesenchyme
Mesenchymal cells	Smooth myoblasts
Angioblasts	Adipocytes

Tissues or organs	
Spinal ganglia	Parasympathetic nervous system
Sympathetic nervous system	Peripheral nervous system
Thyroid gland	Ultimobranchial body
Adrenal gland	Craniofacial cartilages
Craniofacial bone	Tooth papilla
Dentine	Connective tissue
Adipose tissue	Smooth muscles
Striated muscles	Cardiac septa
Dermis	Eye
Cornea	Endothelia
Blood vessels	Heart
Dorsal fin	Brain
Connective tissue of glands (thyroid, parathyroid, thymus, pituitary, lachrymal)	

covery of the quail nuclear marker and its exploitation in chick-quail chimeras, set off an avalanche of research on the neural crest that has played an important role in forging the new synthesis of developmental and evolutionary biology, for the origin of the vertebrate head and the elaboration of craniofacial skeletal and connective tissues is intimately bound up with evolution of the neural crest, placing the neural crest at centre stage in the vertebrate evolutionary play.[24]

10.4.1 Embryonic origin of neural crest

How does the neural crest arise in embryos? The site of origin is clear; it arises in neurula stage embryos in the neural folds at the junction of neural and epidermal ectoderm (Figures 8.5, 10.3). Neural crest and neural, epidermal and placodal ectoderm all arise within the neural folds. This intimate association between those presumptive areas does not, however, arise at neurulation. Presumptive neural crest lies at the boundary of presumptive epidermal and neural ectoderms in the early amphibian blastula and chick epiblast (Figure 10.4).[25]

I used this and other evidence, coupled with the existence of secondary neurulation (s. 10.5) to argue that the neural crest fits the definition of a germ layer, that the neural crest is therefore a fourth germ layer, and that organisms possessing a neural crest – vertebrates – are tetrablastic (Hall, 1997a). The essence of the argument is outlined in Box 10.2 on p. 168.

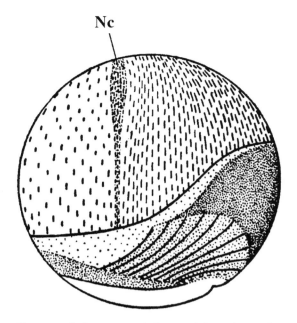

Nc

Fig. 10.4 A fate map of the urodele embryo (late blastula) to show the position of future neural crest (Nc) between epidermal ectoderm on the left and neural ectoderm on the right. Modified from Hall and Hörstadius (1988).

Which events determine that cells at the neural/epidermal ectoderm boundary will become neural crest? Because of their position, and because they produce neurons and ganglia, neural crest cells are regarded as derivatives of neural rather than epidermal ectoderm. Although such neural-crest derivatives as pigment cells and mesenchyme can arise in the absence of neural derivatives, neural tissue is almost always accompanied by neural crest derivatives when induced experimentally. Ectodermal (neurogenic) placodes lying adjacent to neural ectoderm also give rise to (sensory, lateral line) neurons and to some of the receptor cells (neuromasts) of sensory systems. The neural crest, neural plate, neural tube and all differentiated neuronal cells express the same cell surface glycoprotein, the neural cell adhesion molecule N-CAM. Epidermal ectoderm does not.

(Placodes express N- and L-CAM; see s. 11.2.1.)[26]

10.4.2 Induction of neural crest

Does the association of neural crest with neural ectoderm and shared CAM expression patterns, mean that the neural crest (like neural ectoderm) arises as a response of ectoderm to induction? Alternatively, is the neural crest set aside as a determined 'tissue' earlier or later in development?

Models of initiation of the neural crest based on inductive interaction between epidermal and neural ectoderm have been confirmed experimentally in *Xenopus* and the axolotl *Ambystoma mexicanum*. Neural crest forms at the neural-epidermal boundary, whether that boundary is in the normal location *in vivo* or created ectopically by transplantation of neural tube into epidermal ectoderm. Differentiation of mesenchyme, which has been used as the marker for differentiation of neural crest, may not be adequate to distinguish neural crest from mesoderm induction (Box 10.3).[28]

Neural crest arises between neural and epidermal ectoderm precisely because this junction is where neuralizing and epidermalizing influences meet. Thus, when future epidermal ectoderm from a gastrulating amphibian embryo is grafted in place of the future neural crest of a neurula-stage embryo, the grafted ectoderm forms neural folds which respond to the combined 'neurularizing' induction of the chordamesoderm and 'epidermalizing' induction of the lateral mesoderm by differentiating as neural crest cells (Rollhäuser-ter-Horst, 1980). The hierarchy long known for mesoderm (notochord–paraxial–intermediate–lateral plate) may be paralleled by an ectodermal hierarchy (neural–neural crest–[placodal?]–epidermal).

The classic explanation for these associations was strength of induction. The median roof of the archenteron (presumptive notochord) is a stronger inducer and therefore induces neural structures. The lateral

BOX 10.2
THE NEURAL CREST AS A FOURTH GERM LAYER AND VERTEBRATES AS TETRABLASTIC

Evidence for the neural crest as a fourth germ layer and for vertebrates as tetrablastic rather than triploblastic, is discussed in some detail in Hall (1997a). A précis of the arguments is included here.

Ectoderm and endoderm are primary germ layers. They are the only germ layers in diploblastic organisms, and the only germ layers in the unfertilized egg and newly fertilized zygote of triploblastic organisms. Mesoderm is a secondary germ layer, arising from equatorial cells of the amphibian blastula by induction from the adjacent endoderm (s. 10.1). Ectoderm and endoderm are maternal in origin, the products of maternal cytoplasmic control over oogenesis. Mesoderm and neural crest, on the other hand, are the products of inductive interactions after fertilization, i.e. both arise secondarily in the zygote.

Thus the three germ layers recognized for the past almost 180 years, on the dual bases of timing and methods of development, can be replaced by four germ layers, two of which are primary (ecto- and endoderm) and two secondary (mesoderm, neural crest). As might be expected from a secondary layer which is phylogenetically derived, a single evolutionary origin for mesoderm cannot be supported rigorously. Nor does a single developmental mechanism explain the origin of mesoderm during ontogeny. Neural crest, like mesoderm, arises from a primary germ layer after secondary induction.

The integrity of the neural crest as a definitive embryonic layer of pluripotent cells is outlined in section 10.4.1. Clinicians recognize this integrity in neurocristopathies, which are pathological states or syndromes united by the affected tissues and organs either arising from, or being dependent on, the neural crest. Common neuro-cristopathies include neuroblastoma (adrenal medulla, autonomic ganglia), mandibulofacial dysostosis (facial skeleton, ears), and otocephaly (absence of lower jaws, ear defects, heart anomalies).[27]

The neural crest is a vertebrate synapomorphy. If the neural crest is a fourth germ layer, then vertebrates are tetrablastic (Figure 10.5). The evolution of secondary germ layers and evolutionary transformations at the germ-layer stage in developmental mechanisms, such as the origin of secondary neurulation, s. 10.5), illustrate the dynamic nature of early embryonic development and that early development is not immutable to phylogenetic change, even when that change affects something as basic as germ layer organization.

archenteron roof, which contains future lateral mesoderm, is a weaker inducer and therefore induces neural crest (Figure 10.6). This interpretation rests on the threshold for induction of neural tissue being higher than that for neural crest and therefore requiring stronger in- duction. Although there is much experimental evidence for a graded distribution of neuralizing inducer, the explanation for neural-crest induction may, however, be exactly the reverse and lie in differential competence, not differential inductive signaling. Loss of

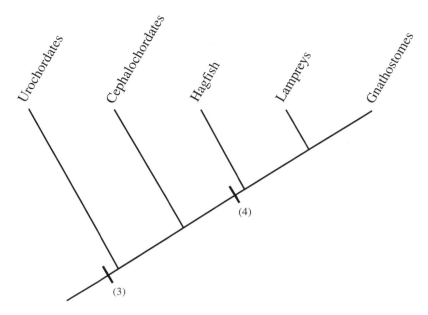

Fig. 10.5 The origin of the neural crest as a fourth germ layer (4) separates triploblastic (3) urochordates/ cephalochordates from tetrablastic craniates (hagfishes, lampreys, gnathostomes). Only extant groups are shown.

ectodermal competence, spread of neural induction along the ectoderm, and formation of placodes in association with weak competence at the boundary, have all been demonstrated.[31]

Induction of an organized nervous system results from the cascade of signals described in section 10.2. **Homoeogenetic induction** is when induced cells function to induce further cells of the same type. Neural induction proceeds homeogenetically in *Xenopus*; additional neural tissue is induced from ectoderm by already-induced neural tissues through a planar signal traveling across the ectoderm, rather than from continued induction from the notochord below.[32]

The outer boundary of the neural plate is determined by loss of competence of the ectoderm to respond to neural induction, rather than by a low threshold of induction at the boundary, i.e., one need not seek a neural-crest inductor, but rather seek altered responsiveness of the ectoderm to neural induction

(Figure 10.6). Kengaku and Okamoto (1993) demonstrated that with increasing ectodermal age, neural differentiation declined and melanophore differentiation increased when *Xenopus* gastrula ectoderm was exposed to bFGF, a finding consistent with altered competence of the ectoderm with age.[33]

Although Mitari and Okamoto (1991) appear to have obtained evidence for separate inductions of neural tube and neural crest in their microculture assay of *Xenopus* early gastrula cells, close-range and/or homeogenetic inductions cannot be ruled out as explanations for their results. Furthermore, they used antibody markers for neurons, melanophores and epidermal cells, but not neural crest markers. Mayor, Morgan and Sargent (1995), using the genes *slug*, *snail* and *noggin* as neural crest markers, claimed that neural crest was induced independently of the neural plate. This finding is at variance with the other studies, alerting us to be cautious in

BOX 10.3
MESENCHYME AS MARKER FOR NEURAL CREST INDUCTION

Production of mesenchyme is not sufficiently diagnostic to identify induced tissue as neural crest. Mesoderm is also a source of mesenchyme. A wave of controversy was initiated with the proposal that mesenchymal cells arose from this ectodermal location, a proposal that violated the entrenched germ-layer theory discussed in section 5.5.3.[29]

Somewhat more than one third of samples of early gastrula ectoderm exposed to basic fibrobast growth factor (bFGF) form neural tissue, some of which is associated with mesenchyme. The presumption is that this is neural-crest-derived mesenchyme but it is not known whether it is chondrogenic. Similar studies using five parts per million bFGF used differentiation of pigment cells as the marker for differentiation of a neural crest phenotype. Again, no mesenchymal derivatives were shown. Cartilage (and neural tissue) was evoked from early gastrula ectoderm of *Rana temporaria* using concanavalin A as the evoking agent. As the starting tissue was embryonic ectoderm, the presumption is that the cartilage is neural crest in origin, but induction of mesoderm from the ectoderm cannot be ruled out. Mancilla and Mayor (1996) and Mayor, Guerroro and Martinez (1997) used expression of the zinc-finger gene *Xslug* to identify neural crest in their induction studies. *Xslug*, expressed in neural crest in neurula stage embryos, is also expressed in mesodermally derived mesenchyme. Ann Graveson has results from experiments undertaken in my laboratory in which neural crest was induced in Japanese quail embryos using chick tissues as the inducers. Chondrocytes formed and could be positively identified as neural crest in origin because of the presence of the quail nuclear marker.[30]

concluding that we fully understand neural crest induction.[34]

10.4.3 BMPs and neural crest induction

Bronner Fraser and her colleagues have identified BMP-4 and BMP-7 as molecules mediating neural crest induction in chick embryos. Genes that are preferentially expressed in the dorsal neural tube, from which neural crest cells arise, initially have a more uniform distribution throughout the neural tube but are inhibited ventrally by such genes as *shh*. Similarly, suppression of BMP inhibits ventralizing actions of BMP in dorsal locations. A further step in acquiring or maintaining a dorsal fate involves interaction with

signals from the adjacent epidermal ectoderm. It is here that BMP-4 and BMP-7 play their roles. Both are expressed in the epidermal ectoderm adjacent to the neural tube and either can substitute for that ectoderm in promoting migration of neural crest from the neural tube and activating such neural crest markers as *slug*.[35]

In the mouse at equivalent stages of neural development, BMP-7 is initially distributed in paraxial and ventral tissues adjacent to the future hindbrain. Over-expression of BMP-7 in this ventrolateral mesoderm promotes dorsalization of the neural tube, enhances expression of *shh* in the floorplate (but not in the notochord) and promotes neuroectodermal growth. These findings by Arkell and

A   STRENGTH OF INDUCTION

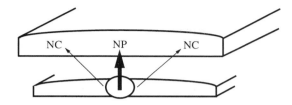

B   DIFFERENTIAL COMPETENCE

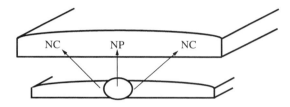

C   SECONDARY INDUCTION

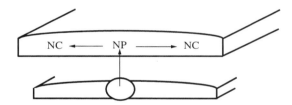

Fig. 10.6 Three alternative models for induction of the neural crest (NC). A. Strength of induction. The notochord is a stronger inducer (large arrow) than the adjacent mesoderm (arrows) so that neural plate (NP) arises in the midline with NC at the edges of the neural plate. B. Differential competence. A lower threshold of response to uniform inductive signals allows NP to form in the midline. C. Secondary induction. NC induction is a two-step process with initial induction of NP by notochord, followed by secondary induction of NC from the already induced neural plate. NC induction could either follow or accompany NP induction.

Beddington (1997) are consistent with and extend the role for BMP-7 demonstrated in the chick. From gene knock-out experiments in mice we now know that BMP-7 also plays a role in the development of the kidneys, eyes and skeleton (Luo *et al.*, 1995).

The ascidian *Halocynthia roretzi* has a homologue of BMP-7 that is expressed in neuroectoderm in spatial and temporal patterns consistent with it playing a role in determination of the boundary of neural and epidermal ectoderm. That ascidians may use a homologue of BMP-7 to segregate neural from non-neural ectoderm sets this pathway in prechordate ancestors.[36]

Dorsalin-1, a BMP-like member of the TGFβ superfamily of growth factors, exerts positive and negative control. *Dorsalin* is expressed in the dorsal neural tube; ventral expression is restricted by notochordal signals. Neural crest is promoted by *dorsalin*, which is mitogenic, enhancing by some 15-fold the numbers of neural crest cells that migrate from isolated neural tubes. *Dorsalin* promotes differentiation of one neural crest derivative (the melanocyte) but inhibits neuronal differentiation, i.e. while the proliferative stimulus is general, differentiative signalling is specific to subpopulations of neural crest cells.[37]

Once neural crest is specified, other growth factors direct the differentiation and migration of neural crest cells. A level of complexity still to be examined in detail is the differential use by neural crest cells of the same molecules, either at different times or at different positions along the neural axis (Hall and Ekanayake, 1991). Thus BMP-4, which promotes migration and differentiation of murine trunk neural crest, induces apoptosis in hindbrain neural crest cells. Later in development, BMP-2 and BMP-4 selectively regulates apoptosis in derivatives of the neural crest in the first and second branchial arches.[38]

10.5 SECONDARY NEURULATION AND THE TAIL BUD

Heads and tails of vertebrate embryos do not develop in the same or even similar ways. Moving from the head to the tail provides an unexpected transition between the cellular and genetic mechanisms used to build body plans (this chapter) and those discussed in Chapter 11 that build organs.

The precisely ordered, three-germ-layer organization, so evident in the gastrula and in the anterior regions of the neurula, is not evident at the caudal end of vertebrate embryos. Indeed, the tail region does not arise by differentiation of primary germ layers but rather from a process of **secondary neurulation** more akin to how organs such as limbs and the heart develop.

Suggestions of two major phases of embryonic development have been in the literature for some time, although not all workers see the same phases.

Wilhelm Roux distinguished two periods in embryonic development: an early period of self-differentiation, characterized by cleavage and organ formation that was independent of function, and a later period characterized by organ differentiation and growth that depended upon function (s. 6.8).

Holmdahl (1925a–c) distinguished two different phases: *primärer Körperentwicklung* and *sekundäarer Körperentwicklung* (primary and secondary body development). To him, germ layers formed during primary body formation. The caudal region developed without organization into germ layers during secondary body development. Such a divergence between cranial and caudal development leads us to ask when during ontogeny primary and secondary developmental processes are initiated and how they relate to the germ layers.

Issues such as these came to the attention of the embryological community in the 1880s, when Kölliker described how the most caudal part of the nervous system arose, not from neural ectoderm, but from mesoderm. In 1931 Bijtel, utilizing vital dyes to label the caudal region of developing frog embryos, confirmed that the most caudal somites that form the tail arise from the mesodermal medullary plate. Indeed, it is now known from the transplantation of labelled grafts, production of chimaeric

embryos, and from culture of tail mesenchyme, that caudal nerve, muscle, vascular and skeletal cells all originate from a common mass of mesenchyme known as the **tail bud**. Transformation of mesenchyme into an epithelial organization is a feature of development of the tail bud in secondary neurulation.[39]

Secondary neurulation characterizes all vertebrate embryos, and is secondary ontogenetically, but not phylogenetically. Members of all vertebrate classes display primary and secondary neurulation. In some species, notably *Xenopus*, caudal neurulation does bear a greater similarity to cranial neurulation than in other species, but recall that *Xenopus* is perhaps the most highly derived model organism used in developmental biology today (s. 8.1).

Tail buds develop following inductive interactions between mesenchyme and epithelium, a mode of differentiation seen in many other organs (some of which are discussed in the next chapter), and quite unlike primary differentiation of germ layers. Limb buds are surmounted by an ectodermal ridge (s. 11.4). So are tail buds. Just as epithelium and mesenchyme interact inductively during limb bud development, so too do tail bud epithelia and mesenchyme. These conclusions are based on experimental studies and analyses of mutants such as *vestigial tail* (*vt*) or *Brachyury* (*T*) mice which lack the specialized tail ridge and *repeated epilation* (*Er*), in which both limb and tail ridges are abnormal. Tail bud development is defective in such mutant embryos.[40]

Caudal regions and tail buds therefore develop secondarily as organs and not by primary differentiation of germ layers. One might therefore expect independent evolutionary changes in caudal and cranial regions in vertebrate evolution, which is precisely what is seen. Development of selected other organs and organ systems (sense organs, limbs, teeth) and the epithelial–mesenchymal interactions that direct their development, are the topics of the next chapter.

10.6 ENDNOTES

1. For overviews of the epigenetic and hierarchical nature of much of vertebrate development, see Hall (1978, 1994b,c), Nieuwkoop, Johnen and Albers (1985), Slack (1985), Maclean and Hall (1987), Gurdon (1987, 1988, 1989) and Langille and Hall (1989).

2. For the labelling studies on which these conclusions are based, see Slack, Dale and Smith (1984), Cooke and Smith (1987) and J. C. Smith (1987).

3. For important milestones in our understanding of induction of mesoderm in *Xenopus*, see Nieuwkoop (1977), Nieuwkoop *et al.* (1985), Godsave, Isaacs and Slack (1988), Slack *et al.* (1984), Slack and Isaacs (1989), J. C. Smith (1989), Gerhart *et al.* (1989, 1991), Kimelman, Christian and Moon (1992) and Sasai and De Robertis (1997). See Chen and Grunz (1997) for a recent study of prepatterning of *Xenopus* ectoderm before mesoderm induction and Stennard *et al.* (1997) for recently discovered molecular markers of mesoderm induction.

4. For the role of these growth factors in mesoderm induction, see J. C. Smith *et al.* (1990), Asashima *et al.* (1990), van den Eijnden-Van Raaij *et al.* (1990), Ariizumi and Asashima (1995a,b), Hainski and Moody (1996), Jones, Armes and Smith (1996) and McDowell *et al.* (1997).

5. See Lemaire (1996), Sasai and De Robertis (1997) and Taira, Saint-Jeannet and Dawid (1997) for other genes involved in the head and trunk organizers and in induction of ventral mesoderm. See Nieuwkoop (1987) for an account (recorded in an interview on his 70th birthday) of his seminal views on the interactive, inductive (epigenetic) nature of vertebrate development. See Gerhart (1997) for an *in memoriam* evaluation of Nieuwkoop's contributions.

6. See Kao, Masui and Elinson (1986), Kao and Elinson (1988) and Elinson and Kao (1993) for the lithium studies, and Sharpe, Pluck and Gurdon (1989) and Sato and Sargent (1990) for the expression studies.

7. For the role of activins in amphibian and avian mesoderm induction, see Green and Smith (1990), Mitrani *et al.* (1990), Thomsen *et al.* (1990), Cooke and Wong (1991), Ariizumi and Asashima (1995a,b), Hainski and Moody (1996), Jones *et al.* (1991, 1996) and Seleiro, Connolly and Cooke (1996).

8. See Christian *et al.* (1991) and Sokol *et al.* (1991) for *Wnt* as an inducer of body axis, and Glinka *et al.* (1997) for inhibition of *Wnt* and BMP by *Cerberus*.

9. For the roles of *Cerberus*, *Lim-1*, *Otx-1* and *goosecoid* in rostral head development, see Acampora *et al.* (1995), Shawlot and Behringer (1995), Ang *et al.* (1996), Bouwmeester *et al.* (1996) and Bally-Cuif and Boncinelli (1997). For an overview of the emerging evidence that anterior endoderm plays a role, perhaps a primary role, in head induction, see Bouwmeester and Leyns (1997).

10. Appropriately, the classic paper by Spemann and Mangold appeared in the centennial volume of the journal established by Wilhelm Roux, the founder of modern experimental embryology. Hilda Mangold, who performed the experiments, died in September 1924 from severe burns received when a heater in her kitchen exploded. Spemann's autobiographical (1938) and Hamburger's first-hand account, published 50 years later, discuss early discoveries in embryonic induction.

11. See Hawley *et al.* (1995), Xu *et al.*, (1995), Jones *et al.* (1996) and Onichtchouk *et al.* (1996) for studies on *Xenopus*, and Nakaido *et al.* (1997) for studies on *Danio rerio*. For ability of activin to establish concentration gradients by diffusion through tissues, see McDowell *et al.* (1997).

12. See Jacobson and Sater (1988) for an overview of temporal and spatial aspects of induction and s. 23.1 for a discussion of developmental time.

13. For discussions of inductive interactions positioning organs and setting limits to their size, see Medawar (1954), Gurdon (1988, 1989, 1992), Hall (1984b,c, 1988a, 1990a) and Hall and Miyake (1992, 1995b).

14. See Lumsden (1991) for a summary of the role of the notochord and Lumsden and Krumlauf (1996) for current understanding of how the neural axis is patterned.

15. For the studies establishing rhombomeres in the chick hindbrain, see Fraser, Keynes and Lumsden (1990), Sundin and Eichele (1990), Lumsden and Wilkinson (1990), Wilkinson (1990) and Wilkinson and Krumlauf (1990). See Lumsden and Krumlauf (1995) and Köntges and Lumsden (1996) for recent summaries, and for independent behaviour of rhombomeres

that extends to rhombomere-specific matching of the origin of connective tissue sheaths of specific muscles and muscle attachment to specific bones. For the presence of pro-rhombomeres in the mammalian hindbrain (localized by the positions of the preotic and otic sulci) and the relationship between pro-rhombomeres and rhombomeres, see Ruberte, Wood and Morriss-Kay (1997).

16. See Northcutt (1995), Lumsden and Krumlauf (1996), Joyner (1996) and Kuratani (1997) for reviews of patterning of the developing nervous system. See Lee *et al.* (1997) and Urbanek *et al.* (1997) for the genes involved in determining the mid/hindbrain boundary. For a study indicating that patterning the nervous system is based on different combinations of signals in the zebrafish than in amphibians, see Woo and Fraser (1997).

17. The involvement of RA in specification of neural and mesodermal tissues is based on studies by Durston *et al.* (1989), Cho and De Robertis (1990), De Robertis, Oliver and Wright (1990), Green (1990), Boncinelli *et al.* (1991), Ruiz i Altaba and Jessell (1991) and Sasai and De Robertis (1997).

18. See L. Z. and N. D. Holland (1996), Katsuyama *et al.* (1995) and Shimeld (1996) for the studies on RA.

19. For nomenclature of homeobox and *Hox* genes, see Hill *et al.* (1989), Scott (1992) and the encyclopaedic compendium compiled and edited by Duboule (1994a). For a listing of *Hox* genes complete as of December 1995, see Stein *et al.* (1996). For *Msh* genes from mouse, zebrafish and the ascidian *Ciona*, see P. W. H. Holland (1991).

20. For homeotic transformation of murine vertebrae, see Kessel and Gruss (1990, 1991), Kessel, Balling and Gruss (1990) and Kessel (1991). See Hemmati-Brivanlou, Stewart and Harland (1990) and Chuong *et al.* (1990) and Duluc *et al.* (1997) for homeobox genes and A-P patterning.

21. For the role of *Sonic hedgehog* in notochordal induction of sclerotomal mesoderm, see Johnson, Riddle and Tabin (1994), Pourquié *et al.* (1996) and Lumsden and Graham (1995). For *Cyclops*, see Macdonald *et al.* (1994, 1995). The gene *Paraxis*, a member of a novel family of basic helix-loop-helix proteins, has been identified as an important regulator of somite formation in chick embryos by Barnes *et al.* (1997).

22. See Monsoro-Burq *et al.* (1996) and Pourquié *et al.* (1996) for BMP as a lateralizing mesodermal signal, and Fan *et al.* (1997) for the role of *Wnt* genes in specification of the dermoyotome.

23. See Hall and Hörstadius (1988) and Hall (1997a) for the discovery of the neural crest.

24. Weston (1963), Johnston (1966) and Le Douarin (1969, 1974) are the pioneering studies on neural crest development in the chick. For analyses of the neural crest and vertebrate head development, see Schaeffer and Thomson (1980), Le Douarin (1982), Gans and Northcutt (1983, 1985), Northcutt and Gans (1983), Maisey (1986), Hall and Hörstadius (1988), M. M. Smith and Hall (1990, 1993), Hanken and Thorogood (1993) and the chapters in Maderson (1987) and Hanken and Hall (1993a). For analyses of the skeletogenic neural crest, see Lumsden (1987, 1988), Hall and Hörstadius (1988), M. M. Smith and Hall (1990, 1993), Smith *et al.* (1994), Graveson, Smith and Hall (1997).

25. See Rosenquist (1981), Brun (1985) and Couly and Le Douarin (1990) for fate maps of amphibian and avian neural ectoderm. Similarly, the neurectoderm (medullary plate) in the marsupial *Sminthopsis macroura* consists of epidermal and neural ectoderm (Tam and Selwood, 1996; Cruz, Yousef and Selwood, 1996). Germ layers retain lability in avian embryos for longer than was thought from extrapolation of studies from amphibians to birds. Garcia-Martinez and her colleagues (1997) demonstrated that prospective neural plate ectoderm could substitute for mesoderm and *vice versa* as late as the early neurula stage.

26. See Webb and Noden (1993), S. C. Smith, Graveson and Hall (1994), Northcutt, Brändle and Fritzsch (1995) and Northcutt (1996) for placodes, Edelman (1983) and Levi, Crossin and Edelman (1987) for N-CAM, and Holtfreter (1933), Raven (1935) and Holtfreter and Hamburger (1955) for experimental studies demonstrating the association of neural tissues with neural crest. Dale and Slack (1987) produced a complete fate map for all the cells of the 32-cell stage *Xenopus* blastula. Extension and refinement of such maps to later stages should enable neural crest and neural crest cell derivatives to be mapped back to individual blastomeres. With such knowledge we should be able to investigate the intransigent problem of the state of determination of individual neural crest cells in the neural tube.

27. See Hall and Hörstadius (1988) and Opitz *et al.* (1988) for discussions of neurocristopathies.

28. For models of neural crest induction, see Hall and Hörstadius (1988) and Graveson (1993). A two-step model for *Xenopus* neural crest in which lateral mesoderm begins induction of neural crest during gastrulation, followed by interaction between neural plate and epidermal ectoderm at the end of gastrulation, has been proposed by Mancilla and Mayor (1996).

29. See Hall and Hörstadius (1988) and Hall (1997a) for discussion and amplification of the neural crest in relation to the germ-layer controversy.

30. For induction of the neural crest in *Ambystoma* and the action of FGF, see Moury and Jacobson (1990), Tiedermann *et al.* (1994) and Kengaku and Okamoto (1993). See Mikhailov and Gorgolyuk (1988) for the studies with concanavalin A. For induction of neural crest in birds, see Bronner Fraser (1995a,b).

31. For studies on thresholds of induction, see Raven and Kloos (1945), Holtfreter and Hamburger (1955) and Saxén and Toivonen (1962).

32. See Albers (1987), Servetnick and Grainger (1991a) and Mancilla and Mayor (1996) for homeogenetic neural induction and progressive loss of competence.

33. For neural crest and boundary conditions, see Sharpe (1990), Yamada (1990), Nieuwkoop and Albers (1990), Kengaku and Okamoto (1993) and Mancilla and Mayor (1996).

34. In more recent studies, Mayor *et al.* (1997) have demonstrated a dependence on FGF for neural crest induction and possible interaction with a neural inducer such as *noggin.*

35. For involvement of BMP-4 and BMP-7 in neural crest induction in the chick, see Selleck and Bronner Fraser (1995), Dickinson *et al.* (1995) and Liem *et al.* (1995).

36. For suppression of BMP dorsally and BMP-7 in ascidians, see Hammerschmidt *et al.* (1996) and Miya *et al.* (1996).

37. See Basler *et al.* (1993) and Kingsley (1994b) for the role of *Dorsalin* in specification of neural crest.

38. For BMP-4 and apoptosis of midbrain neural crest cells, see Graham *et al.* (1994). For similar roles in neural crest derivatives in the branchial arches, see Barlow and Francis-West (1997) and Ekanayake and Hall (1997).

39. See Kölliker (1879, 1884, 1889) and Bijtel (1931) for the origins of caudal neural ectoderm. For mechanisms responsible for formation of tail buds in vertebrate embryos, see Schoenwolf (1977), Schoenwolf and Nichols (1984), Newgreen (1985) Schoenwolf, Chandler and Smith (1985), Griffith, Wiley and Sanders (1992), Presley, Norder and Slipka (1996), Catala *et al.* (1997) and Hall (1997a).

40. For descriptions of caudal neurulation in *Xenopus,* see Pasteels (1943), Gont *et al.* (1993) and De Robertis *et al.* (1994). For the supporting literature on epithelial-mesenchymal interactions in the tail bud and for the mutant mice, see Grüneberg (1956), Searls and Zwilling (1964), Amprino, Amprino-Bonetti and Ambrosi (1969), Grüneberg and des Wickramaratne (1974), Salzgeber and Guénet (1984), Herrmann (1991), Kostovic-Knezevic, Gajovic and Svajger (1991), Gajovic and Kostovic-Knezevic (1995) and Hall (1997a). For the role of the zinc finger gene *Manx* in notochord and tail formation, see also s. 22.2.3.

11

BUILDING ORGAN SYSTEMS

'But ways need to be found to describe, measure and analyse variance in developmental *processes* because these are the ultimate determinants of form.'

Nijhout, 1990, p. 445

Inductive interactions come in hierarchical cascades. As discussed in the previous chapter, inductive interactions coupled with threshold responses initiate the differentiation and morphogenesis of single tissues (nervous system, neural crest), and integrate the differentiation and morphogenesis of primary organ systems within embryos. How such interactions build tissues and organ systems through secondary (epithelial–mesenchymal) interactions is the topic of this chapter.

11.1 EPITHELIA AND MESENCHYME

From gastrulation onwards embryonic cells are organized either into sheets of connected cells as epithelia or meshworks of isolated cells as mesenchyme (Figures 5.3, 6.4, 11.1). Both epithelial and mesenchymal cells secrete and deposit extracellular matrices (ECM). The epithelial ECM is secreted as a basement membrane upon which polarized and coupled epithelial cells sit (Figure 11.2). Mesenchymal ECM takes the form of peri- and extracellular matrices surrounding otherwise isolated and unpolarized cells (Figure 11.1). Epithelial and mesenchymal cell states are not immutable. Epithelial cells can reorganize their extracellular matrices and transform into mesenchymal cells, as occurs when neural crest cells emigrate from the epithelial neural tube (Figure 10.3).[1]

In virtually all tissues and organs of vertebrate embryos, differentiation and morphogenesis is initiated through interaction between epithelia and mesenchyme. Such **epithelial–mesenchymal interactions** are often reciprocal, one setting the stage for the next, that for the next, and so on. I used the term **epigenetic cascades** for the sequential, spatial and temporal inductive interactions that lead to the differentiation and/or morphogenesis of particular cells, tissues and organs.[2]

Few epigenetic cascades are known in any detail, however. Of those that are, development, regionalization and coordination of mesoderm, neural ectoderm and neural crest along the primary embryonic axis were discussed in Chapter 10. Sense organs, limbs and teeth are discussed below. Kidney induction in the newt (s. 11.6) illustrates that inhibitory as well as stimulatory signals characterize these interactions as they characterized mesoderm induction. Development of the craniofacial region, discussed in section 12.1, shows how adjacent tissues and organs, once formed, continue to interact with one another in a hierarchical manner. Part Four contains examples of modification of epigenetic cascades in the generation of morphological diversity over the course of evolution.

A further level of epigenetic control that appears as development unfolds, consists of **functional interaction and integration** between adjacent tissues and organs, for example between nerves and muscles or between the muscular and skeletal systems. Such

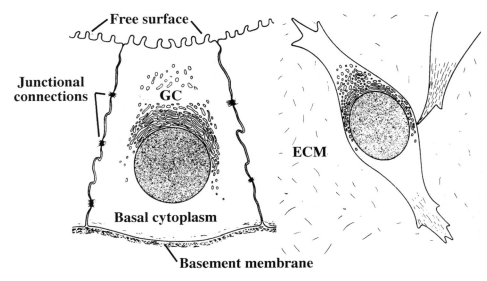

Fig. 11.1 A diagrammatic representation of the differences between epithelia and mesenchyme; see text for details. Modified from Hay (1982).

interactions provide the developmental integration so essential for functional morphology (Chapters 12 and 17). Thresholds of interactions are important, as demonstrated by the requirement for a critical mass of cells before a skeletal element can develop. (Cellular condensations as fundamental units of morphology are discussed in section 20.4.2.) Hierarchical and sequential functional interactions are therefore a hallmark of vertebrate embryonic development.[3]

11.2 CELLULAR MECHANISMS UNDERLYING EPITHELIAL–MESENCHYMAL INTERACTIONS

Because of the practicalities of experimental design and the interests, capabilities, and biases of individual experimenters, epigenetic control is often studied at a single level or with a particular type of control in mind. Consequently, the underlying mechanisms are often represented as single classes of regulatory processes; witness the central roles assigned by different researchers to extracellular matrices, matrix-mediated interactions, cell and substrate adhesion molecules, positional infor-

mation, or selective cell affinity. What appears to be a single level of explanation (e.g. matrix-mediation as the cellular basis for an interaction) may appear sufficient merely because it is the last step in an epigenetic cascade. Cells have histories. The following two sections provide a flavour of current thinking about cellular mechanisms underlying epithelial–mesenchymal interactions. The genetic and molecular bases of these interactions, which reside to a large degree in growth factors, are discussed using two examples – limb and tooth development – in sections 11.4 and 11.5.[4]

11.2.1 Cell adhesion molecules

Over the past decade, a group of cell-membrane molecules with specific spatial and temporal distributions in early embryos has been identified. These are the adhesion and junctional molecules, of which there are three classes:

1. cell adhesion molecules (**CAMs**, cadherins), which account for cell-to-cell adhesion and maintenance of cell associations. (Hox

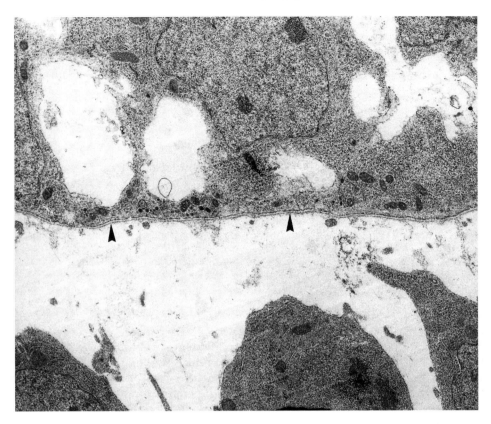

Fig. 11.2 A transmission electron micrograph of an epithelial–mesenchymal junction demonstrating the basal lamina (arrows) and the approximation of mesenchymal cells (below) and mesenchymal cell processes to the basal lamina.

genes also regulate cell adhesion and therefore morphogenesis [s. 11.4.5]);

2. substrate adhesion molecules (**SAMs**), which adhere cells to substrates such as extracellular matrices. SAMs facilitate cell separation; and
3. cell junctional molecules (**CJMS**), which bind adjacent cells allowing cell-to-cell communication.[5]

Two of the most studied CAMs are neural cell adhesion molecule (N-CAM) and liver cell adhesion molecule (L-CAM). Both appear early in development and are localized in reproducible spatial and temporal patterns. The early chick blastoderm expresses N- and L-CAM. With the onset of primary embryonic induction, N-CAM is preferentially expressed in chordamesoderm, neural plate, and subsequently in neural ectoderm and mesenchyme. As noted when discussing origins of the neural crest in section 10.4.2, N-CAM is not seen in epidermal ectoderm, which preferentially expresses L-CAM. Some structures, such as placodes, express N- and L-CAM.

In his morphoregulatory hypothesis, Edelman (1988) gave expression of CAMs and SAMs a regulatory role in embryonic induction. The jury is still out, however, on whether CAMs and SAMs indicate early changes in induced cells, or are the signal molecules that trigger interactions. Current evidence

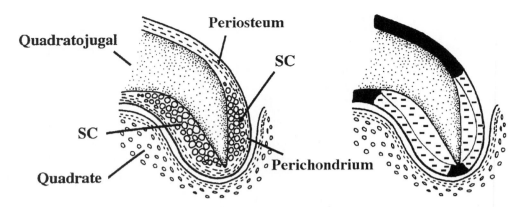

Fig. 11.3 On the left is the articulation of the quadratojugal with the quadrate from a chick embryo incubated for 12 days. At the joint surface, the periosteum has transformed into a perichondrium and deposited secondary cartilage (SC). On the right is the distribution of N-CAM at the same stage. Periosteal cells are positive for N-CAM (black), but N-CAM is downregulated in perichondrial cells and secondary cartilage ($-$).

suggests that their expression reflects the responses of groups of cells to embryonic induction and not inductive signalling to other cells, except indirectly by modifying the behaviour of induced cells.

CAMs are involved, however, in determination of cell fate later in embryonic development:

- N-CAM is down-regulated in the switch of periosteal cells on membrane bones from osteogenesis to secondary chondrogenesis (Figure 11.3).
- N-CAM expression is maintained if chondrogenesis is prevented by embryonic paralysis, in which case the periosteal cells continue along an osteogenic pathway of differentiation (Fang and Hall, 1995).
- As a molecular signal, N-CAM modulates a physical stimulus, embryonic movement, into the ability of these cells to enter an alternate pathway of differentiation.[6]

11.2.2 Matrix-mediated interactions

For some two decades extracellular matrices have been viewed as cellular mediators of interactions.

We have learned a great deal about how the extracellular matrix signals changes in such fundamental cellular processes as proliferation and differentiation. Two important concepts are the cell and its extracellular matrix as the functional unit, and the role of the cell surface and the peri- and extracellular matrices in mediating inductive interaction between a cell and its environment. Components of the matrix cluster integrin receptors on the cell surface, which in turn activate cytoplasmic kinase-mediated pathways that signal through the nuclear matrix to effect changes in chromatin conformation and gene expression.[7]

Two examples of the differentiation of skeletal tissues illustrate some of the unrecognized subtlety underlying these interactions: differentiation of scleral cartilage in response to pigmented retinal epithelium, and of mandibular membrane bone in response to mandibular epithelium. Both are mediated by epithelial extracellular matrices localized within basement membranes. Mesenchymal cells are closely associated with the epithelial basement membrane in spatial and temporal patterns consistent with a role in what are termed short-term, **matrix-mediated** interactions (Figures 11.2, 11.4).[8]

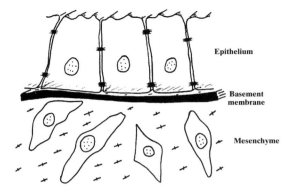

Epithelium

Basement
membrane

Mesenchyme

Fig. 11.4 A diagrammatic representation of junctionally connected epithelial cells in a sheet, resting on a basement membrane, and isolated mesenchymal cells surrounded by extracellular matrix.

Differentiation of many skeletal elements is based on matrix-mediated interactions and mediated by growth factors (ss. 11.4, 11.5). Not all skeletogenic interactions are matrix-mediated, however. The differentiation of scleral bones in response to scleral epithelial papillae is a long-range (300–500 μm), **diffusion-mediated** interaction. Is differentiation of mandibular membrane bones and scleral ossicles therefore based on fundamentally different interactions? Since osteogenesis can be evoked from scleral mesenchyme by mandibular epithelium and from mandibular mesenchyme by scleral epithelium, it probably is not.[9]

Although the cellular bases of these two interactions differ, both epithelia may produce similar signalling molecules that are trapped in mandibular epithelial basement membrane, creating a matrix-mediated interaction, but which diffuse through the scleral epithelial basement membrane creating a diffusion-mediated interaction; there is considerable structural and molecular heterogeneity between basement membranes. If so, then the differences between the two events may reside in the structural organization of the epithelial basement membranes rather than in different

signalling molecules (i.e., the differences would be epigenetic).[10]

Alternatively, different signalling molecules may be produced by scleral and mandibular epithelia, the two mesenchymal cell populations displaying competence to respond to either molecule, a concept known as **double assurance** (the ability to respond to more signals than the one normally encountered during development). In either case (and the two are not mutually exclusive), the interactive and regulatory nature of vertebrate developmental processes ensures normal development and the ability to respond to, or compensate for, mutational, temporal or other changes in the signals controlling development.[11]

Although such heterotypic interactions elicit differentiation, the shape of the bone (bone morphogenesis) is specific to the mesenchyme, irrespective of the source of the epithelium that triggers differentiation. Mandibular mesenchyme forms rods of typical mandibular bone when induced by scleral epithelium, while scleral mesenchyme forms flat scleral ossicles when induced by mandibular epithelium. Other organs behave similarly. Differentiation in epithelial structure and function along a salivary gland tubule reflects properties of the underlying mesenchyme. Mammary gland epithelial responsiveness to testosterone – loss of the epithelial connection in male foetuses – results from binding of testosterone to receptors on the mesenchyme and signalling to the epithelium to elicit the 'epithelial' hormone response. The cumulative, temporally and spatially regulated effects of heterotypic cell-cell and cell-matrix interactions in tissues and organs constitute an important mechanism directing vertebrate differentiation and morphogenesis.[12]

11.3 INDUCTION OF SENSE ORGANS

Neural ectoderm forms the three major regions of the brain – fore-, mid- and hindbrain. Further inductions initiate the major sense

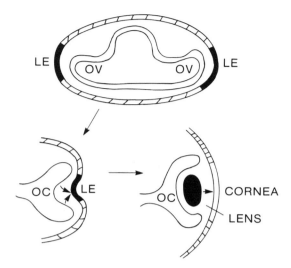

Fig. 11.5 The sequence of inductive interactions in the differentiation of the lens and cornea shown as diagrammatic transverse sections through the developing forebrain (top) and eye (below). Left and right optic vesicles (OV) develop from the forebrain and approach the lens ectoderm (LE). As the optic vesicles transform into optic cups (OC) they signal to the lens ectoderm (arrows, bottom left) to elicit differentiation of the lens from lens ectoderm (bottom right) which in turn signals to the overlying epithelium (arrow, bottom right) to evoke the differentiation of the transparent cornea.

organs, the nose, eyes and ears, associated with fore-, mid- and hindbrain respectively.

11.3.1 Eyes

The forebrain develops into two regions, the centrally located cerebral hemispheres and the peripheral optic lobes (Figure 11.5). The optic lobes grow toward the ectoderm to become optic vesicles, which transform into the optic cup, in which several distinct types of cells differentiate – neural retina, pigmented retinal epithelium, iris. Once contact with ectoderm is made, the optic vesicle induces ectoderm to bud off as a lens vesicle which subsequently rounds up and differentiates into the lens. (An unusual feature of this induction is that it per-

sists into adult life, maintenance of the differentiated state of the lens requiring constant induction.) In turn, the lens acts inductively upon overlying ectoderm to transform what would otherwise be opaque epidermis into transparent cornea (Figure 11.5). This sequence of inductions, in which one induced structure becomes an inducer to elicit the formation of another structure, which in turn becomes the inducer of a third structure, is an exquisite example of cascading inductive interactions. Genetic control of eye development is discussed in section 21.7.3 and by Graw (1996).[13]

Although what has just been described is the 'textbook version' of induction of vertebrate eyes, there are many species of anuran and urodele amphibians, and some species in other groups, in which the lens forms without inductive influences from the optic cup. Prior interaction of lens epithelium with endoderm and/or mesoderm is sufficient to elicit lens formation in such species. There are genera containing pairs of species (*Ambystoma maculatum* and *A. mexicanum*, for example) in which the lens is induced by the optic vesicle in one species but not in the other; the evolutionary significance of the different developmental mechanisms involved in the production of homologous structures in closely related species is discussed further in Chapter 21. In species of frogs such as *Rana esculenta*, the lens differentiates without induction from the optic vesicle, but the optic vesicle nevertheless retains the ability to induce a lens, the clear indication being that embryonic induction can be modified quite readily during phylogeny. For further analysis of lens development, see sections 21.7.2 and 24.1.3.[14]

11.3.2 Ears

The development of the primordia of the ear, especially the protective cartilaginous otic capsule, provides an example of cooperation between different inductors in the formation

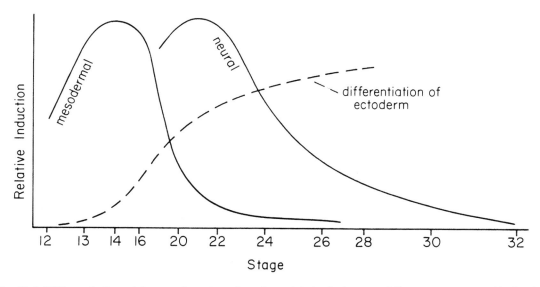

Fig. 11.6 Differentiation of the ectoderm into the otic vesicle in *Ambystoma* follows two waves of induction, one mesodermal with a peak at stages 14–15, and one neural with a peak at stages 21–22. Stages 12–16, neurulation; stage 21, early tail-bud embryo; stage 32, late tail-bud embryo. Redrawn from Yntema (1955).

of a single structure. As documented by Yntema for the urodele *Ambystoma punctatum*, cranial ectoderm in the region of the hindbrain is induced to form an otic vesicle in two major steps. During early to mid-neurulation, cranial mesoderm acts inductively upon cranial ectoderm. This induction wanes during mid-neurulation. A second induction from the hindbrain then acts upon the same ectoderm to complete induction of the otic vesicle (Figure 11.6). Once formed, the otic vesicle acts inductively upon adjacent mesenchyme to initiate chondrogenesis to form the otic capsule (see s. 12.1. for how such interactions control morphogenesis and growth).

Timing and association of these inductive interactions is facilitated by the temporally regulated development of cellular contacts (including gap junctions) that enable hindbrain and ectoderm only to communicate from the mid-neurula stage onwards, i.e. during the induction. Timing is also coordinated by acquisition of competence by the ectoderm.

Ectoderm from the early neurula does not respond to inductive influences from the hindbrain of late neurulae; ectoderm from embryos beyond neurulation no longer responds to neurula mesoderm.[15]

In chick embryos, further inductive interactions occur between the differentiated otic capsule and adjacent mesenchyme that forms the foot plate portion of the columella; the remainder of the columella differentiates without induction from the otic capsule. Once formed, the columella induces already differentiated cartilage cells of the otic capsule to dedifferentiate and transform into fibroblasts to form the annular ligament, which attaches the foot plate to the otic capsule (Figure 11.7). This intricate series of interactions ensures that the connecting ligament forms precisely where and when it is required, and demonstrates that, in addition to controlling differentiation, inductions can control dedifferentiation and redifferentiation during embryonic development as they do during regeneration (s. 21.3).

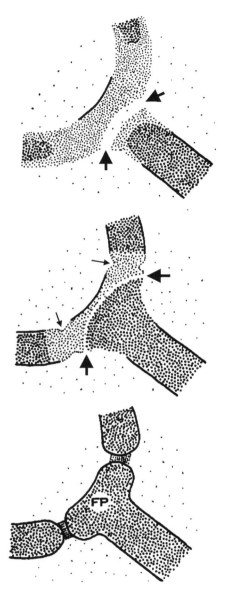

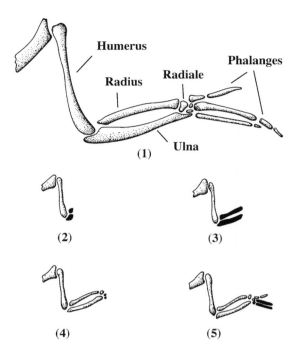

Fig. 11.8 (1) The skeletal elements of the avian wing. (2–5) The appearance of the skeletal elements after removal of the AER at progressively later stages between stage 18 or 2.5 days of incubation (2), and stage 26 or 5-days of incubation (5). The earlier that the AER is removed, the more distal the limb skeletal deficiencies. Modified from Hall (1978).

Fig. 11.7 Major stages in the interaction between the columella and otic capsule to form the annular ligament during development of the avian middle ear. Top, the rod-like columella and U-shaped otic capsule are not yet apposed. Middle, chondrocytes that will dedifferentiate to form the annular ligament are now evident (small arrows) and the two cartilages still separate (large arrows). Bottom, the annular ligament has formed and anchors the footplate (FP) of the columella to the otic capsule. Modified from Jaskoll and Maderson (1978).

11.4 DEVELOPMENT OF VERTEBRATE LIMBS

Vertebrate limbs, especially of developing chick embryos, have been a favourite model system in developmental biology ever since the 1940s, when John Saunders and Ed Zwilling demonstrated that:

- removal of the ectodermal cap or apical ectodermal ridge (AER) from limb mesenchyme truncates limb development (Figures 11.8, 11.9);
- the earlier the ectoderm is removed, the greater the truncation; and
- that the *wingless* mutant fails to develop wings because of arrested development of the AER.[16]

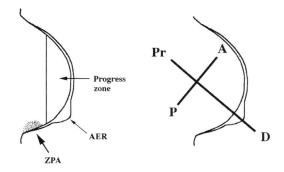

Fig. 11.9 On the left is shown an early embryonic chick limb bud with the apical ectodermal ridge (AER), progress zone of distal mesenchyme, and zone of polarizing activity (ZPA). On the right are shown the principal limb axes, the proximo-distal (Pr-D) axis that passes through the AER, and the antero-posterior (A-P) axis. The dorso-ventral axis runs from the viewer into the page.

The two-fold demonstration of signalling between limb ectoderm and mesoderm, and of the fact that this signalling has a genetic basis, ushered in decades of research on secondary induction. The results presented below are primarily from recent studies on developing chick limbs, with additional studies from gene knock-out experiments in mice. Given that birds are highly derived vertebrates it may be risky to extrapolate mechanisms of wing development to limbs of other vertebrates, although it does appear that the mechanisms initiating limb buds are highly conserved in tetrapods. Indeed, they share elements of control with fish fins, from which they arose evolutionarily (s. 16.2).

Tetrapod limb buds appear as paired lateral swellings in the position of the future fore- and hindlimbs. Soon thereafter a thickening of the ectoderm signals the formation of the AER (Figure 11.9). Outgrowth ceases if the AER is removed (Figure 11.8). A zone of undifferentiated cells lying within some 300 μm of the AER also begins to differentiate if the AER is removed. The AER is therefore involved in limb outgrowth and in maintaining a distal zone (known as the **progress zone**) in

which cells fail to differentiate and from which they exit to differentiate and contribute to limb outgrowth (Figure 11.9). To treat summarily a vast amount of literature, the time that cells spend as undifferentiated cells in the progress zone determines whether they form proximal limb skeleton (those cells that exit early) or more distal elements (those cells that exit progressively later). An AER-mesenchymal interaction therefore controls one of the three limb axes, the proximo-distal (P-D) axis that extends from pectoral or pelvic girdle, through humerus or femur to the distal phalanges of the digits (Figures 11.9).

The D-V axis, the least understood of the three limb axes, and the one which imposes pattern onto the developing muscles and tendons, is maintained by signalling between dorsal and ventral limb-bud ectoderm. The third axis, the A-P axis, specifies digits in their anterior-posterior sequence and is regulated from a centre at the posterior boundary where limb bud meets flank. Because it polarizes the limb axis this area is known as the polarizing region or **zone of polarizing activity** (ZPA; Figure 11.9).

Thus, mesenchymal cells, wherever they lie in developing limb buds, are exposed to three orthogonal signalling systems. The molecular bases of these systems are now quite well understood. Cheryll Tickle, who is responsible for much of the innovative work on limb development, has summarized the state of knowledge. Her review (1996) should be consulted for literature supporting the following analysis.

11.4.1 AER signalling

AER signalling is mediated by fibroblast growth factors. Ectopic application of FGF-2 to limb mesoderm functions in the same way as the specialized AER; it maintains normal mesodermal gene expression and permits limb development. In a dramatic illustration of the controlling roles of FGFs in limb initiation, beads soaked in FGF-1, -2, or -4 and

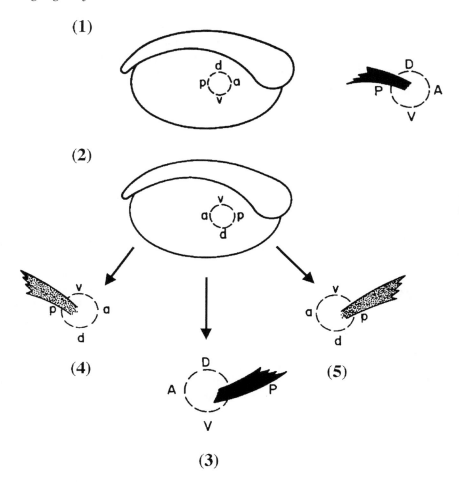

Fig. 11.10 A summary of the classic experiments by Harrison (1918) on determination of the axes of limb fields. (1) The major axes of the limb field in an *Ambystoma* neurula (anterior to the right) with the orientation of the limb in black. (2) Rotation of the limb disc through 180° and reimplantation reverses the A-P but not the D-V limb axis (3) because the A-P axis is determined early in development. (4, 5) The expected result if the D-V and not A-P axis had been determined early (4) or if both axes had been determined before rotation (5). Modified from Hall (1978).

implanted between future wing and leg buds to the flanks of chick embryos, induce an ectopic limb. Indeed, either a wing or leg can be elicited from the same population of mesenchymal cells by FGF treatments that modify patterns of expression of Hox genes. Recent studies by Tanaka and colleagues, in which ectopic limbs were evoked by grafting the D-V boundary of the limb bud, reinforce the role of the limb-bud ectoderm in specifying an AER and initiating limb-bud development.[17]

11.4.2 Dorsal signalling

Signalling from dorsal ectoderm is associated with the *Wnt-7a* gene, which encodes a cysteine-rich glycoprotein. In part this conclusion comes from knock-out experiments; mice from which *Wnt-7a* has been deleted have a double ventral pattern in the distal portions of their limbs. This result is consistent with the normal distribution of *Wnt-7a* along the dorsal and not the ventral ectoderm. *Radical fringe*, a homologue of the *Drosophila* gene *fringe* (which establishes the dorsoventral border of the fly wing), directs formation of the D-V boundary of the limb bud in chick embryos, providing a molecular explanation for the classic experiments by Ross Harrison (1918) on establishment of limb axes in *Ambystoma* (Figure 11.10). The AER forms within ectoderm at the border between cells that do, and cells that do not express *Radical fringe*.[18]

11.4.3 ZPA-signalling

The best evidence for a cascade of signals in limb development comes from analysis of A-P signalling that emanates from the ZPA. It has been known for some time that placing a ZPA into the anterior region of the limb bud leads to a mirror image duplication of the digits. This is consistent with a gradient signal based in the ZPA specifying posterior skeletal structures close to the ZPA and anterior structures farthest from the ZPA.

Signalling from the ZPA involves RA and *Sonic hedgehog*. RA applied to the anterior margin of the limb bud mimics grafting a ZPA to that region. Application of *shh* duplicates digits, and *shh* is induced by RA and localized in the ZPA. *Shh* and FGF-4 are induced by signals from the ZPA and maintain the activity of the ZPA.[19]

Finally, *shh* is produced by the chick notochord, floor plate and Hensen's node (s. 10.3). Chiang *et al.* (1996) demonstrated a major role in embryonic patterning when they inactivated *Shh* by targeted mutagenesis in the mouse. Midline structures such as the notochord and floor plate failed to form. Consequently, structures such as the ventral neural tube and spinal column that depend on these tissues for induction, also failed to form.

Notochord, floor plate and Hensen's node can function as a ZPA and up-regulate *Msx-1* and *shh* when grafted into the anterior face of a limb bud. Such results are consistent with *shh* as an important gene patterning early development in numerous organ systems.[20]

11.4.4 Linkage of the three signalling systems

The three signalling systems just described, each specifying an individual axis and a diversity of tissue types (skeleton, muscle, tendon), may be linked by their signalling molecules. Signalling in the ZPA is maintained by signals from AER and dorsal ectoderm as *shh* and FGF-4 form a feedback loop. Shape of the limb bud, an important element of pattern formation, appears to be regulated, at least in part, by BMP-2 and BMP-4. As might be predicted, mutants with wide limb buds and polydactyly such as *talpid*[3] express transcripts of BMP-2 much more widely across the limb bud, and FGF-4 more widely throughout the apical ectoderm, than limb buds of wild-type embryos. The zinc finger transcription factor *slug* is expressed in the ZPA, progress zone and interdigital area, a dynamic pattern of expression suggestive of a role in coordinating the three signalling systems. *Slug* expression depends on signals from the ZPA but is independent of *shh*. The recent analysis by Buxton *et al.* (1997) demonstrates that *slug* is regulated by FGF-4 and associated with transition of undifferentiated cells of the progress zone to differentiated chondrogenic and connective tissue cell (but not muscle) cells.[21]

As indicated in section 11.4.2, *Radical fringe* couples dorsoventral and AER in a system in which the AER forms at the dorsoventral boundary of *Radical fringe* expression. Before the AER forms (stage 14 in the embryonic chick), *Wnt-7a* is expressed in a restricted zone at the dorsal side and *Engrailed-1* (*En-1*) at the ventral side of future limb ectoderm. Expression of *Radical fringe* in the dorsal ectoderm at stage 15 establishes a boundary of gene

Table 11.1 Similarity of signalling genes involved in specification of the axes of vertebrate and *Drosophila* limbs[1]

Axis	Drosophila	Vertebrate
A-P	Hedgehog	Sonic hedgehog
	Decapentaplegic	BMP-4/2
P-D	Apterous	
	Fringe	Radical fringe
	Serrate	Serrate-2
D-V	Wingless	Wnt-7a

[1] See Shubin, Tabin and Carroll (1997) for a summary.

expression. Cells at the boundary express neither *Wnt-7a* nor *En-1*, but do express FGF-8. The implication is that *Radical fringe* downregulates the former genes to establish a pattern of gene expression required to initiate an AER and therefore couples D-V positioning with formation of the AER.

As will have become apparent from this discussion, the genes directing all three axes of vertebrate limbs have astounding parallels to the genes specifying the same axes in *Drosophila* limbs (Table 11.1). The implication that these genes, and perhaps their role in axial signalling existed in the common arthropod–vertebrate ancestor and may be even older, is borne out by recent studies demonstrating that all the *Hox* genes found in insects are also found in myriapods and onychophorans, but that the anterior expression boundaries of trunk *Hox* genes lie at quite different levels in different groups of arthropods (Grenier *et al.*, 1997).

11.4.5 Mutants, *Msx* genes and cell death

The tissue-level response of limb mesenchymal cells to inductive signalling is to produce skeletal, muscular and connective tissues in precisely patterned arrays. The response at the molecular level, especially to signals from the AER and ZPA, is to express *Msx-1* and *Msx-2* in the distal mesenchyme

and *Hoxa* and *Hoxd* genes in specific arrays throughout the mesenchyme. Overlapping expression boundaries of *Hox* genes have been mapped in developing limb buds; they are discussed in section 16.2.4 in the context of the fin → limb transition and the origin of tetrapods.[22]

I document the effect of mutations on *Msx-1* and *Msx-2* gene expression in limb development by drawing attention to three of the studies by Caroline Coelho and her colleagues at the University of Connecticut (Coelho *et al.*, 1991, 1993a,b). Their model system is the developing chick limb bud. Expression of high amounts of *Msx-2* in the AER and of *Msx-1* in the subjacent limb mesoderm indicate regional specificity of limb ectoderm. Ectoderm from the anterior, proximal border of the bud (where bud meets flank, and the site of the anterior necrotic zone of cell death that sculpts the anterior border of the bud) induces high expression of both *Msx* genes in the mesenchyme. Ectoderm from the posterior proximal border (the site of the ZPA and of the posterior necrotic zone of cell death that sculpts the posterior border of the bud) induced neither. Ectodermal control of region-specific expression seems to be one aspect of ectoderm-mesoderm signalling.

In a further analysis, Coelho and colleagues examined the natural polydactylous mutants *talpid²* and *diplopodia-5*, which have greatly expanded and thickened AERs and additional digits. Both mutants lack zones of cell death; suppression of cell death is a partial explanation for the expansion of the width of the limb bud that allows additional digits to form. Expression of *Msx-1* is expanded in these mutant AERs, suggestive of a role for *Msx-1* in promoting limb outgrowth via the AER. In neither mutant is *Msx-1* expressed at the anterior or posterior proximal borders of the limb buds, consistent with expansion of the limb bud following suppression of cell death.

In a third analysis, this group used the avian mutant *limbless*, in which limb buds arise but do not form an AER and so fail to continue to develop. *Msx-1* but not *Msx-2* is reduced in *limbless* mesoderm. In contrast, *Msx-2* is reduced dramatically in *limbless* ectoderm, suggestive of *Msx-2* signalling to the AER and of regulation of mesodermal *Msx-1* by ectodermal *Msx-2*.

Absence of zones of cell death in polydactylous mutants facilitates the development of polydactyly. *Msx-2* regulates formation of these zones, which sculpt the limb borders in normal embryos. So too do BMP-2 and BMP-4. Overexpression of the BMP receptor in chick embryos blocks signalling of BMP-2 and BMP-4 and suppresses cell death in anterior and posterior necrotic zones and between the digits. Interdigital cells maintained *in vitro* respond to BMP protein by activation of cell death (apoptosis). Implanting BMP-2 or BMP-7 *in ovo* induces cell death in limb mesenchyme but not in limb ectoderm or in differentiating cartilage cells; the action is specific for undifferentiated mesenchyme. Amanda Barlow and Philippa Francis-West (1997) and we (Ekanayake and Hall, 1997) have shown that BMP-2 and BMP-4 also induce cell death in chick mandibular arches in a dose-dependent manner. With high doses mandibles fail to form; with lower doses, mandibular skeletal structures are either reduced in size or undergo abnormal morphogenesis. Patterns of expression of *Msx-1*, *Msx-2* and FGF-4 are also modified.[23]

Exciting days are ahead as the signals and responses during limb development are teased out and as these analyses are extended to more genes and more species, allowing comparative analyses of limb formation and evolution. Such studies in fish are revealing the developmental mechanisms involved in the fin → limb transition that accompanied the origin of the tetrapods (s. 16.2).

11.5 INDUCTION OF MAMMALIAN TEETH

Teeth are composite organs made up of two tissue types. Enamel forms the 'pearly white', mineralized covering of the teeth, primarily above the gums. Dentine, a different mineralized tissue, forms the roots and interior of the teeth. Dentine and enamel are essentially extracellular matrices; odontoblasts synthesize and deposit dentine, ameloblasts synthesize and deposit enamel.

Ameloblasts arise from buccal epithelium at the site of tooth formation. Odontoblasts differentiate from cells that arise in the embryonic neural crest. Preodontoblasts migrate as mesenchymal cells into the developing oral cavity and along the oral epithelium. Odontoblasts differentiate after a series of interactions with cells of the future enamelforming epithelium.[24]

We do not know the nature of the signal that enables neural-crest-derived mesenchymal cells to stop migrating and settle down against the buccal epithelium at precisely localized positions where teeth will develop. Once in position, an intricate series of reciprocal interactions leads to the differentiation of odontoblasts and ameloblasts and deposition of dentine and enamel, producing fully differentiated and morphologically complex teeth (Figure 11.11). The cascade unfolds as follows:

1. Neural-crest-derived dental mesenchyme induces adjacent oral epithelium to proliferate (interaction 1 in Figure 11.11).
2. Dental epithelium triggers dental mesenchyme to proliferate and develop as a dental papilla (interaction 2).
3. The dental papilla induces dental epithelium to form an enamel organ (interaction 3). No further tooth development ensues if enamel organ and dental papilla are separated at this stage.
4. If left in association, the enamel organ induces cells of the dental papilla to differentiate as preodontoblasts that differentiate into odontoblasts (interaction 4).
5. Odontoblasts and the predentine deposited by them induce preameloblasts of the enamel organ to complete their differentiation into ameloblasts and to deposit enamel (interactions 5 and 6). Again, no further

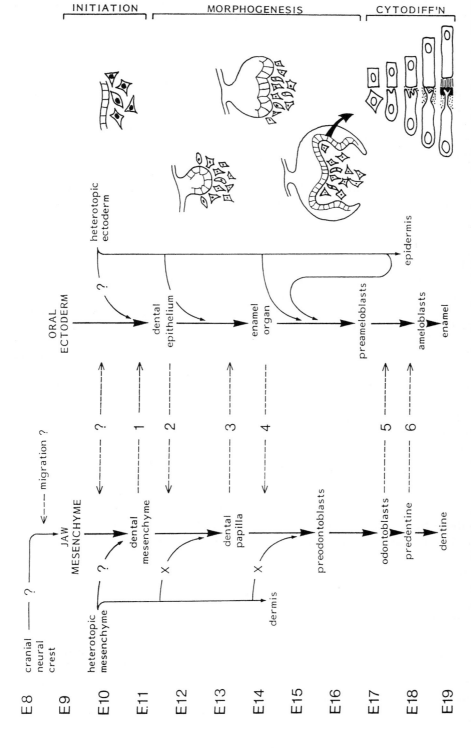

Fig. 11.11 The sequence of epithelial-mesenchymal interactions involved in the development of teeth in the embryonic mouse. E8–E19 represent age in days post-conception. The major stages of tooth development are shown diagrammatically on the right. Solid arrows represent cell or tissue transformations in dental mesenchyme or epithelium; dashed arrows represent epithelial mesenchymal interactions 1 to 6. (1) Dental mesenchyme induces oral epithelium to become dental epithelium. (2) Dental epithelium induces dental mesenchyme to condense as a dental papilla, which in turn (3) induces dental epithelium to form an enamel organ. (4) The enamel organ then induces cells of the dental papilla to differentiate into preodontoblasts and odontoblasts. (5) Odontoblasts then induce preameloblasts to differentiate into ameloblasts. (6) Predentine induces ameloblasts to deposit enamel. Reproduced from Lumsden (1987).

tooth development takes place if preamelo-blasts and preodontoblasts are separated at this stage.

Separating the components at various stages arrests tooth development at those stages, illustrating that each stage represents a threshold or boundary. Additional signals are required before tooth development can proceed to the next stage.

11.5.1 Growth factor signalling, homeobox genes and tooth differentiation

Unravelling the molecular basis of such complex interactions has not been straightforward. Nevertheless, substantial progress has been made. Growth factors play major roles as mediators of differentiation and morphogenesis of mammalian teeth. This conclusion is based on localization of growth factors and their receptors *in vivo*, and experimental application or blocking of growth factors *in vivo* and *in vitro*. As might be expected in a situation where signalling between epithelium and mesenchyme is reciprocal, growth factors in both tissues are involved, as they are in limb development.[25]

At the initiation of tooth development, BMP-2, BMP-4 and FGF-8 are expressed in the dental epithelium. These growth factors are mitogenic, stimulating proliferation of dental mesenchymal cells as Vainio *et al.* (1993) demonstrated so elegantly. Dental mesenchyme maintained apart from epithelium but in the presence of these growth factors, behaves as if dental epithelium were present, i.e. these growth factors mimic the initial action of dental epithelium on mesenchyme. Dental epithelium up to the bell stage (Figure 11.11), and/or these growth factors, up-regulate a variety of genes in dental mesenchyme, including genes for growth factors (TGFß-1 and BMP-4), homeobox genes (*Msx-1* and *Msx-2*) and transcription factors (*Notch -1, -2,* and *-3*). Other homeobox-containing genes such as *Distalless-1* (*Dll*) and *Dll-2* also are present in dental mesenchyme.

Once dental mesenchyme is activated by epithelium, growth factors such as BMP-3, BMP-4, FGF-3, LEF-1, and activin $-\beta$A signal back to the epithelium. The functional role of these genes has been demonstrated *in vivo*; reduced activin function is accompanied by lack of development or greatly reduced development of tooth buds; tooth buds in *Msx-1* knock-out mice fail to develop beyond the bud stage and can be rescued by such upstream signals as BMP-4. Extracellular matrix molecules also may be involved: TGFβ-1 and BMP-2 induce odontogenesis *in vitro* when applied in association with the EDTA-soluble fraction of dentine proteins, although it is not yet clear whether the matrix serves as carrier or signal.[26]

Msx-1 and *-2* and BMP-2 and *-4* are similarly regulated in the epithelial-mesenchymal interactions that govern mammary gland development and play important roles in limb and craniofacial development. *Msx* genes are therefore emerging as important regulators of various situations involving cell-to-cell signalling, and might be expected to play important roles in evolutionary modification of morphology. BMPs also are emerging as major players in other aspects of embryonic development. Homozygote mice deficient in BMP-2 fail to survive because of defective development of extraembryonic membranes and the heart. BMP-4 and BMP-7 are the epidermal signalling molecules responsible for the induction of neural crest from the dorsal neural tube (s. 10.4.3) and BMP-4 is responsible for induction of lateral somitic mesoderm (s. 10.3).[27]

11.5.2 The enamel knot and tooth morphogenesis

Tooth primordia are initiated by growth factors and homeobox-containing genes involved in epithelial-mesenchymal signalling. What about tooth morphogenesis? How are the complex yet characteristic, species-specific shapes of mammalian teeth generated? This question, which has intrigued specialists in

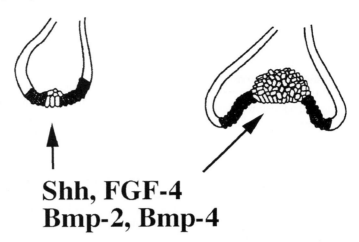

Shh, FGF-4
Bmp-2, Bmp-4

Fig. 11.12 The enamel knot (white cells) in a developing tooth germ express *sonic hedgehog* (*Shh*), FGF-4 and BMP-2 and -4. Modified from Jernvall (1995).

many fields for generations, now seems closer to resolution. The **enamel knot**, a long-known, little-understood, and often argued-about portion of the dental epithelium, appears to hold this year's answer to tooth morphogenesis.

The enamel knot (more correctly, the primary enamel knot because secondary knots subsequently develop), appears at the early cap stage in the centre of the dental epithelium facing the mesenchyme of the dental papilla (Figure 11.12). Its appearance coincides with initiation of the cusps of the teeth. As the name suggests, it is no more than a knot of epithelial cells. Cells of the enamel knot do not divide and are transient, so transient that their existence was denied as often as studied.[28]

Recent studies on patterns of gene expression have confirmed that the enamel knot does exist, and that it displays a specific association of gene products otherwise known only from embryonic regions that control patterning. These are the notochord, which regulates D-V patterning of the neural tube, and the ZPA and AER of developing limb buds, which between them pattern the A-P and P-D axes of the developing limb skeleton. A common array of genes (*shh*, BMP-2, -4, and -7 and FGF-4) is expressed in the enamel knot, notochord, ZPA

and AER. Expression of these genes during tooth development immediately precedes formation of the primary cusps. Additional cusps appear at the early bell stage, preceded by secondary enamel knots. They too express FGF-4 and possibly the other genes.[29]

FGF-4 is of especial interest: FGF-4 up-regulates *Msx-1* and *shh*; implanting FGF-4 provides a sufficient signal to induce a new limb to form (s. 11.4). The enamel knot controls growth and patterning of the teeth via a signalling pathway based on FGF-4, which is a potent stimulator of mesenchymal and epithelial proliferation. This is paradoxical. FGF-4 is found in regions of high proliferation. Cells of the enamel knot do not divide, yet have high levels of FGF-4. The explanation is that the knot, an island of stasis in a dividing epithelial sea, sends its proliferative signal to adjacent cells. Once the tooth is patterned, the enamel knot is removed by apoptosis (as is the AER once its patterning and growth promoting job is done; Vaahtokari *et al.*, 1996b). FGF-4 inhibits apoptosis and so FGF must be removed before the enamel knot can be inactivated.

Mammalian teeth are patterned along the tooth row from incisors to molars. This

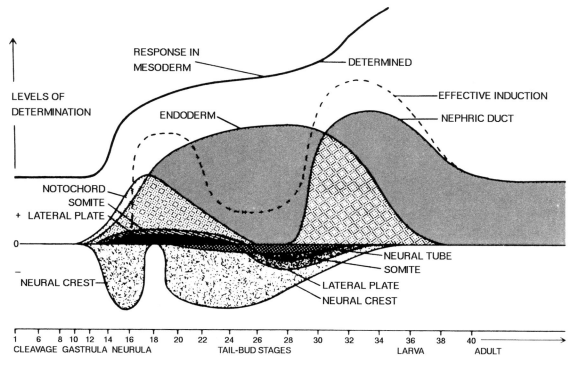

Fig. 11.13 A summary of the stimulatory (+) and inhibitory (−) interactions involved in differentiation of the kidney in the newt *Taricha torosa*. These interactions, which result in increasing determination of mesoderm as nephrogenic, are plotted against developmental stage. Notochord from gastrula to tail bud stages, endoderm from gastrula to larva, and nephric ducts from tail bud stage to adult are major stimulators of kidney development. Neural crest is a major inhibitor of kidney development. Somitic and lateral plate mesoderm are stimulatory early in development (gastrula to tail bud) but inhibitory thereafter (tail bud stage to larva). Reproduced from Jacobson (1987).

patterning also may be regulated by a growth factor, in this case by epidermal growth factor (EGF), which along with its receptor (EGF-R) is found in dental epithelium and mesenchyme at tooth initiation. EGF, or retinoids acting via EGF, modifies the pattern of the tooth row. Indeed, teeth can be induced in the otherwise toothless diastema of the rodent jaw following culture in EGF. Tooth type is modified (molars are replaced by incisors) following exposure of the developing tooth row to retinoids which activate EGF. Clearly, here is a mechanism ideally suited to modification over evolutionary time as the basis for the enormous range of tooth morphologies found among extinct and

extant mammals (s. 17.2.1). Studies on other species are awaited with eager anticipation.[30]

11.6 INDUCTION OF NEWT KIDNEY

A short section on the epigenetic cascades involved in induction of the kidney of the newt *Taricha torosa* (primarily based upon the work of Etheridge, 1968) is included to illustrate that not all the interactions in a cascade are positive and promote differentiation and morphogenesis. Inhibitions also function in epigenetic cascades.

In kidney development, a complex cascade of inductive interactions begins in the early gastrula and continues late into larval life. As is clear from Figure 11.13, which summarizes the major results of Etheridge's experiments, endoderm early in development (from gastrula to larval stages), and the nephric tubules later in development (from tail bud to larval stages), exert the greatest stimulation on kidney induction. Notochord between the gastrula and mid-tail-bud stage also acts as a positive inducer. Somitic and lateral plate mesoderm (not shown in Figure 11.13), initially stimulate but then inhibit kidney induction. The neural tube plays a minor inhibitory role that continues throughout the tail-bud stages of embryonic development. Neural crest exerts the greatest inhibitory influence during embryonic and larval development (Figure 11.13).

A little-studied aspect of such epigenetic cascades that may be of considerable importance in integrating tissues during development, is the inhibition of other cell types by neural crest cells. Etheridge (1972) showed that neural crest suppresses lens, heart and kidney development at different stages and times. Woellwarth (1961) demonstrated neural crest inhibition of lens formation in *Triturws alpestris*.[31]

Inductive control of tissue and organ development consists neither of simple, one-step interactions, nor only of cascades of sequential, stimulatory interactions. If kidney induction in the newt is typical of inductions in general, then epigenetic cascades represent a complex interplay of inhibitory and stimulatory interactions from one embryonic region to another. What we see as 'induction' and almost always discuss as a positive, switching-on process, is, in reality the result of sequential and perhaps interacting switching-on and switching-off processes, which on balance, and over time, 'induce' a particular tissue or organ. Mesoderm induction certainly fits this pattern, indeed is paradigmatic for it (s. 10.1). There is enormous scope for ontogenetic and phylogenetic tinkering with such a process; witness the loss of induction of the lens by the

optic cup in the species discussed in section 11.3. Other examples of this important developmental basis for morphological change during evolution are discussed in Part Four and in Chapter 24.

In addition to initiating organogenesis, interactions between individual tissues and organs coordinate and stabilize their development and facilitate their functional integration. Such interactions, and the associated phenomena of developmental canalization and deviations from stability (fluctuating asymmetry), are taken up in the following chapter. They provide another bridge from embryos in development to embryos in evolution.

11.7 ENDNOTES

1. For basic studies on transformation of epithelial to mesenchymal organization, see Greenburg and Hay (1982) and Hay (1982, 1990). See the papers in Newgreen (1995) for examples and analyses of the mechanisms involved.

2. For development of the concept of epigenetic cascades, see Hall (1983a, 1990a) and Hall and Hörstadius (1988). Thomson refers to 'cascades of decisions [that] form a developmental hierarchy' (1991a, p. 112). For reviews and historical aspects of epithelial–mesenchymal interactions in various vertebrate tissues, see Hall (1982a, 1984b,d, 1988a, 1990a,b, 1994b,c), Nieuwkoop, Johnen and Albers (1985), Maclean and Hall (1987), Sanders (1989), Smith and Hall (1990, 1993) and Gilbert (1994a).

3. For studies on inductive interactions between muscular and skeletal systems, including responses of the avian tibiotarsus and fibula to natural or experimentally induced limb elongation, see Müller (1986, 1989), Müller and Streicher (1989), Hall and Herring (1990), Streicher (1991), Streicher and Müller (1992) and Fang and Hall (1997). For the importance of such interactions in functional morphology, see Dullemeijer (1974, 1991), Hilderbrand *et al.* (1985), D. B.Wake and Roth (1989) and M. H. Wake (1990, 1992b). For further discussion of hierarchical epigenetic interactions during embryonic development, see Hall (1983a, 1990a), Chandebois and Faber (1983), Nieuwkoop (1987, 1992), Nieuwkoop *et al.*

(1985), Maclean and Hall (1987), Gurdon (1987, 1992) and Langille and Hall (1989).

4. For extracellular matrices and matrix-mediated interactions, see Hall (1982a, 1984d, 1988a, 1994b,c), Maclean and Hall (1987) and Ettinger and Doljanski (1992). For cell and substrate adhesion molecules, see Edelman (1988). See Wolpert (1981) for positional information and Steinberg and Poole (1982) for selective cell affinity.

5. For the discovery and elucidation of cell adhesion molecules, see Edelman (1986, 1988) and Takeichi (1987).

6. For paralysis as an experimental approach to interactions within the musculoskeletal system, see Hall (1984b, 1985a, 1986), Hall and Herring (1990), Bertram *et al.* (1997). For N-CAM as a molecular mediator of mechanical influence on periosteal cells, see Fang and Hall (1995, 1997).

7. For elaboration of the extracellular matrix as an essential element of our minimal definition of a cell, see Bissell and Barcellos-Hoff (1987), Maclean and Hall (1987), Gilbert (1992), Ettinger and Doljanski (1992) and Ashkenas, Muschler and Bissell (1996). For signalling between integrins, cytoskeletal filaments and the nuclear membrane, see Maniotis, Chen and Ingber (1997).

8. For reviews of extracellular-matrix-mediated skeletogenesis, see Hall (1982a, 1984d, 1988a, 1994b,c), Thorogood and Smith (1984), Van de Water and Galinovic-Schwartz (1986), Maclean and Hall (1987) and Sanders (1988, 1989) and the chapters in Hall (1994e,f). For extracellular matricial signalling in other organ systems such as developing mammary glands, see Ashkenas *et al.* (1996) and Robinson and Hennighausen (1997).

9. See Hall (1981) and Pinto and Hall (1991) for induction of scleral ossicles by mandibular epithelium, and as a diffusion-mediated, epithelial-mesenchymal interaction. For mediation of these interactions by growth factors, see Hall (1988a, 1994b,c), Coffin-Collins and Hall (1989), Hall and Coffin-Collins (1990), Hall and Ekanayake (1991), Xue, Gehring and Taira (1991), Frenz *et al.* (1992, 1994), Le Douarin *et al.* (1994), Watanabe and Le Douarin (1996) and Ekanayake and Hall (1997).

10. For structural and molecular heterogeneity of basement membranes, see Hay (1982), Hall (1982b), Hall and Van Exan (1982), Hall and MacSween (1984), Dziadek and Mitrangas

(1989), Leblond and Inoue (1989) and Sanders (1989).

11. The term *doppelte Sicherung* for the concept of double assurance was first applied to embryonic induction by Hans Spemann in relation to his studies on lens induction. It was adopted from engineering, where the term referred to the excess structural capacity built into designs as a safety factor; see Hamburger (1988) for the history of the term and concept, and section 6.6.1 for excess structural capacity in organisms.

12. See Hall (1981, 1988a,b, 1989), Noden (1983) and Thorogood (1993) for specification of morphogenesis of bone and cartilage within the mesenchyme. For salivary and mammary glands, see Bernfield and Banerjee (1982) and Kratochwil and Schwartz (1976).

13. The 1955 reviews by Yntema and Coulombre provide excellent entries into the older literature on induction of amphibian and avian sense organs. Henry and Grainger (1990) and Grainger (1992) contain analyses of recent work; Henry and Mittelman (1995) discuss the persistence of lens-inducing ability into adult life.

14. See Filatow (1925), Jacobson and Sater (1988), Saha, Spann and Grainger (1989) and Hall (1990b) for discussions of divergent mechanisms of lens induction.

15. For cellular aspects of these interactions, see Model, Jarrett and Bonazzoli (1981) and Hall (1982a, 1989).

16. The classic literature on limb development is summarized in *Developmental and Cellular Skeletal Biology* (Hall, 1978) and in *The Development of the Vertebrate Limb* (Hinchliffe and Johnson, 1980). See Niswander (1997) for a recent summary of what can be learned from the study of limb mutants.

17. For the role of FGFs in limb induction, see Fallon *et al.* (1994) and Cohn *et al.* (1995). For induction of either wing or leg from the same population of flank mesenchyme and concomitant changes in expression of *Hoxb-9*, *Hoxc-9* and *Hoxd-9*, see Cohn *et al.* (1997). See Tanaka *et al.* (1997) for induction of limbs at the D-V boundary.

18. See Parr and McMahon (1995) for *Wnt-7a*, Laufer *et al.* (1997) and Rodriguez-Ezteban *et al.* (1997) for *Radical fringe* and positioning of the AER at the dorsoventral boundary, and Niswander (1997) and Zeller and Duboule

(1997) for overviews of these and other genes involved in early stages of limb bud development. Signals from somitic and lateral somatopleural mesoderm also are involved in determination of D-V polarity of the limb bud (Michaud, Lapointe and Le Douarin, 1997).

19. Evidence for involvement of RA in signalling from the ZPA in the developing limb bud comes from the experiments of Riddle *et al.* (1993), Johnson, Riddle and Tabin (1994) and Niswander *et al.* (1994).

20. Evidence for up-regulation of *Msx-1* and *shh* by FGF-4 comes from the studies by Niswander *et al.* (1994), Kostakopoulou *et al.* (1996) and Li *et al.* (1996), and see the review by Tickle (1996).

21. Linkage between the signals involved in the three primary limb axes is based on the work of Francis *et al.* (1994), Laufer *et al.* (1994) and Francis-West *et al.* (1995). For the distribution and function of *slug*, see Ros *et al.* (1997) and Buxton *et al.* (1997).

22. For analyses of the overlapping expression boundaries of *Hox* genes in the developing limb bud, see Duboule (1992), Nelson et al. (1996) and Tickle (1996).

23. See Kawakami *et al.* (1996) and Yokouchi *et al.* (1996) for overexpression of BMP receptors and the *in vitro* analyses. See Ganan *et al.* (1996) and Macias *et al.* (1997) for cell death and altered morphogenesis of limb mesenchyme in response to application of BMP *in ovo*. Recent analyses of the function and control of programmed cell death (apoptosis) are Jacobson, Well and Raff (1997) and Nagata (1997). See Winograd *et al.* (1997) for the role of Msx-2 in determining the balance between cell survival and cell death in craniofacial development.

24. For the origin of odontoblast precursors in the neural crest, see Lumsden (1987) and Imai *et al.* (1996). For summaries of mammalian tooth development, see Lumsden (1987, 1988), Slavkin *et al.* (1990), Ruch (1995), Thesleff (1995), Thesleff *et al.* (1995a,b) and Thesleff and Sahlberg (1996).

25. See Thesleff and Sahlberg (1996) for a summary of the growth factors involved in tooth induction. See Kolodziejczyk and Hall (1996) for an analysis of growth-factor receptors and signalling.

26. Signalling from dental mesenchyme to epithelium by BMP and FGF is based on studies by Béque-Kirn *et al.* (1992, 1994), Satokata and Maas (1994), Matzuk *et al.* (1995), Chen *et al.* (1996) and Mitsiadis *et al.* (1997). BMP signals, in part, through activation of a transcription factor, lymphoid-enhancer-binding factor 1 (LEF-1; Kratochwil *et al.*, 1996).

27. For up-regulation of homeobox-containing genes during tooth and mammary gland development, see Jowett *et al.* (1993), Mitsiadis *et al.* (1995), Weiss *et al.* (1994), Weiss, Ruddle and Bollekans (1995) and Phippard *et al.* (1996). See D. Davidson (1995) for a general overview. See Zhang and Bradley (1996) for deletion of BMP-2 and Hogan (1996a,b) for comprehensive reviews of the role of BMPs in development.

28. The enamel knot and tooth morphogenesis are discussed by Butler (1956), MacKenzie, Ferguson and Sharpe (1992), Thesleff and Sahlberg (1996) and Vaahtokari *et al.* (1996a).

29. The expression in common of *shh*, BMP-2, -4 and -7 by enamel knot, notochord, ZPA and AER is based on the studies of Francis *et al.* (1994), Jernvall *et al.* (1994), Laufer *et al.* (1994), Francis-West *et al.* (1995), Jernvall (1995), Lyons, Hogan and Robertson (1995), Kostakopoulou *et al.* (1996) and Vaahtokari *et al.* (1996a).

30. For studies on EGF and retinoids in tooth development, see Kronmiller (1995), Kronmiller, Nguyen and Berndt (1995) and Hu *et al.* (1992).

31. For discussions of the inhibitory role of neural crest on induction of such organs as lens, heart and kidneys, see Jacobson (1966, 1987), Jacobson and Sater (1988) and Hall and Hörstadius (1988).

12

INTEGRATING ORGAN SYSTEMS, DEVELOPMENTAL CANALIZATION AND ASYMMETRY

'Mechanisms for ensuring developmental stability are as crucial for maintaining the continuity of a species as the basic information for encoding gene products in the DNA itself.'

Wilkins, 1997, p. 261

Tissues and organ systems in vertebrate embryos develop as a result of cascades of inductive interactions that activate the developmental programme for that particular tissue or organ. Although we know a great deal about inductive events for some tissues, organs and regions of individual embryos, we cannot yet produce a complete epigenetic blueprint for a single embryo. It is therefore important to identify further interactive cascades, to understand their cellular, molecular and genetic bases, and to incorporate them into modern approaches to functional morphology on the one hand, and models of morphological change in development and evolution on the other.[1]

Following initiation of organogenesis, interactions with adjacent tissues increasingly direct morphogenesis and growth, and, in some instances, differentiation, as in the formation of secondary cartilage on dermal bones (section 19.7.2). The mechanisms underlying these interactions are diverse and vary with time, from organ to organ and in concert with population, ecological and environmental regulators.

In this chapter I discuss the integrative and integrating component of embryonic development, using aspects of amphibian craniofacial development as an example (s. 12.1). Among other model systems discussed in the book are developmental and structural adaptation to dwarfing (miniaturization) in wild populations, phenotypic plasticity/trophic polymorphisms (Chapter 18), and the development of pharyngeal jaws in cichlid fishes and secondary jaw articulations in birds (elaborated in ss. 16.5 and 17.2.2). I round out the chapter by considering how stability is maintained through developmental canalization (s. 12.2) and how deviations from stability ('developmental noise'), as measured by fluctuating asymmetry, can be used to analyse the mechanisms that produce stability (s. 12.3).[2]

12.1 INTERACTIONS AT THE TISSUE LEVEL: AMPHIBIAN CRANIOFACIAL DEVELOPMENT

A complex series of interactions involving four primary cell populations (neural crest, pharyngeal endoderm, lateral and prechordal plate mesoderm), produce the mesenchyme and epithelia that form the teeth, cartilages, bones, loose connective tissue and epithelia of the buccal cavity and associated structures.

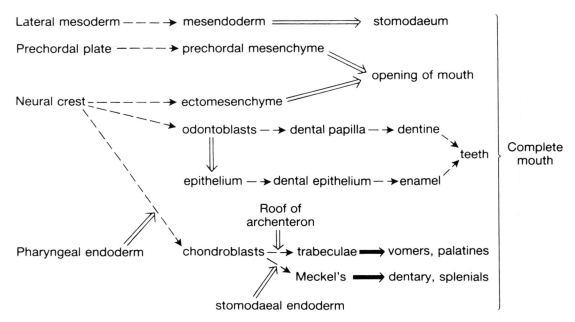

Fig. 12.1 A summary of the epigenetic cascades involving mesoderm, neural crest cells and epithelia in the production of cartilage, bone and teeth in the oral region of the European urodele *Pleurodeles waltl*. Dashed arrows indicate differentiation of a cell or tissue type. Open arrows indicate inductive interactions. Solid arrows indicate epigenetic interactions between tissues. Based on the work of Cassin and Capuron (1979) and reproduced from Hall and Hörstadius (1988) by permission of the Oxford University Press.

12.1.1 Tissue interactions

Our appreciation of the sequences of interactions involved in the integration of cartilage, bone and tooth development in the craniofacial skeleton comes in large part from studies undertaken in the late 1950s and '60s of the frog *Discoglossus pictus*. Figure 12.1 summarizes some of these interactions.[3]

A cascade of inductive interactions begins as pharyngeal endoderm induces mandibular neural crest cells to form the suprarostral and infrarostral cartilages of the jaws. Pharyngeal endoderm (and cartilage, once formed) induces stomodaeal ectoderm of the future mouth cavity to form teeth, beak and oral papillae (Figure 12.2). In the absence of cartilage differentiation, the mouth, oral papillae, teeth and keratinous beak all fail to form.

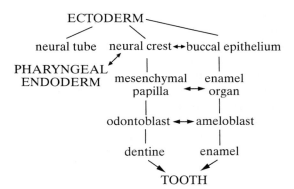

Fig. 12.2 The major sequences of differentiation of neural crest and buccal epithelium in forming odontoblasts and ameloblasts in mammalian tooth formation (single arrow). Induction interactions are shown by double arrows. Modified from Hall (1975).

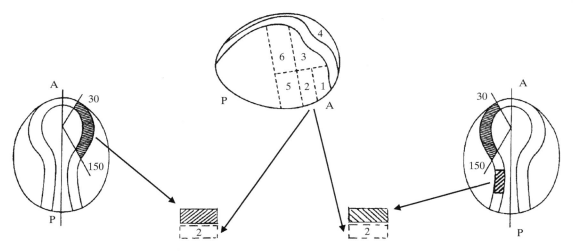

Fig 12.3 A schematic representation of experiments in amphibians in which neural crest (shaded in the neurulae seen from the dorsal surface) is recombined with endoderm (1–6 in the neurula seen in lateral view) before being raised in culture. The recombinations shown are between cranial neural crest which lies between the 30° and 150° marks (left) or trunk neural crest below 150° (right) and endoderm from region 2. See Graveson, Smith and Hall (1997) for further details of these experiments.

12.1.2 Mutations and the craniofacial region

Because regions such as the craniofacial are so tightly integrated developmentally, alteration in a single key early process can affect the entire region to a greater or lesser degree depending on the process(es) affected. One example introduced in Box 10.2 is those neural crest deficiencies manifested as neuro-cristopathies. Migration is a key process in the early development of neural crest cells. In amphibians, anomalous migration leads to failure of development of the organs of the mouth, abnormally placed teeth, fused cartilages, or induction of teeth ectopically in epithelial tissue.[4]

Some mutations affect only a subpopulation of cells, or have a more local action. The *premature death* (*p*) mutation in *Ambystoma mexicanum* (studied and maintained by John Armstrong and his colleagues in Ottawa) affects a subpopulation of neural crest cells that produce craniofacial cartilage and contribute to the heart. Other neural crest deriva-tives form normally in *p*-mutants; neural-crest-derived cartilages do not.

Combining neural crest with known inductive epithelia and establishing the recombined tissues in culture is an effective means of localizing the site of action of mutant genes or of identifying regionally specific populations of neural crest cells (Figure 12.3). Although branchial endoderm is deficient, pharyngeal endoderm (the cartilage inducer) is not; mutant endoderm induces neural-crest-derived cartilage and heart to develop from wild-type embryos. Competence of the chondrogenic subpopulation of neural crest cells is defective in (*p*) mutants.[5]

As might be anticipated, differentiation into the many cell types that arise from the neural crest is controlled by a range of molecules acting during development. Genes such as *Dlx* exert specific actions on craniofacial cartilages of the first and second branchial arches which arise from hindbrain neural crest but have little effect on midbrain-derived chondrogenic crest cells. I discuss other genes, such as *Pax-3* and *Msx-1*, in the context of differentiation of branchial arch and dental tissues in sections 11.5 and 16.1.

Another recently identified gene, *Xbap*, the *Xenopus* equivalent of *bagpipe*, which is expressed only in muscle of the midgut in *Drosophila*, plays a specific role in neural-crest-derived cartilages. *Xbap* is expressed in midgut muscles in *Xenopus*, a remarkable example of conservation of patterns of gene expression. *Xbap*, however, is also expressed in the neural-crest-derived precursors of the basihyobranchial, palatoquadrate and perhaps Meckelian cartilages (Newman *et al.*, 1997). It appears that in the transition to vertebrates, the old role for *bagpipe* in patterning the midgut was conserved and a new role – patterning cartilages – has been added. Such re-utilization of genes in new roles is emerging in the mid-1990s as an important mechanism for the evolution of developmental processes; see sections 9.4 and 15.2 for further examples.

Amphibians are not ideal animals for genetic screening and generation of mutations affecting craniofacial development, but zebrafish are. Induced mutagenesis has revealed numerous neural crest mutants in zebrafish. One, *chinless* (*chn*), has now been described in some detail. *Chinless* mutants lack neural-crest-derived cartilages and mesodermal-derived muscles in all pharyngeal arches, although cartilage and muscle precursor cells are present. Given the effects on neural crest and mesoderm, it appears that *chinless* acts downstream of neural crest cell initiation and specification to inhibit the interactions required for neural crest cell differentiation. Isolation of mutations of such genes will allow regional control of neural crest populations and craniofacial development to be uncovered. Indeed, mutations in *Dlx-1* and *Dlx-2* in mice provide evidence that both genes play important and overlapping roles in regulating P-D patterning of skeletal and soft tissues of the first and second branchial arches.[6]

12.1.3 Growth and morphogenesis

In amphibians, as in other vertebrates, skull morphogenesis is controlled in part by inductive interactions. Elsewhere I discuss studies in which exchange of nasal vesicles between *Ambystoma tigrinum* and *A. punctatum* produces nasal capsules intermediate in size between those typical of the two species (Hall, 1983b, 1998c). When the nasal placode is removed from embryos of *Rana sylvatica*, *R. pipiens* and *Triturus alpestris* before stage 19, all elements of the skull form, but skull shape is abnormal and connections between adjacent elements are misplaced.

In another example, removal of the otic vesicle from frogs and salamanders prevents the otic capsule from forming. This is not a species-specific interaction; frog otic vesicle can induce salamander otic capsule formation. A classic series of studies are those by Medvedeva on inductive interactions between the nasolacrimal duct and several skull bones (the prefrontal, lachrymal, septomaxilla) that form late in larval life or during metamorphosis in urodeles. These bones fail to form, or are small or incomplete following removal of the duct or its placodal precursor. Similarly, the vomer, palatine, dentary and splenial bones do not form unless the trabecular and Meckelian cartilages have already formed. While each element develops under its own cascade of interactions, all elements interact and so are integrated.[7]

Of interest, in terms of inductive cascades, is the effect of the annular tympanic cartilage on the epithelium; the cartilage induces and maintains the epithelial tympanic membrane in frogs. In *R. pipiens*, contact of the epithelium with the cartilage during larval life results in epithelial dedifferentiation and transformation into the tympanic membrane; continued contact with the cartilage after metamorphosis maintains the epithelium as a membrane. The stage of differentiation of the cartilage affects its ability to induce. (You will recall a conceptually similar example involving the annular ligament from s. 11.3 and illustrated in Figure 11.7). This is true not only for tympanic cartilage and tympanic membrane formation, but also for the columella and lamina propria portion of the membrane. The quadrate and suprascapular can also induce a tympanic membrane in *Rana palustris*.[8]

Another promising model system for exploring mechanisms of integration during ontogeny is exogenous application of hormones and the association/dissociation of chondrogenesis from osteogenesis and differentiation from morphogenesis in anuran cranial development. Morphological differences are established early in embryonic cell populations rather than at the tadpole stage of the life cycle, bind organs together in allometric relationships that remain so constant through metamorphosis of many anurans that species can be discriminated even in the smallest post-hatching tadpoles. These examples are considered in Chapters 15–17 in the context of whether evolutionary change is initiated by key innovations or through integrated changes of tissues and organs whose development is coupled.[9]

12.2 DEVELOPMENTAL CANALIZATION

Developmental processes proceed along only a limited number of defined pathways. Deviations from those pathways are controlled genetically and developmentally by threshold reactions. **Canalization** or developmental buffering is the process whereby the more invariant characters of an organism are channeled into restricted sets of developmental pathways – the valleys in Waddington's epigenetic landscape (Figure 12.4).

Canalization was developed as a concept by Waddington, and has more recently been defined as:

'. . . developmental reactions, *as they occur in organisms submitted to natural selection*, are in general canalized. That is to say, they are adjusted so as to bring about one definite

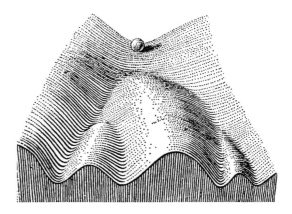

Fig. 12.4 The epigenetic landscape as devised by Conrad Waddington. Embryos (or cells) are represented by the ball which is channelled into one valley just as embryos or cells are channelled (canalized) along developmental pathways. Modified from Waddington (1975).

end-result regardless of minor variations in conditions during the course of the reaction.'
Waddington, 1942, p. 563

'the existence of developmental pathways that lead to a standard phenotype in spite of genetic or environmental disturbances.'
King and Stansfield, 1985, p. 54

Lerner saw genetic homeostasis as resulting from canalization of development. Sewall Wright saw universal pleiotropy in a similar light.

Canalization results from the action of selection upon the intrinsic genetic and epigenetic variability that allows organisms to respond to change, whether that change is genetically or environmentally initiated. Canalization allows unexpressed genetic variability to be built up in the genotype, even when it is not expressed in the phenotype and is therefore not directly subject to selection (ss. 19.4, 19.5). In part, canalization is based on the redundancy of genes in developmental programmes and the ability to compensate for mutations, and even deletions, of individual genes within cascades (s. 20.4.4).[10]

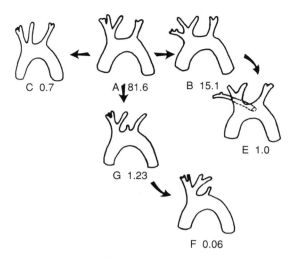

Fig. 12.5 The six types of variability found in 3000 rabbit aortic arches with frequency of each type shown as a percentage. Patterns A and B account for 96.7% of the variability. Reproduced from Calow (1983) based on data from Sawin and Edmonds (1949).

Canalization is expressed phenotypically as directed pathways of development leading to discrete morphologies occupying only a portion of available morphospace. The more highly canalized a character, the less it varies from individual to individual. Characters that define *Baupläne* or phylotypic stages are highly canalized. In vertebrate gastrulae/neurulae and germ-band stage insects, variability is limited to the basic patterns discussed in sections 8.3 and 8.4. Although variability is greater for other aspects of morphology, it is nevertheless restricted. In rabbits, for example, we find only six different patterns of aortic arches (Figure 12.5). Almost 97% of the variability is accounted for by two patterns, A and B in Figure 12.5.[11]

Other examples, considered in the following section, are variability in:

1. tarsal and carpal bones in a population of the salamander *Plethodon cinereus* at the extreme edge of its geographical range, and in

a large number of individuals from a single population of *Taricha granulosa* in California;
2. skeletal patterns in seven subspecies of the plethodontid salamander *Ensatina eschscholtzii* thought to be undergoing speciation; and
3. phalangeal formulae in geographically isolated Italian populations of *Triturus carnifex*.

12.2.1 Tarsal and carpal patterns in salamanders

Hanken (1983) documented eight carpal and five tarsal patterns in a Nova Scotian population of the red-backed salamander, *Plethodon cinereus*, at the edge of its geographical range (Figure 12.6). All involve reduction in carpal or tarsal numbers through fusions or failure of primordia to separate. Given the diversity of developmental mechanisms by which these elements can be reduced – failure to form, failure to separate from an adjacent element(s), fusion with an adjacent element(s), production of an additional element(s) – the limited numbers of patterns reflect canalization of development.

Some patterns are common, others are rare, representing 1% of the sample. Almost 90% of carpal variability is accounted for by just three patterns, with 70% of that variability in one pattern alone. Ninety-six per cent of tarsal variability is accounted for by just two pat-

Fig. 12.6 The five types (I–V) of variability of tarsal bones in a population of the salamander *Plethodon cinereus* at the edge of its geographical range. The frequency of the five tarsal patterns is: I (71%), II (25%), III (2%), IV and V (1% each). Patterns of fusion of tarsal elements are shown by Arabic numerals, e.g. elements 1 and 5 fuse in deriving pattern III from pattern I. c, c1, centrale and centrale 1; d1–5, distal tarsals 1 to 5; f, fibulare; 1, intermedium; t, tibiale. Reproduced from Hanken (1983) with permission of the publisher.

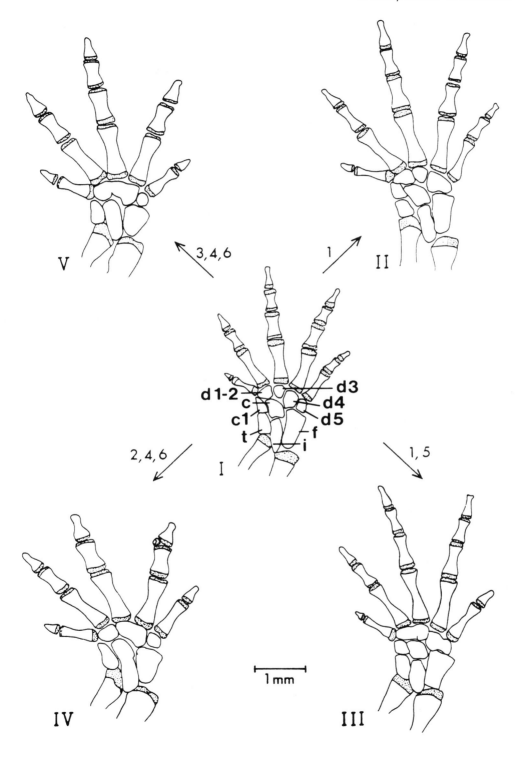

terns. Three of the carpal and two of the tarsal patterns each only represent 1% of the population. The adaptive significance of these patterns is not clear. Each may be equally adaptive, the restricted range of patterns reflecting the dynamics of the developmental processes producing the individual structures. This appears to be the case in a salamander, the polymorphic ring species *Ensatina eschscholtzii* studied by Frolich (1991) and D. B. Wake (1996c), in which skeletal patterning is conservative but proteins and colour patterns are not.

Following Emerson (1988a,b), who applied explicit tests of constraint to limb and girdle elements in frogs, Vogl and Rienesl (1991) applied tests for developmental constraint to Hanken's data set. They were not able to resolve whether constraint and/or stabilizing selection explained the conservatism of patterns. Rienesl and Wagner (1992) continued the search for developmental constraints in an examination of 384 limbs from *Triturus cristatus*, *Triturus marmoratus* and their natural hybrids. Frequencies of carpal and tarsal variation were the same in the two species and in the hybrids, although digit variation was significantly higher in the hybrids. Twelve patterns of tarsal fusion are theoretically possible, but only three are found. In a population at the edge of the geographical range, one pattern was even more predominant. Rienesl and Wagner concluded that these patterns were constrained in the two species, and that neither hybridization nor geographical isolation provided sufficient pressure to override that constraint. Their data support constraint (canalization) as an important element determining skeletal variation.

Pattern variation in skeletal morphology of adult urodele limbs, once identified, turns out to be much more common than anticipated. A sudden freeze which killed over 500 individual salamanders of the species *Taricha granulosa* from a single population, provided Neil Shubin, David Wake and Andrew Crawford with an enviable data base. Indi-

viduals with variable tarsal patterns were twice as common as those with variable carpal patterns (19% *vs* 9%), but the diversity of carpal patterns was greater. An important outcome of this study, possible only because of the large sample size and the authors' knowledge of amphibian evolution, was the discovery of patterns that are expressed in other extant or fossil salamanders, or which can be related to particular models of salamander limb development.

The eight patterns of carpals or tarsals described by Shubin, Wake and Crawford (1995) in this population represent the entire diversity of carpal and tarsal patterns seen throughout the urodeles. Two of the five bilaterally symmetrical patterns represent the atavistic reappearance of the ancestral plesiomorphic character state. The most common of these, the appearance of an additional element termed 'm' or 'm3', reconstructs the central postaxial region of the tarsus seen in the outgroup *Trematops milleri* and in the primitive cryptobranchid urodele *Andrias*. [This element was designated 'm3' by the Russian herpetologist and evolutionary biologist Ivan Schmalhausen (1917) in a study of the salamander *Ranodon sibiricus*. Because of its central position within the tarsal elements, m3 displays much evolutionary variation.]

The other three bilaterally symmetrical patterns duplicate patterns found in more derived urodeles; i.e. they anticipate evolutionary change. That patterns can be assigned as ancestral (atavistic) or derived ('forward looking') speaks to constraints on the design of the limb and to the utility of such population studies for recovering information about possible as well as past evolutionary change.

In the context of loss of characters, atavisms have usually been seen as reflecting past history, rather than as providing information about future evolutionary change (s. 17.5). For Shubin *et al.* (1995) and Shubin and Wake (1996) atavisms provide a basis for an analysis of the developmental bases of structural evolution, intraspecific variability and homology. In their words:

'Bilateral patterns of variation in *Taricha* both restore ancient structures and "anticipate" derived conditions that arise in parallel within highly nested taxa. These regularities suggest that the same processes that underlie the expression of atavistic characters are involved in the origin of evolutionary novelties.'

Shubin *et al.*, 1995, p. 882[12]

Pacces Zaffaroni *et al.* (1996) examined two geographically isolated populations of *Triturus carnifex* from northern and central Italy. Skeletal variation was 36% and 15% of forelimbs and 55% and 23% of hindlimbs in northern and central populations respectively. The basis for the variation was primarily fusions of two elements and variations on the phalangeal formula in the northern population, with no predominant patterns in the central population; see Box 16.1 for problems assigning phalangeal formulae and numbering digits. Variation in the hindlimb skeletons was much lower than forelimb skeletal variation in both populations. Such variable patterns in limbs, individuals, and populations speak to the likelihood of a variety of causative agents influencing skeletal patterns.

12.2.2 Skeletal patterns in wild, laboratory-reared and regenerating salamanders

Skeletal patterns are known to be more variable in regenerated limbs of animals maintained in the laboratory than in those in the wild. Indeed, regenerated amphibian limbs display a greater variety of patterns than do all limbs of all tetrapods. Fusion of elements in salamanders from natural populations (s. 12.2.1) is lateral, i.e., across the carpus or tarsus. Fusions during regeneration, however, are proximo-distal. Consequently, the patterns that result when limbs of *Plethodon cinereus* are amputated and allowed to regenerate, are not those seen in the wild. Naturally occurring

patterns are variations of limb development independent of regeneration. The patterns seen after regeneration are mechanical consequences of regenerating a limb in a terrestrial environment.[13]

Diet may be a contributing factor in laboratory animals. Martinez and colleagues demonstrated defective or failed ossification resulting from altered collagen metabolism in tadpoles of hatchery-reared *Rana pereri*. In part, this may be because of the prolonged cartilaginous phase associated with skeletal regeneration; Libbin and his colleagues documented that the skeleton in regenerating forelimbs of *N. viridescens* was still cartilaginous 270 days after onset of regeneration. In a subsequent paper, they saw this persistence as a reflection (they say recapitulation) of the prolonged retention of the cartilaginous phase and consequent slow replacement of cartilage by bone during urodele limb development. Such a prolongation of the cartilaginous phase in a functioning regenerating limb would certainly provide ample opportunity for fusions. Ossification is also delayed in laboratory-reared amphibians, demonstrating that different constrained patterns occur in different developmental and ecological conditions.[14]

12.3 FLUCTUATING ASYMMETRY AND DEVELOPMENTAL NOISE

I now want to consider a form of asymmetry – fluctuating asymmetry – which may be the antithesis of canalization and stability.

We expect bilaterally symmetrical organisms to be just that – identical on their right and left sides. Symmetry is an important consequence of canalization and developmental stability. Mechanisms promoting symmetry contribute to canalization.

Expectations of symmetry arise from our perception that development of symmetry is an error-free process. Embryos are selected for study based on staging tables that reinforce the constancy and symmetry of development

(s. 23.2). As organisms or their paired parts are normally symmetrical, we conclude that developmental processes are tightly controlled, even to the extent that separate left and right structures such as forelimbs develop in synchrony when nothing in their individual development integrates their development. If we are surprised when symmetry is not perfect, it is a reflection, in microcosm, of differences in the approaches of developmental and evolutionary biology. In the former, variation is often assumed not to exist; in the latter, variation is the stuff of life.

Deviations from symmetry, if random or non-directional, can be used as an index of deviation from the regularity, stability, or canalization of development. Such deviations, known as **fluctuating asymmetry**, are distinct from normally occurring **directional asymmetry** (as in the left and right portions of the mammalian heart), or **anti-symmetry** (as in the left and right claws of lobsters), or indeed as can occur as directional asymmetry between elements expected to be symmetrical such as left and right mandibles or femora. Although associated with variability (as is any biological process), directional asymmetry is unlikely to yield information about mechanisms underlying developmental stability. On the other hand, fluctuating asymmetry will.

Any bilateral character or characters that exist in pairs can be quantified for analysis of fluctuating asymmetry, but it is important to determine that asymmetry is neither biased nor directional. Teather (1996), who analysed tarsus, beak, and primary feathers in nestling tree swallows for fluctuating asymmetry, was surprised to find directional asymmetry in all three, with peak asymmetry at different times for each character. Distinguishing directional from fluctuating asymmetry is not easy. Nor is detecting the developmental basis of fluctuating asymmetry. Several examples illustrate these statements.

The molecular bases of directional asymmetry have begun to emerge with studies that indicate that the maternal gene *Vg-1* (from the TGF-β superfamily of growth factors) initiates left-right axis formation in *Xenopus*, and that *cSnr* (a zinc finger protein of the *Snail* family) is expressed only in mesoderm on the right-hand side and is involved in the characteristic handedness of looping of the heart in chick embryos. (A pathway involving *sonic hedgehog* and *nodal* is also involved in heart looping.) Altering the normal expression of *Vg-1* at the 16-cell stage in *Xenopus* or of *cSnr* at the head-process stage in the chick, produces embryos with randomized situs, reversing the normal left-right asymmetry of the heart and viscera.[15]

Analyses of developmental pathways using fluctuating asymmetry have considerable power:

1. Fluctuating asymmetry is an index of developmental noise, or how well development is canalized or buffered.
2. It can be increased through selection, as Leamy and Atchley (1985) demonstrated in skeletal features of rats selected for gain of body weight.
3. Heterozygosity, hybridization and developmental abnormalities all affect fluctuating asymmetry.
4. Acquired resistance to insecticides alters fluctuating asymmetry, as described in section 12.3.3 for Australian blowflies exposed to dieldrin or diazinon.
5. Ridley (1993) suggests that fluctuating asymmetry could be used as a test of the stability of the developmental programmes of Burgess Shale animals as an aid to resolving the controversy surrounding how different types of animals arose.[16]

12.3.1 Comparative studies

In one of the first comparisons across a diversity of taxa, Leigh Van Valen (1962) investigated a metric character (the number of microchaetae in *Drosophila melanogaster*) and parameters of cheek and upper teeth in deer mice (*Peromyscus leucopus*) and the extinct horse (*Griphippus* [*Hipparion*] *gratus*) respectively. Finding little asymmetry and a general lack of variation for buffering capacity, Van Valen raised the question of whether fluctuating asymmetry is more likely to reflect developmental stability at the individual or population levels, an issue that is still with us (Leamy, 1993). Demonstrations, for example by Soulé (1967) in 20 populations of lizards, that individuals are asymmetrical for many characters, and that island populations are more asymmetrical than land-locked populations, argue that mechanisms that affect individuals spread through populations.

Three decades later, Benedikt Hallgrimsson (1995), a student of Van Valen's, addressed the possible relationships between fluctuating asymmetry and maturational time in 22 dimensions of the teeth and cranial bones of six primate and three other mammalian species. He considered two explanations for the fluctuating asymmetry that acumulated during ontogeny:

1. **functional asymmetry**, in which explanations are sought in factors external to the organs under study but consistent with known mechanical influences on bone growth; and
2. a **morphogenetic-drift** hypothesis based on mechanisms of growth intrinsic to skeletal development.

Separating these two explanations is not easy. As demonstrated by computer models of bone growth, and an analysis of motor activity and the effects of paralysis, mechanical influences direct bone growth from early in ontogeny. Such controls as rates of cell division, cell enlargement, and matrix deposition, previously thought to be intrinsic, may be under extrinsic control from early in ontogeny. Any accumulation of fluctuating asymmetry with time during skeletal development, as detected by Hallgrimsson, may reflect the increasing role of extrinsic factors or functional asymmetry.[17]

12.3.2 Degeneration and the Mexican cave fish

Degeneration has been a subject of study in evolutionary biology since 1880, when Lankester wrote *Degeneration: a Chapter in Darwinism*, in which he argued that natural selection, in addition to better fitting organisms to their environment, could reduce the complexity of an organism. Animals, such as parasites, with degenerate adults or larvae might be expected to show increased asymmetry as embryos. A greater symmetry early in development is exemplified in situations where paired organs degenerate, become vestigial or non-functional, as occurs with sense organs in parasitic and cave-dwelling species.

Langecker, Schmale and Wilkins (1993) compared development of the degenerating eyes of the Mexican cave fish, *Astyanax fasciatus*, with individuals from the Rio Teapao, the form from which the cave fish descended. Cave-dwelling fish have small, rudimentary eyes buried under the skin. Eye development is similar in both river- and cave-dwelling forms until hatching (which occurs some eighteen hours after fertilization), including expression of the gene for the visual pigment opsin in the eye rudiment. During the second day, cell death in the retina limits further development and growth of the eyes in the cave dwellers. There is little variation in morphogenesis or size between left and right eyes during this regressive phase, although adult eyes vary considerably. As in Hallgrimsson's study, fluctuating asymmetry has a cumulative effect, increasing with developmental time and organ maturation. Perhaps not unexpectedly, maximal asymmetry appears during the growth phase, suggesting external control at these stages.

It is not possible, however, to determine the balance of intrinsic and extrinsic control over fluctuating asymmetry by pinpointing the time when maximal asymmetry appears alone. Although detecting heritability of fluctuating asymmetry is not easy, other studies discussed below point to genetic control over developmental stability.[18]

12.3.3 Genetic bases of fluctuating asymmetry

McKenzie and Yen (1995) investigated asymmetry in strains of the Australian blowfly *Lucilia cuprina* that were either resistant or susceptible to the organophosphate insecticide dieldrin. They measured left and right bristle numbers in flies subjected to environmental factors such as altered temperature and larval density, in the presence or absence of various concentrations of dieldrin. Dieldrin-susceptible strains showed increased fluctuating asymmetry in response to temperature and larval density. Furthermore, their asymmetry scores were positively correlated with dieldrin concentration.

In a continuation of these studies, Batterham *et al.* (1996) and Davies *et al.* (1996) studied the resistance gene *Rop-1* and its proposed modifier *Scalloped wings* (*Scl*). Resistance to diazinon is controlled by allelic substitution at the *Rop-1* locus. Twenty years after introduction of diazinon, many blowfly populations are close to fixation for the resistant allele (*R*). Batterham and colleagues argue that *Scl* is the blowfly homologue of the *Drosophila* gene *Notch*, which encodes a large transmembrane protein involved in cell adhesion. Altered cell adhesiveness in resistant populations and the instabilities of morphogenesis and growth that follow changes in cell surface properties, is a likely cellular mechanism mediating fluctuating asymmetry of bristle numbers.

These and similar studies (for example, the demonstration by Lacy and Horner (1996) of thresholds for fluctuating asymmetry in inbred populations of the Australian long-haired rat, *Rattus villosissimus*, and of Leamy *et al.* (1997) of the small number of quantitative trait loci affecting fluctuating asymmetry of murine mandibles) demonstrate that:

- environment and genotype both influence fluctuating asymmetry, as indeed they influence developmental canalization;
- environmental responses are genotype-dependent (Chapter 19); and
- the role of embryonic development in evolutionary concepts such as fitness and adaptation can be approached through analyses of developmental stability (Chapter 20).[19]

12.4 ENDNOTES

1. For discussions of inductive cascades and the evolution of morphology, see Gould (1977), Hall (1982a, 1984b, 1990b) and Alberch (1985). See M. H. Wake (1990, 1992b) and Atchley and Hall (1991) for incorporation of hierarchical thinking into functional morphology and quantitative genetics models respectively. See Maze *et al.* (1992) for a similar approach in plant biology.

2. For literature on miniaturization, see Hanken (1984), Hanken and Hall (1993b) and Hanken and Wake (1993). For paralysis as an experimental approach to interactions within the musculoskeletal system, see Hall (1984d, 1985a, 1986), Hall and Herring (1990), Bertram *et al.* (1997) and Fang and Hall (1995, 1997).

3. For analysis of the literature on interactions leading to development of the mouth and craniofacial region in tadpoles of *Pleurodeles waltl, Triturus alpestris* and *Discoglossus pictus*, see Cassin and Capuron (1979) and Hall and Hörstadius (1988). See Seufert and Hall (1990) and Clemen (1978, 1979) for studies with *Xenopus* and *Salamandra* respectively. For interactions at the tissue level in rodents and fish, see Lumsden (1987, 1988), Huysseune (1983, 1989), Huysseune and Verraes (1987) and Huysseune, Sire and Meunier (1994). For a broad overview of these issues, see Hall and Hörstadius (1988).

4. Effects of anomalous migration were demonstrated in the re-analysis by Levy, Detwiler and Copenhaver (1956) of the histological specimens from a study reported by Detwiler and Copenhaver in 1940.

5. For studies on the *premature death* (*p*) mutant in *A. mexicanum*, see Mes-Hartree and Armstrong (1980) and Graveson and Armstrong (1990, 1996).

6. For *chinless* (*chn*) mutation and *Dlx* in the zebrafish, see Schilling, Walker and Kimmel (1996) and Ellies *et al.* (1997) respectively. For a list and discussion of 109 mutants that affect zebrafish branchial arches and 48 that affect craniofacial development – an extraordinary resource for developmental biology – see Schilling *et al.* (1996), Piotrowski *et al.* (1996), Neuhauss *et al.* (1996) and Schilling (1997). See Qiu *et al.* (1995) for studies with *Dlx-1* and *-2* in mice.

7. See Medvedeva (1986a,b) and earlier studies cited in those papers for inductive interactions involving the nasolacrimal duct. Hall (1991a) documents in some detail tissue specificity of interactions involving amphibian otic vesicle. For discussions of the none too conclusive literature that placodal derivatives induce lateral line canals and/or associated dermal bones, see Hall and Hanken (1985a), Webb and Noden (1993) and Wonsettler and Webb (1997).

8. These classic studies on the tympanic cartilage by Helff (1927, 1934, 1940) cry out for re-analysis using modern molecular approaches.

9. See Strauss and Altig (1992) and Hall and Miyake (1995b) for early establishment of morphology in embryonic cell populations. On exogenous application of hormone and dissociation of aspects of skeletogenesis, see Emerson (1987), Hanken and Hall (1988a,b), Davies (1989) and Trueb and Hanken (1992).

10. See Waddington (1957a), especially pp. 19–41 for developmental pathways and threshold reactions; Waddington (1942, 1959) and Waddington and Robertson (1966) for canalization; and Lerner (1954) and Wright (1968) for genetic homeostasis and universal pleiotropy respectively. If internal selection or molecular drive operate within the genotype, genetic variability would undergo a process of sifting (Whyte, 1965; Dover, 1982), and/or initially unrelated genes coding for similar molecules could be incorporated into individual cascades (Cooke *et al.*, 1997). See Scharloo (1991), Wagner, Booth and Bagheri-Chaichian (1997) and Wilkins (1997) for recent perspectives on the genetics underlying canalization.

11. See Edmonds and Sawin (1936) and Sawin and Edmonds (1949) for limited numbers of aortic arch patterns in rabbits.

12. See Hall (1998c) for a review of these and other skeletal patterns in amphibians, and Gollmann (1991) for basipodial patterns typical of other taxa in the Australian anurans *Geocrinia laevis*, *G. victoriana* and their hybrids. Neoteny is one mechanism by which ancestral patterns can be reinstated in salamanders, as illustrated by the pattern of the skull in *Triturus alpestris* (Rocek, 1996).

13. See Goodwin (1984, pp. 114ff), Scadding (1991) and Martinez *et al.* (1992) for differences in regeneration between animals in the laboratory and in the wild. See Hanken and Dinsmore (1986) and Dinsmore and Hanken (1986) for analysis and explanations of skeletal patterns in regenerating salamander limbs, and Vaglia, Babcock and Harris (1997) for patterns of salamander tail regeneration throughout the life cycle.

14. See Libbin *et al.* (1988, 1989), Trueb (1985), Neufeld (1985), Hall and Hanken (1985b) and Hall (1998d) for studies relating to cartilage in regenerating newt limbs.

15. See Hyatt, Lohr and Yost (1996) and Isaac, Sargent and Cooke (1997) for the studies on *Vg-1* in *Xenopus* and *cSnr* in the chick. See Levin (1997) for an overview of left-right directional asymmetry in vertebrate embryos and for lists of mutants with left-right asymmetry and genes that affect symmetry. See Kraak (1997) for directional asymmetry.

16. For literature on fluctuating asymmetry and developmental buffering, see Ludwig (1932), Waddington (1957a), Van Valen (1962), Palmer and Strobeck (1992, 1997), Parsons (1990), Hallgrimsson (1993, 1995), Markow (1994) and Lacy and Horner (1996). For some of the initial studies demonstrating affects of heterozygosity, hybridization and developmental abnormalities on fluctuating asymmetry, see Soulé's (1979) study on 15 populations of side-blotched lizards (*Uta stansburiana*). See Ferguson (1986) and Adams and Niswander (1967) for analyses of slower growth and associated increased developmental instability in strains of rainbow trout (*Salmo gairdneri*).

17. See Hall and Herring (1990), Herring (1993a,b), Hallgrimsson (1993), Carter, Van der Meuleun and Beaupré (1996) and Bertram *et al.* (1997) for mechanical influences early in ontogeny.

18. See Wilkens (1988) for variation in eye size during ontogeny, and Whitlock (1996) for genetic control of fluctuating asymmetry.

19. See McKenzie and Batterham (1994) and Freebairn, Yen and McKenzie (1996) for genotype by environment interactions in fluctuating asymmetry.

PART FOUR
EMBRYOS IN EVOLUTION

13

INNOVATION, NOVELTY AND THE ORIGIN OF MULTICELLULARITY

'A comprehensive theory of organismic form is needed that relates embryonic development to longer-term processes of evolution, with the rationale that phenotypic structural change must originate through changes in ontogeny, and the two processes must be linked.'

M. H. Wake, 1992b, p. 315

Against the background presented in the last four chapters of development as hierarchical, integrative and epigenetic at all levels from origins of germ layers, through body plans and embryos, to organs and tissues, I now turn to how change, especially morphological and functional change, occurs in evolution and whether our knowledge of development as hierarchical and interactive allows us to 'explain' the developmental basis of such evolutionary change. In the past, most evolutionary biologists have not seen epigenetics (or even a hierarchical approach) as a necessary component of the mechanisms of evolutionary change. That situation has changed dramatically.

Chapters 13 and 14 treat key innovations and integrated changes in development and their roles in two major evolutionary transitions: the origins of multicellularity and the origins of the Metazoa. The same approach is taken to the origin and diversification of chordates and vertebrates (Chapter 15) and to evolution within the animal kingdom (Chapters 16 and 17).

13.1 DEVELOPMENTAL EVOLUTIONISTS AND EPIGENETICS IN EVOLUTION

In the 20th century, a number of embryologists and evolutionary biologists, collectively re-ferred to by Wallace Arthur (1988, pp. 56–9) as 'developmental evolutionists', attempted to incorporate developmental, integrative and epigenetic elements into evolutionary theory. The most notable attempts were those of Ivan Schmalhausen in the USSR, Conrad Waddington in Great Britain, Rupert Riedl in Austria, and Sven Løvtrup in Sweden. Although they did not collaborate, the approaches taken by these individuals have much in common. Wallace Arthur identified four common themes in their work:

1. development is hierarchical;
2. macromutations play a rare but important role in evolutionary change;
3. canalization of development is evolution-arily relevant; and
4. the environment is not just a sieve but interacts with developmental processes.

[I do not concur with Arthur's view that Waddington advocated macromutations, even on rare occasions. Nor, I suspect, would Waddington; see s. 19.2.]

Their ideas were propounded in a number of important books:

- *Factors of Evolution: the Theory of Stabilizing Selection* (Schmalhausen, 1949).

- *The Strategy of the Genes; New Patterns in Genetics and Development; The Nature of Life* (Waddington, 1957a, 1962a,b).
- *Order in Living Organisms* (Riedl, 1978).
- Epigenetics – a Treatise on Theoretical Biology; The Phylogeny of the Vertebrata (Løvtrup, 1974, 1977).

There is a remarkable parallel in the ideas developed by Waddington and Schmalhausen, especially amazing when we consider that the two men were isolated by geography, language and the politics of their day. The pivotal contributions by Waddington and Schmalhausen on canalization of development, genetic assimilation, stabilizing selection and norms of reaction are discussed more fully in section 7.2, 12.2, 18.3 and in Chapter 19.

Schmalhausen viewed developmental interactions such as gradients or embryonic inductions as providing the regulatory control over much of development, using as evidence, his experimental studies on tissue and organ transplantation in amphibian embryos. Variation in these interactions and their susceptibility to genetic change provided the genetic raw material upon which selection could act. To him, the mechanism of evolutionary change resided in development: 'In the last analysis, all new differentiations are due either to changes in characteristics and concentrations of the morphogenetic substances or to changes in reactivity of tissues and in their morphogenetic effects' (Schmalhausen, 1949, p. 32).

Løvtrup and Riedl also formulated epigenetic approaches and mechanisms for morphological change. Løvtrup stresses macromutations, emphasizing major leaps from one form to another, essentially as Geoffroy had done 150 years earlier. Løvtrup maintains that laws specific to biology cannot be reduced to a chemical basis, a disturbingly vitalistic view, and also advocates recapitulation. The distance of his views from the mainstream may be gauged by the title of his 1987 book, *Darwinism: the Refutation of a Myth*.[1]

Others also stress the importance of epigenetics. Victor Hamburger, a pioneer of 20th-century developmental biology, argued that epigenetics provides 'the missing chapter of evolutionary biology' (1980, p. 108). E. O. Wilson concluded that 'the key to macroevolutionary synthesis is epigenesis' (1981, p. 71). Pere Alberch and his colleagues documented the importance of epigenetic control as a basis for morphological change, while Sue Herring argued that 'the study of epigenetic mechanisms is the missing link between molecular genetics and functioning organisms' (1990, p. 408). Structuralists such as Webster and Goodwin invoke epigenetics as the basis for an approach to the maintenance and evolution of biological form. Evolutionary biologists such as John Maynard Smith and Eörs Szathmáry place considerable emphasis on an epigenetic inheritance system whereby states of gene activity are transmitted through cell division (Box 13.1). They see transition from limited heredity (in which only few states exist and are transmitted), to unlimited heredity (where very many states are transmitted), as crucial for major evolutionary transitions.[2]

In *Order in Living Organisms* (1978), Riedl developed the concept of **epigenetic burden** for the canalization that comes with extensive interactions between parts of developing embryos. Despite difficulties in the individuality of its language and assumptions of a catholic background knowledge, his book merits close analysis. Riedl saw vertebrate structures such as the notochord, neural crest, mesoderm and medullary canal as expressing the greatest epigenetic burden because they have persisted unchanged for hundreds of millions of years. These are the structures upon which the greatest number of developmental structures and processes depend. In part this is 'merely' because they arise early in development. More importantly, it is because of the interactive and hierarchical nature of the epigenetic cascades characteristic of vertebrate development (discussed in the previous two chapters) that flow

BOX 13.1
EPIGENETICS AS REGULATION NOT REPLICATION

In their 1995 book, Jablonka and Lamb concluded that what they term epigenetic inheritance systems (Chapter 7) are less stable, more sensitive to the environment (and they claim, therefore more adaptive), more directed, and have more predictable variation, but more limited alternate states, than genetic inheritance. They see epigenetic inheritance systems as forces for evolutionary change and as contributing to speciation and hybridization. Their closely documented and argued book is worthy of study by those interested in these inheritance systems.[3]

Jablonka and Lamb (1995, 1997) take epigenetic inheritance systems beyond Weismannian continuity of the germ line and separation of soma from germ line. The majority of phyla, and all plants, fungi and protists, lack a separate germ line; somatic inheritance is at least theoretically possible in such organisms (see Table 2.1 and s. 9.1.1). But Jablonka and Lamb go further to embrace a Lamarckian dimension even for metazoans in which germ line and soma are separate. Relationships between epigenetics, environment, replication and selection do not require adoption of a Lamarckian dimension, which is one dimension too many. I have commented on their Lamarckian approach to epigenetic inheritance elsewhere. My critique was that in metazoans, epigenetics functions as a regulator not as a replicator.[4]

The need is to integrate environment, genetics and epigenetics with the environment neither as replicator (an entity from which copies are made) nor interactor (an entity that interacts with its environment to promote differential replication and on which selection acts). The environment does not generate nor epigenetics replicate heritable variation in metazoans. Both can selectively evoke existing genetic variability, and epigenetics activates the determined states of cell lineages. Organization of the egg, not embryonic inheritance, is epigenetic.[5]

from them. Thus, the greater the epigenetic interactions, the greater the burden and the more conservative and long-lasting the features. Günter Wagner (1989a,b) discusses these as '**epigenetic traps**', a term that derives from the concept that the epigenetic nature of development constitutes a constraint on evolutionary change.

Features of early ontogeny which are in general more widespread (see s. 8.4 for exceptions) carry a greater burden than features of later ontogeny, which are often (although not always) more specialized. Epigenetic burden relates to the characters of *Baupläne* and con-

served or phylotypic stages. The implication of burden and epigenetic traps as constraints is that they vary with the importance to the body plan of the particular developmental structure or process. Wagner is careful to document that the epigenetic traps that bound variation arise from selection acting on genetic variability. I attempted to take the same care in section 6.8.1 when defining the processes operating when *Baupläne* are laid down. Key innovations and integrated changes utilize the epigenetic control of development to turn epigenetic burden or epigenetic traps to new ends.

13.2 KEY INNOVATIONS AND INTEGRATED CHANGE

Evolutionary biologists often discuss morphological change in the context of key innovations and/or integrated change, and it is in that context that I proceed. In Chapter 6, I raised three possible paths for the origins of *Baupläne*. Levinton (1988) distinguished three mechanisms that correspond to these paths. The paths and mechanisms are:

1. as **integrated units** – Levinton's saltational hypotheses, in which the whole organism, or major components of the organism, change as a unit; akin to the hopeful monsters proposed by Richard Goldschmidt (1933, 1940);
2. through single **key innovations** that set the stage for subsequent changes – Levinton's independent blocks hypotheses or mosaic evolution in which blocks of features evolve independently and slowly become integrated over time; and
3. by **gradual accumulation** of multiple (initially uncorrelated) features that are correlated over time – correlated progression hypotheses in which components become sufficiently interrelated that they evolve without losing those relationships.

Hopeful monsters are more often likely to be hopeless than hopeful. The epigenetic thesis developed in the previous chapters argues against the evolution of independent blocks except at the outset of metazoan evolution, when mosaic evolution would have been an important mechanism. Levinton argued, as I would, for gradual and correlated accumulation of features. He supports this view primarily with data on the origin of mammalian genera provided by Kemp (1982) and including gradual increase in the size of the jaw muscles, nature and complexity of the teeth, and alteration in the jaw articulation and postcranial skeleton. Following Lauder (1981), Levinton emphasized that even though features are tightly correlated in the current

members of a group, they need not have arisen simultaneously or in a highly correlated way. Given the strongly hierarchical nature of ontogeny, this is especially true for features of embryonic development. Conserved phases of embryonic development, although tightly correlated with one another in a causal hierarchical sequence now, arose sequentially, causal links being established one by one as development evolved and *Baupläne* arose (ss. 8.3, 15.1, 16.1.1). New features appear against the background of history (the constraints imposed by ancestral morphology), and in the context of adaptation to changing environments. Once evolved, they became highly canalized (s. 12.2).

Oft-cited examples of key structural, physiological, and/or behavioural innovations (nine of which are discussed herein, most in Chapter 16) include:

1. multicellularity and the origin of Metazoa and Metaphyta from unicellular organisms (ss. 13.5, 14.2);
2. the neural crest, dorsal nerve cord, tissue interactions, and the origin of the vertebrates (s. 15.1);
3. jaws and the transformation of jawless to jawed vertebrates (s. 16.1);
4. lungs and limbs and the origin of tetrapods from their aquatic ancestors (s. 16.2);
5. flight, feathers, and the evolution of birds from archosaurian reptiles (s. 16.3);
6. the shell and the origin of turtles (s. 16.4);
7. the development of pharyngeal jaws and speciation in cichlid fishes (s. 16.5);
8. changes in jaw suspension and middle ear ossicles and the origin of mammals (ss. 17.3.1, 24.2);
9. wings and the origins of insects (s. 16.6);
10. spiral cleavage and the origin of torsion in gastropods (s. 16.7);
11. constantly growing incisors and the origin of rodents;
12. larval versus adult instars and the origin/radiation of insects;

13. segmentation and the origin of annelids;
14. an exoskeleton/epicuticle and the origin of arthropods;
15. homeothermy and the origin of mammals;
16. enlarged brain size and the diversification of mammals;
17. internal fertilization, development of the amniotic (cleidoic) egg, and the origin of terrestrial amniotes from aquatic amphibians;
18. upright posture, toolmaking and language in the origin of humans from anthropoid apes;
19. evolution of placentae and origins of viviparous reptiles and placental mammals; and
20. flowers, insect/bird pollination, and the origin of angiosperms in the Cretaceous.

Although novelties are expressed most obviously as the appearance of new characters, novelty also embraces:

- loss of characters;
- alterations in shapes or numbers of characters;
- change in position; and/or
- change in function of a character.

The bones of reptilian lower jaws that transformed into middle ear ossicles in mammal-like reptiles and mammals are new characters, of different shapes, in a radically new position and serving a new function, sound transmission. They are also transformed lower jaw elements (ss. 17.3.1, 24.2.1). Innovations also may lead to novelties that bound, rather than promote change, or, as in the sabre teeth of the sabre-toothed tiger, that are innovations only in the short term.[6]

13.3 EVOLUTIONARY NOVELTY AND SPECIATION

Darwin dealt at length with key adaptations although he did not use that term. The first to use it may have been Miller, in 1949, but G. G. Simpson and Ernst Mayr developed the concept and emphasized the importance of key anatomical novelties that trigger a burst of subsequent evolution (by which they meant speciation). Since their work in the '50s, the notion of key innovations has been coupled to evolutionary novelty and/or speciation, although higher-level taxa are characterized not by special characters, but by groups of species that descended from a common ancestor possessing those characters.[7]

Mayr defined an evolutionary novelty as 'any newly arisen character, structural or otherwise, that differs more than quantitatively from the character that gave rise to it' (1960, p. 351). This definition could just as easily stand for key innovations, provided that 'character' is taken to include physiological and behavioural characters, and 'otherwise' to include ecological innovations. In 1982 Mayr qualified his definition to include any phenotypic change that permits an organism to perform a new function. Although Mayr specifically excluded the origin of new taxa from his discussion of evolutionary novelties (the processes, although often correlated, are sufficiently different to be regarded as separate), such innovations are vital for subsequent evolution. George Lauder, on the other hand, specified that key innovations could *only* be identified on the basis of associated speciation.[8]

Similarly, evolutionary biologists such as Futuyma and Cracraft tie novelty to new features of taxa, novelties becoming every shared-derived character of a taxon. Gerd Müller and Günter Wagner explicitly tie morphological novelties to homology: a novelty is neither homologous to any other structure in the same organism, nor to any ancestral structure. In essence their approach is to identify a novelty as a character state and then investigate and separate mechanisms responsible for generating the novelty from those responsible for maintaining and/or fixing it in the population. Such an approach integrates developmental and population-level analyses.[9]

Determining whether a structure is a shared derived character, homologue, or unique novelty, is not always easy. The 12 genera of the South American electric fishes of the family Apteronotidae all share a unique dorsal filament originating on the posterior half of the back. The dorsal filament is a novelty; no other fishes possess it. Yet, despite quite extensive analysis of these fishes and their closest neighbours, we still do not know the function of the dorsal filament, nor whether it is a modified adipose fin or a uniquely derived structure that evolved with the Apteronotidae. Knowledge of the ontogeny of the filament should help, as we could determine whether it is a homologue of the adipose fin of other fishes (Franchina and Hopkins, 1996).

Levinton (1988) saw a key innovation as having to satisfy two criteria: appearance of other character changes in a larger number of derived taxa, and key functional significance. Although he tied key innovations to speciation, he did not see them as necessarily causing speciation; 'a key innovation is necessary, but not sufficient for a subsequent radiation' (p. 305). A key innovation provides an opportunity, not a guarantee.

13.4 EVOLUTIONARY NOVELTY AND ADAPTATION

Key innovations have been linked with adaptation and the exploitation of new environments since Simpson's 1953 paper. Innovative changes and adaptive radiation that are analysed as morphological differences can reveal changes that transcend taxonomic changes or strikingly different ecology. Morphological changes in molar teeth of Eocene ungulates (s. 17.2.1), and post-Paleozoic crinoids (Foote, 1996) illustrate changes that transcend taxonomy or ecology respectively. Decoupling morphology from taxonomic status in such analyses permits us to tease out whether morphological innovation is, or is not, tied to generation of new taxa.

Stanley, Futuyma and Eldredge all argue that models of the origin of higher taxa are theories of the origin of adaptations, not of the origin of taxa *per se*. Simpson, who discussed key innovations in the context of key mutations (a mechanism now discounted), and fixation by Schmalhausen's stabilizing selection (s. 19.5), clearly saw key innovations as 'the ticket of entry' into a new adaptive zone; i.e. key innovation as a causal mechanism for the adaptive radiation that follows speciation (but see Mayr's comment above). Such an approach is especially operational in situations where the phylogeny of a group is insecure, but where a key innovation characterizes all members of the group. This is exactly the situation in bats (Order Chiroptera).[10]

With 927 species, bats, after rodents, are the second largest order of mammals. There are two suborders: megabats (Megachiroptera) and microbats (Microchiroptera). All bats fly, but the wings of the two groups differ, especially in the shoulder girdle and number of nails modified as claws. Megabats have claws on both thumb and index finger (digits I and II), microbats only on digit I. All microbats, but few megabats, navigate using echolocation. Relationships between micro- and megabats are sufficiently unresolved that they may each be more closely related to other mammals than to one another; i.e. bats may be diphyletic. If so, then the possibility that flight in mammals evolved twice has to be entertained. The biomechanical coupling of flight with echolocation in microbats (and the constraints placed on body size) argues either for a single origin in an echolocating ancestor, or for evolution of an adaptive complex of innovations in microbats not found in megabats. Resolution of the origins of such key innovations as flight in bats will require advances in our knowledge, not only of the mechanical, structural, developmental, physiological, and adaptational bases of mammalian flight, but also of phylogenetic relationships.[11]

It may be easier to identify key innovations in the evolution of embryonic stages – especially those phylotypic features that characterize generation of *Baupläne* (s. 6.4) – than in other transitions. I explore this idea in the context of the origins of multicellularity in the balance of this chapter, and in relation to the origins of the Metazoa in Chapter 14.

13.5 EVOLUTION OF MULTICELLULARITY

This section should be read along with section 14.2 on the origin of the Metazoa. This is not because all multicellular organisms are metazoans. They are not. Plants, fungi and many prokaryotes are multicellular. But multicellularity was a prerequisite for the origin of metazoans and consequently of metazoan ontogeny.

13.5.1 Origins

The origin of multicellularity was not the first step on the road to ontogeny, although it is often argued that only multicellular animals and plants (Metazoa and Metaphyta) have an ontogeny. Multicellularity evolved in several lineages of prokaryotes, for example, the Volvocales (*Chlamydomonas*, *Volvox*) and flagellated green algae, many of which consist of colonies or multicellular individuals with complex life cycles. Indeed, multicellularity arose on six independent occasions in the Volvocales. The smaller forms consist of as few as four and as many as 64 cells embedded in a common matrix. The largest forms, such as *Volvox*, are composed of thousands of cells. Bell and Koufopanou (1991) argue that the major limitation to multicellularity among the protists is their mode of reproduction. A rigid cell wall within which multiple fission occurs, favours the evolution of multicellularity in these groups; release of cells through a more flexible cell wall would not. Repeated division within a rigid cell wall – a precondition for the evolution of multicellularity within these groups – differs from multicellularity in the Metazoa, which do not have rigid cell walls. Plants, with their rigid cell walls, may be closer to the multicellular prokaryotes in how they attained multicellularity.

13.5.2 Advantages of multicellularity

The advantages of multicellularity are many:

1. organisms increase in size beyond the surface-to-volume limits set for single cells;
2. specialization into distinct cell types each with its own function(s) replaces intracellular specialization, which is limited to (by) cytoplasmic domains;
3. advantages accrue from segregation of cytoplasmic factors, cell membrane and microenvironmental specialization, and inductive cell-to-cell interactions;
4. feeding becomes more efficient;
5. dispersal is facilitated;
6. protection from predators and from environmental changes is afforded; and
7. sexual reproduction is facilitated.

Multicellularity has arisen repeatedly and development is a direct consequence of multicellularity:

'Because the increase in size has, under certain conditions, adaptive value, there has been a repeated and persistent trend towards a multicellular condition. If the organism is to be multicellular and its variation system unicellular, the direct consequence will be a development. To put the matter succinctly, development is the result of sex and size.'

Bonner, 1958, p. 11

In a more recent analysis, Michod (1997) demonstrated that sexual reproduction maintains higher levels of cooperation and lower levels of change when the fitness covariance of an organism overcomes any tendency of cells to leave a cooperative (multicellular) unit. To gain the advantages that accrue from genetic recombination, gametes have to be single cells; to gain the advantages of specialization and to overcome surface to volume constraints, organisms need to be multicellular. A life cycle that starts from single-celled gametes, followed by division into a multicellular stage, provides both advantages and in so doing generates development.

Development is the compromise between recombination of parental genes and specialization as an individual.

13.5.3 Some cellular mechanisms

The origin of multicellularity would have required cells that normally reproduced by fission and then separated into individual unicellular organisms, to develop mechanisms either to stay together or to come together (fail to separate or aggregate). Aggregation would produce an organism with at least two genetic constituents, essentially a colony. Failure to separate would produce an organism whose cells had identical genetic constitutions, essentially a blastula. One could also imagine a 'hybrid' situation in which cells which failed to separate combined with other cells to produce a new hybrid organism. Some multicellular organisms such as *Volvox* arise by staying together. Others, such as myxobacteria (slime bacteria), myxomycetes (slime molds), dictyostelid amoebae, and some ciliated protozoa, arise by aggregation. Cellularization of a syncytium, another theoretical possibility, appears not to have been evoked.

Could the same cellular properties control cell fusion and failure of cells to separate? Fusion requires the evolution of cell adhesion molecules or ionic coupling. Failure to separate requires the development of a mechanism to keep coupled cells together, again via cell adhesion molecules or ionic coupling. It may well be that the same molecular processes regulate failure to separate and fusion, the essential differences being in the timing of expression of cell membrane molecules. Cell adhesion, substrate adhesion (CAMs, SAMs) and cell junctional molecules that could have mediated such events were introduced in section 11.2.1. Failure of cells to separate would be characterized by early and continued expression of CAMs and cell junctional molecules and lack of (or reduced) expression of SAMs, so that CAMs exceeded SAMs. Fusion of separate cells would require

downregulation of SAMs and re-expression of CAMs and cell junctional molecules.

This dichotomy of possibilities – fusion or failure to separate – is based on properties of the cell membrane. A different dichotomy, one that is intracellular and tied to cell division, involves one pathway in which a daughter cell is maintained as a stem cell (the lineage pathway), and a second (the fission pathway) in which both daughter cells divide and in which cell-to-cell interactions generate cell diversity (Wolpert, 1990). Given the requirement for differential division, generation of stem cells, unequal distribution of cytoplasmic information and sophisticated cell-to-cell interactions required in Wolpert's model, it is best viewed as operating at a later evolutionary stage than the failure to separate versus fusion model. It is a model for the origins of early germ line segregation (s. 9.1.1).

13.5.4 Variability within cell lineages

In his superb analysis of the evolution of individuality and multicellularity, Buss (1987) argued for mechanisms that reflect interaction between selection at the level of the individual and selection at the level of the cell lineage, and emphasized initial selection of variants within that somatic environment. This concept raises the fundamental issue of where the variability lies that allows development to respond to natural selection so as to facilitate phenotypic change.[12]

As Norman Maclean and I argued in *Cell Commitment and Differentiation* in 1987, specification of cell fate is a stochastic process. Fate maps are not canonical. Blastomeres not only produce different cell types, but the products of individual blastomeres can vary from embryo to embryo within a single species. Variability in the fate of individual blastomeres within a single cleavage-stage embryo is an important but little-explored stochastic mechanism. Documented in *Xenopus* and in *Danio rerio*, such individual variability reflects the fact that the fate of these cells is not fixed early in development, that cells take up new positions during development, and that those new positions create situations for the cell-to-cell interactions and inductions that determine cell fate. Davidson used these concepts in his scenario of the origins of the Metazoa (s. 14.2.5). It is critically important to use a population approach to embryonic development to obtain information from multiple embryos of single species so that estimates of variability can be placed on apparently canonical fate maps.[13]

13.5.5 Transcriptional regulation

Transcriptional regulation did not arise with eukaryotes but was conserved in the transition from prokaryotic to eukaryotic cells. The key innovation of multicellularity is the ability to regulate where and when transcription occurs within multicellular embryos and organisms. In groups such as arthropods there is enormous divergence of body plans but conservation of the clusters of *Hox* genes that regulate body-plan organization (s. 9.4). This presents a strong argument for regulatory control of genes downstream from *Hox* genes as important elements in diversification. Indeed, that the diversity of segments in arthropods evolved without any increase in *Hox* gene number strongly implicates regulatory control as an important mechanism in body plan diversification (Grenier *et al.*, 1997). Regulatory control could be based on any one or combination of:

1. changes in *Hox* genes;
2. changes in upstream regulators of *Hox* genes;
3. changes in cofactors of *Hox* genes;
4. changes in how *Hox* gene products interact with downstream target genes such as N-CAM; and/or
5. changes in target genes.

Despite much searching, comparatively little is known about genes that act downstream of *Hox* genes. What is known is mostly from studies on *Drosophila*, in which some 19 downstream genes have been identified. These include structural genes (β3-tubulin, centrosomin), a cell adhesion molecule (connectin), and regulatory molecules (transcriptional factors and cell signalling molecules such as *Dll*, *decapentaplegic* (*dpp*), *wingless*, and *forkhead*). Such target genes are also highly conserved; *Dll* is expressed along the P-D axes of annelid parapodia, onychophoran lobopodia, ascidian ampullae and echinoderm tube feet, strongly supporting the origin of such genetic mechanisms in an early common ancestor. The homeobox genes *Dll*, *engrailed* and *orthodenticle* have now been shown by Christopher Lowe and Grey Wray to be involved in specification of the echinoderm body plan in ways that diverge from their use in bilateral metazoans. Furthermore, most target genes are regulated by several *Hox* genes, which, of course, makes determination of pathways even more difficult. Most target genes also function early in development before they come under *Hox* gene control, which makes research into their regulation and the unique roles played by *Hox* genes formidable tasks indeed.[14]

With multicellularity came a conservation of genetic mechanisms that might appear to be a genetic constraint, but that actually facilitated generation of additional cell specialization, especially as different cells came together and interacted. Within distinctive cell lineages genes became linked into networks or cascades, a different cascade for each cell lineage, furthering diversification. John Gerhart captured the essence of the consequence of such linkages in his summation of a Royal Society symposium on the cell in evolution:

'The cell's capacity to discriminate conditions outside and within itself, its capacity to generate conditions to which other cells respond, and its capacity to give individualized responses, all these arise out of contingency and linkage.'

Gerhart, 1995, p. 334[15]

Multicellularity facilitated increasing complexity and the beginnings of the metazoan taxa, the topic of the next chapter.

13.6 ENDNOTES

1. For development of his position, see Løvtrup (1974, 1982, 1983, 1984a,b, 1987, 1988).
2. For an epigenetic approach to morphological evolution, see Alberch *et al.* (1979), Alberch (1980) and Alberch and Alberch (1981). For expositions on the structuralist position, see Webster and Goodwin (1981), Goodwin and Trainor (1983), Ho (1984) and Goodwin (1994). For epigenetic inheritance systems, see Maynard Smith (1990), Szathmáry and Maynard Smith (1995) and Hall (1997c).
3. For recent literature evaluating epigenetic inheritance, see Jablonka and Lamb (1989, 1995, 1997), Maynard Smith (1990), Holliday (1991), Gruenert and Cozens (1991), Paro (1993), Szathmáry and Maynard Smith (1995), Russo *et al.* (1996) and Hall (1997c).
4. For critiques of the Lamarckian aspects of the approach of Jablonka and Lamb, see Walsh (1995), Hall (1997c) and the other papers in the issue of *J. Evol. Biol.* in which my paper was published. For epigenetics as regulator not replicator, see Maynard Smith (1990) and Hall (1997c).
5. See Dawkins (1976, 1978) and Hull (1980, 1981, 1988) for the replicator/interactor concepts and Hall (1997c) for more detailed development of these arguments.
6. Trueb (1996) is a good source for a recent discussion of novelties in relation to an analysis of the evolution of the skeletal system of pipid frogs, of which *Xenopus* is the most well-known example. See Eldredge (1989) for innovations that place bounds on evolutionary change.
7. See Simpson (1953, 1959), Mayr (1954, 1960) and Eldredge (1985) for development of the concept of key adaptations. For discussions of innovations, see the papers in the symposium volume edited by Nitecki (1990).
8. For the obligatory association of key innovations and speciation, see Lauder (1981) and Lauder *et al.* (1995).
9. See Futuyma (1986) and Cracraft (1990) for types of novelties. For an approach to the study of novelty, see G. B. Müller (1989, 1990, 1991, 1994) and see pp. 243ff in Müller and Wagner (1991) for further details of morphological novelty and homology.
10. For the association of key innovations with adaptation rather than speciation see, for example, Stanley (1981, p. 94), Futuyma (1986, p. 336) and Eldredge (1989). See Amundson (1996) and Rose and Lauder (1996a) for a history and evaluation of the concept of adaptation.
11. See Honeycutt and Adkins (1993) and Pettigrew (1991) for divergent views on the relationships among bats and of bats to other mammals; Thewissen and Babcock (1992) for independent evolution of flight in micro- and megabats; and Arita and Fenton (1997) for biomechanical coupling of echolocation and flight in microbats.
12. Gilbert (1992) has provided a critique of the concept of cell community in relation to the pioneering ideas of August Weismann. See Michod (1997) for an analysis of cooperation, conflict and selection in organisms without a germ line.
13. See Dale and Slack (1987) and Kimmel *et al.* (1990) for fate maps of *Xenopus* and *Danio rerio*.
14. For an overview of downstream targets of *Hox* genes, especially in *Drosophila*, see Graba *et al.* (1997). For conservation of *Dll* in metazoan appendages, see Panganiban *et al.* (1997). For involvement of homeobox genes in specification of the echinoderm body plan, including modification of the symmetry of expression domains and both addition and loss of developmental roles from those seen in bilateral metazoans, see Lowe and Wray (1997) and the accompanying commentary by Davidson (1997).
15. For studies on the conservation or evolution of regulatory control, see Erwin (1993), Warren *et al.* (1994), Carroll (1995) and Gerhart (1995).

14

COMPLEXITY AND THE ORIGIN OF THE METAZOA

'As any analysis of biological systems builds upon cell theory, an evolutionary analysis of development would benefit from describing evolutionary transformations as changes of the properties and activities of individual cells.'

Sommer, 1997, p. 225

In this chapter I examine the origins of the Metazoa against the backdrop of the evolution of multicellularity in the previous chapter and the discussion of the origin of phyla in Chapter 2. While separating this discussion from that on the origins of phyla in Chapter 2 may seem artificial, it is important to remember that the origins of body plans predated the origins of phyla, by perhaps hundreds of millions of years. Twelve chapters is a small gap in comparison.

Moving from unicellular organisms to multicellularity and metazoans immediately raises the problem of complexity and whether the transition from unicellular → multicellular → metazoans represents increasing complexity. From a discussion of complexity in section 14.1, I proceed to the conditions under which metazoans would have arisen (s. 14.2.1), including a discussion of phylotypic stages, zootypes and phylotypes (s. 14.2.2–14.2.3), whether larvae of present-day invertebrates shed any light on metazoan origins (s. 14.2.4), and the likely developmental mechanisms involved (s. 14.2.5).

14.1 COMPLEXITY

Complexity is notoriously difficult to characterize. Complexity is complex. Neville

Willmer's classic analysis, *Cytology and Evolution* (1960), an absolute gold mine of information and ideas, is the starting point for any analysis of the evolution of complexity at the cellular level.

Superficially, determining complexity would seem to be a trivial task. Ask the average person in the street and they would say that worms are simple and humans complex. But there is no agreed way of defining and measuring complexity (not for want of trying), and certainly no agreement that worms are less complex than we are, or that complexity has increased throughout the evolution of life. For instance, complexity in living things has been defined in terms of:

1. the size of the genome and/or the total number of genes;
2. the number of genes encoding proteins;
3. the number of parts, or the variety of morphological or behavioural units;
4. the number of cell types;
5. increased division of labour between parts;
6. the number of interactions between the parts of an organism or in the causal chains producing those parts, variously known as character entrenchment or hierarchical developmental complexity;
7. the length and complexity of the minimal statement required to describe the organ-

ism, an approach that effectively combines the other six.[1]

While biologists disagree on the best measure of complexity, all would agree that complexity increases during ontogeny, and that irrespective of which of the seven measures is used, complexity has increased during at least some portion of the evolutionary history of life on earth. Increasing size is associated with increasing complexity in development. Is this also true in evolution? Confounding the issue is the fact that complexity increases with size independently of phylogenetic relationships, but that different groups of organisms of the same size differ in complexity (Bell and Mooers, 1997). Cope's rule of the tendency of animals to increase in size during evolution embodies such thinking and raises such questions as whether evolution occurs more frequently from smaller rather than larger members of a lineage (Box 14.1).

14.1.1 Complexity and hierarchy

Perhaps the most comprehensive recent approach to complexity is that of Daniel McShea (1996). By separating process from pattern and hierarchical from non-hierarchical structure, McShea orders complexity into two nested types – morphological and developmental. After reviewing the evidence for each, he concludes that the Early Phanaerozoic trend toward increasing hierarchical developmental complexity associated with (causing?) the origin of body plans provides the only good evidence for increasing complexity.

Complexity increases during the ontogeny of multicellular embryos, and development is hierarchical. Consequently, new properties emerge as development unfolds. Simon (1962) provided an elegant exposition of why hierarchical organization is basic to what he termed 'the architecture of complexity', focusing on two main properties of hierarchical organization:

1. Hierarchical systems possess common properties that are independent of the specific content of the system. This theme was developed by Waddington in his attempt to unify genetics and embryology, and is discussed in Chapter 19 in the context of developmental programmes that change without every gene involved in the programme having to change.
2. Hierarchical systems evolve faster than non-hierarchical ones of similar size and complexity. In organisms such as vertebrates, where interactive processes are common and development hierarchical, we expect to see more epigenetic control and increasingly complex interactions as development proceeds, and we do.[5]

14.1.2 Numbers of cell types

Bonner (1988) elected to use number of cell types as an index of complexity and a guide to the relative level of organization of organisms with a particular body plan. Given the subjectivity of deciding what is a cell type (are all blood cells one type or are erythrocytes and the various white blood cells separate cell types?), Bonner's numbers do correlate with our subjective notions of organismal complexity. They also coincide with our understanding of phylogenetic relationships among phyla (s. 2.3). Numbers of cell types as low as 6–12 are found in sponges and cnidarians. Flatworms have 20–30, nemertines 35–40. A number of phyla have around 55 cell types, e.g., molluscs, arthropods, annelids and echinoderms, each of which has a distinctive body plan, and one of which (echinoderms) is based on distinctly different axes of symmetry from the rest. At this level, any close correlation between number of cell types and organismal organization breaks down.

Bell and Mooers (1997) estimated numbers of cell types for representatives of many phyla of animals, plants, fungi, green algae, amoebae, ciliates and brown seaweeds. Their use of the number and diversity of cell types in the

BOX 14.1
COPE'S RULE AND THE EVOLUTION OF NEW LINEAGES

Working in the 1870s and '80s, Edward Drinker Cope proposed an evolutionary rule that was based on theories of embryological development. We know it as Cope's rule that animals tend to increase in size during evolution. A leading American contributor to evolutionary theory, Cope saw evolutionary laws firmly grounded in the regular, stepwise changes of embryology rather than in chance random variations.[2]

Evolution often occurs most rapidly in animal species of small size. Recent investigators have sought explanations for Cope's rule in the dynamics of the interplay between size, generation time, speciation, extinction, or optimality for energy utilization. A specific concern is whether new taxa preferentially arise from small organisms. Some investigators claim to have found evidence.

Stanley (1973) propounded the idea that new lineages are founded from organisms of small size. In their analysis of 342 Caenozoic planktonic species of foraminifera, A J. Arnold *et al.* (1995) saw the size increase embodied in Cope's rule as a consequence of the greater adaptive responsiveness of small species with short generation times. They see size as correlated with evolutionary change, not causing that change. The ontogenetical feature of shorter generation time, not adaptation related to body size *per se*, results in patterns of increasing size during evolution. Such studies affirm the validity of Cope's rule.[3]

David Jablonski's 1996 and 1997 analyses of all the known marine bivalves and gastropods from the Upper Cretaceous Gulf and Atlantic coastal plains of North America – a total of 1086 species in 300 genera, representing 191 lineages – did not. In part this reflects Jablonski's application of Cope's rule in the strict sense of the combination of net increase in body size and loss of small-sized species. This coupled trend was no more common than either an increase in size range or net decrease in size over 16 million years. It is arguable whether this strict interpretation can be applied apart from a context in which phylogenetic relationships between lineages that show the two changes are compared. Origination of new lineages from small-sized species on the one hand, and advantages of increased size in established lineages on the other, can only be teased apart in a fine-grained phylogenetic analysis. Even these most comprehensive data sets assembled to test Cope's rule might be inadequate to support a conclusion that the rule is invalid, or that macroevolutionary patterns do not have their origin in microevolutionary processes. It is important to remember that neither Cope's rule, nor any other 'rules' of evolution, operate in isolation.[4]

nematode *Caenorhabditis elegans* as the standard (27 cell types within the categories of epithelia, nervous tissue, mesoderm, intestine, glands, excretory tissue, ovary and endothelium) immediately raises the question of an appropriate standard. A difficulty arises if complexity is not uniform but is related to numbers of cell types. The use of *C. elegans* (or any other species) as a standard, would result in underestimation of cell numbers for (more complex) organisms with a larger number of cell types. Thus, the three vertebrates used in their analysis (two fish and the mouse) yielded cell numbers of 112, 116 and 102, when others identify many more. In his comprehensive overview of complexity in relation to metazoan evolution, McShea (1996) suggested that an approach based on estimating numbers of different types of cells is not without its problems. His criticism is based, not on the standard selected, but on the difficulty of determining numbers of cell types in organisms with very few types. In fact, numbers of cell types can be determined with greater ease in organsms with few, than they can for those with many. This is certainly true for vertebrates, especially mammals, where estimates range as high as 210 for human cell types, but where some workers estimate twice that number if all neuronal and other subtypes are treated as separate types.[6]

Therefore, it is not easy to determine whether numbers of cell types in different phyla represent the level of cellular complexity required to generate their body plans, or for different organisms to function. Numbers of cell types may not reveal anything fundamental about organisms. James Valentine and his colleagues (1994) tested for these issues in an analysis of increase in morphological complexity in metazoans. By plotting the time of origin of body plans against number of cell types these authors demonstrated that the upper limits of complexity increased from the earliest metazoans (such as sponges) to the vertebrates, and that early rates of increase were high in comparison to changes within the vertebrates.

While it is tempting to correlate number of genes or genome size with numbers of cell types as an index of the increasing complexity of development required to direct organisms with larger numbers of cell types, genome size bears no direct correlation to perceived complexity. Insects such as *Drosophila*, with eight homeobox genes and some 50 cell types, are certainly as developmentally complex as mammals which have 39 known *Hox* genes

and cell types numbered in the hundreds. Neither number of genes nor number of cell types appear to correlate directly with complexity.

A further difficulty with using number of cell types as the metric, is that cellular interactions increase with increasing cellular diversity and complexity. The more cell types and the more differentiated they are, the greater the potential for cell-to-cell interactions and generation of additional cell types. It may be better to measure complexity by the number and/or types of inductive interactions between cells.

Larsen (1992) compared taxa in three groups: those in which cells self-differentiate after autonomous specification (what she calls self-governing), those in which cellular interactions are required to generate cell diversity, and those that use both strategies. Because of the difficulty of finding enough species in which experimental analyses of cell and tissue interactions have been investigated, and because self-governing species tend to lack mesenchyme, Larsen used presence or absence of mesenchyme as a surrogate for interactions (Table 14.1). Interestingly, autonomous specification of cell fate as a strategy does not appear to limit speciation; arthropods and nematodes with autonomous specification are the most speciose phyla.

Complexity is a difficult concept to characterize.

Table 14.1 Phyla listed on the basis of whether their cells are self-governing (self-differentiating), derive from cell or tissue interactions, or use both mechanisms of cell specification[1]

Self-governing	Interactions	Both
Nematoda	Porifera	Mollusca
Arthropoda	Cnidaria	Bryozoa
Brachiopoda	Ctenophora	Urochordata
Cephalochordata	Platyhelminthes	Chordata
	Annelida	

[1] Following Larsen (1992).

14.2 ORIGINS AND DIVERSIFICATION OF THE METAZOA

The problem of how to represent phyla and phyla grouped into super-phyla has persisted as a problem for a long time (s. 2.3.1). Considerable uncertainty exists as to the best criteria to use at this highest level of taxonomic organization. It is axiomatic that when two phyla split they possessed the same body plans (unless a key morphological innovation drove the split). Compilation of evidence from the fossil record, from molecular analyses, and from development itself, indicates that it is unlikely that the origins of major metazoan body plans coincided with the separation of metazoans into lineages. Valentine, Erwin and Jablonski (1996) critically analysed three likely models:

1. maximal time interval between the splitting of metazoan lineages and the origin of body plans, with all body plans arising around 530–540 mya;
2. minimal time interval between the splitting of metazoan lineages and the origin of body plans, with body plans arising progressively over some tens of millions of years; or
3. rapid diversification of lineages and body plans without any temporal separation.

Molecular phylogenetic analysis should be able to distinguish between these alternatives as additional data become available, although, as discussed in the context of divergence rates of molecular sequences in section 2.3.2, just how best to use such data remains at issue.

14.2.1 Conditions required

The origin of metazoan development was a necessary but not sufficient precondition for the origin and diversification of metazoans. In 1993 Doug Erwin evaluated three requirements:

1. the existence of eukaryotic cells;
2. environmental changes; and
3. key innovations.

His approach was a phylogenetic analysis of fossil, molecular (16S and 18S rRNA) and developmental evidence.

Eukaryotic cells arose some 1.5 or more billion years ago, metazoans some time before 600 mya. This enormous gap appears to rule out the evolutionary origin of eukaryotic cells as a sufficient condition to establish metazoan development. Indeed, as Erwin demonstrates, only three of 23 multicellular eukaryotic lineages evolved complex development.

The primary environmental agent proposed as key to the origin of the Metazoa (and to the radiation of sulphide-oxidizing bacteria about half a billion years earlier) is increase in atmospheric oxygen, thought to be prerequisite to such key evolutionary events as increasing size and the origin of the fibrous protein collagen.[7]

Erwin's analysis is consistent with environmental triggering of successive innovations, each building on preceding innovations as thresholds or boundary conditions are crossed. Diversification of cells, ability to produce extracellular matrices and their products, elaboration of epithelia, mesenchyme, epithelial-mesenchymal interactions and tissue types are obvious steps along the way. Some of these steps (specifically the origins of epithelia, mesenchyme and the ability to produce extracellular matrices) have been proposed as deep-rooted synapomorphies for the Kingdom Animalia. In the future it should be possible to determine the sequential origins of such features as the evolution of collagen, cell migration, such molecules as the integrins required for cell-matrix signalling, epithelial-mesenchymal transformations, and elaboration of components of extracellular matrices.[8]

14.2.2 Phylotypic stages

The existence of common embryonic stages within members of individual taxa (usually phyla), the topic of section 8.3, prompted a search for whether a stage can be so conserved that it typifies a phylum. The explicit presumption is that the basic body plan or *Bauplan* characteristic of the phylum is laid down at that stage; hence its evolutionary conservation and conservatism (s. 6.8.1). Seidel (1960) called such a stage in insects *Körpergrundgestalt*, the basic body-plan stage. Such a stage was designated phyletic by Cohen in 1979, phylotypic by Sander in 1983, and a phylotype by Slack, Holland and Graham in 1993.[9]

Von Baer's embryonic stages and Haeckel's conserved phylogenetic stages show that such ideas were present even at the origins of evolutionary morphology and evolutionary embryology. Haeckel identified the gastrula as a universal metazoan stage equivalent to the metazoan ancestor, the Gastrea, although as Anderson and Richardson remind us, Haeckel was less than rigorous in his interpretation when depicting conserved and conservative stages during embryogenesis, combining features of different stages in single figures. Harking back to von Baer and Haeckel, William Ballard identified a phylotypic stage in vertebrates, the **pharyngula** – essentially an early, post-neurula 'larva' with paired pharyngeal slits and initiation of the basic vertebrate organ systems. Anderson identified the blastula as the phylotypic stage of the phyla Annelida and Arthropoda. Sander identified the germ-band stage as the insect phylotypic stage (s. 8.4.1). This larval stage, with its segmented head, thorax and abdomen, is sufficiently differentiated to possess three thoracic segments, 8–11 abdominal segments, antennal buds, and segments from which the mouth parts develop.[10]

Haeckel, Anderson, Ballard and Sander all had the same intent: the search for a phylotypic stage is a search for the physical embodiment of the link between development and evolution. Ballard's summary of the pharyngula as the vertebrate phylotypic stage exemplifies this position:

'The pharyngula exhibits the basic anatomical pattern of all vertebrates in its simplest form: a set of similar organs similarly arranged with respect to a bilaterally symmetrical body axis, possessing chiefly the characters that are common to all the vertebrate classes . . . One sees in them (the pharyngulas of vertebrates) epidermis but no scales, hair or feathers; kidney tubules and longitudinal kidney ducts are there, but no metanephros; all the little hearts have the same four chambers and there is at least a transient cloaca; there are no middle ears, no gills on the pharynx segments, no tongue, penis, uterus, etc. Basically just vertebrate anatomy, unobscured by the vast array of characters that appear later in development to distinguish the various classes, orders and families.'

Ballard, 1981, p. 392

As discussed in section 8.4, development before the phylotypic stage can be quite divergent. Development after the phylotypic stage is equally divergent. In a pioneering analysis, Medawar (1954) categorized development before and after the stage at which vertebrates least differ from each other as **developmental convergence** and **divergence**. In placing these ideas into a modern context 40 years later, Duboule used the metaphor of the phylogenetic egg-timer, the narrow waist representing a constraint between more variable phases.[11]

Within a phylum, early development converges toward and later development diverges from the phylotypic stage. For Medawar, vertebrate embryos least differed from each other at the late-neurula stage (Ballard's pharyngula). Embryologists have long discussed pre-gastrula development in such forward-looking terms as 'presumptive ectoderm' and 'future fates' of the blastula, as I do in Chapter 10. Post-gastrula embryos, on the other hand, are described in backward-looking terms – from which part of the gastrula did a tissue arise; how are cells specified during gastrulation? It is no accident that we focus on the gastrula and the gastrula/neurula transition as the turning point in vertebrate development.

Anderson (1973) and more recently but for

similar reasons Richardson (1995) advocated the notion of a **phylotypic period** to emphasize the natural variability seen during embryogenesis, the dissociation of timing between parts (a concept formalized by Needham [1933, 1942]), and the importance of shifts in timing of development that produce heterochrony as an evolutionary process (s. 24.1).

On the basis of an extensive analysis of comparative embryology, Richardson and colleagues argue that equivalent stages occur earlier or later in development in related species. As they so carefully document, stages are conserved, time of appearance of the stage is not. Recognition of conserved stages and conserved suites of characters is important; it forms the basis of attempts to 'stage' precisely embryos irrespective of when during ontogeny individual embryos reach that stage (see Miyake *et al.* [1996a,b] and the discussion in section 23.2). I do not, however, see that variability in the time of appearance of the phylotypic stage, indicative of temporal not morphological shifts in evolution, requires abandoning the concept of phylotypic stages. Variability in timing of appearance of the phylotypic stage is amply demonstrated by germ-band-stage embryos in insects (s. 8.4.1). Rather, the dynamics of development would be better reflected were we to emphasize that time of appearance of phylotypic stages is in itself subject to evolutionary change, without in any way negating our ability to recognize the stage or the suite of characters that typify the stage. Indeed, recognition of phylogenetic suites of characters is central to understanding how organisms arose and are modified over the evolutionary history of the lineage; see Hall (1997d) for further discussion.

In the following section I discuss whether phylotypic stages arose as integrated stages, or represent a key innovation that led to the origin of the phyla. Against this background we can evaluate the dramatic proposal made by Slack, Holland and Graham in 1993 that the minimal definition of an animal can be extracted from the conserved pattern of homeotic genes in what they call the zootype. From this gene-based definition of an animal flow gene-based definitions of phylotypic stages.

14.2.3 Zootypes and phylotypes

What might the first metazoan have looked like?

Based on a close analysis of patterns of embryological development throughout the Metazoa, Valentine (1997) sees the common protostome-deuterostome ancestor as having the following embryonic features:

- an egg in which a small amount of yolk was evenly distributed;
- total and radial cleavage;
- bilateral symmetry;
- a larval mouth formed apically as in deuterostomes;
- ectoderm, mesoderm and endoderm; and
- a pseudoceolome.

But, what of earlier metazoans?

Gould (1986) saw a parallel between homeobox genes and Geoffroy's homology of insect and vertebrate segments. Slack, Holland and Graham (1993) used the accumulated evidence on specification of body plans by *Hox* genes to propose a modern equivalent of Geoffroy's universal type. This **zootype** displays a particular spatial pattern of gene expression (*labial* anteriorly, *Abd-B* posterially, and *Ubx* centrally) which Slack and his colleagues see as the minimal genetic definition of an animal.

The ontogenetic stage at which the zootype is more evident is the phylotypic stage – the vertebrate pharyngula, the insect germ-band stage, the fully segmented, ventrally closed leech embryo. Miller and Miles (1993) presented evidence from the coral *Acropora formosa* for the inclusion of the pair-rule gene *even-skipped* (*eve*) in the definition of the zootype. *Eve* lies adjacent to *Antennapedia* in the coral and in vertebrates, and appears to specify patterning of structures of the posterior body axis.

Slack and his colleagues reasoned that if the zootype represents fundamental features embodied in all metazoans, then the next level of genes controlled by the zootype code should be representative of phyla and/or super-phyla. Phylotypic stages for mice, *Xenopus*, amphioxus, *Drosophila*, nematodes and leeches were defined on the basis of the expression boundaries of particular *Hox* genes as follows:

1. the *Antennapedia-Ultrabithorax-abdominal-A* group for mice (mammals?), *Drosophila* (flies?), nematodes and leeches;
2. the *Abdominal-B* complex for *Xenopus* (amphibians?); and
3. *proboscipedia* for amphioxus.

Minelli and Schram (1994) defined a series of **phylotypes** corresponding to other gene complexes that are downstream from the homeobox genes and likely to control more specific aspects of metazoan body plans. This attempt to define controlling elements of phyla and super-phyla in sequences of regulatory genes complements that employed by Emile Zuckerkandl, and Eric Davidson and colleagues in their attempts to understand the origins of the Metazoa in terms of developmental regulatory mechanisms. Davidson's scenario is outlined in section 14.2.5.

Minelli and Schram list 11 general principles they consider important for a theory of pattern formation in macroevolution and as an adjunct to process-oriented theories of microevolution. Scott Gilbert, John Opitz and Rudy Raff (1996) develop this theme further in their integrative approach to a synthesis of micro- and macroevolution. Below are listed those principles that pertain to the zootype and phylotypic stages; a number have been introduced already:

1. Cellular organization (syncytial or cellular blastoderm) is secondary when considering the zootype.
2. Cell lineage also may be unimportant – the zootype establishes pattern not type.
3. Whether or how segmentation occurs is of secondary importance (s. 9.4).
4. The zootype is determined by maternal and zygotic genes and not by maternal control alone (s. 7.4.1).
5. Establishing the posterior axis stabilizes patterns that are size-dependent. Lack of terminal control is part and parcel of the lack of individuality seen in colony formation or the ability to add additional units (strobilization).
6. Specific body boundaries (hot spots or nodes representing combinations of organ-specifying genes) take priority over boundaries between traditionally recognized body regions.
7. A minimal size is required to generate such patterns as the A-P axis during development and regeneration.
8. Pattern, whether A-P, D-V, or of hot spots, is stabilized as individual structures develop.

The genetic and developmental bases of some of these principles are incompletely understood. Some are controversial; the use of combinations of genes rather than homologous structures to represent positional homology (number 6 above) unites convergent structures as homologous. This will trouble many, and so it should (see section 21.7.3 for the reasons). Nevertheless, such principles serve as a backdrop against which to organize our approach to the evolution of development and to show how microevolutionary developmental processes translate into macroevolutionary patterns within the Metazoa.

For Minelli and Schram, as for Slack and colleagues, expression boundaries of *Hox* genes define 'animal'. Genes downstream from the *Hox* complex define phyla, super-phyla or perhaps grades of organization which bear a strong resemblance to the embryological archetypes established by Von Baer on entirely different criteria (s. 5.1). Strong anterior and posterior controlling regions are found in the members of several phyla and in successful grades of organization, including lophophorate and deuterostome phyla. With these additional levels of control, Minelli and Schram distinguished five phylotypes:

1. a **cyclotype**, seen in groups with radiate *Baupläne* such as Cnidaria (*Hydra*) and Ctenophora, consists of the *Hox* complex plus a polar coordinator. This phylotype comes closest to the unmodified zootype. How linear control of the zootype was modified to give radial symmetry in these organisms is not known;

2. a **platytype**, as seen in flatworms, consists of the *Hox* complex plus genes controlling a mid-body region. A-P, D-V and left-right axes are preserved. A mid-body 'hot spot' (an organizing region) is associated with the mouth and/or reproductive structures. Absence of effective posterior control allows strobilization, as seen in cestodes;

3. a **malacotype**, seen in molluscs and nemerteans, is based on the *Hox* complex plus a posterior terminator control. The latter, not seen in the cyclotype, provides effective terminal control over development;

4. an **arthrotype**, seen in arthropods, consists of the *Hox* complex plus serial-selector controls; and finally

5. a **trimerotype**, seen in the Hemichordata (*Balanoglossus* and its allies), consists of the *Hox* complex combined with anterior and posterior regional controllers. Given the strength of anterior control over development in hemichordates and chordates, the trimerotype may subsume other yet unrecognized phylotypes.

Others have sought to define the minimal genome or most essential portions of the genome in the Cambrian animal ancestor. For Ohno (1996), such genes included a gene for lysyloxidase to cross-link collagen, genes for haemoglobin to transport oxygen, the *Pax-6* gene for development of sense organs, and *Hox* genes to specify the zootype. From an examination of the Burgess Shale fossils, Ohno concluded that *Hallucigenia sparsa* came closest to the hypothetical Cambrian 'pananimal'. With a body composed of six or seven identi-

cal segments, each bearing identical pairs of ventral and dorsal simple appendages, simplified differentiation along the A-P axis, and D-V and L-R symmetry (Figure 3.4), *Hallucigenia*, constructed so simply, is notwithstanding its phylogenetic antiquity, highly derived. The common ancestor is likely to have been even more simply constructed than *Hallucigenia*, perhaps more like *Kimberella* and akin to a gastrula (s. 2.2.2c).

Jacobs (1990) placed his emphasis on how, as unsegmented and segmented body plans arose and diversified, selector genes such as those of the *Hox* series led to the elaboration of morphogenetic fields and the ability to generate serially repeated morphogenetic fields. Recent evidence supports these ideas: *Her-1*, the zebrafish equivalent of the *Drosophila* pair-rule gene *Hairy*, is expressed in paired bands within zebrafish paraxial somitic mesoderm; *Ch-1* (*chairy*), the chick equivalent of *Hairy* is expressed at the posterior border of the somites as they segregate from paraxial meosderm; *engrailed* is expressed in posterior compartments of each segment in *Drosophila* and in the posterior half of each of the first eight somites in amphioxus. Such conserved patterns of expression have been used as evidence for similarities in mechanisms of segmentation across the animal kingdom, for segmentation in the common ancestor of insects and chordates, and for that ancestor as the primitive bilateral animal, *Urbilateria*.[12]

14.2.4 Larvae

'No questions . . . are of greater importance for the embryologist than . . . the secondary changes likely to occur . . . in the larval state.'
Balfour, 1880a, p. 381

The approach to metazoan origins used by Eric Davidson and colleagues is based on identification of the basic developmental regulatory mechanisms within embryonic and larval metazoans, and extension of that knowledge to the evolutionary origins of those mechanisms. The arguments were elaborated in three papers of incredible erudition and compass by Davidson, and a fourth paper by Davidson, Kevin Peterson and Andrew Cameron. An understanding of larvae is crucial to their approach.[13]

Davidson took the 19th-century classification of larvae as primary and secondary and molded it into a modern approach to metazoan origins. Before considering his approach I provide a brief overview of primary and secondary classes of larvae. An important caveat is that, because of the existence of larval adaptations, only some larvae provide information useful for reconstructing phylogenetic histories and evaluating evolutionary relationships. This concept of **caenogenesis** was introduced by Ernst Haeckel in 1866.[14]

In the halcyon days of marine biology of the late 19th century, naturalists such as Johannes Müller (1801–58) identified a vast diversity of marine larvae in plankton samples. How could this diversity be rationalized? What did it reveal about animal relationships and origins? One way to capture the flavour of the important position assigned to larvae at that time is to see how they were treated in the first comparative embryology textbook, Francis Balfour's two-volume *Treatise on Comparative Embryology* (1880/1). In the 'department of Phylogeny' Balfour saw the aim of embryological research as:

'(1) To test how far Comparative Embryology brings to light ancestral forms common to the whole of the Metazoa. Examples . . . the ovum itself, supposed to represent the unicellular ancestral form of the Metazoa: in the ovum at the close of segmentation [cleavage] regarded as the polycellular Protozoon parent form: in the two-layered gastrula, etc., regarded by Haeckel as the ancestral form of all the Metazoa.
(2) How far some special embryonic larval form is constantly reproduced in the ontogeny of the members of one or more groups of the animal kingdom; and how such larval forms

may be interpreted as the ancestral type of those groups.
(3) How far such forms [larvae] agree with living or fossil forms in the adult state; such an agreement being held to imply that the living or fossil form in question is closely related to the parent stock of the group in which the larval form occurs.
(4) How far organs appear in the embryo or larva which either atrophy or become functionless in the adult state, and which persist permanently in members of some other group or in lower members of the same group.
(5) How far organs pass in the course of their development through a condition permanent in some lower form.'

Balfour, 1880–81, Volume 1, pp. 4–5[15]

Balfour set himself three tasks: describe larvae, discuss their origin, and comment on their relationships to one another. His definition of larvae and view of their importance in evolutionary studies, especially relationships and origins, are straightforward. Larvae are those animals: 'born in a condition differing to a greater or less extent from the adult . . . No questions . . . are of greater importance for the embryologist than . . . the secondary changes likely to occur . . . in the larval state' (Balfour, 1880a, p. 381).

Today we assume that 19th-century embryologists all followed von Baer and Haeckel in regarding early development as immutable – as not subject to any change either ontogenetic or phylogenetic. Not so. A hundred years ahead of his time, in studies undertaken as an undergraduate at Trinity College, Cambridge, Balfour perceived the hand of natural selection early in embryonic stages:

'I see no reason for doubting that the embryo in the earliest periods of development is as subject to the laws of natural selection as is the animal at any other period. Indeed, there appear to me grounds for the thinking that it is more so.'

Balfour, 1874, p. 343

'The principles which govern the perpetuation of variations which occur in either the larval or the foetal state are the same as those for the adult state. Variations favorable to the survival of the species are equally likely to be perpetuated, at whatever period of life they occur, prior to the loss of the reproductive powers.'

Balfour, 1880a, p. 381

Balfour thus had a perceptive view of the operation of natural selection, survival of the fittest, and inheritance of variation, and specifically addressed these topics in the vice-presidential address to Section D of the BAAS in August 1880. To paraphrase the conclusions of that address: only the most favourable variations persist and they persist as secondary changes and adaptations in larvae. This view would be echoed by Walter Garstang in his presidential address to section D 40 years later, and by Davidson in his modern approach to these problems.[16]

Balfour cites three lines of evidence for his emphasis on the strength of selection at early developmental stages:

1. variation of yolk content in closely related species and the effect of yolk content on development;
2. different modes of cleavage; and
3. suppression of embryonic stages in freshwater species and their retention in marine species.

In the absence of a larval stage, development is likely to be abbreviated, direct, and associated with increased amounts of yolk in the egg. Balfour even uses the term '**direct development**' for ontogeny from which the larval stage has been lost. We still use the phrase for species that have lost the larval stage and contrast this condition with **indirect development**, in which the larval stage is retained. A remarkable finding of the recent work on direct development discussed in Chapter 22 is that embryonic organization can vary as much between direct- and indirect-developing species in the same genus as it does between the members of a particular phylum.

Balfour offered an interesting argument for why development that includes a larval stage is more likely to repeat ancestral history than is direct development. Production of a larva necessitates that organs are always retained in a functional state to permit the larva's free and independent existence. Therefore (and even though secondary adaptations occur), larval development is a closer representation of the unabbreviated ancestral history than is the abbreviated ontogeny of direct development. Abbreviation obscures ancestral life history more than does secondary adaptation, or as Balfour expressed it: 'There is a greater chance of the ancestral history being lost in forms which develop in the egg; and masked in those which are hatched as larvae' (1880a, p. 383).

In discussing larvae, Balfour is concerned with understanding which types most resemble the ancestors of the phyla. To this end he distinguished two kinds:

1. **primary larvae** as modified ancestral forms that have existed as free larvae 'from the time when they constituted the adult form of the species' (1880a, p. 383); and
2. **secondary larvae**, introduced secondarily into the life history of a species that previously developed directly.

Within these two types Balfour identifies no fewer than 33 larval forms with even more unnamed varieties.

Balfour set the primary larva, the **planula** (the ancestral form of coelenterates) apart from all other larval forms, which he regarded as secondary, recognizing pilidium, echinoderm, trochosphere, tornaria, actinotrocha and brachiopod larvae. Secondary larval adaptations were thought to arise from changes in larval life, changes in the order of appearance of structures, or to be related to the struggle for existence.

Balfour ends his lengthy discussion of larval types with some phylogenetic conclusions.

They hinge on the central notion that groups that share a common larval type are 'descended from a common stem' (Balfour, 1880a, p. 405). Common larval types are used to deduce the form of a common ancestor for all animal groups above coelenterates, i.e. for triploblastic animals. This is a radially symmetrical, medusa-like organism. From this basis, in what Balfour acknowledges are conclusions 'of a highly speculative character' (*ibid.*, p. 407), he derives the basic bilaterally symmetrical larval type (of which the Pilidium larva is the closest approximation), and the more highly differentiated larval types. For Balfour this was the way ahead if embryology was to unravel phylogenetic relationships, ancestors, and evolutionary history:

'The majority of these conclusions are undoubtedly of a highly speculative character, but while they cannot be regarded as part of our stock of embryological knowledge, they may, nevertheless, serve to indicate an important line for continued embryological research. A thorough histological investigation of the larval forms dealt with in this essay will be likely to lead to valuable results.'

Balfour, 1880a, p. 407

Davidson has provided the thorough histological investigation called for by Balfour.

14.2.5 Developmental mechanisms

In Davidson's scenario, metazoans would initially have resembled modern, marine, microscopic invertebrate larvae. The few embryos that are known from the Early Cambrian fit this concept (s. 3.1). Whether early metazoans were similarly constructed is contentious.

Davidson recognizes three types of embryogenesis, conveniently, though not very descriptively, called types 1, 2 and 3, and delineated on how cell fate is specified, how and when embryonic axes are determined, and such features as cleavage patterns and regulative ability (Table 14.2).[17]

(a) Type 1 embryogenesis

Davidson recognizes four suites of developmental characters in type 1 embryogenesis:

1. invariance of cleavage within species;
2. one axis preformed during oogenesis, the second organized after fertilization;
3. specification and appearance of distinct cell types before gastrulation; and
4. autonomous and conditional specification of founder cells for cell lineages (Table 14.3).[18]

As a reminder, in autonomous specification, cell fate is specified on the basis of internal constituents, usually inherited maternally. Two examples are *bicoid* in *Drosophila* and the specification of structures as anterior rather than posterior, and myogenic determinants in ascidian cell lineages. Autonomous specification is independent of interaction with other cells. Conditional specification is just the opposite. It is establishment of cell fate through interaction(s) with adjacent cells.

(b) Type 2 embryogenesis

Embryos with type 2 embryogenesis do not display invariant cell lineages (Tables 14.2, 14.3). Furthermore, cells from different lineages can contribute to the same tissues or structures, i.e. cell lineage does not equal cell fate. (Cell lineage does not always equal cell fate in type 1 embryogenesis either, although cases where cells from different lineages contribute to a single structure are infrequent.) Extensive cell migration establishes the new cell-to-cell associations required for conditional specification.

(c) Type 3 embryogenesis

Type 3 embryogenesis and long germ band insects represent a special case. Delayed cellularization prolongs the phase during which maternal gene products pattern em-

Table 14.2 The three major types of embryogenesis recognized by Davidson (1991)

Type 1
- Found in most invertebrate taxa. Most information is known about the nematode *C. elegans*, ascidians, molluscs and echinoderms (chiefly sea urchins)
- Invariant patterns of cleavage specify cell lineages
- Cell lineage an important means for spatial organization
- Cell lineage is important in specification of cell fate; cell lineage normally equivalent to cell fate
- Maternal cytoplasmic factors important in specification of cell fate
- As a consequence of the two previous features, cell specification occurs *in situ*, i.e. in the absence of extensive cell movement
- Both autonomous and conditional[1] mechanisms specify cell fate with some lineage founder cells specified autonomously at the embryonic poles
- One embryonic axis specified during oogenesis, the second axis specified after fertilization
- Regulation possible
- Selective gene expression occurs early in embryogenesis

Type 2
- Found in vertebrates. Most information is known about *Xenopus laevis*, the domestic fowl and the mouse
- Variable patterns of cleavage
- Cell lineage plays a minor role; structures are composed of cells from different lineages; cell lineage does not automatically equal cell fate
- Diffusible morphogenes important in specification of cell fate
- Maternal factors primarily active in cleavage stages
- Cell migration is prominent and important
- Primarily conditional specification of cell fate
- Either one embryonic axis specified during oogenesis, or axis unspecified until later in development
- Regulation of large portions of the embryo possible
- Selective gene expression occurs later in embryogenesis, usually after gastrulation

Type 3
- Found in long germ band insects such as *Drosophila*
- Embryo initially syncytial
- Adult cell lineages set aside in imaginal discs
- Spatial patterns of gene expression generated in a syncytial embryo by both maternal and zygotic genes
- Maternal factors establish body axes
- Cellularization of the embryo a late event
- Both autonomous and conditional specification of cell fate
- Both A-P and D-V axes specified during oogenesis
- Regulation limited or absent
- Selective gene expression throughout embryogenesis

[1]Autonomous specification is by internal cellular constituents and is independent of interactions with adjacent cells. Conditional specification requires interaction with adjacent cells; see text.

bryos. Insects such as *Drosophila* take full advantage of such products to produce their derived embryos (Table 14.3). Onset of gene activity specific to individual cell types coincides with transition from maternal to zygotic gene control (s. 7.4.1).

(d) The scenario

The embryogenesis seen in advanced insects and vertebrates is derived; their larvae are secondary. Such larvae are unlikely to yield information pertinent to initial conditions. Some

Table 14.3 Examples of cell lineages that are specified autonomously or conditionally in embryos representing the three types of embryogenesis

Species	Autonomous	Conditional
Type 1		
Caenorhabditis elegans (nematode)	Gut, germ line, posterior body muscles	Pharyngeal muscles
Ilyanassa (mollusc)	Gut, heart	Eye, foot shell gland
Stronglyocentrotus (echinoderm)	Primary mesenchyme	Gut, ectoderm
Type 2		
Xenopus (frog)	Endoderm of gut, ciliated ectoderm	Brain, mesoderm
Danio rerio (zebrafish)	None	Central and peripheral nervous systems, mesoderm, notochord, pigment cells
Mus (mouse)	None	All embryonic lineages
Type 3		
Drosophila (insect)	Germ cells	Neuroblasts, portions of cuticle and ectoderm

Taken from data compiled in Davidson (1990).

invertebrates display maximal indirect development. Such groups have larvae that bear no resemblance to the adults and arise from totally separate cells (imaginal discs in *Drosophila*) with developmental programmes that are distinct from those that produce adult structures. Such larvae illustrate 'the almost brutal independence of the developmental process by which the adult body plan is generated' (Davidson, Peterson and Cameron, 1995, p. 1320).

Most invertebrate phyla have type 1 embryogenesis and share common mechanisms for specification of cells and cell lineages. Clusters of genes encoding cell autonomous regions are expressed from cleavage onwards. Primary larvae derived from type I embryogenesis are proposed as representing basic metazoan organization.

The first essential element of Davidson's scenario is that larvae of extant indirect-developing marine invertebrates (type 1) are surrogates of the Precambrian genetic regulatory systems that formed the basis for the evolution of metazoan body plans.

The second essential element translating genetic regulatory systems into cellular activity is **set-aside cells** and the development of hierarchical, regulatory, developmental programmes of the types discussed in Chapters 8 and 15. These programmes specify different morphogenetic regions physically based in lineages of cells as domains of expression of transcription factors, and are consistent with conservation of larval programmes as ancestral. Each lineage of set-aside cells evolved its own hierarchical programme of gene expression.

Set-aside cells with their own patterns of expression of upstream transcription factors provided the cellular and genetic bases for the diversification of metazoan body plans. Elaboration of upstream regulatory processes within these cell lineages or set-aside cells provides the basis for morphogenetic diversification. Once bounded regions are established within embryos, the potential for rapid specialization and evolution of novel cells and arrangements of cells is in place. Conservation of genes specifying basic body plans across the Metazoa and the elaboration of genetic control for subsequent differentiation and morphogenesis, are consistent with this scenario. Indeed they form the basis for it.

Davidson and his colleagues do not neglect to treat how selection would have been sharpened once set-aside cells arose, a topic tackled by Buss in his analysis of selection between cell lineages and the evolution of multicellularity (s. 13.5.4). Phylogenetic analyses and the precise time scales that can now be placed on the Precambrian, the Cambrian and their boundaries, indicate that metazoans may have arisen over as little as 25 million years of the over 600-million-year history of their existence (Figure 2.11). Selection on the most successful combinations effectively channeled body plans to those represented by phyla that persist today.

14.3 ENDNOTES

1. See Cavalier-Smith (1985) for genome size; Maynard Smith and Szathmáry (1995) and Szathmáry and Maynard Smith (1995) for genes encoding proteins; Szathmáry and Maynard Smith (1995), McShea (1991, 1996), Donoghue and Sanderson (1994) and Minelli (1996a) for the number of parts; Bonner (1988), Valentine, Collins and Meyer (1994) and Bell and Mooers (1997) for the number of cell types; Bonner (1988) and Szathmáry and Maynard Smith (1995) for division of labour; Wimsatt (1986) and McShea (1991, 1996) for character entrenchment/hierarchical developmental complexity, and Hinegardner and Engleberg (1983) for the minimal description concept. Converting the minimal description into bits of information, the latter authors estimate that to describe a bacterium would take 250 pages of the Encyclopaedia Britannica.

2. See Cope (1887, 1896) for his development of size increase in evolution and Rainger (1985) for an in depth evaluation of Cope's contributions.

3. See Shea (1992), Hanken and Wake (1993) and Shea and Bailey (1996) for recent analyses of the evolutionary consequences of small body size.

4. For other recent tests of Cope's rule, see Newell (1949), Stanley (1973), Hallam (1990) and Damuth (1993).

5. See Waddington (1940, 1942) for hierarchical control of development, and Pettersson (1996) for an analysis of types of hierarchies, complexity and evolution. Epigenetic developmental programmes and their inherent stability in the face of loss of individual genes involved in the programme are elaborated and extended in Hall (1983a, 1990b), Sachs (1988), Verraes (1989), Salthe (1993), Rollo (1994), Valentine and May (1996) and Wilkins (1997).

6. See Alberts et al. (1983), pp. 941–7 for detailed listings of numbers of cell types and sub-types.

7. See Canfield and Teske (1996) and Thomas (1997) for increased atmospheric oxygen as a factor in the radiation of sulphide-oxidizing bacteria and the Cambrian explosion of metazoans; see Garrone (1978) for evolution of collagen.

8. See Morris (1993) for evolution of cellular organization and extracellular matrices.

9. Cohen and Massey (1983), and Cohen (1993) trace the concept of the phylotypic stage to the American embryologists E. B. Wilson and E. G. Conklin early in the 20th century.

10. See Anderson (1973), Richardson (1995) and Richardson et al. (1997) for Haeckel's loose use of comparative stages; Ballard (1976, 1981), Anderson (1973) and Sander (1976, 1983) for delineation of these phylotypic stages, and Hall (1997d) for an analysis of the veracity of phylotypic stages.

11. The double-coned, narrow-waisted egg-timer is an old device. In a paper published in the same month (April) that he was appointed Regius Professor of Natural History at Edinburgh, Edward Forbes used a double cone to illustrate periods of greatest generic types in the 'Neozoic' and Palaeozoic with a narrow zone of minimal production of types in between (Forbes, 1854, and see Box 5.1).

12. See P. W. H. Holland and Williams (1990), M. Müller et al. (1996) and L. Z. Holland et al. (1997) for *hairy* and *engrailed* in *Drosophila*, zebrafish and amphioxus, and De Robertis and Sasai (1996), Kimmel (1996) and De Robertis (1997) for *Urbilateria* and discussions of segmentation. *Chairy* is expressed repetitively with an endogenous periodicity of 90 mins, which is precisely the time taken for a pair of segments to segment from paraxial mesoderm (Palmeirim *et al.*, 1997).

13. See Zuckerkandl (1994), Davidson (1990, 1991, 1993) and Davidson *et al.* (1995) for

developmental approaches to larvae and metazoan origins.

14. Haeckel (1866) introduced caenogenesis for those situations in which recapitulation of phylogeny in ontogeny was obscured because of larval adaptations or the displacement of embryonic or larval stages in time or space during ontogeny. The modern usage tends to follow de Beer (1958) who restricted caenogenesis to larval adaptations without any reference to recapitulation.

15. I have omitted the detailed examples Balfour provided for each of these five points, not all of which he necessarily accepted. Balfour devotes the second part of Chapter 13 of Vol. 2 to larvae, publishing the same material as a memoir in the *Quarterly Journal of Microscopical Science* (Balfour, 1880a).

16. See Balfour (1880b) for his address to the British Association for the Advancement of Science. In *The origin and evolution of larval forms*, Garstang argued 'that on the whole a larval evolution has taken place more or less parallel to that of the adult evolution, but subject to conspicuous deviations' (1929, p. 77). See Hall and Wake (1998) for a modern evaluation of larval evolution.

17. Strathmann (1993) is a good source for views on the origins of marine larvae. Fortey *et al.* (1997) and references therein, consider the likelihood that the first metazoans were small and planktonic.

18. See Davidson (1991) for the delineation of these three types of embryogenesis, and Goldstein and Freeman (1997) for details of the timing of axis specification across the animal kingdom.

15

CHORDATE AND VERTEBRATE ORIGINS AND DIVERSIFICATION

'Gene duplications, which free up the old or the new copy of a gene, or group of genes, to take on a new function are possibly one of the major forces of molecular evolution that can lead to the evolution of new function and novelty . . . Alternatively, the regulatory control of the expression of genes is also likely to be involved, and may be an even more important force in evolution than gene duplications.'

Meyer, 1996, p. 329

Chordate and vertebrate origins can be approached at three levels – the origins of distinctive morphology, the developmental programmes producing those morphologies, and the molecular controls of those programmes. In sections 3.1.4 and 5.5.5 I discussed the discovery of the distinctive morphology of chordates, hemichordates and cephalochordates and the possible presence of a cephalochordate (*Pikaia*) in the Burgess Shale fauna. In this chapter I concentrate on causal developmental sequences and the origin of basic chordate and vertebrate features (s. 15.1) and the role played by *Hox* genes in the origins of the chordates, the vertebrate head and axial patterning in vertebrates (ss. 15.2–15.5).

15.1 CAUSAL SEQUENCES IN DEVELOPMENT

Langille and Hall (1989) summarized arguments for the sequence of fundamental developmental processes involved in the separation of the protochordates from a deuterostome invertebrate ancestor and the origin of the chordates. The evidence is briefly summarized here and in Figure 15.1; see Chapter 10 for basic developmental information on germ layers, neural and mesodermal induction/regionalization and the neural crest.

Deuterostome invertebrates possess the three germ layers, ectoderm, mesoderm and endoderm (s. 8.3). Specialization of the mid-dorsal mesoderm as chordamesoderm and then notochord, and coupling of the notochord to the ectoderm in neural induction, are the minimal innovations required to produce the first protochordate body plans (P1 in Figure 15.1).

Primary embryonic induction, in which the notochord induces overlying ectoderm to become neural ectoderm, thus forming the primary embryonic axis, is a fundamental chordate characteristic. The embryonic notochord has been retained throughout chordate evolution, during which its role as a supporting rod became subsumed by the vertebral column in adults (in apparent recapitulation of an ancestral condition), precisely because the notochord plays such an important inductive role in development. Such structures are retained for ontogenetic, not phylogenetic reasons, and as part of the somatic programme regulating tissue interactions (Mayr, 1994).

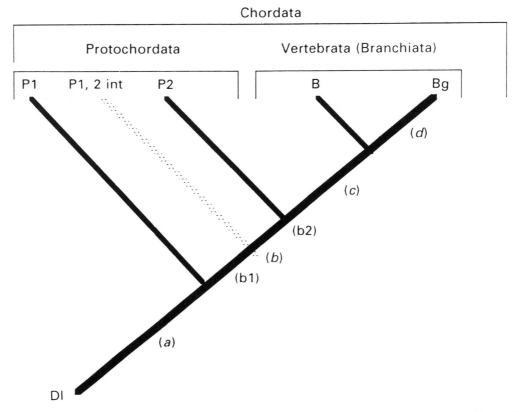

Fig. 15.1 Development events associated with the evolution of the vertebrates from a deuterostome invertebrate ancestor. The developmental events are (a) origin of chordamesoderm, notochord and primary embryonic induction of the neural tube; (b) regionalization of the neural tube into fore-, mid- and hind brains and spinal cord. b1 and b2 represent two steps in these processes; acquisition of neural induction and development of forebrain (b1), and acquisition of regionalization of the remainder of the central nervous system (b2). (c) origin of the neural crest and epidermal placodes leading to the Branchiata (B). (d) modification of the first visceral arch into jaws, and the origin of the gnathostome (jawed) vertebrates (B_g). B, vertebrates (craniates) with branchial arches (jawed and jawless); B_g, jawed vertebrates; DI, deuterostome invertebrate ancestor; P, protochordates (urochordates). See text and Langille and Hall (1989) for details. Reproduced from Langille and Hall (1989) with the permission of the publisher, Cambridge University Press.

The second set of chordate innovations consists of A-P regionalization of the primitive nervous system into brain and spinal cord (P1, 2 int, P2 in Figure 15.1), and of the brain into major regions (s.10.2.2). The origin of neural crest from neural ectoderm as a fourth germ layer along with the differentiation of ectodermal placodes, allowed increasing so-phistication of the nervous system and the elaboration of sense organs and connective and skeletal tissues. The result was increasing cephalization and further specialization of the craniate head. Craniates (Craniata) include hagfishes and vertebrates. Vertebrates (Vertebrata) include lampreys, gnathostomes and extinct agnathans, but exclude hagfishes,

Fig. 15.2 An inclusion diagram to show that Craniata includes hagfishes and vertebrates, while Vertebrata excludes hagfishes but includes lampreys, agnatha and gnathostomes.

which lack vertebrae (Figure 15.2). Although experts still argue, lampreys and gnathostomes are more closely related to one another as vertebrates than either is to hagfishes (Figure 15.3).

Evolution of branchial arches characterized the Agnatha, the first (jawless) vertebrates (B in Figure 15.1). Modification of the elements of the first (most anterior) pair of branchial arches into jaws ushered in the gnathostomes or jawed vertebrates (Bg in Figure 15.1 and see s. 16.1).[1]

This is an abbreviated summary of the data compiled from causal developmental sequences. The important point in the present context is that an evolutionary series of innovations can be deduced from an analysis of causal sequences of embryonic development because sequences are so highly conserved. As we summarized the scenario in 1989:

'However, once a fundamental (i.e. causal or necessary) sequence or event is recognized through developmental analysis, the elemental part(s) of the sequence (a particular type of embryonic tissue, a key epigenetic interaction, a sequence of steps) can be designated as a developmental character. This character can then be sought in two or more groups in an attempt to test for phylogenetic links, since those which display the character(s) which allow key developmental innovations will possess the same conserved, fundamental, developmental sets of instructions in their ontogeny. Although the numbers of such developmental characters . . . may be fewer than

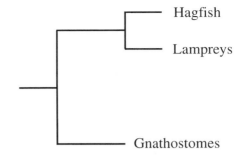

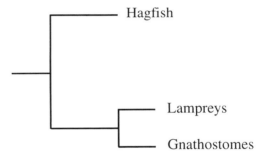

Fig. 15.3 It has been proposed that lampreys are the sister group of hagfishes in a single agnathan clade (top) or the sister group of gnathostomes (bottom).

the large lists of morphological or physiological characters . . . the strength of the developmental character is that it represents a developmental sequence that is shared between the linked groups.'

Langille and Hall, 1989, p. 74

15.2 THE EVOLUTION OF *HOX* CODES IN CHORDATES AND CHORDATE ALLIES

Another way to study chordate origins is now available in phylogenetic information contained within gene sequences, especially *Hox* gene sequences. Organization of *Hox* and homeotic genes into common gene clusters provides strong evidence that the common ancestor of arthropods and chordates possessed a *Hox* gene cluster.

Vertebrate homeobox genes with substantial sequence similarity to such *Drosophila* genes as *Ultrabithorax* and *Antennapedia*, and with a patterned distribution along the body axis (s. 15.5), are represented by the *Hox* series of helix-turn-helix transcription factors. A dramatic demonstration of conservation of the patterning roles of *Hox* genes across the animal kingdom is that mouse *Hoxb-6* transfected into *Drosophila* eggs elicits development of legs in place of antennae (Malicki *et al.*, 1990). *Hox* genes are organized as homeobox clusters on four chromosomes in vertebrates. Mice have four clusters of 39 *Hox* genes, lampreys may have two or three clusters, teleosts as many as five. The order of *Hox* genes in individual homeotic gene clusters has a surprising parallel to the anterior-posterior order of the body segments in which they are expressed.[2]

Homeotic transformations of mouse vertebrae by modification of *Hox* genes (either by creating *Hox*-mutant mice, or following application of retinoic acid [RA]) are dramatic illustrations of the patterning function of the *Hox* code. A further dramatic illustration of the patterning function of *Hox* genes is in regeneration of limbs from the stumps of amputated tadpole tails treated with RA. This homeotic transformation duplicates the limb territory on the tail. Even more astounding is respecification on the tail stump of a second caudal body region complete with duplicated pelvic girdle.[3]

Ruddle, Kappen and colleagues model *Hox* gene clusters as duplications of a primordial gene cluster in the common ancestor of chordates and arthropods. They speculate that duplication of the homeobox gene clusters was important in initiating novel aspects of the chordate body plan at the time of the origin of the chordates. In *Evolution by Gene Duplication* (1970), a book that was way ahead of its time, Ohno proposed that evolution of gene duplication and evolution by gene duplication are key innovations. Gene duplication permits subsequent specialization and structural and functional divergence. An exciting question is how gene duplication allowed diversification of chordate features – how duplication 'overcame a genetic constraint to elaboration of the chordate body plan' (P. W. H. Holland and Garcia-Fernàndez, 1996, p. 392).

Evolutionary changes in gene regulation, supported by Akam (1995), is an alternative (but not necessarily mutually exclusive) mechanism. In fact, at least five possible mechanisms are available for modification over evolutionary history:

1. change in the number of *Hox* gene clusters;
2. change in the number of genes per cluster;
3. altered functions of individual genes;
4. altered regulation of upstream or downstream expression; and
5. increased complexity of interactions between genetic programmes.

Sufficient information has now accumulated that we can begin to understand changes in *Hox* genes that underlie the origin of the hemichordate, urochordate, cephalochordate and chordate phyla.

15.2.1 Hemichordates

As members of a phylum closely related to the chordates, hemichordates (acorn worms) are obviously of considerable interest (Figure 2.8). Three of the *Hox* genes in *Saccoglossus kowalewski* are homologous to genes of the anterior cluster of mammalian *Hox* genes, a finding consistent with, and further reinforcing, the close phylogenetic relationship between hemichordates and chordates. For further discussion, see section 15.3.

15.2.2 Urochordates

Urochordates (tunicates) comprise three groups: ascidians, larvaceans (Appendicularia) and thaliaceans (doliolids, salps and pyrosomes).

Initial experiments by Tung (1934) indicated that notochord does not induce neural ectoderm in ascidians. Such a fundamental difference in development of so basic an embryonic structure would influence our thinking on whether urochordates are a chordate subphylum or separate phylum. This apparently important exception to the rule was, in fact, simply a consequence of insufficient information. Further studies by Tung and Nishida demonstrated that the notochord does induce neural development in ascidians, as indeed it does in amphioxus (s. 15.2.3). Ascidians and chordates share other fundamental features of embryonic development: the neural plate in the ascidian *Halocynthia roretzi* develops in the same way as it does in vertebrates; the ascidian notochord is induced following interactions between ventral blastomeres before the 64-cell stage; and the precursor of the spinal cord plays a role in induction of sensory pigment cells, as in vertebrates.[4]

On first analysis using PCR primers, urochordates appeared to possess only a single *Hox* gene. This is surprising, for the chordate ancestor is posited to have had multiple *Hox* genes. It now appears, from studies using genomic and cDNA libraries, that ascidians do have more than one *Hox* gene. This change in position reflects, in part, growing awareness of the extreme sensitivity of the PCR approach and its consequent susceptibility to false results from even the most minor contaminants. It also reflects the unusual sequence divergence of ascidian *Hox* genes.

Other patterning genes found in urochordates and chordates are also similar. The ascidian *Halocynthia roretzi* has a Pax gene (*HrPax-37*) that patterns the neural tube, as do *Pax-3* and *Pax-7* in vertebrates. The reiterated distribution of *HrPax-37* along the neural tube, the structure of the tube, and the presence of serially repeated cilia pairs, provide clues to the origins of early segmentation. Wada *et al.* (1996) saw in the expression of *HrPax-37* in the dorsal epidermis, a similarity to *Pax-3* expression in the vertebrate neural crest, possibly adding one piece to the puzzle of the evolu-tionary origin of the neural crest from dorsal ectoderm.

The presence of a BMP-7-like gene and its possible role in setting the boundary between neural and epidermal ectoderm in the ascidian *Halocynthia roretzi* (s. 10.4.3) speaks to another common control system that would have been present in the urochordate/chordate ancestor. So does shared regulation by RA of *Hox* genes in urochordates, cephalochordates and vertebrates (s. 10.2.2 and see below). Tunicate *Hox* genes also are up-regulated by RA, implicating a common ancestral role for RA in axis determination and neural tube specification.[5]

Presence in the common ancestor of genes shared between urochordates and vertebrates is emerging as an important theme in gene and chordate evolution.

15.2.3 Cephalochordates

The cephalochordate (lancelet) *Branchiostoma* (amphioxus, Figure 3.5b) has been of interest to zoologists ever since Kovalewsky recognized that it possesses a notochord and was therefore allied to the chordates (s. 5.5.5). For 150 years, students of embryology and vertebrate anatomy have studied amphioxus as the closest living representative of the vertebrate ancestor. Cephalochordates resurfaced in a symposium held at the IVth International Congress of Vertebrate Morphology in 1994 (published under the editorship of Carl Gans *et al.*, 1996) and in two recent books on vertebrate origins: *Before the Backbone*, by Henry Gee (1996) and *Life's Splendid Drama*, by Peter Bowler (1996). A close relationship to chordates is borne out by an analysis of the processes underlying amphioxus development; phylogenetic analyses using morphological characters and 18S RNA confirm cephalochordates as the closest sister group to the vertebrates, as shown in Figures 2.5 and 2.8.[6]

The question of whether amphioxus has one or two *Hox* gene clusters has been resolved in favour of one cluster with at least 12 genes.

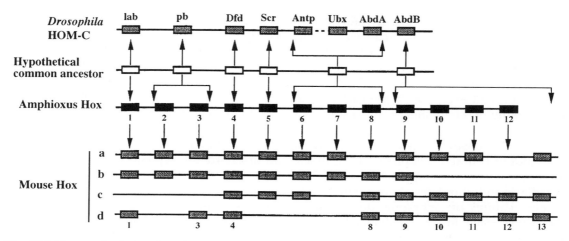

Fig. 15.4 Proposed relationships between homeotic gene clusters in the hypothetical common ancestor of chordates and arthropods; homeotic genes in *Drosophila* (*lab, pb, Dfd, Scr, Antp, Ubx, AbdA, AbdB*); *Hox* genes in amphioxus (*AmphiHox-1–AmphiHox-12*); and *Hoxa–Hoxd* in the mouse. Modified from Holland and Garcia-Fernàndez (1996).

Designated *AmphiHox-1* to *-12*, they are arranged in the same order as the corresponding mouse *Hox* genes (Figure 15.4). As in vertebrate and tunicate *Hox* genes, RA up-regulates *AmphiHox* genes.

Organization of *AmphiHox* genes is consistent with:

- duplication of the ancestral *Hox* cluster after vertebrates diverged from cephalochordates (*Msx* genes also were duplicated at the origin of the vertebrates; there are three *Msx* genes in the mouse and zebrafish, one in *Drosophila* and one in the ascidian *Ciona* [P. W. H. Holland, 1991]);
- amphioxus retaining the organization of *Hox* gene clusters assumed for the common chordate ancestor; and
- our century-old view of amphioxus as the most primitive chordate; see s. 5.5.5.[7]

One word of caution: Amphioxus is neither a 500-million-year-old chordate nor the vertebrate ancestor. As an extant species with a long history, amphioxus itself has evolved, even if morphological stasis characterizes its evolution. Developmental control mecha-

nisms also evolve. While the single *Hox* gene cluster of amphioxus is as postulated for the craniate ancestor, other genes in amphioxus are not present in ancestral patterns. *Brachyury* (*T*), a gene involved in mesoderm and notochord formation in vertebrates, is a good example. [For a complete list of homeobox and other developmentally important genes cloned from the several species of *Branchiostoma* to the end of 1995, see P. W. H. Holland (1996).]

The ascidian *Ciona intestinalis* regulates *Brachyury* in the notochord using a regulatory element related to the vertebrate *Notch* signalling pathway. A role for *Brachyury* in notochord formation therefore predates the origin of the vertebrates. Detection of a homologue of *Brachyury* in the secondary mesenchyme of the sea urchin *Hemicentrotus pulcherrimus* indicates that the gene had an even more ancient role predating the origin of the chordates.

Amphioxus has two *Brachyury* genes with remarkably similar patterns of expression to the single vertebrate gene. Amphioxus *Brachyury* is expressed in mesoderm of gastrulae and in somitic and notochordal tissue of neurulae. Therefore, *Brachyury* was duplicated independently in amphioxus **after** the divergence of the vertebrates. To reiterate, notwithstanding our notions of amphioxus as an ancestral vertebrate, amphioxus did evolve. This should alert us to be cautious when assigning primitive ancestor status to amphioxus. Peter Holland (1996) has explored this issue in relation to the primitiveness of the amphioxus body plan in a paper that addresses also an alternative (and old) notion that amphioxus is degenerate (s. 5.5.5), a notion Holland rejects.[8]

Transcriptions factors from the *fork head/ HNF-3* family have now been described from amphioxus (Shimeld, 1997). *HNF-3* genes are found in the dorsal lip of amphibian embryos, Hensen's node of avian embryos, and in the notochord and floor plate. Amphioxus has two hepatic nuclear factor (*HNF-3*) genes, *AmHNF-3-1* and *AmHNF-3.2*, which are expressed in the presumptive organizer and notochord, as is *Brachyury*. Shimeld interprets a column of *AmHNF-3-1*-expressing cells in the ventral midline of the neural tube as evidence for a floor plate in amphioxus. Even if not a latent homologue for the floor plate, these data support a role for *HNF-3* in neural development that precedes the separation of vertebrates and cephalochordates, reinforcing the conclusions based on *Hox* gene expression; see s. 15.3.

15.2.4 Vertebrates

Within the craniates, the hagfish *Eptatretus stoutii* appears to have four *Hox* gene clusters, while among 'lower' vertebrates, the lampreys *Petromyzon marinus* and *Lampetra fluviatilis* may have between two and four. These findings again are consistent with duplication of the single gene cluster at the origin of the vertebrates.

A number of laboratories are beginning to clone and sequence *Hox* genes from teleost fishes. The zebrafish (*Danio rerio*), puffer fish (*Fugu rubripes*) and killifish (*Fundulus heteroclitus*) may each have four or more *Hox* gene clusters. Although this is the same number of clusters as in mice and humans, different genes may have been lost in different groups of fishes and in mammals. Indeed, a very recent study indicates that the puffer fish has lost as many as 11 *Hox* genes, the mouse only three, and that only three of the puffer fish gene clusters (*Hoxa–c*) match those in the mouse. A pattern of four *Hox* gene clusters in a common ancestor with differential loss within different classes of vertebrates is emerging. Analysis of more species is required to verify this pattern and reveal whether cluster duplication has occurred in some lineages.[9]

15.3. AMPHIOXUS, *HOX* GENES AND THE ORIGIN OF THE CRANIATE HEAD

Although not a degenerate chordate, amphioxus is the least derived and closest relative to the vertebrates. Expression patterns of *AmphiHox* genes are thus of considerable interest, and have provided a new source of data for the age-old problem of the origin of the craniate and then vertebrate head. Further information will come from determining whether amphioxus contains homologues of genes that govern head development in vertebrates, from analysis of conserved functions of such genes, and from a search for elements of a 'protoneural crest' in amphioxus.

Nicholas Holland and his colleagues (1996) see three features of amphioxus dorso-lateral ectoderm suggestive of neural crest cells:

1. topography – ectodermal cells at the lateral border of the neural plate;
2. migration of epidermal cells over the neural tube; and
3. expression of *Dll* in pre-migratory and migratory epidermal cells.

At this stage of our understanding, these are no more than suggestions of a protoneural crest in amphioxus. That we did not have even these suggestions a few years ago, bodes well for future understanding.

AmphiHox-3 is expressed in the nerve cord except most rostrally, a pattern of expression essentially identical to *Hox-3* in the developing vertebrate brain. In the first investigation of molecular aspects of amphioxus development, Peter Holland and his colleagues (1992, 1994) used the limits of *AmphiHox-3* expression as a guide to the homology of regions of the nerve cord in amphioxus and the vertebrates. They concluded that the vertebrate hindbrain is homologous to an extended region of the amphioxus nerve cord. These conclusions could shed new light on the origins of the craniate head as the elaboration of a pre-existing region, rather than a novel anterior addition to a pre-existing invertebrate head. Such studies do, however, raise the fundamental issue (discussed below and in Chapter 21) of how homology should be defined and detected.

The arguments for how to detect homology are that rostral expression limits of *Hox* genes allow identification of boundaries of homologous morphological structures. This approach is sound **provided that** the *Hox* genes have not changed their function during evolution. Axial patterning of *Hox* expression, in which homologous genes are expressed at different axial levels in different vertebrate taxa but retain their function in specification of somites as cervical, thoracic, lumbar, etc., is consistent with a constant role for the genes even though expression boundaries diverge between taxa (s. 15.5).[10]

Holland and his colleagues extend such constancy of role to the *Hox* and *AmphiHox* genes expressed in developing vertebrate and amphioxus nervous systems. *Hox* genes from groups 1 and 3 have rostral expression boundaries in the hindbrains of chicks and mice: *Hoxa-1, b-1* and *d-1* at the boundary of rhombomeres three and four; *Hoxa-3, b-3* and *d-3* at the more caudal boundaries of rhombomeres four and five. By projecting these expression boundaries onto the amphioxus nerve cord, and relating them to ex-

pression boundaries of *AmphiHox-1* and *Hox-3*, Holland and Garcia-Fernàndez (1996) argue that the limits of *Hox-1* and *Hox-3* (rhombomeres three to five) represent only a small portion of the amphioxus nerve cord. Consequently, they argue that the vertebrate hindbrain with eight rhombomeres (as seen in chick and mouse) is equivalent to an even more extensive region of the amphioxus nerve cord than circumscribed by *Hox* or *AmphiHox-1* to *-3*. L. Z. Holland *et al.* (1997) discuss implications, for whether amphioxus has any homologue of the vertebrate hindbrain, of the question of temporal and spatial expression pattern of the *engrailed* gene (*AmphiEn*) in amphioxus cerebral vesicles late in development.

The conclusions reached from expression patterns of *Hox* and *engrailed* are congruent with fine-grained neuroanatomical analysis of the amphioxus cerebral vesicle at the rostral tip of the nerve cord, for the cerebral vesicle contains cells homologous to those of the vertebrate diencephalon. It appears that an extensive nerve cord with an anterior specialization in the form of a single cerebral vesicle was elaborated into the hindbrain and diencephalon. Structural, developmental and genomic data thus combine to illuminate major aspects of the elaboration of the head at the outset of craniate evolution.[11]

The following list summarizes the evolutionary and developmental processes emerging from studies of *Hox* gene clusters:

1. Amphioxus and the common chordate ancestor has (had) a single *Hox* gene cluster.
2. Vertebrates have multiple *Hox* gene clusters that arose with the origin of the craniates.
3. Gene duplication in vertebrates is not unique to *Hox* genes but is seen in other regulatory genes such as *engrailed*, *Msx*, *Otx*, *MyoD*, the globin gene family, and insulin-like growth factor (IGF). Gene duplication was associated with the origin of

the vertebrates; for each of these genes examined, amphioxus has one, while jawed vertebrates have two or more.

4. Gene duplication continued after the separation of amphioxus from the vertebrates; *Brachyury* is an example of a gene that is duplicated in amphioxus but not in the vertebrates.[12]

5. *Hox* genes used to pattern the nerve cord in the craniate ancestor assumed new roles as mesodermal, ectodermal and neural crest derivatives arose and were elaborated. That the nerve cord is the only site in both amphioxus and ascidian embryos with regional *Hox* gene expression, is consistent with diversification of roles in craniates.

15.4 *HOX* CODES AND THE VERTEBRATE HEAD

As outlined in section 15.1, formation of the branchial region of vertebrates is by concerted and coordinated actions and interaction between neural crest cells, branchial arches and the developing brain, especially the hindbrain. The delineation by Paul Hunt and his colleagues of a *Hox* code responsible for patterning the branchial regions is a major advance in our understanding of how the head is patterned.[13]

Using knowledge of the expression boundaries in the rhombomeric segments of chick hindbrains as a basis, Hunt and colleagues demonstrated identical limits of anterior expression of *Hoxb-1* to *Hoxb-4* at particular rhombomere boundaries. Expression patterns are intrinsic and autonomous; they are maintained even if rhombomeres are exchanged or transplanted to another site. Similar expression boundaries are detected in surface ectoderm, cranial ganglia, migrating neural crest cells and the mesenchyme of the branchial arches of early mouse embryos (Figure 15.5). The branchial *Hox* code has now been demonstrated in human development by Vieille-Grosjean *et al.* (1997). There is a high

degree of conservatism of *Hox-1* to *Hox-4* early in development and differential down-regulation of *Hox-3* later in development.

As neural crest cells migrate from the hindbrain they express a combination of *Hox* genes appropriate to the rhombomeres from which they emerge. This pattern of expression continues as neural-crest-derived mesenchyme populates the branchial arches. Indeed, branchial arch ectoderm expresses the same combination of *Hox* genes as the mesenchyme that invades it, strongly suggesting a *Hox* code within ectoderm before the epithelial-mesenchymal interactions discussed in sections 11.2 and 11.5. Couly and Le Douarin (1990) demonstrated that the facial ectoderm of chick embryos (much of which arises from the neural folds of the forebrain) is patterned into regions (**ectomeres**) corresponding to neuromeres. *Hox* gene expression in hindbrain, neural crest and branchial arches into which neural crest cells migrate represents a fundamental unity of mechanism patterning this region of the head.[14]

The demonstration by Qiu *et al.* (1997) of differential expression of *Distal-less* genes in branchial arch ectoderm in the mouse, and specific arch deficiencies in embryos carrying mutations of *Dll*, reveals another level of the genetic cascade governing branchial arch development and specification. Expression boundaries of *Hox* and *Dll* pattern the branchial arches in orthogonal A-P and P-D directions, corresponding to the embryonic A-P axis and the temporal pathway of neural crest cell migration.

Experimental confirmation that *Hox* genes do indeed pattern the head and branchial region comes from a number of studies of mouse embryogenesis in which particular *Hox* genes have been disrupted following homologous recombination in embryonic stem cells and introduction into the germ line. (See the next section for a discussion of similar experiments demonstrating homeotic transformations of vertebral type.) One caution with such

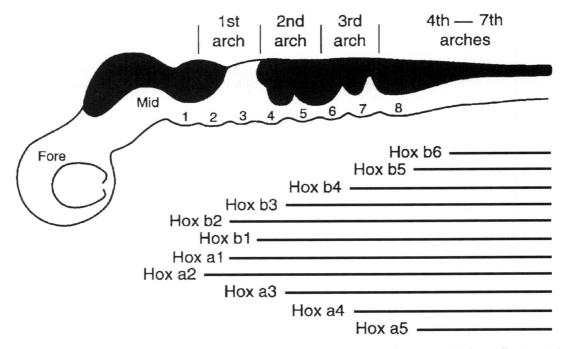

Fig. 15.5 A lateral view of the vertebrate nervous system (Fore, forebrain; Mid, midbrain; 1–8, rhombomeres of the hindbrain) with populations of neural crest cells (black) and locations of the seven branchial arches shown. The Hox code is illustrated by the most rostral expression boundaries of *Hoxa1–a5* and *Hoxb1–b5* (and the most caudal expression boundary of *Hoxb1*). Modified from Northcutt (1996).

experiments, elegant as they are, is that normally occurring functional overlap between *Hox* genes and the ability of one gene to compensate for inactivation of another, may mean that loss of a gene will not result in an altered phenotype, even though the gene is involved in the production of that phenotype, an idea pioneered by Waddington. Section 16.2.4 contains a detailed example from limb development.

Hoxa-1 and *Hoxa-3* play especially important roles in the neural crest and its mesenchymal derivatives and/or in the tissue interactions that induce them. These roles, plus the absence of multiple *Hox* clusters in hemichordates (which lack neural crest cells) suggest that the acquisition of new or expanded gene functions at the outset of vertebrate evolution, was instrumental in permitting the origin and diversification of neural crest cells that facilitated development of the vertebrate head. One or two examples of studies disrupting *Hox* genes are now discussed.

15.4.1 Disruption of *Hoxa-1*

Homozygotes in which *Hoxa-1* has been disrupted display defects that can be traced to alterations in rhombomeres 4 to 7 and the most rostral expression boundary of *Hoxa-1*. Defects were not confined to the hindbrain but included failure of associated cranial nerves and ganglia to develop and malformation of the inner ear and bones of the skull derived from paraxial mesoderm. Because *Hox* genes function in a variety of tissues and because tissues interact during ontogeny, care must be exercised in assigning the primary site of gene action to a single tissue.

15.4.2 Overexpression of *Hoxa-1*

Overexpression of *Hoxa-1* after injection of RNA into zebrafish eggs alters anterior hindbrain growth and changes the fate of neural and neural crest cells arising in rhombomere 2. Meckel's cartilage and the palatoquadrate of the mandible (first branchial arch) fail to form, while second arch hyoid cartilages are enlarged and partially duplicated; the mandibular arch skeleton in the chick has a composite origin from neural crest arising from midbrain and rhombomeres 1, 2 and 4 of the hindbrain.[15]

15.4.3 Disruption of *Hoxa-2–Hoxa-4*

Disruption of *Hoxa-3* in mice produces a different suite of defects centred on the neural crest, with defects in the pharyngeal arches, thymus, thyroid, hyoid bone and hyoid arch skeleton. While such changes are not equivalent to homeotic mutations, homeotic transformations can be elicited if boundaries of homeobox genes are expressed more rostrally. As one example, disruption of *Hoxa-2* induces defects in the branchial region corresponding to the rostral extent of expression of *Hoxa-2*. Neural crest mesenchyme derived from rhombomeres which normally populate first and second branchial arches, fails to produce normal second arch derivatives but does produce first arch structures. The result is a homeotic transformation of the entire second arch skeleton into a first arch skeleton. As a second example, the normal rostral level of expression of *Hoxd-4* is at the level of the first cervical somite. When *Hoxd-4* is expressed more rostrally, the occipital bones (which lie rostral to the cervical vertebrae) are transformed into bones that mimic cervical vertebrae.[16]

In an alternative approach Paul Hunt and Gérald Couly and their colleagues found that surgical removal of the rhombencephalic neural crest in chick embryos was followed by normal branchial arch development and normal expression patterns of *Hoxa-2*, *a-3*, and *b-4*. Regulation is the ability of embryos to compensate for diminished numbers of cells assigned to a specific fate by redirecting other cells to that fate. Regulation from the neural epithelium at the surgical site compensated for neural crest cells removed by replacing them and regenerating the ablated *Hox* code.[17]

15.5 AXIAL PATTERNING IN VERTEBRATES

An elegant demonstration within the vertebrates of the correlation between shifts in positions of *Hox* genes, specification of vertebral type, and positioning of paired appendages along that body axis, is the expression studies of 23 chick and 16 mouse *Hox* genes undertaken by Annie Burke and her associates (1995). Gaunt (1994) came to similar conclusions but with a much smaller data set.

Axial organization, such as transitions between body regions (cervical-thoracic, lumbar-sacral) and the position of the limbs along the axis, tracks patterns of *Hox* gene expression. Thus, even though *Hoxc-5* has its most anterior level of expression at different somite levels in chicks and mice (somites 17–18 in the chick, somites 10–11 in the mouse), expression in both species coincides with the level at which the forelimb arises and the brachial plexus innervates the forelimb. Similarly, the anterior limit of expression of *Hoxc-6* maps to the boundary between somite pairs 11–12 in mice, 18–19 in chicks, 20–21 in geese and 3–4 in *Xenopus* (Figure 15.6). Despite these different transitions with respect to somite numbers in the different species (and vertebrate classes), *Hoxc-6* coincides with the transition between cervical and thoracic vertebrae in all species. Other *Hox* genes show similar patterns. *Hox-9* genes map to the thoracic-lumbar boundary in the chick and mouse, *Hox-10* to the lumbar-sacral boundary, and *Hox-12* to the sacral-caudal boundary.

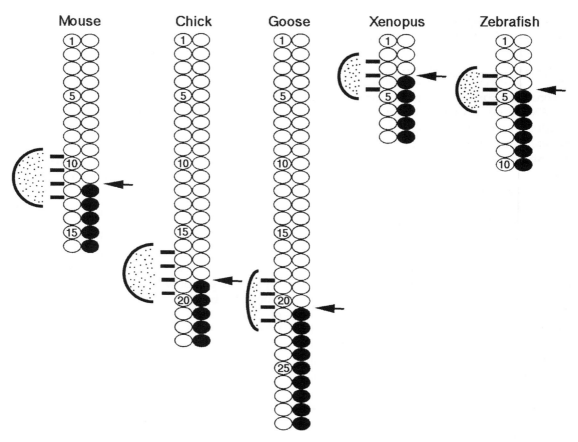

Fig. 15.6 Anterior expression boundaries of *Hoxc-6* depicted in relation to pairs of somites (1–29) in five vertebrate species. Although anterior boundaries of expression differ with respect to somite levels (somites 11–12 in the mouse, 18–19 in the chick, for example), they coincide with the level of the spinal nerves of the brachial plexus (black bars) that innervate the limb buds (curved lines, stippled) in all species. Adapted from Burke *et al.* (1995).

Do the same transitions occur in fish, i.e. do they antedate tetrapods? It appears that they do, for in the zebrafish the anterior expression boundary of *Hoxc-6* lies between somite pairs 4 and 5, a level which coincides with the most posterior segment innervating the pectoral girdle, equivalent to the first thoracic vertebra of tetrapods.

Therefore vertebral type and limb position track boundaries of *Hox* gene expression. Combinations of *Hox* genes specify vertebrae to particular regions of the body axis: *Hoxa-1*

and *Hoxa-3*, *Hoxb-1* and *Hoxd-4* characterize the atlas. These four genes plus *Hoxa-4* and *Hoxb-4* specify the axis, and so forth. As one of many examples of combinatorial action, I discuss the role of *Hoxd-4* in specification of vertebral type.

15.5.1 Combinatorial codes

During normal embryogenesis in the mouse, *Hoxd-4* is expressed in several tissues including the developing spinal cord and the sclerotomal mesoderm from which the vertebrae develop. In heterozygous and homozygous mice in which *Hoxd-4* is mutated (using a combination of embryonic stem cells and a site-specific recombination system), the expression boundary is altered and the second cervical vertebra is transformed into a first cervical vertebra. On both counts – recognition of vertebral type and altered expression boundaries – the transformation is regarded as homeotic, the patterning of one unit having been transformed into that of the adjacent unit (Horan *et al.*, 1995a).

An anterior shift in *Hox* gene expression is therefore followed by posterior transformation of the affected vertebra. These transformations were determined from subjective analysis of vertebral shape and therefore vertebral type. A close association between changing vertebral shape along the body axis and the number of *Hox* genes specifying particular vertebrae has now been demonstrated using more sophisticated measurements of vertebral shape (Johnson and O'Higgins, 1996). Such associations open up the real possibility of understanding the genetic specification of shape and morphogenesis.[18]

Penetrance of the mutant phenotype was incomplete and variably expressed on two different genetic backgrounds. Horan and colleagues reasoned that compound *Hox* mutations might show greater penetrance and more complete transformation of vertebral type. They therefore created compound mutants of three paralogous group 4 genes, *Hoxa-4*, *Hoxb-4* and *Hoxd-4*. Double mutant embryos showed almost complete transformation of the second to the first cervical vertebra (Horan *et al.*, 1995b).

That different double mutants display slightly different phenotypic patterns argues for functional redundancy between these *Hox* genes. Indeed, even more vertebrae are transformed in triple mutants; vertebrae as rostral as the fifth cervical vertebra transform into the first in a dose-dependent response to gene deletion. Significantly, when combinations of

Hox genes are disrupted by targeted mutations, the increased penetrance and expression of phenotypic change is concentrated at transitions between vertebral types. Chen and Capecchi (1997) convincingly demonstrated a similar phenomenon with *Hoxa-9* and *Hoxb-9* in an analysis of rib and sternal changes in mice.

Thus there is serial organization of homeotic genes and a hierarchy in their control and expression. Denis Duboule is a strong advocate for a functional hierarchy among *Hox* genes as a way of silencing a gene without repressing transcription (see section 16.2.4 for additional evidence). If such a mechanism is of widespread importance in development and evolution, then investigating patterns of expression of mRNA may not be an adequate way to map functional genes.[19]

15.6 ENDNOTES

1. For experimentation on and discussions about the developmental and evolutionary relationships of placodal and neural crest ectoderm, see Northcutt (1992, 1996), Graveson (1993), Webb and Noden (1993), Collazo *et al.* (1994), Northcutt *et al.* (1994), S. C. Smith *et al.* (1994), Northcutt and Brändle (1995) and Northcutt *et al.* (1995).

2. For organization of mouse *Hox* genes into four clusters and for their ordering along the chromosome, see Kappen *et al.* (1989a,b, 1993), Schughart *et al.* (1989), De Robertis *et al.* (1990), Gaunt and Singh (1990), Ruddle *et al.* (1994), Walsh (1995) and Meyer (1996).

3. Included in the analyses by Kessel and Gruss (1990, Figure 2) and Kessel (1992, Figure 12) is a summary of the anterior expression boundaries of 27 homeobox-containing genes in the developing murine neural tube, the *Hox* code in mesodermal segments, and the RA-induced transformations in that code. P. W. H. Holland and Garcia-Fernàndez (1996) and Sharman and Holland (1996) present comprehensive overviews and references documenting the evolution of *Hox* codes in chordate origins. See Holland and Hogan (1988), De Robertis *et al.* (1990), Wilkinson (1990) and Wilkinson and Krumlauf (1990) for overviews of basic studies

on *Hox* genes. For homeotic transformation of amphibian tail blastema into limb primordia, see Mohanty-Hejmadi *et al.* (1992) and Maden (1993). For respecification of a second caudal body region, see G. B. Müller *et al.* (1996). For an overview of control of limb regeneration in amphibians, see Brockes (1997). See Géraudie and Ferretti (1997) for the role of RA in inducing cell death in regenerating fins and limbs.

4. See de Beer (1958) for the influence of early studies on urochordate induction on taxonomy. Reverberi *et al.* (1960), Tung *et al.* (1962) Nishida (1991), Nakatani and Nishida (1994) and Nakatani *et al.* (1996) demonstrated neural or notochord induction in ascidians and amphioxus. See Hall (1983a), Satoh (1994) and Satoh and Jeffery (1995) for how this induction compares with neural induction in vertebrates.

5. For evidence for a single *Hox* gene cluster in urochordates, see Ruddle *et al.* (1994), P. W. H. Holland *et al.* (1994) and Sharman and Holland (1996). See Wada *et al.* (1996) for expression of *HrPax-37* in the ascidian, Serbedzija and McMahon (1997) for *Pax-3* as essential for neural-crest-cell development, and Hall and Hörstadius (1988) for a discussion of the evolution of the neural crest. See Crowther and Whittaker (1992, 1994) for cilia patterns; see Sasai *et al.* (1995) and Wilson and Hemmati-Brivanlou (1995) for studies with BMP-7. For up-regulation of tunicate *Hox* genes by RA, see Shimeld (1996).

6. See Lacalli *et al.* (1994), Lacalli (1996b) and Presley *et al.* (1996) for embryonic development of cephalochordates, and Wada and Satoh (1994) and Peterson (1995) for the 18S RNA analysis.

7. For evidence that amphioxus has a single *Hox* cluster that is upregulated by RA, see Holland *et al.* (1992), Garcia-Fernàndez and Holland (1994), Holland (1996) and Shimeld (1996).

8. The *Ciona* studies are those of Corbo *et al.* (1997). See P. W. H. Holland *et al.* (1995) and Terazawa and Satoh (1995) for *Brachyury* in amphioxus, Harada *et al.* (1995) for *Brachyury* in sea urchins. See s. 5.5.5, Maienschein (1994), Gee (1996) and Bowler (1996) for histories of changing ideas on the primitiveness of amphioxus.

9. For studies on *Hox* genes in fishes, see P. W. H. Holland and Garcia-Fernàndez (1996), Misof *et al.* (1996) and van der Hoeven *et al.* (1996). For the complete sequence of *Hox* genes in the

Japanese puffer fish, see Aparicio *et al.* (1997) and P. W. H. Holland (1997). For similar patterns in *Msh* genes, see P. W. H. Holland (1991).

10. See P. W. H. Holland and Garcia-Fernandez (1996) and N. D. Holland (1996) for homology between amphioxus and vertebrates, and Burke *et al.* (1995) for axial patterning and *Hox* codes.

11. See Hunt *et al.* (1991a–d) and Hunt and Krumlauf (1992) for expression boundaries of *Hox* genes in the vertebrate hindbrain; Holland and Garcia-Fernàndez (1996) for projection of these boundaries to the amphioxus nerve cord; and Lacalli *et al.* (1994) and Lacalli (1996b) for structural homology of elements of amphioxus and chordate nervous systems. For exercising caution in such interpretations, especially for the possibility that the same developmental signals may be used to generate different structures, see G. B. Müller and Wagner (1996) and Abouheif (1997).

12. For duplication of regulatory genes in vertebrates, see Atchley *et al.* (1994), Sharman and Holland (1996) and Holland and Garcia-Fernàndez (1996). For shared expression of *engrailed* between posterior compartments in each segment of *Drosophila* and the posterior half of each of the first eight somites in amphioxus, see L. Z. Holland *et al.* (1997). For duplication of *engrailed* during the evolution and diversification of barnacles, see J.-M. Gilbert *et al.* (1997).

13. For the *Hox* code, see Hunt *et al.* (1991a–c), Kessel and Gruss (1990, 1991), Hunt and Krumlauf (1992) and Kessel (1992).

14. Cephalic (placodal ectoderm) in the axolotl is also derived from cells that arise in the neural folds (Northcutt, 1996). Initially thought to reflect a transfer of the *Hox* code from neural ectoderm → neural crest → branchial arches → ectoderm, the similarity of the expression boundaries does not reflect simple transfer of a *Hox* code; separate enhancer elements are present in *Hox* gene clusters in the neural tube and neural crest.

15. See Köntges and Lumsden (1996) and Miyake *et al.* (1996b) for skeletal structures with a composite origin within the mandibular arch (Meckel's cartilage, mandibular membrane bones) or from two branchial arches (middle ear ossicles).

16. For studies on disruption of *Hoxa-1* or *Hoxa-3*, see Lufkin *et al.* (1991), Chisaka and Capecchi (1991) and Alexandre *et al.* (1996). For homolo-

gous recombination of *Hoxd-4* and *Hoxa-2*, see Lufkin *et al.* (1992), Gendron-Maguire *et al.* (1993) and Rijli *et al.* (1993).

17. See Scherson *et al.* (1993), Hunt *et al.* (1995) and Couly *et al.* (1996) for regulation of the avian neural crest, Snow and Tam (1979) for regulation of the mouse neural crest, and Maclean and Hall (1987) and Gilbert (1994a) for general discussions of regulation.

18. For practical approaches to the analysis of shape, for the utility of morphometrics to evolutionary studies and for a caution on shape and homology, see Bookstein *et al.* (1985) and Bookstein (1991, 1994).

19. See Duboule (1992, 1994b) and Duboule and Morata (1994) for functional hierarchies among *Hox* genes.

16

TRANSITIONS IN ANIMAL EVOLUTION

'It cannot be said that this ancient extinct bird [*Archaeopteryx*] goes far towards connecting birds with reptiles: but in the possession of separate claw-bearing fingers, a long bony tail and teeth, in the apparent want of a beak, it does come nearer to lizard-like reptiles than does any other known bird.'

Lankester, 1909, p. 239

Although the features that characterize each group of organisms are readily identifiable, there is no conclusive evidence that their acquisition was the key innovation responsible for the origin of the group. Obviously the Metazoa could not have arisen without multicellularity, but the key innovation, if there were one, may well have had to do with cell division or cell surface properties, changes that enabled multicellularity but that were neither multicellularity *per se*, nor selected for as multicellularity (s. 13.5). This approach sees multicellularity as consequential, not causal. Gradual improvement in function and the availability of new habitats are just as likely initiators of change as key innovations. Often change in structure/function and habitat proceed hand-in-hand, as in the proposed origin of book-lungs from book-gills in the transition from aquatic to terrestrial scorpions in the Carboniferous (Jeram, 1990), insect wings from limbs or gills, limbs from fins, or feathers from scales.[1]

In this chapter I provide further examples of transitions in animal evolution. Invertebrate examples include the evolution of wings in insects, torsion in gastropods (ss. 16.6, 16.7) and eyespot patterns in butterfly wings (s. 20.5). For vertebrates, examples include the

origin of jaws and transformation of jawless to jawed vertebrates, the transition from fish fins to tetrapod limbs, the origins of birds and feathers, evolution of turtle shells, and the development of pharyngeal jaws in cichlid fishes (ss. 16.1–16.5, and see sections 11.5 and 17.2.1 for the evolution of specialized features of mammalian teeth). Other examples, such as the neural crest and the origin of the vertebrates, were covered in Chapter 15, while even more examples are covered in Bob Carroll's 1997 book *Patterns and Processes of Vertebrate Evolution*, which may be read with much profit.

16.1 JAWLESS → JAWED VERTEBRATES

Cope (1889) divided vertebrates into agnatha (jawless) and gnathostomata (jawed). The earliest known vertebrates of the Cambrian Era are jawless fishes possessing a series of gill arches, each with a skeletal component supporting the respiratory organs. The fossil record of these jawless fishes is very complex and subject to several interpretations. In the traditional scenario of the transition from jawless to jawed vertebrates, the most anterior of these gill arches was extended anteriorly to a position supporting the mouth and buccal cavity, carrying its cartilaginous supporting

A

B

C

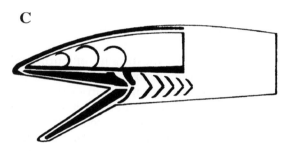

Fig. 16.1 Proposed transformations of the first of the serial gill arches of a jawless agnathan (A) into jaws in gnathostomes (B), which are then braced by the skeleton of the second gill arch (C). Modified from du Brul (1964).

element with it. The dorsal portion of the arch formed the palatoquadrate, the skeleton of the upper jaw; the ventral portion formed the mandible, the skeleton of the lower jaw (Figure 16.1). The necessary elements were thus present in the original gill arch and functioned in respiration, but were co-opted for a new function (feeding) in this new position. No new gill arch had to be constructed, no 'new' skeletal elements had to evolve. The traditional view is that enhanced feeding was the driving selective force, but whether feeding, predation, or respiration drove this evolutionary change is uncertain.[2]

16.1.1 Developmental evidence

Evidence that jaws are modified gill arches comes in part from development (the mandibular arch is the most anterior of a series of pharyngeal arches), in part from homology of the skeletal elements, and in part from the equivalent developmental origins and inductive interactions responsible for their formation (s. 15.1).

Although this is the traditional textbook view, it is unclear how jaws arose, especially considering suspicions that agnathan gills may not be homologous with the gills of jawed vertebrates (gill lamellae lie on the inside of the skeleton in lampreys but outside in gnathostomes; Schaeffer and Thomson, 1980). An alternative scenario is based on the velum, which pumps the pharynx in lampreys. Like the jaws of gnathostomes, the velum is derived from the mandibular arch and may be the agnathan homologue of the jaws. These alternative scenarios emphasize how critical it is to establish homologies of the structures and organisms being considered, before proposing an evolutionary scenario.[3]

The key innovation in the agnathan-gnathostome transition could be regarded as behavioural, functional/ecological, structural, developmental or size-related, depending more on individual inclination than on evidence for the change itself. Arguments for each of the four areas are:

1. functional/ecological: increasing toughness of potential food and the need to develop a more robust food-gathering apparatus;
2. structural: availability of the first visceral arch with its skeletal elements and associated musculature;
3. developmental: the transformational capabilities of the mesenchymal and skeletal tissues of the arch (Hall, 1975); and
4. growth-related: the capability of the arch and its components to grow anteriorly.

None of these could have been effective in isolation. To isolate one as 'the' key innovation

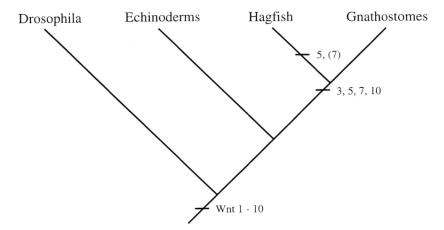

Fig. 16.2 The genes *Wnt-1–Wnt-10* were present in the common ancestor of arthropods and deuterostomes. *Wnt-3, -5, -7* and *-10* were duplicated in craniate ancestors. *Wnt-5* and perhaps also *Wnt-7* underwent separate duplications on the line to hagfishes. Based on data in Sidow (1992).

is certainly misleading and other innovations (the origins of pectoral fins, transformations of pelvic fins, elaboration of the forebrain) were taking place in these same organisms.

However they arose, jaws, once in place, provided a key innovation for the 500-million-year radiation into every conceivable and even some inconceivable habitats, allowing jawed vertebrates to capture and utilize an extraordinary range of prey and foodstuffs. In section 16.5, I discuss an example of a subsequent addition to jaw structures – pharyngeal jaws in cichlid fishes – and the ecological exploitation this addition permits.[4]

16.1.2 Molecular evidence

Comparative studies of structure and development can be augmented by analyses of sets of genes or even of individual genes. With the rapid growth in comparative molecular biology over the past seven to ten years has come knowledge of molecules for which sequences and partial sequences are now known for a sufficient number of organisms that they can be used for broad-based phylogenetic analyses and/or for analysis of key events

accompanying or accounting for the origin of individual groups of organisms.

Sidow (1992) used 55 partial sequences of the ten *Wnt* genes then known in an investigation of the initial diversification of jawed vertebrates. *Wnt-1* to *Wnt-10* are ancient, having diversified before the last common ancestor of arthropods and deuterostomes some 600 or more mya. A round of duplication of *Wnt-3, -5, -7* and *-10* occurred early in the lineage leading to craniates (Figure 16.2). Subsequent amino acid replacement in *Wnt* genes was four times lower in jawed vertebrates than in agnathans and echinoderms. The implication is that *Wnt* genes were serving an important function(s) regulating development in the 100 or so million years associated with the origin of jawed vertebrates. A further implication is that *Wnt* plays or played more important roles in the development of jawed than jawless vertebrates, although the hagfish *Eptatretus stoutii* has several *Wnt* genes. Sidow generalizes from the expression of *Wnt* in limb buds, the neural crest, and central nervous system, to argue that diversification of *Wnt* genes facilitated generation of cell types and developmental pathways characteristic of vertebrates.

An excellent example illustrating the utility and potential pitfalls of analysing homology on the basis of patterns of gene expression (ss. 15.2, 21.7.3), is expression of the segment polarity gene *engrailed* in the lamprey *Lampetra japonica* (N. D. Holland *et al.*, 1993). As in other vertebrates, *engrailed* is expressed at the midbrain–hindbrain boundary and in mesoderm of lamprey mandibular arches. At this level of analysis the expression patterns suggest conservation of *engrailed* across the vertebrates. Later in development (by the eyespot stage), expression of *engrailed* is restricted to one particular set of muscles, the velothyroideus muscles that activate the velum.

As discussed above, homology of the lamprey velum is unclear. Two of the jaw muscles in zebrafish, the levator arcus palatini and dilator operculi, express *engrailed* at an early condensation stage in their development. While accepting the difficulties of using expression patterns of isolated genes to establish homologies of the structures expressing those genes, Holland and his colleagues favour the view that the velothyroideus muscles in the lamprey are homologues of the two *engrailed* positive muscles in zebrafish, and by implication, in all teleosts. The inference – that the lamprey velum is the homologue of the gnathostome jaw and that jaws arose from the velum – illustrates the dilemma.

Are we justified in inferring the origin of vertebrate jaws on the basis of the expression pattern of one gene? Most of us would answer with a resounding no! Even a comparative analysis of *engrailed* expression across the teleosts, while it would reveal the generality of the pattern, would not resolve the fundamental difficulty of whether the pattern of expression of a single gene is an adequate basis to establish structural homology and/or to infer evolutionary origins. Even analysis of patterns of diversification of many genes might not be instructive, given that multiple genes would change when a new pathway was co-opted. Patterns of gene expression in conjunction with more traditional measures of homology (position, innervation, musculature and so forth) are much more likely to reveal patterns of origination. Determination of *engrailed* expression in parallel with analyses of homology based on muscle condensations and development in primitive fishes (as undertaken by Miyake, McEachran and Hall (1992a) for the gill arches in the highly derived batoid fishes) may provide a base from which to tackle such fundamental problems as the origin of jawed vertebrates. Combined molecular, developmental and phylogenetic analyses of such developmental regulatory genes hold great promise.[5]

16.2 FINS → LIMBS

The belief that tetrapod limbs arose from modified fins has a long history, reviewed by Richard Hinchliffe and David Johnson in their now classic text, *The Development of the Vertebrate Limb* (1980) and in a more recent NATO symposium, *Developmental Patterning of the Vertebrate Limb* (Hinchliffe *et al.*, 1991).

16.2.1 Palaeontological evidence

Paired lateral projections which were not appendages, and which consisted of spines and plates as modifications of dermal armour, first appeared in vertebrates in Silurian and Devonian jawless ostracoderms. Such exoskeletal projections are not homologous with paired fins and limbs: they were ectodermal (neural-crest-derived) and not mesodermal, stabilizers rather than locomotory organs.

Flexible anterior and/or posterior sets of fins are first found in various Devonian fishes. Just as median unpaired dorsal and ventral fins preceded paired fins, so pectoral preceded pelvic fins. Structural evidence, essentially based on position and connections as proposed by Geoffroy, makes the case for the transition from fins to limbs; pectoral and pelvic fins transformed into fore- and hindlimbs

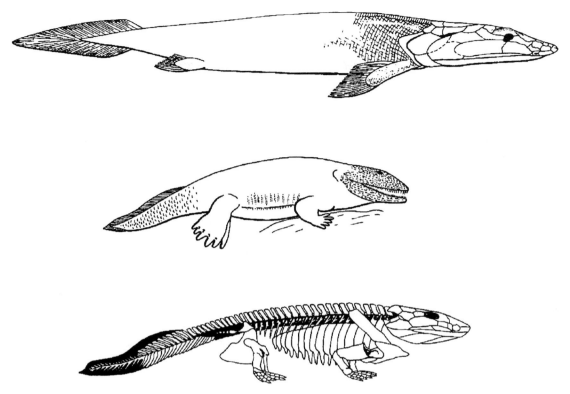

Fig. 16.3 Reconstructions of *Panderichthys*, a sarcopterygian fish from the Devonian (top) and *Ichthyostega* (middle, bottom), an early, partly aquatic tetrapod. Modified from Janvier (1996a) and Waterman *et al.* (1971).

when a group of Devonian lobe-finned fishes ventured onto land. As Michael Coates and Jennifer Clack have documented from the new fossil evidence, this major ecological transition from water to land was neither accompanied by, nor perhaps even facilitated by, a simultaneous transformation of the fins.[6]

Panderichthys and related sarcopterygian fishes from the Devonian appear to lie closest to the first tetrapods. *Ichthyostega* and *Acanthostega*, two of the first tetrapod genera, were partly aquatic (Figure 16.3). Their pectoral fins, like those of earlier sarcopterygian fishes, are composed of a single proximal element and two more distal elements regarded as homologous with the humerus, radius and ulna of the tetrapod limb (Figure 16.4). Elements equivalent to wrist/ankle elements (carpals/

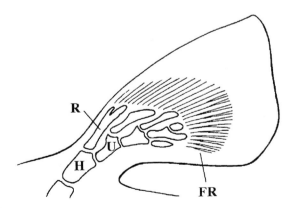

Fig. 16.4 A transitional stage between fin and limb with proximal elements that can be identified as humerus (H), radius (R) and ulna (U); more distal elements; and distal fin rays (FR). Modified from Shubin (1995) after Steiner.

tarsals) are present in *Ichthyostega* and *Acanthostega*, as they were to varying extents in earlier sarcopterygians. Significantly, although elements equivalent to proximal phalanges may be present, no distal elements equivalent to the digits of the tetrapod limb are present (Figure 16.4). They appear to be a tetrapod invention. See Box 16.1 for problems assigning phalangeal formulae and numbering digits.

According to one long-standing view, lateral fin folds set the stage for the evolution of fins and then tetrapod limbs. According to this theory, known from its original proponents as the Thacher–Balfour or Thacher–Mivart–Balfour fin-fold theory of the origin of the paired fins, paired fins share common elements with the median unpaired fins in that both are derived from folds of the body wall and supported by skeletal elements. As demonstrated from the fossil evidence, however, pectoral and pelvic fins did not evolve simul-

taneously among primitive fishes. It is unclear whether this necessarily argues against their origin in a continuous fin fold; the separation need not have been contemporaneous. That no extant vertebrates possess such a continuous lateral fin fold that can be related to orthodox fins, may be a more compelling reason against the fin-fold theory.[8]

16.2.2 Developmental evidence

Fins and limbs do develop in similar ways at the cellular and molecular levels.

A complex of interactions between mesenchyme and ectoderm regulates proliferation, outgrowth and patterning along the three limb axes (s. 11.4). Fins also begin development as a core of mesenchyme surmounted by an ectodermal thickening. Most extensively studied in the trout, this 'pseudoAER' or fin fold does not remain as a prominent projection but rather folds onto itself and increases in

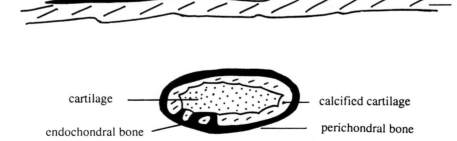

Fig. 16.5 The dermal exoskeleton is based on dentine underlain with bone and overlain with enamel. The endoskeleton is based on cartilage which, both ontogenetically and phylogenetically, is first surrounded by perichondral bone and then invaded and replaced by endochondral bone.

BOX 16.1
PHALANGEAL FORMULAE AND NUMBERING DIGITS

Assigning phalangeal formulae and homology to tetrapod digits is a controversial undertaking. Hinchliffe (1985) and Cooke (1990) have addressed the issue of digit nomenclature, an especially difficult problem when digits are lost (as they are in birds and many amphibians), when palaeontological records demonstrate that the first tetrapods were polydactylous, and/or if the pentadactyl limb evolved independently in such groups as amphibians and reptiles. Padian (1992) proposed a standardized scheme for phalangeal formulae for tetrapods that takes into account homology and the pentadactyl basis of digit number. Thus, the phalangeal formula for a vertebrate with four digits and three phalanges per digit would be 0-3-3-3-3 (a notation that clarifies that digit I is absent) rather than 3-3-3-3 which does not.[7]

If individual digits can indeed be homologized, Padian's scheme for phalangeal formulae clarifies digit loss and digit homology. However, opinions vary as to whether individual digits can be homologized (s. 21.2). This is important; presence of digits defines the tetrapods (Thomson, 1991b), although the existence of species with fins anteriorly and limbs posteriorly complicates such a neat and crisp definition, as does the recent finding of a sarcopterygian fish with radials encased in fin rays (Daeschler and shubin, 1998).

length. Dermal fin rays, which make up the distal fin elements in fishes, develop within these ectodermal folds. Mesenchyme of the fin rays is of neural crest origin. It is presumed, but not yet proven, that the lepidotrichia and actinotrichia that make up the fin rays themselves are also of neural crest origin. With respect to the tissues that form them, the dermal exoskeleton is based on bone and associated dentine (± enamel). The endoskeleton (which may be of neural crest or mesodermal origin) is based in cartilage which may either persist as a permanent cartilage, or be replaced by being surrounded by (perichondrial) bone or invaded by (endochondral) bone (Figure 16.5).[9]

The proximal fin endoskeleton of teleosts, elasmobranchs and chondrichthyes is mesodermal. In teleosts, bony fin rays are the major component of the fins; the proximal endoskeletal is a minor component. The presence of neural crest derivatives in fins is a fundamental difference from tetrapod limbs, which lack any neural crest (dermal) elements, and in

which the endodermal skeleton has expanded to include formation of digits. From such descriptive analyses comes the surmise that the transition from fins to limbs involved:

1. elimination of the dermal fin rays;
2. elaboration of the existing proximal endoskeleton;
3. elaboration of the distal endoskeleton for wrist, ankle and the proximal phalanges; and
4. formation of new distal endoskeletal elements (the distal and fourth carpal) as digits.

Developmentally, these changes are essentially the loss of the neural crest component (dermal skeleton) and a take-over by the mesoderm-derived endoskeleton.

In 1991 Peter Thorogood proposed, and in 1995 Sordino and colleagues elaborated, a model of altered timing to explain these shifts in the fin → limb transition. Thorogood related development of the endoskeleton to the degree and timing of folding of the ectodermal fold. Alteration in timing of appearance of the apical fin fold would prolong the time during which mesenchyme was in association with the ectodermal proliferative signal. The consequent extension of signalling from ectoderm to distal mesenchyme would promote formation of additional distal mesenchyme. Such a timing switch or heterochrony, when it changes an evolutionary pattern (s. 24.1), would explain how the autopod of the limb bud was extended in the fin-limb transition, but does not explain how that mesenchyme came to form novel limb structures such as digits.[10] [The term autopod may be unfamiliar to some readers. Three terms are used to describe skeletal elements in different regions of tetrapod limbs. Stylopod refers to the single proximal elements – humerus in forelimb and femur in hind. Zeugopod refers to the paired elements immediately distal to the stylopod – radius and ulna in the forelimbs, tibia and fibula in the hindlimb. The most distal region, the autopod, consists of wrist or ankle elements (carpals and metacarpals or tarsals and metatarsals) and digits (phalanges).]

16.2.3 Polydactyly

The story has become even more interesting because of some surprising palaeontological discoveries associated with digits in the first tetrapods.

1. The first tetrapods were 'polydactylous'. They were not built on the pentadactyl (five-digit) plan of later tetrapods. *Ichthyostega* had seven digits in the hindlimb (fossils with forelimbs have not been found); *Acanthostega* had eight digits in fore- and hindlimbs.
2. The concept of a primitive pentadactylous limb no longer stands.
3. Digits arose at a stage when animals like *Acanthostega* were facultatively aquatic.
4. The number of digits may have been reduced independently in amphibians and reptiles.

If the original tetrapod condition was of limbs with more than five digits, then 'polydactylous' is not the appropriate term for the early tetrapod pattern. The early tetrapods were not polydactylous, for the term indicates a condition with more than the typical number of digits for the taxon. Modern tetrapods with a maximum of five and sometimes fewer digits show a reduction from the original condition. They are oligodactylous with respect to the original condition.[11]

16.2.4 Molecular evidence

We still do not know how fins acquired the distal chondrogenic condensations necessary to produce the tetrapod novelty of an autopod with the original (but transformed) sarcopterygian carpals and/or tarsals, to which digits were added. Molecular evidence from limb development contributes to our understanding of this aspect of the fin → limb transition.

Hox genes position limbs along the body axis and play regulatory roles in limb bud development (ss. 11.4, 15.5). Similarities in cellular development of fins and limbs have prompted investigators to ask whether *Hox* genes play the same role in fin as in limb development. We now know that they do. Can transitions in a *Hox* code between fins and limbs explain the formation of one from the other, especially the formation of the skeleton of the new digits? Perhaps yes, perhaps no. It certainly appears, at least from knowledge of zebrafish development, that unpaired median fins do not express the *Hox* genes (*Hoxa* and *Hoxd*) that pattern paired appendages. If this is the ancestral condition, then *Hox* genes had to be co-opted to pattern paired appendages.

Similar expression boundaries of *Hox* and homeotic genes are found in fin and limb buds. Admitedly not many species of fish have been sampled; for that matter, not many species of tetrapods have been studied either. Nevertheless, the similarities are very close, as the following list indicates:

1. Expression patterns of *Hoxc-6* and the *Hoxd* and *Hoxa* clusters in the fin buds of the zebrafish *Danio rerio* are equivalent to those seen in the limb buds of amphibian, avian and mouse embryos.
2. *Sonic hedgehog* is expressed in a zone of polarizing activity (ZPA) in fin and limb buds.
3. *Sonic hedgehog* is inducible by RA in zebrafish pectoral fin, as it is in tetrapod limb buds. (RA also induces apoptosis in regenerating fins and limbs.)
4. *Msx* is expressed in the distal limb mesenchyme in fin and limb buds.[12]

There have been several proposals to explain the transitions between fin and limb buds and fin and limb skeletons using these expression patterns, but it may be that the genetic signals are encoded not in expression patterns, but in regulatory sequences or cascades. Tabin and Laufer (1993) proposed more distal (ectopic) activation of *Hoxa* and *Hoxd* clusters as one way to re-specify distal mesenchyme in the fin → limb transition, but more recent data does not support activation of *Hox* genes in a new distal domain. Sordino and colleagues (1995) developed a scenario from the initial expression of *Hoxd-9* to *-13* and *Hoxa-9* to *-13* in posterior and proximal limb regions, followed by progressive expression more anteriorly and distally within tetrapod limb buds, but restricted anterior expression in fin buds

of *Danio rerio*. More detailed analysis of patterns of *Hox* gene expression in avian limb buds confirm and expand such a hypothesis (Nelson *et al.*, 1996).

Support for differential extension of a prior, proximal domain accompanied by differential regulation comes from three lines of investigation. One is detailed fate-mapping of the chick wing bud and the demonstration that cells already expressing *Hoxd-13* 'fan out' into the distal bud. The second is the cloning of 23 *Hox* genes from the embryonic chick and detailed analysis of their deployment in three phases during limb bud development. The third is targeted mutations in mice of *Hoxa-11* and *Hoxd-11* which result in loss of the radius and ulna. Indeed, targeted deletion of *Hox* genes in the clusters *Hox-9* to *-13* is accompanied by elimination of successively more distal limb elements, indicating selective control over limb development (Table 16.1).[13]

Eliminating *Hoxd-13* interferes with limb size rather than limb patterning. In chick limb development, *Hoxd-13* regulates growth at the growth-plate stage while *Hoxd-11* acts at the prior cartilage condensation stage. Yokouchi and colleagues (1995) misexpressed *Hoxa-13* in the entire chick limb bud using a replication-competent retroviral system. Misexpression of *Hoxa-13* (like *Hoxd-13*) dramatically slowed growth and differentiation of more proximal limb cartilages (radius/ulna, tibia/fibula).

Table 16.1 Progressively more distal control of limb skeletal elements by progressively more 5' Hox genes of the *Hox-9* to *Hox-13* clusters[1]

Limb skeletal element		Hox gene
Forelimb	*Hindlimb*	
Pectoral girdle	Pelvic girdle	*Hox-9*
Humerus	Femur	*Hox-10*
Ulna/radius, proximal	Tibia/fibula, proximal	
Carpals	Tarsals	*Hox-11*
Distal carpals	Distal tarsals	*Hox-12*
Digits and phalanges	Digits and phalanges	*Hox-13*

[1] Based on data summarized in Davis *et al.* (1995).

These skeletal elements displayed substantial arrest of chondrocyte growth and differentiation. This change was interpreted as a homeotic transformation of the proximal cartilages to a more distal type, producing short cartilages characteristic of the carpus or tarsus. Over-expression of other *Hox* genes does not produce this phenotypic shift. The cartilage cells in *Hoxa-13*-deficient embryos are more adhesive, fail to separate, and so produce smaller elements. Even if not a homeotic transformation (and it is difficult to identify limb elements as proximal or distal when size is perturbed), this study links cell-surface responsiveness with overall skeletal patterning in an unexpected way and suggests a genetic mechanism for transformation of proximal into distal elements in the fin → limb transition.[14]

Favier *et al.* (1996) demonstrated effective functional cooperation between non-paralogous *Hox* gene products in developing mouse limbs and differential control over fore- and hindlimbs. Targeted disruption of *Hoxa-10* had minimal effect on the forelimb skeleton, but affected proximal elements in the hindlimb. Mice made double mutant for *Hoxa-10/Hoxd-10* showed a synergistic response with exaggerated defects in the forelimb. A further analysis, targeting mutation of *Hoxd-12* resulted in relatively minor defects in the distal forelimb skeleton: reduced length of metacarpals and phalanges and malformation of one distal carpal. Mutation of *Hoxd-13* produced a more pronounced phenotypic effect on growth and eliminated many of the metacarpals and phalanges. Again, overlapping domains of single classes of *Hox* genes provide a potential mechanism for evolutionary transformations.

Davis and Capecchi (1996) and Zákány and Duboule (1996) assessed interactions between *Hoxd-11, -12, and -13* by interbreeding mutant strains. Interaction between these three 5' *Hox* genes was demonstrated by the presence of phenotypes (including an additional postaxial digit) not seen with mutations of individual

genes. The homozygous mice resulting from this triple inactivation had small digit primordia, disorganized cartilage and impaired skeletal growth, suggesting involvement of this gene in early prechondrogenic condensations. Similarly, targeted mutation of *Hoxa-9* and *Hoxb-9* produces more severe rib and sternal defects (including increased penetrance and increased expression of the fusions), than does mutation of either gene alone (Chen and Capecchi, 1997). Such studies provide further support for an interactive and synergistic role for *Hox* genes in patterning the limb skeleton and a mechanism for changes in the fin → limb transition.[15]

Sequential spread of expression across buds in the transition from fins → limbs is likely to be sufficient to produce additional mesenchyme distally, and even to specify that the elements that develop are distal rather than proximal, but not to specify them as limb elements. Sordino and colleagues emphasized that delayed folding of the ectoderm in the fin bud would permit prolonged signalling and mesenchymal proliferation, and allow expression boundaries of *Hoxd* genes to shift more anteriorly and distally. Such a proposal effectively combines mechanisms of altered timing and modified gene expression. An overlapping regulatory cascade of *Hoxa-9–13* to *Hoxd-9–13*, operating along the P-D axis, may be a sufficient signalling system to explain reduction of the exoskeleton and distal extension and transformation of the endoskeleton.

16.2.5 Limb patterning

Our understanding of how prechondrogenic condensations are patterned into skeletal elements in tetrapod limbs increased dramatically with the work of Pere Alberch and his colleagues in the 1980s. They identified branching and segmentation as the two major processes involved in subdividing and extending condensations and patterning developing skeletal elements. The resulting bifurcations establish primary limb axes con-

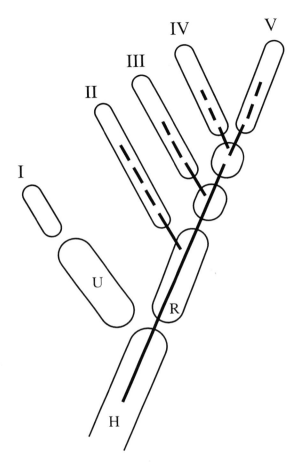

Fig. 16.6 Proximally, the primary limb axis runs proximo-distally from the humerus through the radius. Distally, the axis runs antero-posteriorly through the digits. H, humerus; R, radius; U, ulna; I–V, digits I–V.

sisting of humerus → ulna → digital arch in the forelimb, and femur → fibula → digital arch in the hindlimb (Figure 6.1). Proximally the axis runs proximo-distally, but distally, the digital arch shifts preaxially and so runs from posterior to anterior, except in (some?) salamanders, where the digital arch extends antero-posteriorly (Figure 16.6).[16]

Arguments based on modification of branching patterns of skeletal condensations cannot offer a sufficient explanation for the origin of tetrapod digits, for they do not ad-

dress how altered branching would produce novel structures. Postaxial branching that produces digits is the branching of already established chondrogenic condensations. Branching of proximal condensations, followed by growth and extension of the condensations distally and anteriorly, would produce novel distal condensations as the positional shift of a developmental precursor. Such branching, growth and extension are not sufficient to produce novel structures without a second, probably *Hox*-gene-based, mechanism to transform the newly positioned proximal condensations into novel distal elements, and perhaps a third mechanism to establish joints. This may be a situation where the fossils will help to reveal the answers, just as they revealed ancestral polydactyly and the precedence of pectoral over pelvic fins.

A potential molecular basis for the subdivision of condensations into individual skeletal elements is now available with the unexpected finding by Storm and Kingsley (1996) that a BMP-like molecule named GDF-5 (growth differentiation factor-5) is expressed in transverse stripes across many developing skeletal structures of embryonic mice in positions that correspond to sites of joint formation. Furthermore, null mutations of GDF-5 disrupt formation of over 30% of the synovial joints in the limb and change the patterns of elements of the digits, wrists and ankles. Other growth differentiation factors (GDF-6 and GDF-7) also are expressed in joint regions. It will be of great interest to explore the distribution and any possible role of GDF-5 and other GDFs in the earlier branching and segmentation events in limbs, and to search for their expression and/or role in fin development.[17]

16.3 BIRDS AND FEATHERS

Perhaps nothing in vertebrate history, indeed, nothing in the history of palaeontology, is as dramatic as the discovery of *Archaeopteryx* and the origin of birds, flight and feathers. This story was told in detail in 1985 in the proceedings of an international conference on *Archaeopteryx* edited by Hecht *et al.*, and updated in 1996 by Alan Feduccia in *The Origin and Evolution of Birds*, and again in 1997 by Bob Carroll in *Patterns and Processes of Vertebrate Evolution*. Although palaeontological evidence is revealing the origins of birds, neither palaeontological nor developmental evidence yet speaks more than a whisper about the origins of feathers.

16.3.1 Palaeontological evidence

Everyone knows something of the origins of birds and flight because of the fame of *Archaeopteryx*, the Upper Jurassic 'missing link' between reptiles and birds; see Lankester's opinion in the epigraph for this chapter.

A single fossil feather was discovered in 1861. Soon thereafter, two virtually complete fossils with feathers and wishbones (the fused clavicles or furcula that characterize birds) were unearthed, and Huxley (1868) made the dramatic announcement that *Archaeopteryx* was precisely what Geoffroy had hoped for and Darwin's theory had predicted – an intermediate between two major vertebrate groups, reptiles and birds. Had feathers not been preserved in these specimens, without question, *Archaeopteryx* would have been classified as a reptile, for the only non-reptilian feature in the skeleton is the furcula.

Huxley thought dinosaurs were the most likely reptilian ancestors of birds. His theory lay fallow for more than a century until John Ostrom disinterred it and advocated bipedal theropod dinosaurs as the ancestors of birds. Bipedality was already established in dinosaurs by the Middle Triassic; *Sinosauropteryx prima*, from Liaoning province in northeastern China, lived some 120–140 mya and was bipedal.[18]

As might be expected from knowledge of the embryonic development of skeletal systems and epidermal appendages such as scales and feathers (see below), and as verified from the fossil record, the transition to birds was gradual. Different features altered at different rates. A newly discovered fossil, *Unenlagia comahuensis*, intermediate between theropod dinosaurs and birds, illustrates beautifully the mix of primitive theropod and derived avian features found in such 'missing links'. Unfortunately, it is not known whether *Unenlagia* was feathered or possessed anything resembling protofeathers, although it was able to fold its forelimbs in a most bird-like manner.[19]

Recent discoveries of fossil birds with feathers and more specialized wings and pectoral girdles than *Archaeopteryx*, but with dinosaur-like pubic bones and ribs, indicate both how rapidly birds diversified and how they retained a mosaic of reptilian and avian characters. Although only a little younger than *Archaeopteryx*, these Chinese and Spanish fossils had already diversified into several lineages. One such fossil is *Sinornis santensis*, a sparrow-sized, 135 million-year-old bird from the Lower Cretaceous of China, described by Paul Sereno and Rao Chenggang in 1992. Indeed, as new and more complete avian fossils appear – for example the Upper Jurassic/ Lower Cretaceous *Confuciusornis* from northeastern China described by Hou *et al.* (1996) – we are increasingly able to evaluate the range of innovations associated with the origin of flight and the origin and radiation of modern birds that began in the later Mesozoic. How feathers arose is much more problematical.

16.3.2 Developmental evidence

If the first bird hatched from a reptile egg, did the first feather arise from a reptile scale? The answer from embryological evidence is yes, or at least, consistent with yes. Given that birds evolved from reptiles, the only alternative to feathers arising from reptilian scales is an independent origin of feathers as epidermal appendages. The fact that many modern birds have scales on their legs in positions where other birds develop feathers is consistent with a close relationship between scales and feathers.[20]

The transition from reptilian scales to avian feathers is not as dramatic as it might first appear, especially when we consider that common initial steps are found in the embryonic development of scales (whether fish, reptilian or avian), feathers, hairs, teeth and indeed of glands (Figure 16.7). All arise through epithelial-mesenchymal interactions in which the initial steps promote epithelial and/or mesenchymal proliferation to facilitate folding of the epithelium and/or aggregation of mesenchyme into a placode (Figure 16.7). Indeed, epithelia from different tissues, even from different species or classes of vertebrates, can be exchanged and the initial steps in these interactions initiated, as demonstrated in interspecific recombinations between tissues from reptilian and avian embryos. Feathers or scales as alternate differentiation states, can be evoked from chick epidermis in experimental recombination between epidermis and feather- or scale-forming mesenchyme.[21]

Like scales, feathers are epidermal specializations based on keratins: α and β-keratins in scales; β-keratin in feathers, in which α-keratin is limited to the adjacent epidermis. Maderson (1972) proposed a relatively simple set of steps through which the scale could transform into a feather, essentially involving consolidation of β-keratin-forming cells to a centre as a placode which sank into the epidermis as a follicle to which muscles attached.

Major issues remain in connection with this innovation, including the question of how the secondary and tertiary stages of feather development and structure (shafts, barbs, barbules) were added to the developmental programme; the parallel with the problem of the addition of tetrapod digits to a pre-existing fin skeleton will be appreciated. Did feathers evolve for flight or for continuation of a function such as insulation already exercised by reptilian scales? Fossils remain mute on these points, despite the vociferousness of palaeontologists.

Feathered dinosaurs have not yet been found, although a 120 my old specimen housed in the Geology Museum in Beijing has a row of what appear to be appendages along the dorsal midline. Although no account of the specimen has been published in the refereed scientific literature, popular accounts attest to intense debate over whether these are appendages, and if they are, whether they are feathers, protofeathers, a skin frill as seen in some lizards, or something entirely different. Unfortunately, development does not speak to these questions as eloquently as it speaks to us about the origin of limbs, and thus far, palaeontology is also mute on whether flight arose in tree-dwelling reptiles which used feathers to aid in gliding or in ground dwellers such as dinosaurs without a gliding intermediate.[22]

16.4 TURTLE SHELLS

All tetrapods except turtles carry their ribs on the inside of the pectoral or pelvic girdles and limbs. Evolution of the turtle shell (carapace) involved a shift in this fundamental arrangement (Figure 16.8). [Although boid snakes such as pythons are limbless, they possess vestigial pelvic girdles and often a vestigial femur. Boids may also represent an exception to the rule of pelvic girdles outside the ribs (Raynaud, 1985).]

In turtles, each trunk vertebra corresponds to and is intimately associated with a dermal shield or osteoderm, a dermal ossification that arises in the skin independently of the endoskeleton (M. M. Smith and Hall, 1990). The carapace consists of vertebrae and ribs as endochondral elements from the axial skeleton, and fused osteoderms as dermal elements. Expansion and fusion of the ribs to one another and to the dermal ossifications in the skin produce the carapace outside the ribs. As structure (shell), by position (outside the ribs) and by process (fusion between elements), the carapace is novel.

When did this novel skeletal structure evolve and how does it develop? What devel-

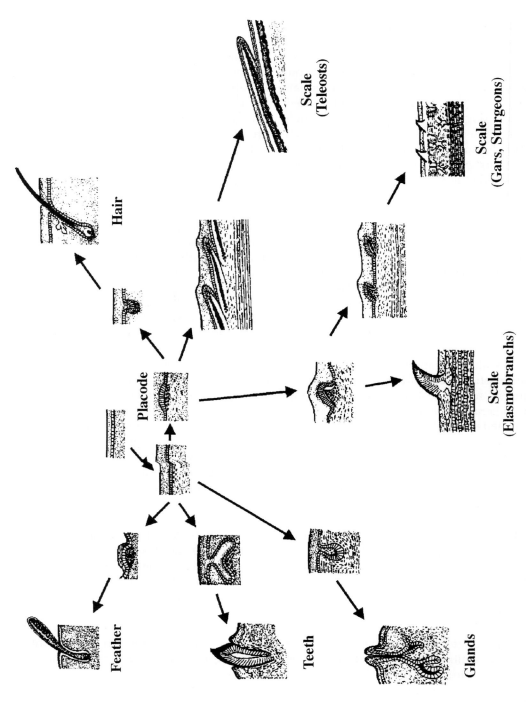

Fig. 16.7 Interactions between epithelia and mesenchyme give rise to feathers, teeth and glands, and, after placode formation, to hair and the various scales found in fish. Modified from M. H. Wake (1979).

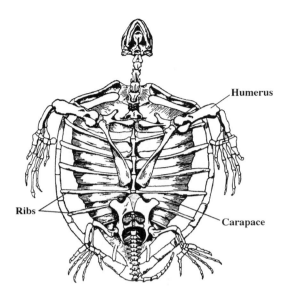

Fig. 16.8 A ventral view of the skeleton of the turtle *Chelydra* to show the relationships between the carapace, ribs, fore- and hindlimbs. Modified from Young (1958).

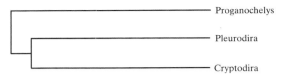

Fig. 16.9 Turtles consist of two monophyletic groups, the Pleurodira and the Cryptodira, with *Proganochelys* representing the nearest sister group. Based on the analysis of 39 characters in 14 families by Gaffney, Meylan and Wyss (1991).

opmental changes are required to place ribs outside the limbs where they can contact and fuse with dermal elements? Did other body parts change in concert or was evolution of the shell a key innovation allowing subsequent diversification?

Gaffney, Meylan and Wyss (1991), using 39 characters for 14 suprageneric categories of living and extinct turtles, confirmed that there are two major phylogenetic groups among living turtles:

1. Pleurodira, freshwater forms that retract their necks under the shell using a sideways movement; and
2. Cryptodira, terrestrial and aquatic forms that retract the neck with an S-shaped flexure (Figure 16.9).

A third more plesiomorphic clade includes Triassic and Early Jurassic taxa.

Relationships of higher categories of turtles may be well-resolved; turtle ancestry is much less well-resolved. Ancestors have been sought in numerous groups of tetrapods, but turtles are now generally thought to have arisen from a primitive amniote group, the procolophonids, of which the Upper Permian South African genus *Owenetta* is the oldest and most completely known. Procolophonids and testudines (turtles) form a monophyletic group on the basis of nine shared skull characters. It appears, however, that skull and shell evolved quite independently.[23]

The oldest known turtles are *Proganochelys* from the Upper Triassic of Germany, *Palaeochersis talampayensis* from the Upper Triassic of Argentina, and *Australochelys africanus* from the Lower Jurassic of South Africa. These specimens are known from skulls and, for *Proganochelys*, also from post-cranial skeletons. *Australochelys* and *Proganochelys* share many primitive features. *Australochelys* (which is younger) is more advanced, possessing such features as a fused basipterygoid articulation, a middle ear partially enclosed laterally, and a temporal roof extending posterior to the opisthotic. *Proganochelys* is presumed to have been partially aquatic; *Australochelys* is from an arid, terrestrial environment.[24]

Given the paucity of postcranial characters in more primitive forms that might represent intermediate conditions, how do we determine when and how the shell arose? I discuss two approaches, one palaeontological, one embryological.

16.4.1 Palaeontological evidence

Fossil evidence for the evolution of turtle shells is minimal. Lee (1993, 1996) used evidence from pareiasaurs (large, herbivorous anapsid reptiles) for clues to the progression that may have also led to the origin and elaboration of turtle shells. In a study emphasizing post-cranial modification, he argues that pareiasaurs are the nearest relatives or sister group of turtles. Some dwarf, heavily armoured forms paralleled turtles in having elaborate (but quite different) dermal armour (derived only from osteoderms) and short, fused presacral vertebrae. Lee hypothesizes that shells arose gradually by correlated progression of several organ systems. The following sequence describes the acquisition of features described in pareiasaurs (but remember that pareiasaur shells are not turtle shells, which form from ribs as well as from dermal elements):

1. The most primitive arrangement and the earliest evidence for dermal armour in the tetrapod fossil record is the presence of small, isolated osteoderms on the dorsal midline. The postulated function of these osteoderms is not protection but postural; they provide additional area for insertions of axial muscles.
2. Osteoderms increased in size and spread over the dorsal body surface. Although most stayed separate, some fused over the pectoral and pelvic girdles.
3. Osteoderms fused in later forms to form solid shells.

It is unclear whether fusion of osteoderms with vertebrae and ribs in turtles required a fully fused shell of osteoderms or occurred before osteoderm fusion was complete. Whichever the case, Lee argues for extrapolation of the sequence from pareiasaurs to turtles, in part because the dermal armour in two other groups of reptiles (archosauromorphs and placodonts) was elaborated in the same sequence. He also uses pareiasaurs to follow the reduction in vertebral number and expansion and widening of the ribs associated with production of an inflexible body. With inflexibility came ankylosis of vertebrae and ribs and then, in turtles, fusion of ribs and vertebrae with osteoderms; or so goes the scenario. Such gradual, small-step, correlated progression is just what is expected in organisms where organ systems interact inductively during development and growth, and functionally at maturity.

Lee interpreted the morphology of the anterior body of pareiasaurs as consistent with migration of the turtle shoulder girdle posteriorly into the rib cage, reversing the normal tetrapod arrangement. Burke demonstrated just such an event using developmental evidence.

16.4.2 Developmental evidence

Three stages in the development of the carapace in the European turtle *Chelonia midas* are shown in Figure 16.10. As determined in the snapping turtle *Chelydra serpentina*, the margin of the carapace appears as a ridge on either side of the body. This carapacial ridge consists of a mesenchymal core overlain by thickened epithelium as seen in limb buds (s. 11.4). A unique relationship between this carapacial ridge and the rib primordia sets the stage for development of the shell.

Rib primordia migrate dorsally following the expanding carapacial ridge until they terminate in the margin of the carapace. In all other vertebrates the ribs move ventrolaterally and so remain medial to the pectoral girdles (Figure 16.11).

Burke postulated a causal connection mediated by epithelial-mesenchymal interaction between the carapacial ridge and rib mesenchyme. In all other vertebrates, somitic cells that form the ribs migrate ventrally so that the ribs are positioned on the inside of the pectoral girdle. In turtles, alteration in the pathway of migration of somitic cell populations places rib-forming cells outside the pectoral girdle close to the dermis, where epithelial-mesenchymal interactions modify

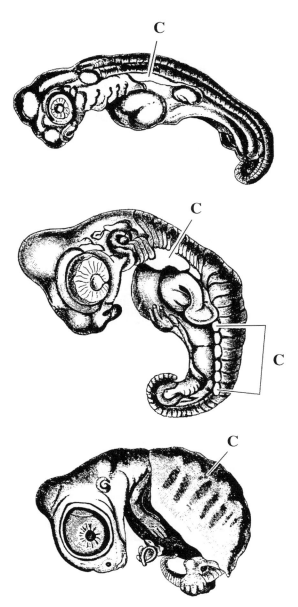

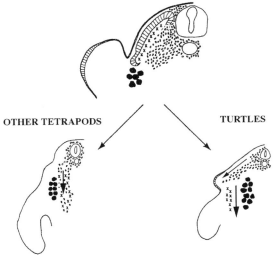

Fig. 16.11 In tetrapods other than turtles myogenic (x), scapular (black) and rib (small black) cells migrate ventrally in association with the body wall. In turtles, the rib cells (small arrow) migrate laterally while scapular cells migrate medially, leaving the ribs lateral to the pectoral girdle. Adapted from Burke (1989a).

Fig. 16.10 Three stages in the development of the carapace (C) in the European turtle, *Chelonia midas.* Modified from Balfour (1880–1, Vol. 2).

skeletogenesis to allow the ribs to fuse, forming the shell. The result is a novel *Bauplan.* Although the epithelial-mesenchymal interaction has yet to be demonstrated experimentally, the fact that important phases of skeletal development (including both sequences and patterns of chondrogenesis) are decoupled in the turtle *Chelydra serpentina*, demonstrates the requisite plasticity of skeletogenesis.[25]

An essentially similar scenario of inductive flexibility explains the repositioning of pouch rudiments which form internal cheek pouches in some mice and squirrels but external pouches in pocket gophers and kangaroo rats. The evolutionary origins of external cheek pouches and turtle shells are two examples of **heterotopy**, the topic of section 24.2.

16.5 PHARYNGEAL JAWS AND SPECIATION IN CICHLID FISHES

Predation, competition and ecological diversity in some African and South American lakes have produced extraordinary species diversity and niche exploitation in fishes of the family Cichlidae. Just three African lakes (Victoria,

Malawi and Tanganyika) contain >500, >500 and >150 endemic species, i.e. species found nowhere else. Even very small lakes contain endemic species. Barombi Mbo, with an area of $4.15\,km^2$ contains 11 species. Lake Nabugabo, which lies near Lake Victoria, measures only $2 \times 3\,km$, and yet contains five endemic species; Lake Bermin ($0.6\,km^2$) contains nine species, and so forth. Analysis of mitochondrial DNA of cichlids from African lakes provides strong evidence that the species within each lake are monophyletic. Analyses of cichlid species flocks from the Cameroon lakes Barombi Mbo and Bermin support monophyly and sympatric speciation facilitated by ecological diversification within each of the lakes. Indeed sympatric speciation was a driving force for cichlid radiations.

16.5.1 Speciation

Cichlid speciation has been (and continues to be) very rapid, most of it occurring over just a few hundred thousand years. Lake Victoria, with perhaps as many as 500 endemic species, is only 750 000 years old, and Lake Nabugabo with five endemic species is only 4000 years old, establishing a rapid rate of speciation. If Lake Victoria dried up completely some 12 400–14 000 years ago, as has been suggested, and if no refugia remained, then recolonization and speciation was very rapid indeed.[26]

Cichlids have especially malleable body shapes, skull forms, and tooth and gut structures. Morphological specialization is especially evident in the feeding apparatus, reflecting a tremendous diversity of diets and a degree of ecological specialization that may be unparalleled among vertebrates. The elaboration of the pharyngeal bones as masticatory pharyngeal 'jaws' to process food, freed the jaws proper (the premaxillae and mandibles) to become specialized for prey capture, allowing cichlids to adapt to the great diversity of African lake microhabitats (Figure 16.12). Specializations include feeding on plankton,

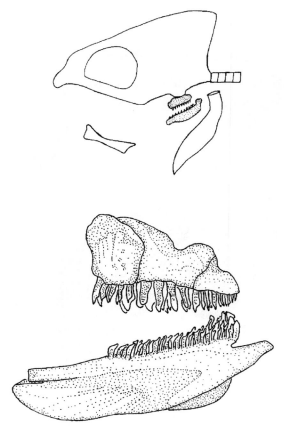

Fig. 16.12 Top: The skull and jaws of a cichlid fish to show the location of the pharyngeal jaws and the articulation of the upper pharyngeal jaw with the skull. Bottom: Detail of the pharyngeal jaws of *Astatotilapia elegans*. Upper figure modified from Liem and Greenwood (1981). Lower figure modified from Ismail, Verraes and Huysseune (1982).

detritus, insects, insect larvae, other fish and fish embryos, along with grazing on algae or crushing snails.[27]

Specialization is extreme. One species rasps scales from the caudal fins of other fishes. Another takes scales only from the left side of its prey. Another eats the eyes of other species. Yet another sucks eggs and embryos out of the mouths of one of its mouth-brooding relatives! Behavioural attributes such as mouth-brooding, short generation times, production of many young, and aggressive defence of territory, play prominent roles in cichlid success and speciation.[28]

16.5.2 An evolutionary avalanche

The ability of cichlids to adapt so extensively and rapidly in what has been called an 'evolutionary avalanche' is attributed by Liem (1974) to both prospective adaptive zones and a unique morphological key innovation involving pharyngeal jaws. This key innovation and environmental opportunity have gone hand in hand, and rapid isolation has been facilitated by behavioural traits.

The evolutionary novelty identified by Liem, which I expand on below, has three facets:

1. the development of a joint between the lower pharyngeal jaws;
2. a shift in insertion of the fourth levator externi muscles; and
3. the development of a joint between the upper pharyngeal jaws and the base of the cranium (Figure 16.12).

It might be argued that three such drastic changes constitute much more than a single key innovation, as indeed they do at the level of structural change. Liem, however, refers to it as a 'specialized, highly integrated innovation' and views the entire cichlid pharyngeal jaw apparatus as the key innovation. He documents plasticity in the formation of the basipharyngeal joint, incorporating knowledge gained from experimental studies in other vertebrates of the ontogeny of joints and their responsiveness to biomechanical factors. Periosteal and perichondrial cells at joint surfaces are remarkably adaptive, with the ability to modulate cellular differentiation between osteo- and chondrogenesis (s. 17.2.2). Because of the sliding of the two bony surfaces against one another, simple contact between the upper pharyngeal jaws and the cranial base would initiate the development of a cartilaginous pad between the two bones (point 3 above). The

development of such joints is a simple and rapid ontogenetic process.[29] Liem views these shifts as rapid and as minor genetically and developmentally. Thus:

'It is known . . . that the development of such a simple contact between two bones into an amphiarthrosis and the subsequent change from an amphiarthrosis to a diarthrosis is a relatively simple and rapid ontogenetic process controlled by minor genetic changes . . . The conversion of the preexisting elements into a new and significantly improved cichlid adaptive complex of high selective value may have evolved in rapid steps under influence of strong selection pressure acting on the minor reconstruction of the genotype which is involved in evolutionary changes of the pertinent ontogenetic mechanism.'

Liem, 1974, pp. 425, 434

Is it possible (or profitable) to attempt to isolate 'the' key innovation in such highly integrated changes involving so many tissues and organs in such a complex region of the organism? It may not be possible to reconstruct the historical events, but identification of 'the' key innovation would greatly facilitate experimental analysis of cause and effect sequences such as whether the differentiative potential of skeletogenic (points 1 and 3 above) or myogenic tissues (point 2) led to the morphological innovation just described.

Upper and lower pharyngeal jaws and the visceral skeleton are coupled structurally, mechanically and functionally in such fishes as centrarchids (sunfishes), but decoupled in cichlids. Liem (1974, 1980) argued that a shift of insertion of the fourth levator externus muscle facilitated by adaptive alterations in the lower pharyngeal jaws was the primary innovation. In a recent comparative analysis of cichlid and centrarchid fishes, Galis and Drucker (1996) argue that the shift in insertion is secondary to a decoupling of the epibranchials from the upper pharyngeal jaws. The independence of movement of upper and lower pharyngeal jaws allowed greater biting force, which was reinforced by a subsequent shift in muscle insertion. In both these scenarios the key innovative event (or sequence of events) had its origin in the developmental plasticity of the skeleto-muscular system and the ability to decouple previously coupled elements.

Galis and Drucker see four ways in which such novel structures could arise:

1. excess structural capacity;
2. decoupling of developmental pathways;
3. variability in developmental pathways; and/or
4. phenotypic plasticity.

Cichlids display all four.

Schaefer and Lauder (1996) tested for decoupling of the morphologically complex jaw articulation region of loricarioid catfishes. Their prediction that clades with decoupled systems would exhibit greater morphological variability than outgroups without decoupling, was confirmed in 10 of 12 comparisons. The two exceptions could be explained by lack of a tight association between morphological and functional evolution. Such rigorous phylogenetic tests hold great promise for revealing the nature of the integration of morphologically complex regions. Liem (1991a) undertook an analysis based on comparisons of the similar development between oviparous pomacentrid and viviparous embiotocid fishes, and he and George Lauder developed the phylogenetic methodology discussed below for such studies.

Meyer (1990) argues that these different morphologies represent intermediate steps in sympatric speciation. According to this view, formulated by West-Eberhard from studies on phenotypic plasticity (s. 18.2), developmental plasticity is an important element driving morphological change.

16.5.3 Phylogenetic methodology

Lauder and Liem (1989) proposed a phylogenetic methodology for testing the role of key innovations as causative agents in the initiation of morphological novelty. This methodology, the culmination of work by two of the more innovative researchers in functional morphology, builds on their studies on form and function in fishes, of which Liem's (1974) study on the origin of pharyngeal jaws in cichlid fishes is an exceptionally fine example. They provided a six-step methodology:

1. Define the key innovation or morphological novelty (the two terms are used interchangeably).
2. Propose the consequences of possession of the innovation, i.e. propose a hypothesis to test.
3. Map the innovation onto the group being studied and onto the appropriate outgroup.
4. Characterize the innovation in the groups under 3, i.e. quantify the form or function associated with the innovation.
5. Perform appropriate statistical analyses on the data generated.
6. Repeat the analysis on other groups.

This and similar approaches have been used to analyse other aspects of form and function. Carrier (1991) developed a conceptually similar approach for the identification of evolutionary constraints and applied it to the constraints that the dual functions of running and breathing impose on hypaxial muscle architecture in some lizards. Zweers used similar approaches in his analyses of multiple roles associated with several measures of functional performance in feeding mechanics in birds. In his analysis of relationships within parasitic platyhelminths, Rohde (1996) emphasized the importance of selecting homologous characters before undertaking any analysis of key innovations and adaptive radiations.[30]

16.6 THE ORIGIN OF INSECT WINGS

A major innovation in the successful speciation and radiation of insects in the Carboniferous was the evolution of wings. But did such innovative appendages arise from transformation of existing structures or *de novo* as novelties? Two major (and many minor) hypotheses for their origin have been proposed: wings either developed by modification of limbs or as new structures independent of the limbs. Both scenarios are considered briefly, although the jury is still out on which (or which other) scenario best explains the origin of wings.

16.6.1 Wings evolved as modified limbs

According to the modified limb hypothesis, wings evolved from a segment of the legs of wingless (apterygote) insects. Given that limbs are ventral and wings are dorsal appendages, origination from limbs involves extensive migration around the body. Unfortunately, little more can be said about the likely changes between limbs and wings. However, from fossil evidence we know that the first winged (pterygote) insects had multiple pairs of wings, perhaps involving all thoracic and abdominal segments. Modern insects, which have one or two pairs confined to the thoracic segments, could have evolved from this primitive condition by reduction in numbers of pairs of wings.

Given what is known of how homeobox genes control segment identity in insects, and given that the primitive condition was one of repeated pairs of wings along the body, modern insects may have come by their reduced number of wing pairs by suppression of homeobox genes, a possibility investigated by Sean Carroll and his colleagues in 1995. Their novel results involved a close investigation of *Antennapedia* in a winged insect (*Drosophila*) and the wingless thysanuran *Thermobia domestica*. Wings and halteres (the modified second pair of appendages in *Drosophila* that transform into wings in such mutants as *Bithorax*; Figure 9.4) develop on the second and third thoracic segments. Although development of the segments is dependent on *Antp*, development of wings and halteres is not. Other homeotic genes (*Sex combs reduced* [*Scr*], *Ultrabithorax* [*Ubx*] and *abdominal-A* [*abd-A*]) control wing initiation and patterning. Carroll and colleagues propose that wings arose without input from homeotic genes. Subsequent modification of *Scr* led to the elimination of wings on the prothorax, while evolution of *Ubx* or *abd-A* eliminated wings on insect abdominal segments and on the mesothorax of dipterans. The latter possibility is strengthened by the recent demonstration that ectopic expression of *engrailed* in *Drosophila* transforms the anterior, but not the posterior, compartment of halteres to wings through negative regulation of *Ubx* by *engrailed*.[31]

Homeotic genes such as *Ubx* do control segment identity (s. 9.4.1). Gibson and Hogness (1996) identified polymorphisms of *Ubx* in *Drosophila*, with phenotypic effects on thoracic segment three (transforming it into thoracic segment two). Their strategy was to use artificial selection for differential sensitivity to the induction of phenocopies of the *Bithorax* phenotype using ether vapour as the environmental trigger (as Waddington had done in the first experiments on genetic assimilation of Bithorax; s. 19.1). Increased sensitivity to ether correlated with loss of expression of *Ubx* in the imaginal disc for the third thoracic segment. There is also potential for regulation through modification of the 120 kb regulatory region of *Ubx*. Polymorphism in regulatory genes that encode transcription factors contributes genetic variation for morphological traits and increases the ability of the organism to respond to environmental signals by using previously unexpressed genetic potential to alter morphological patterns, a topic discussed further in Chapter 19.

16.6.2 Wings from 'gills'

There are several variants of the second proposal, which follows from the hypothesis that wings did not evolve from segments of the legs. In one, wings are novel appendages unrelated to any pre-existing structures, limb or otherwise. In another, proposed by William

Carpenter in the 19th century, wings evolved from gills. (Gill-like structures are found on trunk segments or as lobes of limbs of some arthropods.) Averof and Cohen (1997) investigated this proposal by searching for genes in a crustacean, the brine shrimp, *Artemia francoscana*, that serve functions specific to wing development in winged insects. They found two genes – *pdm*, which is thought to be involved in initiation of wing primordia in *Drosophila*, and *apterous*, which determines D-V patterning of *Drosophila* wings. Both are associated with the 'respiratory limb' in *Artemia* – *pdm* at specification of the limb buds, *apterous* with a dorsal expression pattern. Unfortunately for the 'wings from gills scenario' other genes involved in patterning appendages are not shared between *Artemia* and *Drosophila*.

The approach, based on shared patterns of expression, assumes conservation of function of these genes between wingless and winged organisms. This is no more parsimonious than the assumption that genes in a common ancestor were independently modified in crustaceans and insects for roles in appendage development and is an excellent example of the difficulty of using either patterns of gene expression or shared developmental pathways as the basis for determination of homologies (s. 21.7). Our current inability to resolve the origins of insect wings is precisely because we cannot recognize the homologues of wings in wingless ancestors; see Jockusch and Nagy (1997) and Chapter 21 for elaboration of this theme.

16.7 TORSION IN GASTROPODS

A key innovation need not be structural; it could be genetic, cellular, developmental, physiological, behavioural or ecological.

As an example of a genetic change, a mutation has been proposed as the key innovation in the origin of directionality of torsion in gastropods. Normally, single gene mutations of large effect are lethal. It is therefore of interest to examine the arguments marshalled for a

mutation in a single gene resulting in the production of something as fundamental as directionality of early development and body organs. (See section 12.3 for further discussion of the genetic bases of symmetry and asymmetry.)

The shells of snails, indeed the whole of the viscera, undergo torsion or spiralling during development. Torsion is but one of the many rearrangements of organs found in molluscs, characterized by Ponder and Lindberg (1997) as visceral rotations, folds and flops. Torsion may be right-handed (dextral) or left-handed (sinistral), is species-specific, and can be traced back to the plane of the first cleavage division of the egg (Crampton, 1894).[32]

Gastropods, in common with annelids, turbellarians, nematodes and all molluscs other than cephalopods, display a spiral pattern of cleavage. If the mitotic spindle in the first cleavage division angles to the right, a right-handed spiral is imposed on embryonic and adult structures. If the mitotic spindle angles to the left, a left-handed spiral is imposed. The mitotic spindle rotates from right to left (or *vice versa*) with each successive cell division. Strictly, this is not spiral but alternating cleavage, for spiral cleavage would have a constant direction of rotation of the spindles. The cleavage pattern is more analogous to standing on the North Pole and alternately turning to the right and the left, than to spinning like a top on top of the earth.

Directionality of torsion is controlled by a single pair of genes through cytoplasmic factors inherited in the egg – an instance of the maternal cytoplasmic control described in section 7.4.1. This was revealed when cytoplasm from the eggs of snails with right-handed coiling was injected into eggs of left-handed coiling snails, transforming them into right-handed coiled individuals. Direction of torsion is under simple genetic control with the potential to produce an innovation of large effect. The key appears to be control of a fundamental and early developmental process by a single pair of genes, and the consequences of that process for subsequent body organization. One can't, however, as Stanley did, leap from a simple genetic basis for the directionality of torsion to an equally simple genetic basis for the origin of torsion.[33]

16.8 CONCLUSION

Stepping back and looking at the larger picture of innovations and macroevolution, the conclusion from this overview of key innovations and transitions between taxa is that although features that qualify as novel adaptations can be identified, it is not easy (and may be impossible) to equate such novel features with key innovations. Therefore, we need to avoid the trap of equating possession of a novel adaptive feature (feathers in birds, shells in turtles, torsion in snails) with that feature having triggered the origin of a taxon. In practice, however, it is often difficult to discuss one without the other, especially when the innovation is as dramatic as the origin of flight and development of feathers.

Although evolution may appear to proceed from threshold to threshold, each threshold event is but one in a chain extending back to the immediate ancestors of the group, from them to their ancestors, and so on. Even a structure that is synapomorphic for a taxon need not have been the key innovation that initiated that taxon. There is therefore no valid reason for restricting key innovations to speciation, although there is every reason for associating key innovations with adaptive shifts, which might then have facilitated speciation. Evolution and duplication of the homeobox in homeotic genes and the generation of basic body plans is a prime example.

Given the interactive nature of development, the potential for ontogenetic repatterning, and the integrated nature of changes in interrelated systems, a key innovation would have to be rapidly integrated into the existing functional and structural complexity of the organism for the change to persist. It may be much more profitable to focus on the developmental and functional bases of integrated change (as seen in pharyngeal jaws) and ontogenetic repatterning than to seek individual key innovations. Integrated change and ontogenetic repatterning are the topics of the next chapter.

16.9 ENDNOTES

1. For discussions of these aspects of key innovations, see Stebbins (1973), Jaanusson (1981), Lauder (1981, 1982), Liem and Wake (1985), Stiassny and Jensen (1987), D. B. Wake and Larson (1987), Lauder and Liem (1989), Meyer (1990), Zweers (1991) and the papers in Nitecki (1990).
2. Mallatt (1996), who has amassed all the relevant evidence on whether feeding or respiration drove the evolution of the jaws, favours respiration.
3. See Hall (1975), Langille and Hall (1988a,b, 1989) and Mallatt (1996) for summaries of the origin of jaws; and Alberch and Kollar (1988), Northcutt (1990) and Thomson (1993) for evaluations of head segmentation and the origin of the jaws. Forey and Janvier (1993, 1994) and Mallatt (1996) discuss the velum and/or homologies between jawless and various jawed vertebrates.
4. For the range of structural and ecological opportunities created by the evolution of jaws, see Gans and Northcutt (1983), Northcutt and Gans (1983) and the chapters in Hanken and Hall (1993a).
5. For further elaboration of the expression of *engrailed* in amphioxus and whether amphioxus has a homologue of the vertebrate midbrain, see L. Z. Holland *et al.* (1997). See Forey and Janvier (1993) for homology of the velum to jaws, Hatta *et al.* (1990) for *engrailed* expression in the zebrafish muscles and N. D. Holland (1996) for a recent evaluation of *engrailed* expression.
6. For additional analyses of developmental and palaeontological evidence for the fin → limb transition, see Hinchliffe *et al.* (1991), Coates and Clack (1990, 1991), Coates (1994) and Vorobyeva and Hinchliffe (1996).
7. See Coates (1991, 1994) and Hinchliffe (1991, 1994) for the possibility of independent evolution of the pentadactyl limb. For determination of homology of the digits of birds and theropod

dinosaurs (both of which retain only three digits on their wings and forelimbs) in relation to theories of whether birds arose from dinosaurs, see Burke and Fediccia (1997) and Hinchliffe (1997).

8. For studies proposing the fin-fold theory, see Thacher (1877), Mivart (1879) and Balfour (1881). See Hall (1991b) for an evaluation of the fin-fold theory, and R. L. Carroll (1987) and Coates (1994) for pectoral and pelvic fin evolution.

9. See Géraudie (1978) for the structure of the teleost AER, M. M. Smith and Hall (1993) and M. M. Smith *et al.* (1994) for the neural crest origin of fin mesenchyme and the dermal skeleton; and Hall (1997b) for mechanisms of bone formation.

10. For discussions of the cellular transitions required to go from fin → limb, see Hall (1991b), Thorogood (1991), Coates (1994), Hinchliffe (1994), Shubin (1991, 1995), Sordino *et al.* (1995) and Sordino and Duboule (1996).

11. For 'polydactyly' of the first tetrapod limbs, see Coates and Clack (1990, 1991), Coates (1991, 1994), Hinchliffe (1991, 1994) and Thomson (1991b).

12. For the literature on *Hox* gene expression in fin buds, see Akimenko *et al.* (1991, 1994, 1995), Akimenko and Ekker (1995), Molven *et al.* (1990), Riddle *et al.* (1993), Sordino *et al.* (1995), Misof and Wagner (1996) and Misof *et al.* (1996). For induction of *Sonic hedgehog* in pectoral fin buds treated with RA, see Akimenko and Ekker (1995). For RA-induced apoptosis in regenerating fins and limbs, see Géraudie and Ferretti (1997).

13. See Vargesson *et al.* (1997) for fate-mapping of the chick wing bud, Nelson *et al.* (1996) for *Hox* gene expression in the chick limb bud, and Davis and Capecchi (1996), Davis *et al.* (1995) and Rijli and Chambon (1997) for targeted mutations in mice.

14. For *Hox* knock-out experiments and skeletal growth, see Dollé *et al.* (1993), Yokouchi *et al.* (1995), Newman (1996) and Goff and Tabin (1997). Overlapping functions are not seen when other genes are knocked out; TGFβ-1 and TGFβ-3-null mice show no phenotypic overlap (Sanford *et al.*, 1997).

15. Additional examples and analysis of functional cooperation between non-paralogous gene products may be found in Wilkins (1997), who regards this role as blurring the distinction between orthologous and paralogous genes. Examples are known from yeast, *Drosophila, C. elegans* and mice. An important element in regulation of long-bone growth is regulation of the rate of chondrocyte hypertrophy by *Indian hedgehog* and parathyroid hormone-related protein (Vortkamp *et al.*, 1996; Wallis, 1996).

16. For digital arch development in urodeles, see Alberch (1985), Alberch and Gale (1985), Shubin and Alberch (1986) and Shubin (1995). Detailed fate mapping of the chick limb bud by Vargesson *et al.* (1997) is relevant to interpretations of these models, for, at least in the chick, much of the wing arises from the posterior half of the wing bud.

17. See King *et al.* (1996) and Vortkamp (1997) for further discussion of growth differentiation factors and their interaction with BMPs. *Pax* genes are also emerging as major players in determining the condensation phase of organogenesis; see Dahl *et al.* (1997).

18. See Ostrom (1974, 1976, 1994) and Carroll (1997) for detailed reviews of the origin of birds from theropod dinosaurs, and Shipman (1997) for a general, up-to-date overview of recently discovered fossil bird finds. The theory of the origin of birds from dinosaurs, supported by most workers, is refuted by Feduccia (1996), as is the argument that the ancestors of birds were terrestrial bipeds rather than arboreal gliders. Ernst Mayr (1997b) in reviewing Feduccia's book, concurs on both points. Recent reanalysis of the homology of the digits of birds and theropod dinosaurs in which the development of embryonic turtle, alligatoir and avian forelimbs is used to argue that birds retain digits II-III-IV, while theropods retained digits I-II-III, will add new life to debates over the origins of birds, and over the use of developmental evidence of homology to override other classes of evidence (Burke and Feduccia, 1997; Hinchliffe, 1997, and see s. 21.7).

19. See Novas and Puerta (1997) for *Unenlagia comahuensis*, the theropod dinosaur 'missing link'.

20. For discussions of the evolutionary origin of feathers, see Maderson (1972), Regal (1975) and Brush (1993, 1996).

21. See Dhouailly and Maderson (1984) for the development of reptilian scales, and Hall (1978), Dhouailly and Sengel (1983) and Lemus (1995) for recombinations between reptilian and avian epidermal tissues.

22. For popular accounts of the Chinese 'feathered dinosaur', see Morell (1997a,b).

23. See Rieppel and de Braga (1996) and De Braga and Rieppel (1997) for evidence that turtles are diapsid reptiles, and Lee (1997) and references therein for an analysis of recent studies on the relationship of turtles to other reptiles.

24. For information on fossil turtles, see Reisz and Laurin (1991), Fraser (1991), Gaffney and Kitching (1994) and Rougier *et al.* (1995).

25. See Burke (1989a,b, 1991) for development of the shell; and Rieppel (1993b) for dissociation of chondro- from osteogenesis of the turtle skeleton, a decoupling also seen in other vertebrates (s. 24.1.5d).

26. For analyses of cichlid species diversity and rates of speciation, see Trewavas *et al.* (1972), Sage *et al.* (1984), Meyer *et al.* (1990), R. B. Owen *et al.* (1990), Sturmbauer and Meyer (1992) and Schliewen *et al.* (1994). For other species with rapid rates of speciation, see Stanley (1981).

27. For how possession of pharyngeal jaws facilitates the adaptability of cichlids, see Liem (1974), Liem and Greenwood (1981), Stiassny and Jensen (1987), Meyer (1990), Witte *et al.* (1990), Liem (1991a,b), Galis (1993b) and Galis and Drucker (1996). See Ismail *et al.* (1982) and Clemen *et al.* (1997) for the development of pharyngeal jaws.

28. See Meyer (1987, 1990) and Kingdon (1990) for morphological specializations in cichlid fishes.

29. In Chapter 17, I take up plasticity of skeletal tissues in facilitating morphological change. See Hall (1978, 1985a) for syntheses of past studies, and Fang and Hall (1995, 1996, 1997) for recent studies seeking a molecular basis for differentiative plasticity of periosteal and perichondrial cells.

30. For analyses of constraints and form-function analysis of feeding mechanisms in birds, see Zweers (1991, 1992), Zweers and Gerritsen (1997) and Zweers and Vanden Berge (1997).

31. For evolution of insect wings from limbs, see Carroll (1995) and Carroll *et al.* (1995). See Emerald and Roy (1997) for transformation of anterior haltere compartments to wings.

32. Ano-pedal flexure, which allows the visceral mass to elongate, may be a synapomorphy uniting gastropods, cephalopods and scaphopods. See Ponder and Lindberg (1997) for a comprehensive phylogeny of gastropods including developmental characters, and see Page (1997) for a detailed analysis of the onset and progression of torsion in the archaegastropod *Haliotis kamtschatkana*.

33. Sturtevant (1923) and Boycott *et al.* (1930) demonstrated single gene control of directionality of torsion, while Freeman and Lundelius (1982) did the cytoplasm injection studies. Stanley (1979, 1982) proposed that torsion itself, and therefore the gastropods, had a similarly simple genetic basis.

17

INTEGRATED CHANGE IN VERTEBRATE EVOLUTION

'Genes do not directly cause *anything* of immediate phenotypic significance.'

K. C. Smith, 1992b, p. 338

In this chapter I examine integrated change as a force for evolutionary change in morphology. The three contexts in which this discussion takes place are intensification of function (s. 17.2), acquisition of a new function (s. 17.3), and loss and/or reappearance of structures, as exemplified in Dollo's law and atavisms (ss. 17.4.1, 17.5).

The prevailing, adaptationist view is that new features appear because of a **change in function**, and that the feature that changes was pre-adapted for the new function. This presupposes that the structure has the potential to take on the new function under appropriate conditions of selection, and possessed that capability even as the original function was being performed. Bock (1959) formalized this concept as **pre-adaptation**; Mayr (1960) developed it, and Gould and Vrba (1982) investigated it under the term **exaptation**.

Gradual acquisition of new structures or functions could come about through correlated changes in developing systems, pleiotropy, epigenetic interactions and/or intensification or alteration in function. Most evolutionary changes in morphology are probably of these types. Change must, of course be associated with fitness, and both rate and type of change depend on the genetic variance in fitness and/or on traits that are correlated with fitness. Indeed, evolutionary developmental biology integrates develop-

ment, fitness and adaptation (Freebairn, Yen and McKenzie, 1996). In this chapter, I treat integrated change and correlated progression at the level of developmental processes. The well-known topics of genetics, natural selection, correlated response to selection, and multivariate selection theory are not considered here.

Components of functional complexes may be integrated by additive or synergistic interactions, as Sharon Emerson and her colleagues demonstrated so elegantly for integration of the morphological characters accounting for 'flying' in frogs (Emerson and Koehl, 1990; Emerson, Travis and Koehl, 1990). The functional stimulus for such changes could involve intensification of function or acquisition of a new function. Because functionally interdependent structures are tightly coupled epigenetically in development (for the evidence, see Chapters 11 and 12), innovations could arise as secondary by-products of epigenetic change during development, or as side-effects of phylogenetic change in size or developmental timing, as discussed in Chapter 24.

17.1 ONTOGENETIC REPATTERNING AND CORRELATED PROGRESSION

Bolitoglossine salamanders have one of the most, if not the most, specialized feeding systems of any amphibian. The tongue, used as a projectile, is 'fired' rapidly (<50 msec) and with enormous accuracy at the prey, over distances as great as 80% of the length of the body. Depth, distance perception and a highly specialized musculoskeletal feeding apparatus are required for such specialized and effective feeding, which is so accurate that these animals virtually never miss their prey. At the extreme of this specialization is the genus *Hydromantes* from California, Italy and France in which, not only the tongue, but portion of the visceral skeleton is projected outside the mouth using a ballistic mechanism. Changes in feeding in bolitoglossine salamanders from the ancestral condition are associated with the acquisition of direct development, although loss of the larval stage is a necessary but not a sufficient condition.[1]

David Wake and Gerhard Roth have documented evidence from the functional morphology of feeding behaviour in plethodontid and bolitoglossine salamanders for what they term **'ontogenetic repatterning'**, by which they mean 'the establishment of new sets of morphogenetic processes resulting from heterochronic events' (1989, p. 367).[2] Ontogenetic repatterning is not so broad a concept as to embrace any change in development or any addition, deletion or compression of an existing or ancestral ontogeny. Ontogenetic repatterning establishes a new developmental trajectory leading to change that is structurally and functionally integrated. In their 1985 analysis Roth and Wake identified the following repatterning events in modification of bolitoglossine feeding mechanisms:

1. suppression of branchial arch development so that only a single epibranchial element develops fully (although early stages have more than one epibranchial element);
2. extensive elaboration of the tongue musculature required to project the tongue so far and fast;

3. rotation of the eyes to the front of the head from the more lateral position typical of most other amphibians; and
4. development of ipsilateral retinotectal projections for stereoscopic vision, comparable to those seen in mammals.

New networks of interactions are established as existing networks are uncoupled, creating a new set of conditions for the innovation and morphological novelty. Ontogenetic repatterning provides a powerful developmental process upon which selection can act in the introduction of new adaptive innovations.

Fran Irish proposed a shift in timing of development and resulting paedomorphosis associated with the capture of large prey items as instrumental in the evolution of the snake skull from the more generalized skull of their lizard-like ancestors. The changes included reduction of the chondrocranium, elaboration and enhanced ossification of the dermocranium (including complete closure of the wall of the braincase), a mobile joint between the mandibles, and reduced connections between bones of the snout, palate and jaws. Deleting these connections facilitated the increased mobility of the snake skull and jaws so essential for capturing and swallowing large prey items.[3]

A functional/morphological approach involves investigation of the integration of changes in many systems (visual, neural, skeletal, muscular, behavioural) and, consequently, extensive knowledge of such diverse fields as morphology, development, physiology, neurobiology, biomechanics, ecology, systematics and evolution. These tasks are daunting but necessary if we are to uncover mechanisms for the initiation of morphological changes. Studies on pharyngeal jaws (s. 16.5) are a model. Such studies move us away from identification of single key innovations and toward an emphasis on integrated changes and ontogenetic repatterning in interrelated systems.

17.2 INTENSIFICATION OF FUNCTION

Examples of evolutionary change resulting from intensification of an existing function include long-term improvement in the mechanical efficiency of the skeleton leading to the evolution of the artiodactyls; highly modified digging claws and teeth in taeniodont mammals; evolution of the hypocone on mammalian teeth (s. 17.2.1); and such neomorphic features as the divided maxilla of bolyerine snakes and secondary jaw articulations in birds (s. 17.2.2).[4] [The Taeniodontia are an order of mammals found exclusively in western North America as fossils from the Lower Paleocene to the Upper Eocene. They were large and rat-like, with limb and dental features similar to those found in the ground sloths, reflecting convergent lifestyles.]

17.2.1 Hypocones of mammalian molar teeth

Primitive mammalian upper molar teeth are triangular with a cusp at each of the apices of the triangle. The cusps are known as the paracone, metacone and protocone. On at least 20 independent occasions in mammalian lineages since the Caenozoic a fourth cusp has evolved. Situated at the disto-lingual corner of the upper molar teeth, the hypocone doubles the surface area available for crushing food. Hypocones evolved frequently and are only rarely lost, attesting to the advantages associated with converting a three-cusped to a four-cusped molar.

Jernvall (1995) and Hunter and Jernvall (1995) consider that the hypocone fits Mayr's definition of a key innovation precisely: a new structure allowing a new function that opens up a new adaptive zone to the organism. They document various ways in which the hypocone may be produced and argue for the presence of intermediate conditions in extant and fossil mammals – either an enlarged shelf or a small cone not in occlusal contact with the

lower molars. They further document that hypocones evolved in a great diversity of mammals on all the major continents; that Caenozoic mammals with hypocones underwent much greater radiations than did forms without hypocones; and that hypocones are preferentially found in herbivorous mammals.

The hypocone allows invasion and diversification within herbivorous adaptive zones. As a high proportion of generalist feeders possess hypocones, Jernvall and Hunter suggest that hypocones evolve under generalist dietary regimes and are then pre-adapted for development of the more specialized herbivorous diet requiring additional tooth area to crush and grind food.

17.2.2 Jaw joints in snakes and birds

Bolyerine snakes of the species *Bolyeria multicarinata* and *Ceasarea dussumieri* are a subfamily of boid snakes found only on Round Island, near Mauritius in the Indian Ocean. As with all living boid snakes, they possess vestigial hindlimb skeletons and/or traces of pelvic girdles. Their maxillae are divided into anterior and posterior articulating halves; this condition, unique among vertebrates, probably results from failure of fusion of the two centres of ossification from which the maxillae form.[5]

The development of secondary jaw articulations in birds is a particularly nice example of intensification of function driving an evolutionary novelty. With the strong musculature required for the intensified impacts associated with feeding in many birds came the need to increase the surface area of the jaw and skull skeleton for muscle insertion. This was met by the development of a bony projection from the mandible which grew medially toward the skull. If the bony process comes in contact with the skull, the differentiative properties of the osteogenic cells of the bony process (the cells can differentiate into either cartilage or bone) facilitate the development of an articulating surface between the bony process and the skull. Contact between these two previously separate parts of the craniofacial skeleton is sufficient for a new articulation to develop, giving these birds two jaw articulations – the original, primary quadrate-squamosal articulation and the new, secondary mandible-skull articulation. As in adaptations associated with pharyngeal jaws in cichlid fishes, the adaptive potential of periosteal cells responds to functional demand. Walter Bock estimated that two forms of secondary articulation – between the mandible and the basitemporal and between the mandible and the ectethmoid – evolved independently in 15 and three avian lineages respectively. The evolution of these secondary articulations was achieved readily and rapidly through simple shifts in ontogeny, a mechanism which Bock termed **multiple evolutionary pathways**.[6]

If intensification of function initiates their development, then the presence or degree of development of secondary jaw joints should be correlated with intensity of the impact of feeding or prey capture on the mandible, just as the primary jaw articulation reflects intensity of functional demands (Zweers and Vanden Berge, 1997). This is indeed the case. In the plover *Charadrius*, where impact is low, the joint is a syndesmosis of fibres and ligaments without articular cartilage or cavity. The marine skimmer, *Rhynchops nigra,* on the other hand, flies just above the waves, dipping its knifelike lower beak into the water to catch fish. The head can be thrown back with sufficient force to break the horny tip on the bill. It will not be a surprise to learn that *Rhynchops* has the most robust and highly developed secondary jaw articulation consisting of a true diarthrodial joint with articular cavity, articular membrane and dense cartilaginous articular surfaces.

Such changes are integrated. Intensification of function, developmental and functional association of skeletal and muscular systems, differential growth, and the cellular capabilities of osteogenic cells for joint formation, combine to facilitate formation of a new structure whose function is just sufficiently different from that of the original jaw joint that both persist.

17.3 ACQUISITION OF A NEW FUNCTION

Examples of evolutionary change associated with alteration of function or acquisition of a new function include:

1. transformation of the reptilian to the mammalian jaw articulation, and the development and integration of ossicles into the mammalian middle ear;
2. transformation of the first dorsal fin ray into a lure in angler fish;
3. the origin of the thyroid gland from the endostyle;
4. development of the turtle shell from ribs and vertebrae;
5. derivation of teeth from scales in various groups of fishes; and
6. origin of the snake skull from a lizard-like skull.[7]

I discuss the transition from the reptilian to the mammalian jaw joint and the origin of middle ear ossicles in more detail below. For turtle shells and snake skulls see sections 16.4 and 17.1.

17.3.1 The reptile to mammal transition

The transformation of the reptilian to the mammalian jaw articulation, and the concomitant freeing of bones for inclusion as ossicles in the mammalian middle ear, is a resounding example of triumphant morphological analysis. Following Kemp, who used the term specifically for the changes associated with the transition from reptiles to mammals, Thomson termed such transformations and integrated changes in interrelated systems in the ear as 'correlated progression'.[8]

Mammals have only the dentary as a single bone in the mandible. Reptiles have at least

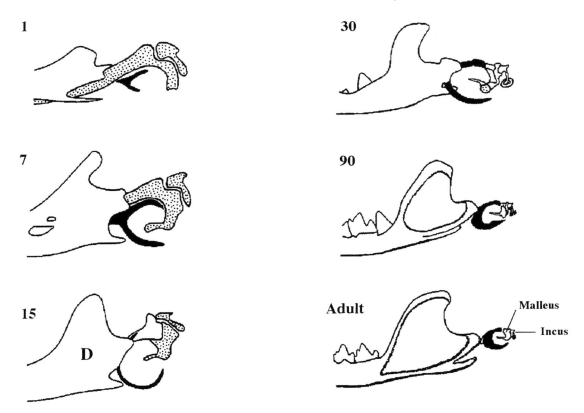

Fig. 17.1 Six stages in development of the posterior end of the dentary (D) in the marsupial *Monodelphis domestica*, from 1–90 days after birth to the adult condition, as seen in lateral views. Meckel's cartilage (stippled) from which the malleus arises, progressively moves from the mandibular arch into the middle ear. The ectotympanic bone (the reptilian angular) is shown in black. Adapted from Rowe (1996b).

seven bones – dentary, articular, prearticular, surangular, angular, splenial and coronoid. Articulation of reptilian jaws is effected through the articular of the lower jaw and the quadrate of the upper jaw. In mammals, articulation is between the dentary of the lower jaw and the squamosal of the skull. Marsupials represent a fascinating intermediate condition. Neonates of the South American possum *Monodelphis domestica* have neither the mammalian dentary-squamosal nor the reptilian quadrate-articular jaw articulations. Acquisition of the mammalian condition can be followed in marsupial embryonic development (Figure 17.1) and also through an excellent fossil record of the reptile → mammal-like reptile -→ mammal transitions.[9]

During evolution of the mammal-like reptiles, the dentary expanded at the expense of the other bones of the lower jaw. With expansion, the dentary came into contact with the wall of the skull, and as happened in birds and cichlid fishes, periosteal cellular reactivity at the contact site initiated the formation of a joint. With the development of this new squamoso-mandibular joint, functional control was removed from the articular-quadrate joint, freeing these two bones to take on new functions (or to be lost if no new function was available for selection; see below).

The articular and quadrate gradually moved posteriorly away from the original jaw articulation. This condition is seen in fossil synapsid reptiles. Ear ossicles, which were present in synapsid reptiles, were co-opted by the developing mammalian middle ear to form the malleus and incus. Had the mammalian hearing system not been available for selection at the same time as the transformation in jaw articulation was taking place, the articular and quadrate would in all likelihood have become vestiges or disappeared in mammals as have the other reptilian jaw bones. They would not have been retained as key players in this adaptive scenario.[10]

Even though the sequence is known in exquisite detail, it is still not clear whether the onset of the mammalian condition was driven by a functional demand for high frequency hearing, by developmental changes in the jaws and jaw musculature, or by differential growth between brain and jaw. This important example of heterotopy is discussed further in section 24.2.1.

17.4 LOSS AND REAPPEARANCE OF STRUCTURES: DOLLO'S LAW

There is a large literature on loss of structures during evolution. Indeed, degeneration, a major evolutionary theme in the years after Darwin wrote *The Origin*, was championed by Ray Lankester and others. Romer (1949) brought together the evidence for the 'reversibility' of evolution (loss of limbs in snakes, loss of the dermal skeleton in tetrapods, and so forth) and for reversal in evolutionary trends such as phyletic size increase/decrease. Henningsmoen (1964) coined the term **zigzag evolution** for repeated patterns of increase or decrease during evolution. Fong and colleagues, in their 1995 review of vestigialization and loss of nonfunctional characters, discuss various well-documented examples of structural degeneration, but also consider vestigial behaviours and genes. Their examples include reduction or loss of digits in

tetrapods; flightlessness and wing reduction in birds and insects; lung loss and larval paedomorphosis in plethodontid salamanders; and loss of sensory structures in cave-dwelling animals and such subterranean dwellers as the mole rat *Spalax ehrenbergi*.[11]

17.4.1 Dollo's law

Dollo's law, named after the Belgian paleontologist Louis Dollo by one of his students, is concerned with reversibility of evolution. Dollo (1893) maintained that organs, other complex structures and indeed whole organisms, cannot return to a condition identical to an ancestral condition. As is so often the case in science, this law was enunciated independently in the same year by the anatomist and zoologist, Hans Gadow: 'What in the course of ages has phylogenetically disappeared cannot again recur' (cited by Berg, 1969, p. 227). Dollo only intended his law to apply to loss of complex structures and entire organs. Nowadays, the Dollo-Gadow law is generally applied to any aspect of reversibility in evolution.[12]

Dollo's law appealed particularly to paleontologists, whose discipline relies so completely on morphology and interpretation of apparent evolutionary trends in morphology. The reappearance of structures lost in ancestors (such as the second lower molar in felids), atavistic structures (such as limbs in snakes, hindlimb skeletons in whales and additional digits in horses; s. 17.5), or phylogenetic character reversals (s. 21.5), affirm that loss of a structure need not necessarily mean loss of the ability to form the structure. Dollo's law thus has substantial embryological implications: traces of developmental stages can be retained in descendants, especially as ancestral structures in early ontogeny.[13]

Dollo's law was tested by Marshall, Raff and Raff (1994) in the light of current knowledge of rates of degradation of genetic information. The premise upon which the test is based is that evolution could not be reversible if genes or developmental programmes rapidly become non-functional after they have been released from the selective pressure associated with maintenance of the organ or structure. If information is lost rapidly (as is the conventional wisdom), a structure could not reappear exactly as it had been in an ancestor. They determined, however, that there was a significant chance that silenced genes or apparently lost developmental programmes could be reactivated if the time scale of loss was of the order of 0.5–6 mya, but not if loss were more than 10 mya. If selection maintains the gene(s) or programme for another function, then even long-silenced genes or programmes have the potential for reactivation. A silenced gene is defined as a gene that is no longer transcribed; i.e. the gene might still be present but inactive, as indeed it must be for an ancestral pattern to be reestablished. It will therefore be important to gather similar data for regulatory gene products, as emphasized by Zuckerkandl (1994) in his approach to parallel evolution via modification of common molecular pathways.

17.5 LOSS AND REAPPEARANCE OF STRUCTURES: ATAVISMS

Atavisms are ancestral characters that occasionally reappear in individuals of descendant species. This is reflected in synonyms for the word, which include 'throwbacks', 'reversions', and 'reproduction of the ancestral type'. Atavisms are traditionally interpreted phylogenetically as the reappearance of an ancestral state. When viewed ontogenetically they provide evidence of the retention of ancestral developmental programmes (see below and s. 21.5).

Situations in which a 'lost' structure 'reappears' tell us a great deal, not only about how the structure was lost, but also which portions of the developmental programme were lost, rather than silenced or modified. Indeed, such atavistic reappearances tell us that loss of the structure cannot have meant loss of the

developmental machinery to produce the structure.

The genetic basis of atavistic characters was established in a large series of breeding experiments on guinea pigs with extra toes, undertaken by Stockard, Castle, Sewall Wright and Scott in the first three decades of this century. These researchers showed how structures reappear either

1. spontaneously in response to a chance mutation that removes a constraint from a developmental programme (ss. 17.5.1, 17.5.2);
2. through artificial breeding; or
3. from experimental manipulation of embryonic tissues, usually involving transplantation of inducers between species (s. 17.5.5).

Surprisingly, structures are often integrated with adjacent tissues and organs as functional components of the organism. Perhaps the classic examples from nature are three-toed horses, hindlimb skeletons in whales, atavistic muscles in passerine birds and prolegs in insects (ss. 17.5.1–17.5.4). Induction of balancers in amphibians is presented in section 17.5.5 as an example of atavisms produced following transplantation of inductively active tissues between species. Other examples, not discussed here, include atavistic muscles in mammals, winged earwigs, molar teeth in Pleistocene *Lynx*, extra nipples, tails and excess hair in humans, dew claws in dogs, an ascending process of the astragalus in bird embryos (present in *Archaeopteryx* and in putative ancestors of birds, the theropod dinosaurs), a vestigial egg tooth in some marsupials, and patterns of bristles in *Drosophila* that are diagnostic for flies in a related family. Atavistic tarsal patterns in the salamander *Taricha* were discussed in section 12.2.1.

Atavisms in basal taxa can point the way to structures that will appear in more derived taxa, as well as telling us what has gone before. Examples of such atavisms at the population level in fishes and urodeles are discussed in sections 12.2.1 and 21.5.[14]

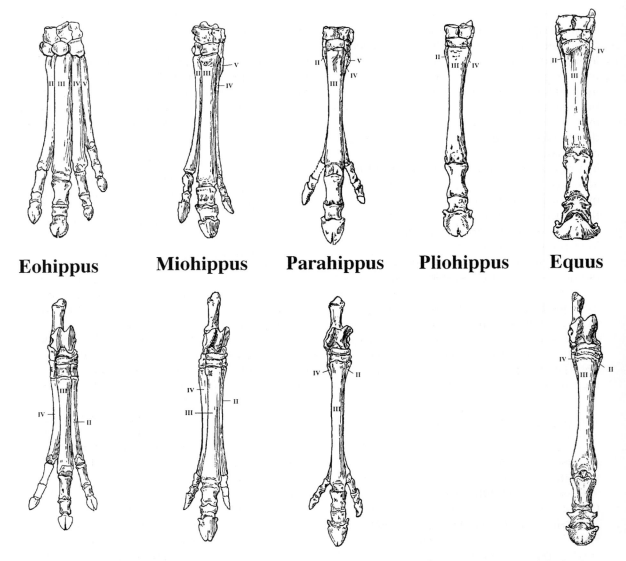

Fig. 17.2 Reduction in the toes from four to one in fore- and hindlimbs (top and bottom rows) of the lineage leading from the Eocene *Eohippus* to the modern horse *Equus* from Pleistocene to now. Digits II and IV are reduced to splint bones. Modified from Gregory (1951).

17.5.1 Three-toed horses

The evolution of modern horses of the genus *Equus* involved numerous changes, including reduction in the number of toes from four in the forelimb and three in the hind of *Eohippus* (Early Eocene) to a single elongated toe in fore- and hindlimbs of *Pliohippus* in the Pliocene and *Equus* (Pleiostocene to now; Figure 17.2). Elongation of the single toe accompanied reduction in the length of humerus and femur and even greater reductions in ulna and fibula, including loss of their distal articulating surfaces. Such extreme reduction in lateral digits was not unique to the horse lineage. A group of ungulates from South America (Order Litopterna) exhibits similar changes.[15]

Embryonic horses have primordia for three toes of equal size and development. During ontogeny the middle toe increases its rate of growth and development, leaving the two lateral toes behind until they are no more than vestiges (splint bones) lateral to the cannon bones (fused metacarpals or metatarsals). But if a mutation allows the lateral toes to continue to grow they can develop as fully formed digits no smaller than the middle toe. Julius Caesar and Alexander the Great prized their war horses with extra toes. The mustangs of the western United States are said to possess extra toes more often than do other breeds. The ability to modulate the growth rate of the lateral toes leads to subsequent and higher order interactions between skeleton, muscles, ligaments, tendons, and the nervous and vascular systems. Extra toes are in contact with the ground and integrated functionally into the musculo-skeletal system.

17.5.2 Hindlimb skeletons in whales

The reappearance of vestigial skeletal elements of the hindlimbs in whales is a second example, although one about which we have less developmental information.

Whales were recognized as mammals by the British naturalist John Ray in 1693 and divided into the two major groups of toothed and baleen whales by Linnaeus in 1758. Following phylogenetic analysis using 12S and 16S mitochondrial ribosomal DNA sequences undertaken with Orti and Meyer in 1993, Milinkovitch (1995) used molecular phylogeny to augment and revise morphological evidence for whale evolution. Cetaceans (whales, porpoises, dolphins) are now regarded as members of a monophyletic group that includes pigs, cows, camels and hippos, although when cetaceans diverged from artiodactyls remains controversial.[16]

Modern whales in the Order Cetacea are all marine but evolved from terrestrial ancestors in the Paleocene or Eocene epochs. Limbed and terrestrial ancestors of modern-day aquatic whales could have ventured into water before flippers evolved, just as aquatic amphibians ventured onto land before tetrapod limbs had evolved fully. All living whales lack hindlimbs and have forelimbs modified as flippers. Before forelimbs were modified as flippers and hindlimbs were lost, aquatic locomotion in whales was probably powered by undulation of the vertebral column rather than by the limbs. Early whales, such as the Eocene *Ambulocetus*, may have used a combination of tail undulation and paddling using pelvic limbs. Evolutionary loss of the skeletal elements of the hindlimb was progressive, beginning with the most distal elements (the digits) and proceeding through loss of more and more proximal elements.

Early forms such as the Middle Eocene archaeocete *Basilosaurus isis*, from Egypt, had relatively complete hindlimb skeletons although the tarsals were fused and the digits reduced to four. Other Eocene forms, such as *Rodhocetus kasrani* from Pakistan, are intermediate between land mammals and fully marine whales. Fossils of *Rodhocetus* lack distal limb elements, but this may be due to lack of preservation. *Rodhocetus* had a reduced femur and modifications of the vertebrae and tail associated with swimming. Other organ systems (for example, the bones of the ear associated with hearing) developed at their own rate early during the transition. Consequently, as in tetrapods, turtles and so many other taxa, the features of modern whales appeared sequentially and not as an integrated package.[17]

The sperm whale (*Physeter catadon*) is the only toothed whale to have a vestigial hindlimb skeleton, although expression is variable, ranging from a single pelvic bone to presence of a nubbin of a femur, rudimentary fibia, fibula, tarsals and metatarsals. Although living baleen whales lack complete hindlimb skeletons as adults they do retain rodlike vestigial pelvic bones or cartilages (which are independent of the axial skeleton), femora and occasionally tibiae that fail to project beyond the body wall. Atavistic hindlimb elements can be surprisingly complete, for example, 12.7 cm long left and right hindlimb elements in a female humpback whale (*Megaptera novaeangliae*); 15 cm diameter elements in a female sperm whale.

17.5.3 Thigh muscles in passerine birds

Passerine birds (Order Passeriformes) are recognized by a variety of characters including absence of two leg muscles, *Musculus iliofemoralis externus* and *M. caudiliofemoralis pars iliofemoralis*. *M. iliofemoralis externus* is readily recognizable from its origin on the iliac crest and insertion on the femur. It is found as a normal element in ducks, falcons and galliform birds (which are phylogenetically distant from passerines) but is absent from kingfishers, hornbills, trogons, toucans, woodpeckers and cuckoos, which are phylogenetically closer to the passerines.

M. iliofemoralis externus has been reported as an atavistic muscle in individuals of the common Hawaiian myna (*Acridotheses tristis*), Rothschild's myna (*Leucopsar rothschildi*), the Australian bowerbird (*Chlamydera nuchalis*), birds-of-paradise (*Epimachus*, *Loria*) and the New Zealand thrush (*Turnagra capensis*). Three patterns of muscle origin and splitting explain the development and/or atavistic reappearance of *M. iliofemoralis externus*; see Hall (1984a) for details. *M. caudiliofemoralis pars iliofemoralis* occurs atavistically in the U.S. fox sparrow (*Passerella iliaca*), the white-breasted woodswallow (*Artamus leucorhynchus*) and in *Aechmorhynchus cancellatus*, a rare shorebird of the Tuamotu Archipelago in the Pacific Ocean.[18]

17.5.4 Insect prolegs

Prolegs of insect larvae may also be atavistic.

Prolegs are abdominal legs found on all lepidopteran larvae and in other groups scattered throughout the insect orders. Adult insects do not have abdominal legs although insect ancestors are thought to have had prolegs. Do prolegs in larvae of extant species represent the retention and expression of an ancestral limb programme that has been lost in some groups or re-expression in some groups of an ancestral limb programme? As our understanding of the genetic basis of proleg formation increases we may be able to address this question. The homeobox gene *abdominal-A* (*abd-A*) suppresses limb formation in *Drosophila* and other insects. Consequently, *abd-A* is expressed in those abdominal segments that lack limbs and not in segments bearing limbs. In lepidopetran larvae, however, *abd-A* is repressed in patches of cells in the four anterior abdominal segments on which prolegs develop. The modified expression pattern and correlation with proleg formation implies a causal connection which, if established, would provide a way to approach prolegs as atavisms.

17.5.5 Induction of balancers in axolotls and frogs

Balancers are paired lateral appendages that lie just behind the mouths of larvae of pond-dwelling salamanders of such families as the Ambystomidae, Salamandridae and Hynobiidae. Covered in mucous, balancers serve as balancing or stabilizing organs until the limbs form at metamorphosis, then the balancers degenerate. Balancers form in the mandibular arch attached to the palatoquadrate cartilage following inductive interactions between the roof of the archenteron, neural plate and overlying ectoderm during neural induction.[19]

Balancers are not found in the axolotl, anurans and some salamanders (e.g. *Ambystoma tigrinum*) but in the 1930s it was discovered that frogs and the axolotl can form these appendages if their embryonic tissues are challenged with ectoderm from embryos

Table 17.1 A summary of experiments demonstrating the ability of axolotl and frog embryos to form balancers if their cranial ectoderm is replaced by salamander ectoderm[1]

Source of ectoderm	Host species	Result	
		Balancers	No balancers
Salamander	Salamander	+	
Axolotl	Axolotl		+
Axolotl	Salamander		+
Salamander	Axolotl	+	
Salamander	Frog	+	
Frog	Salamander		+

[1] See Hall (1984a) for details and literature.

of salamanders such as *Triton taeniatus* that do form balancers. The results of such experiments are summarized in Table 17.1. Salamanders form balancers and axolotls do not when ectodermal transplants are from the same species, confirming the efficacy of the experimental approach.

In reciprocal transplants between axolotls and salamanders, balancers form in axolotls with salamander ectoderm but not in salamanders with axolotl ectoderm, demonstrating:

- that axolotl ectoderm is not competent to respond to salamander inducers;
- that axolotls retain inducers for balancers;
- that axolotls lack balancers because of loss of competence, not loss of inducers; and
- that axolotl inducers can induce balancers from competent salamander ectoderm.

Exactly similar conclusions may be drawn to explain the lack of balancers in frogs; anuran ectoderm is not competent to respond but anurans contain balancer inducers (Table 17.1). [It is an open question whether balancers are homologous with ectodermal adhesive organs (cement glands, suckers) which develop from the hyoid arches of anuran larvae (s. 22.2.1). Salamanders produce adhesive organs, not balancers, if anuran ectoderm is transplanted onto the head (Hall, 1984a). This argues for common inductive interactions underlying both organs, but common inductive

mechanisms do not necessarily prove structural homology (s. 21.7.2 and Hall, 1994c, 1995a).] Indeed, the same conclusions could be drawn regarding experimental induction of teeth in amphibians. Urodele larvae form true teeth (composed of dentine and enamel and located inside the mouth), while anuran larvae produce keratinized teeth on a disc outside the mouth. Salamander larvae produce keratinized teeth if anuran ectoderm is transplanted over the developing mouth.[20]

An interesting natural experiment exists in hybrids between *Ambystoma tigrinum* and *A. texanum*. Hybrids are intermediate in many characters such as size, but closer to *A. texanum* in others. *A. tigrinum* lacks balancers. *A. texanum* has balancers and so do the hybrids (Flick, 1992), indicating retention of balancer-forming potential in both species. The nature of that potential has not been determined experimentally.

17.6 CONCLUSION

The morphological novelties discussed in the last three chapters arise through modification of developing systems. Developmental mechanisms include:

1. temporal and hierarchical activities of sets of genes, as in establishment of the body plan in *Drosophila*;

2. causal and constrained sequences of inductive interactions, as in the origin of the chordates and gnathostomes;

3. developmental plasticity and functional integration between tissues and organs, as in the origin of pharyngeal jaws in cichlid fishes; and

4. ontogenetic repatterning, as seen in feeding mechanisms in salamanders.

Such integrated changes are associated with acquisition of a new function or with intensification of a pre-existing function. In both instances, epigenetic control couples environmental variability to developmental processes. This theme is pursued in the next two chapters.

17.7 ENDNOTES

1. See D. B. Wake and Roth (1989) for feeding specializations in bolitoglossine salamanders, and Deban *et al.* (1997) for ballistic tongue projection in *Hydromantes*. Duellman and Trueb (1986) for a detailed discussion of specialization of the musculature associated with feeding in salamanders; and Reilly and Lauder (1990) for an analysis of the developmental changes associated with tongue projection in the tiger salamander, *Ambystoma tigrinum*.

2. For fuller development of the concept of ontogenetic repatterning, see Larson *et al.* (1981), Roth and Wake (1985), D. B. Wake and Larson (1987) and Wake and Roth (1989). It is still not clear whether heterochrony is indeed the mechanistic basis for this repatterning. Heterochrony must be coupled with the production of developmental variability that selection, change in fitness, genetic drift, etc., can act upon to effect the integrated change (s. 24.1).

3. See Irish (1989) and Rieppel (1993b, 1994) for studies on the development of reptilian skulls, and Caldwell and Lee (1997) for analysis of one of the earliest snakes, *Pachyrhachis problematicus*, a primitive marine Cretaceous reptile from Israel with well-developed hindlimbs and pelvic girdle. Previously regarded as a lizard, *Pachyrhachis* has been reinterpreted as a primitive snake with a skull containing many of the derived features of modern snakes.

4. These examples based on increased function are derived from Schaeffer (1948), Patterson (1949), Bock (1959, 1960), Frazzetta (1970) and Bock and Morioka (1971).

5. See Anthony and Guibe (1952) for the development of the maxilla in *Bolyeria* and *Casarea*.

6. See Bock (1959, 1960), Bock and Morioka (1971) and Hall (1978) for secondary articulations. Padian (1995a) has thoroughly reviewed the form versus function dialectic raised by such studies.

7. For introductions to the literature on middle ear ossicles, see Olson (1959), Crompton and Jenkins (1979), Lombard (1991) and Maier (1990). See Regan and Trewavas (1932) for lures in angler fish, Leloup (1955) for thyroid gland from endostyle, Burke (1989a,b, 1991) for turtle shells, Irish (1989) and Rieppel (1991) for snake skulls, and M. M. Smith and Hall (1990, 1993) for teeth from scales.

8. See Thomson (1966) for correlated progression, and Kemp (1982) and Carroll (1997) for mammal-like reptiles and the transition to mammals.

9. Primitive reptiles have two splenial and one or two coronoids. Early mammals retain the coronoid. Filan (1991), Clark and Smith (1993), K. K. Smith (1994) and Rowe (1996a,b) have studied the development of the marsupial middle ear ossicles most intensively.

10. See Rowe (1996a,b) for the coevolution of the mammalian middle ear and neocortex, and for the recapitulation of the evolutionary history of middle ear ossicle evolution in the ontogeny of living marsupials. (K. K. Smith and van Nievelt [1997] discuss some difficulties associated with assessment of the timing of marsupial maturation used by Rowe in his analyses.)

11. See Fong *et al.* (1995) for a recent discussion of vestiges and the loss of nonfunctional characters, and Tague (1997) for developmental mechanisms that may account for variability in a vestigial pollex (first metacarpal), variability that is similar to that in functional but miniature pollices in primates and evaluation. (Tague is analysing his data to determine whether vestigial metapodials show a higher level of fluctuating asymmetry than nonvestigial metapodials; pers. comm. Robert Tague.) See Collazo and Marks (1994) for the evolution of larval paedomorphosis in the

cave-dwelling plethodontid salamander *Gyrinophilus porphyriticus*. Larval paedomorphosis (perennibranchiation) is a permanent larval condition in which larval features are retained into sexual maturity; see Whiteman (1994) for the multiple mechanisms underlying the evolution of facultative paedomorphosis in salamanders. Larval paedomorphosis is highly correlated with cave-dwelling.

12. Dollo laid out the basis for his law in papers of 1893 and 1922. Gould (1970) reviewed Dollo's law and in 1980 edited a reissue of Dollo's papers.

13. For the application of Dollo's law in palaeontology, see Gregory (1936), Shishkin (1968), Gould (1970) and Rainger (1985, 1989). For the studies on atavisms noted in the text, see Kurten (1963), Hall (1984a, 1995b), Stiassny (1992) and sections 17.5 and 21.5.

14. For overviews of atavisms and discussion of the examples not treated in the text, see Hall (1984a, 1995b), Tintant and Devillers (1995) and Verhulst (1996). For discussions of the genetic basis of atavisms, especially the breeding experiments with guinea pigs with extra toes, and an introduction to the literature, see Castle (1911), Stockard (1930), Hall (1984a) and Provine (1986).

15. For the literature on three-toed horses, see Marsh (1892), Prentiss (1903), Hall (1984a, 1995b) and Thomson (1988, 1991a).

16. For recent studies and evaluation of the phylogenetic relationships of whales, see Shimamura *et al.* (1997) and Milinkovitch and Thewissen (1997).

17. For studies on whale fossils, see Gingerich *et al.* (1990), Milinkovitch *et al.* (1993), Thewissen and Hussain (1993), Gingerich *et al.* (1994), Thewissen *et al.* (1994) and Milinkovitch (1995). For variability of the hindlimb skeleton in sperm whales, see Yablokov (1974) and Hall (1984a). For mechanisms of locomotion in early whales see Thewissen and Fish (1997).

18. For atavistic muscles in birds, see the studies of Zusi and Jehl (1970), Raikow (1975), Raikow *et al.* (1979) and McKitrick (1986). For difficulties associated with identification, homology and phylogeny based on such muscles, see Raikow *et al.* (1990) and McKitrick (1991, 1994). Although relationships between many avian taxa are not well resolved, see Sibley and Ahlquist (1990) for a comprehensive phylogeny and classification of birds. For the model for the origin of *M. iliofemoralis externus* and for its reappearance as an atavism, see Raikow (1975) and Hall (1984a).

19. There are few modern studies on the structure or induction of balancers. They degenerate by a process of autotomy and there is evidence for communication between left and right balancers; time of loss during ontogeny is affected by the absence of regeneration of the contralateral balancer (Kollros, 1940).

20. See Spemann (1938), Hall (1984a), Hamburger (1988) and M. M. Smith and Hall (1993) for discussions of the experiments on balancer and tooth induction in amphibians.

PART FIVE

EMBRYOS, ENVIRONMENT AND EVOLUTION

18

EVOLUTION AS THE CONTROL OF DEVELOPMENT BY ECOLOGY

'A plausible argument could be made that evolution is the control of development by ecology. Oddly, neither area has figured importantly in evolutionary theory since Darwin, who contributed much to each.'

Van Valen, 1973, p. 488

Chapter 8 concentrated on epigenetic control within organismal ontogeny. In Part Three I discussed how that control builds embryos and their organ systems, while in Part Four we saw how the same processes form the developmental bases for major transitions during evolution. In this chapter, I continue this theme by considering epigenetic interactions between species and between species and their environment. Many such interactions are known as **cyclomorphosis** or **seasonal polymorphism**, both of which are examples of **phenotypic plasticity**. Chapter 19 then moves us to genetic assimilation, the other major mechanism by which environmental signals trigger changes in development. Then, in Chapter 20, I present a quantitative genetics model that integrates epigenetic and environmental factors into analyses of morphological change in development and evolution. The generality of the model is supported by its elaboration for mammalian mandibles and eyespots on butterfly wings.

As adaptiveness is central to any meaningful and long-term interactions between ecology and development, I begin with a consideration of adaptation.

18.1 ADAPTATION AND ENVIRONMENTAL CHANGE

Adaptation, which may be defined as 'a structure or function which is appropriate for some particular set of conditions,' or 'the process by which such a structure or function comes into existence' (Waddington, 1957a, p. 145), has been central to theories of evolution since Darwin. Any deviation from the basic plan of a taxonomic category, when associated with special and characteristic modes of life, is an adaptation. A deviation could be a structure, a function, or the process of acquiring the structure or function. Selection operates on phenotypes in the context of adaptation, which maintains or 'improves' conformity between organisms and their environment.[1]

Nineteenth-century embryologists were much concerned with adaptations for embryonic and larval existence (s. 5.5). By and large, 20th-century developmental biologists have not dealt with adaptations and adaptability. An exception, Conrad Waddington, considered three kinds of adaptation as process:

1. **exogenous adaptation**, in which the organism is modified to be better fitted to its

special environmental circumstances. According to Baur (1922) and to Waddington (1957a), truly exogenous adaptation is selection for the capacity to respond developmentally to the environment in an appropriate way;

2. **pseudoexogenous adaptation**, in which an organism exhibits characteristics similar to those of exogenous adaptation, but the characters are hereditary and independent of any particular environmental stimulus. Waddington believed that the genome imposes limits on the degree to which the phenotype could respond to environmental change; and

3. **endogenous adaptation** or functional connections, as of the cornea to vision.

Most current authors, when referring to adaptation as process, use the term in the sense of exogenous adaptation. Population geneticists emphasize the adaptive nature of the relationship of populations to their environment.

In 1949 the American palaeontologist Edwin Colbert endeavoured to show that adaptation could be especially visualized in series of fossils in which progressive adaptation over long periods of time could be related to environmental change. Unfortunately, few fossil lines show progressive adaptation, and when it is absent, the problem of how to infer environmental change arises. Striking exceptions are the studies of bovid speciation and extinctions in sub-Saharan Africa undertaken by Vrba, and similar studies of hominids by Brain. Both these workers correlate speciation and extinction events that parallel adaptive shifts, with global temperature variations that influence rainfall and vegetation. Low temperatures at the end of the Miocene epoch five mya, and again 2.6 mya, were associated with greater rates of speciation of hominids and antelopes than at intervening times when temperatures were rising. Antelope distribution ranges also fluctuated in concert with the climatic and vegetational changes.[2]

To study adaptation from the standpoint of functional morphology (as outlined in Chapter 17), we need to find an approach that will include an environmental component and supplement the mathematical models of the population geneticist. One measure, morphological disparity or morphological distance, has been used in two studies of quite disparate organisms and structures: molar tooth diversity in Caenozoic ungulates (discussed below) and morphological change in post-Palaeozoic crinoids (Foote, 1996). These analyses concentrate on morphological change independent of taxonomic status to reveal ecological components driving morphological change or morphological change that appears to transcend ecological changes.

Jernvall and colleagues (1996) developed an innovative approach to adaptation in their study of molar tooth diversity in Caenozoic ungulates in relation to ecological and climatic changes. Their use of morphological distance follows the methodology introduced by Raup (s. 3.3). Tooth crown types rather than taxonomic units form the basis of their analysis of adaptive radiations during the 50 million years of the Caenozoic. The response of different taxa which share particular adaptive radiations (such as adoption of a herbivorous diet) to such global changes as climatic cooling could be analysed using similar approaches.[3]

Although it has long been thought that slow temperature change facilitates evolution, the notion that abrupt alteration in climate may facilitate periods of rapid evolution is novel. Even deep-sea organisms, such as benthic ostracods, exhibit fluctuations in diversity over time scales of 10 000 to 100 000 years that correlate with glaciation, diversity decreasing as glaciers advance and recovering during interglacial periods (Cronin and Raymo, 1997). Climate also influences evolutionary change over shorter periods. An example is change in cranio-dental characters in the pocket gopher, *Thomomys talpoides*, that track climatic change over 3200 years of the Upper Holocene (Hadley, 1997).

Coope (1979) discovered similar correlations for Caenozoic beetles. Although the morphology of beetle species was remarkably stable during the enormous climatic oscillations of the Quaternary period, their geographical ranges changed enormously: the beetles tracked the tolerable environment across continents, in the process breaking down geographical barriers between populations to allow genetic mixing and prevent speciation.

In cases like this, there are obvious difficulties in identifying adaptive responses. The adaptation may be reflected in morphology, speciation, or alteration of geographical range. As the latter is a physiological and behavioural adaptation, only large-scale study of fossil faunas over considerable geographical areas will reveal it, or allow it to be inferred. The reigning view is that adaptation to the environment occurs by the natural selection of mutations whose direction of change is entirely random with respect to the environment, but which, by chance, give rise to individuals with suitable (i.e. adapted) characteristics. Two further means of adaptation to the environment are discussed at some length: phenotypic plasticity in this chapter, and genetic assimilation in Chapter 20.

18.2 PHENOTYPIC PLASTICITY AND TYPES OF INDUCTIONS

In phenotypic plasticity, which is one class of evolutionary plasticity, 'when the environment is heterogeneous, a single genotype can develop different phenotypes in different environments' (Stearns, 1992, p. 60). Phenotypic plasticity modulates the genetic response of organisms to selection. In terms of heritability, it is the proportion of total phenotypic variance accounted for by genotype by environment interactions.[4]

Epigenetic control directs the building of organisms through inductive interactions (embryonic induction) in which signal and response occur within the same organism. But there are other classes of inductions, including situations, such as cyclomorphosis, in which the products of one organism's genes activate specific developmental programmes in another organism of the same or different species, as in interactions between a predator and embryos of the prey species. There are further situations where the inductive signal originates from a dietary constituent from a food source, or from abiotic sources in the environment such as light and temperature. In other situations, induction by another species or from the environment (either independently or through cooperation) can trigger the same response, either because of independent induction under different circumstances or through cooperative induction.

Therefore, the following classes of induction can be identified:

1. Induction between cells within an individual embryo, larva or adult. This is **embryonic** (interorganismal) induction.
2. Induction between individuals of the same species, as in pheromone-based signalling in insects and density-dependent polymorphism in salamanders. This is **intraorganismal** (intraspecific) induction.
3. Induction between individuals of different species, often between predator and embryos of a prey species. This is **interspecific** induction, with **predator-based** induction as a subclass.
4. Induction initiated by abiotic, environmental signals such as light and temperature. This is **environmental** induction.
5. Inductions involving combinations of an environmental signal and any of the other four classes of induction.[5]

Although these types of induction appear to be quite distinct, all share the ability to activate a developmental pathway in embryos of a responding species that leads to the production of distinctive embryonic, larval or adult morphologies. Fundamentally, they differ only in the source of the inductive signal.

Examples are elaborated in the sections that follow.

18.3 PREDATOR-BASED INDUCTION

The term cyclomorphosis was coined by Lauterborn in 1904 for seasonal polymorphism in plankton. The sense of the term made explicit by Brooks in 1946 is of cyclic or seasonal change in genetically homogeneous organisms. In the present context, cyclomorphosis and seasonal polymorphism are synonyms, whether the morphological changes are in planktonic or non-planktonic species.

In many instances, cyclomorphosis arises in response to inductive stimuli from outside the responding organism, usually produced by a predator. A chemical released by a predator (or obtained from a food source in environmental induction; s. 18.6) evokes a morphological change in the offspring of the prey species (or species consuming the food), switching development into new morphogenetic and differentiative pathways. The genetic component is the ability of a prey species to respond by selective gene expression to activate a new developmental programme. Consequently, alternate or multiple phenotypes can be derived from one genotype in response to different environmental, ecological or inductive signals. This is the ecological equivalent of the phenotype of the gene introduced in section 7.4. Schmalhausen formalized such responses as **norms of reaction**, a concept now virtually synonymous with phenotypic plasticity (see sections 18.2 and 19.5 for Schmalhausen's contributions).[6]

The stimulus that evokes the specific developmental pathway is environmental as far as the responding species is concerned, but involves the production and release of a gene product as far as the predator is concerned. Such interspecific, predator-based responses are not odd rarities. They occur in plants, algae, protozoans, ciliates, rotifers, cladocerans, bryozoans, gorgonians, gastropods, barnacles, insects, frogs, amphibians, reptiles and fish. Interspecific interactions are epigenetic because of the specific response of the prey species; the interaction is epigenetic at the level of developmental response. Other examples, in which a developmental programme is elicited by different signals, include phenocopies and the relationship between genetic assimilation and homeotic mutations (Chapter 19).[7]

In my view seasonal polymorphism provides some of the most compelling evidence for the relevance to organismal ecology of inductive control of development. Perhaps even more importantly, it links evolution, ecology, environment and community-level interactions to individual development. Cyclomorphosis enables organisms to avoid predation, take advantage of seasonal food supplies, compensate for overcrowding, and tolerate and/or avoid extreme environmental variation. Cyclomorphosis is a dramatic illustration of Leigh Van Valen's aphorism, reproduced at the head of this chapter, that evolution is the control of development by ecology. Three examples of predator-based inductions are described below.

18.3.1 Rotifers

My all-time favourite is Gilbert's study of the rotifers *Brachionus calyciflorus* and *Asplanchna brightwelli*, which made the cover of *Science* on March 11 1966.

A. brightwelli is carnivorous and preys upon the much smaller *B. calyciflorus*. In the absence of the predator, *B. calyciflorus* has three pairs of short spines; when predators are present, an additional pair of long, posterolateral spines develops. *A. brightwelli* releases a protein into the water that activates this new developmental pathway in embryos of the prey species, resulting in the differentiation and morphogenesis of new spines; the predator induces these spines in the prey species.

For me, such chemically mediated induction is the essence of both seasonal polymorphism and embryonic induction and is an affirmation of the causal relationships between ecology, development and evolution. Gilbert concluded his 1966 paper, however, by contrasting such interactions with embryonic induction:

'The *Asplanchna*-factor differs from classical embryological inducers in several important respects. First, it is a substance produced by one species which affects the developmental pattern of another species. Second, it exists in effective concentrations and in a free state in the organisms' external environment. Typical inducer substances both form and operate within a single organism and are closely associated with cells or cell layers. Finally, the *Asplanchna*-factor acts prior to cleavage, probably during oogenesis, whereas other inducers appear and exert their influence during or after gastrulation.'

Gilbert, 1966, p. 1236

But given the benefit of an additional 30 years of research into embryonic induction, Gilbert's three points can now be used as arguments for induction. Interspecific induction is but an extension of induction within an individual, just as the sperm-egg interactions comfortably reside within the category of inductive interactions (s. 8.3.1). Embryonic inducers can be diffusible (s. 10.2.2), and inductions do occur prior to gastrulation; mesoderm induction is an obvious example (s. 10.1). Rather than being set apart from embryonic induction, the *Asplanchna-Brachionus* interaction illustrates the range of biological organization over which inductive interactions occur and the essential unity of process operating within and between individuals, even when those individuals come from different species.

The ecological importance of such interactions is clear when we realize that *B. calyciflorus* individuals with the additional set of spines cannot be eaten by the predator. Gilbert subsequently (1980) showed that different quantities of prey elicit different *Asplanchna* morphs and that efficiency of feeding varies from morph to morph. A second rotifer species, *Keratella testudo*, responds to soluble factors released by some ten competitive zooplankton species, including species from other phyla. Additional spines are produced, body size is larger, and survivorship higher, all clear indications of the ecological advantages of cyclomorphosis.

Even more impressive as an indication of the developmental potential and responsiveness within such species is the finding that single clones of *K. testudo* produce spines of different types and sizes in response to signals from different predators. A suite of morphological responses is available, providing a remarkably adaptive plasticity for species that encounter many predators at different seasons. Similarly, copepods that undergo extensive daily vertical migration within the zooplankton and for which vulnerability to predation varies with migration and life-cycle stage, show different ontogenetic patterns adapted to different combinations of predators encountered.[8]

From such studies we see that, not only do such interspecific inductions integrate the life histories of predator and prey, they integrate communities of organisms. An important but little-studied area of evolutionary biology is the origin, development, maintenance and ecological significance of such relationships between predators and prey. Sharon Emerson, Harry Greene and Eric Charnov made a beginning in 1994 when they advocated the application of scaling models used in ecology and functional morphology.

18.3.2 *Daphnia*

The production of 'neckteeth' (elongated spines) in progeny of a cladoceran, *Daphnia pulex*, that is preyed upon by larvae of the phantom midge, *Chaoborus*, parallels the rotifer example, including the release of a diffusible molecule by the predator. Abiotic environmental agents can also evoke alterations in helmet formation; 10 hours exposure to 5μg/l of the carbamate pesticide carbaryl alters helmet form at the second instar.

Seasonal changes in head and tail spines and helmet morphology in various species of *Daphnia* do not involve any change in genotypic frequency – they are genuine polymorphisms. The altered *Daphnia* have ecological consequences themselves, for they induce changes in their own food supply, a green alga, which grows too large to be consumed by the *Daphnia*. These essentially developmental interactions are important elements of the trophic ecology of the species and community.[9]

18.3.3 Carp

One vertebrate example is the development under natural or experimental conditions of a deep-bodied morph of the crucian carp, *Carassius carassius*, in the presence of the predatory pike *Esox lucius*. Divergence in body shape is seen as early as 12 weeks after introduction of the predator. The faster rate of growth and increased drag that the deeper body entails are offset by the final size of the deep-bodied morphs, which are too large to be eaten by the pike. As with the invertebrate examples, polymorphism provides a refuge from predation.

Adaptation to predation is not unique to species exhibiting polymorphism. Various aspects of life history (body size, allotment of resources to reproduction, fecundity, size of offspring and so forth) respond to predator pressure because predators are size-specific in their selection of prey. Thus, guppies (*Poicilia reticulata*) show significant changes in life history parameters in as little as 30 generations of response to altered predation. Polymorphism is a specialized form of such life-history adaptation.[10]

18.4 SYMBIOSIS AND INDUCTION

An elegant demonstration of the inductive nature of associations between species is seen in interaction between the marine luminescent bacterium *Vibrio fischeri* and the squid *Euprymna scolopes*. These squid are luminescent because of the light emitted by symbiotic bacteria housed within epithelial crypts of their light organs. The bacteria invade the developing juvenile light organ through ciliated cells. The symbiont engineers its own entry; ciliated cells die only in the bacteria's presence. Furthermore, epithelial cells of the light organ only increase in size to produce the crypts if bacteria are present. In the absence of bacterial 'infection' the light organs remain in a state of arrested morphogenesis, an elegant example of specific inductive interaction between species (Montgomery and McFall-Ngai, 1995).

18.5 INDUCTIONS BETWEEN INDIVIDUALS OF THE SAME SPECIES

18.5.1 Ants

The entomology literature contains many examples of regulation of development through interactions between individuals of the same species. Such interactions are very highly developed in the social insects (ants, bees, social wasps) with different castes (queen, soldiers, workers) within a single population.

One example that especially appeals to me is regulation of the production of soldier ants in *Pheidole bicarinata*. The number of soldiers produced is regulated by a combination of interactions between individuals and the colony's nutritional status. Adult soldiers in this species release a pheromone which inhibits production of further soldiers by regulating the threshold of larvae to juvenile hormone, the hormone that triggers development of soldiers. Size and allometric relationships of soldier ants are regulated through programming imaginal-disc growth patterns. Hormone levels are sensitive to nutritional status of the colony: a balance of nutritional status and numbers of soldiers determines whether additional soldiers will form. The development of larval bees into workers or queen depends on whether or not they are fed the 'royal jelly' secreted by the worker's salivary glands – an example where nutrition alone determines phenotype.[11]

18.5.2 Cannibalistic amphibians

Some amphibians are polymorphic with respect to whether they metamorphose or remain as permanent aquatic adults.

Less well-known are the cannibalistic morphs of two subspecies of the tiger salamander (*Ambystoma tigrinum nebulosum* and *A. t. mavortium*) and of the New Mexico spadefoot toad, *Scaphiopus multiplicatus*. Diet and overcrowding respectively trigger development of large, aggressive individuals whose morphology and behaviour enables them to cannibalize their smaller siblings. Cannibalistic morphs have enlarged heads, much enlarged jaws and feeding apparatuses (including altered teeth and shortened intestines) enabling them to feed on other amphibian larvae. *Scaphiopus* feeds on crustaceans with high thyroid hormone content, and the cannibalistic morph can be induced by the application of thyroid hormone. The changes are not atypical of metamorphosis, just as the changes in the ants were mediated by juvenile hormone.

A similar example (though one with a much more complex basis) is alteration of the jaw morphology of the cichlid fish *Cichlasoma managuense*, in response to different diets. Genetic variability, coupled with a developmental basis for polymorphism and an ecological releaser, provide plasticity for adaptive morphological novelty.[12]

18.6 ENVIRONMENTAL INDUCTIONS

In terms of a classification of epigenetic factors that can be incorporated into quantitative genetics models of the evolution of development, a distinction needs to be made between situations where the signal is a gene product and situations where it is an environmental stimulus. Rotifers are an example of the former, light-eliciting chlorophyll production in plants of the latter. Such clean distinctions between biotic and abiotic environmental signals are not, however, always possible; for example, predator and environment often interact to elicit the response (s. 18.7). A conceptually similar situation is the artificial activation of eggs by a temperature or pH shock in the absence of sperm; the stimulus may be abiotic, the response is not (s. 8.3.1). The distinction between biotic or abiotic signalling is not critical at the level of analysis of the activation of developmental pathways within a single generation. It is important in the context of models of the heritability and evolutionary stability of such interactions (Atchley and Hall, 1991, and see Chapter 20).

18.6.1 Moths

A fine example of environmental induction is revealed by Erick Greene's study of the geometrid moth *Nemoria arizonaria*, which made the cover of *Science* on 3 February 1989. Caterpillars of this North American and Mexican moth hatch out on several species of oak trees in either spring or summer. Those that hatch in spring feed on oak catkins low in tannin; those that hatch in summer, after the catkins have fallen, feed on oak leaves that have a high tannin content.

Caterpillars are morphologically similar when they hatch but become strikingly different as they develop; they become **seasonal polymorphs**. Spring caterpillars mimic catkins and are therefore referred to as catkin morphs; summer caterpillars mimic oak twigs and are known as twig morphs (Figure 18.1). Catkin morphs have bright yellow skin with many papillae and processes and two rows of stamen-like dots along the midline; twig morphs have grey skin and fewer processes. Greene eliminated temperature and photoperiod as causal factors in morph production and demonstrated that the amount of tannin in the diet controlled morph type. Catkin

Fig. 18.1 The twig morph of the moth *Nemoria arizonaria* on the left is a full sib of the catkin morph on the right. Photographs courtesy of Dr Erick Greene.

morphs developed from 94% of larvae fed catkins, but from only 6% of larvae fed leaves.

Ecological correlates of this remarkable seasonal polymorphism allow caterpillars to overcome the apparent disadvantage of feeding on such a seasonal food supply as oak catkins. Catkin morphs pupate more rapidly, have higher survival to pupation and are larger at pupation than are twig morphs. Furthermore, females that metamorphose from catkin morphs have greater fecundity than do those that metamorphose from twig morphs.

Greene emphasized what may be an unrecognized widespread occurrence of diet-induced developmental polymorphism in the evolution of host specificity and host races, confirming the alternative adaptations theory of incipient speciation put forward by West-Eberhard (1986). West-Eberhard, who hypothesized that the mechanisms in such interactions play a major role in speciation, postulated that the maintenance of different adaptive phenotypes in the same life history stage within a population is the first step to speciation. Having established alternate morphs within a species, reproductive isolation allows speciation from one of the morphs. Given that polymorphisms are developmentally based and induced, her approach places epigenetic control (epigenetic divergence/bifurcation, in her terminology) at the base of phylogenetic bifurcation, linking developmental mechanisms directly to speciation.

18.6.2 Plasticity genes

The genetic basis of environmentally induced responses is now under investigation, especially in plants, with the search for 'plasticity genes', which Pigliucci (1996, p. 169) defines as 'regulatory loci that directly respond to a specific environmental stimulus by triggering a specific series of morphogenic changes'. Alterations in rates of DNA transcription and of RNA translation are known in at least four systems displaying phenotypic plasticity: the response of flowering plants to light, induction of heat-shock proteins in plants and animals, adaptive responses of cyanobacteria to sulphur limitation, and temperature-dependent sex determination in lizards.[13]

18.7 COMBINED ENVIRONMENTAL AND BIOTIC INDUCTION

Environmental and biotic signals can independently elicit the same morphological responses in *Daphnia* (induction of neekteeth) and salamanders (development of cannibalistic morphs), as discussed in sections 18.3.2, 18.5.2. A predator-based induction and an environmental signal can also cooperate to elicit the response, as demonstrated by Michael Bell and his colleagues in an elegant analysis of reduction in the pelvic girdle of the threespine stickleback, *Gasterosteus aculeatus*, in 179 lakes surrounding Cook Inlet, Alaska.

The pelvic girdles are comprised of pelvic spines, posterior and anterior processes, and an ascending branch. The pelvic spines can be elevated and locked upright and so deter predators. Physiologically, burst speed and escape from predation is correlated with other skeletal characters, such as length of vertebrae, vertebral number, and the ratio of abdominal to tail vertebrae.

Pelvic girdle reduction occurs in response to absence of predation by other fishes and low levels of dissolved calcium in the lakes. It occurs in stages. First to go is the pelvic spine, then the posterior process, the ascending branch, and finally the anterior process. Reduction of the pelvic girdles and of other elements of the bony exoskeleton enables sticklebacks to cope with calcium deprivation. By introducing predators into some of the lakes, Bell *et al.* (1993) demonstrated that the physiological response to low calcium levels was contingent on the absence of predators; it does not occur in the presence of predators. Predator and environment interact to provide two levels of signals and two levels of integrated selective force.[14]

18.8 SUMMARY

It is clear from these examples of developmental mechanisms responsible for phenotypic plasticity and from other studies in the literature discussed by Stearns, Dodson and Pigliucci, that one fruitful area for investigation in evolutionary developmental biology is the working-out of inter- and intraspecific causal links between change in transcription or translation of DNA and mRNA, inductive changes in embryonic development, ecological adaptation, evolutionary change and possibly speciation. Inductive control of development provides a mechanism for both integrated short-term (ontogenetic) and long-term (phylogenetic) changes. Epigenetic mechanisms such as embryonic induction act interspecifically to evoke seasonal polymorphisms, amply demonstrating Van Valen's aphorism that 'evolution is the control of development by ecology'.

Genetic assimilation provides an additional mechanism for adaptive change by relating environment and development through selection of inherent genetic variability to produce novel morphologies and behaviours. Genetic assimilation is the topic of the next chapter.

18.9 ENDNOTES

1. Comprehensive discussions of adaptation may be found in Grant (1963), Gould and Lewontin (1979), Burian (1992), West-Eberhard (1992), and the chapters in the books edited by

Loeschcke (1987) and Rose and Lauder (1996b). For a history of the concept and its resurgence, see Amundson (1996) and Rose and Lauder (1996a).

2. See Colbert (1949), Vrba (1983a,b, 1984), Brain (1981, 1983), and the chapters in Vrba *et al.* (1996) for the way in which climatic change can drive adaptation.

3. See Fortelius (1985) for a detailed discussion of the adaptiveness of mammalian teeth.

4. See Stearns (1989, 1992), Gordon (1992) and Via *et al.* (1995) for overviews of phenotypic plasticity.

5. Viviparous reptiles and mammals represent a special case in which the maternal genome and environment exert epigenetic and sometimes inductive control over embryonic and/or foetal development; see ss. 7.2, 7.3, Chapter 20 and Atchley and Hall (1991).

6. For overviews of cyclomorphosis, see Gilbert (1966), Dodson (1989a), Stearns (1989), Roff (1992, 1996), Baldwin (1996) and T. B. Smith and Skúlason (1996). For norms of reaction, see Via and Lande (1985) and Stearns (1989). Although Schmalhausen is most closely associated with reaction norms, the term and essential concept was coined by Woltereck (1909) in relation to studies on *Daphnia*, some of which are discussed in section 18.3.2.

7. Hutchinson (1967) provided an extensive review of cyclomorphosis in plankton, while Black and Slobodkin (1987) reviewed the history of the term and the concepts underlying it. For the studies on which my analysis is based, see Dodson (1989a), Stearns (1989), Harvell (1990), Leclerc and Regier (1990), Brönmark and Miner (1992) and Pigliucci (1996).

8. See Stemberger and Gilbert (1987) and Neill (1992) for adaptive responses of rotifers and copepods to multiple predators.

9. For the genotypic constancy that underlies seasonal polymorphism in *Daphnia*, see Dodson (1989a,b), Lampert and Wolf (1986) and Stirling and McQueen (1987). For the aromatic hydrocarbon carbaryl (1-naphthyl-N-methylcarbamate) as an agent evoking altered helmet morphology and for chemicals released by predators, see Hanazato (1991).

10. See Brönmark and Miner (1992) for response to predators, and Reznick *et al.* (1990, 1997) and Endler (1995) for rapid changes in life history.

11. See E. O. Wilson (1971) for examples of regulation of insect development via interactions between individuals and/or nutrition. For studies on the ant *Pheidole bicarinata*, see Wheeler and Nijhout (1983, 1986), Nijhout and Wheeler (1996) and Nijhout (1998).

12. For cannibalistic morphs of *Scaphiopus*, see Pfennig (1990, 1992), while for *Ambystoma*, see Gehlbach (1969), Collins *et al.* (1980), Collins and Cheek (1983) and T. B. Smith and Skúlason (1996). The response to increased population density was documented experimentally by Collins and Cheek (1983), shown by Maret and Collins (1997) to result from resource competition and to occur in natural populations by Gehlbach (1969).

13. See Schlichting and Pigliucci (1993) and Pigliucci (1996) for analyses of transcriptional regulation in phenotypic plasticity.

14. See Swain (1992a,b) and Walker (1997) for correlations between body shape and size, behaviour and predation in sticklebacks.

19

EVOLUTION, GENETIC VARIABILITY AND THE ENVIRONMENT

'A major gap in our knowledge of internal causes in evolutionary mechanisms is their environmental context. In addition to the genetic origins of particular changes, what are the ecological factors? How are the internal and external environments related? . . . It also brings us to new questions about the inter-relationships of internal processes acting within development and externalist processes acting within populations. To what extent can selection reinforce the causation of change in developmental mechanisms?'

Thomson, 1986, p. 232

The main tenet of Lamarckism remembered today is 'the inheritance of acquired characters', the phrase and law on which so much criticism rightly falls. Lamarckism united environment, genotype and phenotype, with primacy given to the environment. Rejection of Lamarckism and its implications separated the phenotype from the genotype and led to an almost total neglect of the environment as a factor in the production of change. Yet, organisms do adapt to environments during their lifetimes and although such adaptation is not inherited as such, the capacity to respond to the environment is heritable, as amply demonstrated in the last chapter. This capability, coupled with developmental plasticity and pre-adaptation, can lead to long-term evolutionary change. The arguments for epigenetic control of ontogeny (Part Two) and for developmental change in phylogeny (Part Three), were, in part, arguments for inclusion of 'environmental' control. The need to bring the environment back into discussions of genotype–phenotype interactions is the topic of this chapter.

Non-heritable environmental effects, of course, only alter the phenotype of members of one generation; inheritance over multiple generations is the stuff of evolution. One way in which the environment is integrated with developmental and evolutionary plasticity is through the genetically inherited capability of responding to specific environmental signals by producing long-term, stable, phenotypic polymorphism (Chapter 18). Another mechanism, genetic assimilation, is the topic of this chapter.

19.1 GENETIC ASSIMILATION

The essence of Conrad Waddington's concept is that an environmental influence evokes a phenotypic character that is expressed in subsequent generations in the absence of the environmental influence because of the exposure of pre-existing genetic variability to the action of selection. Genetic assimilation has been defined as:

'The process by which a phenotypic character initially produced only in response to some environmental influence becomes, through a process of selection, taken over by the genotype, so that it is formed even in the absence of the environmental influence that at first had been necessary.'

King and Stansfield, 1985, p. 153

In a series of experiments using *Drosophila melanogaster*, undertaken in the '40s, '50s and '60s, Waddington demonstrated genetic assimilation as a real and potentially important evolutionary mechanism. Superficially, genetic assimilation appears to be heritable phenotypic responsiveness to the environment, i.e. acquired inheritance. It is not, although it took a long time for Waddington's experiments and mechanism to be appreciated and accepted for what they are – **genotypic** responsiveness to environmental signals.[1]

Bithorax is a mutant phenotype in which the metathoracic imaginal discs of *Drosophila* larvae develop into an accessory mesothorax. As the mesothorax has wings and the metathorax balancers, bithorax *Drosophila* have two pairs of wings instead of the one pair diagnostic of the Diptera. This dramatic and remarkable change results from a homeotic mutation (Figure 9.4), but bithorax individuals can arise from other genetic bases than a homeotic mutation. Production of such phenotypes was the first experimental demonstration of genetic assimilation.

Flies with the bithorax phenotype (24.5% in the first generation), were produced when Waddington exposed eggs to ether vapour. Adults with this **induced bithorax** phenotype were used to produce the next generation. Eggs from second generation flies were exposed to ether, and bithorax adults from this generation were used as parents for the next generation and so on, for a total of 29 generations. In parallel, individuals that had not been exposed to the ether shock (except in the first generation) were selected for. In the 8th generation of the latter lines, and much to his surprise, a single fly with weakly developed bithorax phenotype appeared. Waddington referred to this as **uninduced bithorax**. Ten such individuals appeared in the 9th generation, others in the 29th. The genetic explana-

tion for the appearance of individuals with this phenotype is explored in the next section.

Uninduced bithorax is neither a unique response nor an aberration. In his 1956 paper Waddington demonstrated that exposing eggs to a brief 40°C temperature shock produced a condition in which a posterior (and sometimes an anterior) wing vein was missing. This induced change mimicked a known mutant, *crossveinless*, just as induced bithorax mimics *Bithorax*.

Waddington selected animals with the crossveinless condition but without further exposure to the temperature shock. The phenotype began to appear in flies from the 14th generation onwards. Milkman (1962) demonstrated that expression of the crossveinless phenotype involved seven loci all distinct from the *crossveinless* locus operating in unshocked animals. This is very important. It indicates that two identical phenotypes have different genotypic bases that can be correlated with the signal eliciting the phenotype; see section 19.2 for further discussion of this point.

Uninduced crossveinless and bithorax flies are both **phenocopies** of known mutations. A phenocopy (a term introduced by Richard Goldschmidt) is an experimentally produced morphological modification which bears a close resemblance, or identity, to a phenotype produced following a mutation. The developmental period that is sensitive to perturbation by the environmental agent often coincides with the time of action of the mutation during development, i.e. phenocopies mimic the time of action of mutants in producing the phenotype. Furthermore, as demonstrated by Richard Goldschmidt using heat shock and *Drosophila*, the same environmental stimulus can produce different phenocopies depending on the intensity of the stimulus applied during the sensitive period of embryonic development. Clearly, the environment can evoke specific genetic and developmental responses to produce specific morphologies. In discussing induction of phenocopies, Walter Landauer

(1958) concluded that considerable genetic variation exists that is only manifest phenotypically in phenocopies. Waddington extended this idea by showing that genetic variation normally only seen in phenocopies, could be selected for until it is expressed in the absence of the specific environmental stress. Causally, a genetically assimilated character is a phenocopy.[2]

19.2 THE GENETICS OF ASSIMILATION

Genetic assimilation could either be the cumulative response to selection for genes which alter a developmental threshold sufficiently that the genetic response occurs in the absence of the environmental stimulus, and/or the result of selection acting upon a chance mutation(s).

The average grade of expression of the phenotype in induced bithorax rose, eventually leading to a large number of **extreme bithorax** types, which were the product of a single dominant gene with recessive lethal effect. Assimilation did not proceed by the selection of many genes or by mutation, but by the fixation of a single major gene. Waddington's data are consistent with the presence of the gene at a low frequency from the beginning of the experiment; there was no evidence for a 'directed induction by the environment of an appropriate mutation' (Waddington, 1957b, p. 245), but such an induction could not be completely excluded by the experimental design used. Further work on *cubitus interruptus* in *Drosophila* by Waddington, Graber and Woolf (1957) showed that if other loci behave in the same way, the mutation rate to isoalleles cannot be great enough for mutation at any one locus to play an important role in response to selection in these experiments.[3]

Assimilated bithorax, on the other hand, is polygenic, and so could not have been derived by chance mutation. Genes other than homeotic genes were involved (Waddington, 1956, 1957b). Homeotic genes are restricted to chromosome 3 (s. 9.4). In assimilated bithorax,

multiple genes distributed on all chromosomes, and a recessive, maternally acting, X-linked gene are involved. There are therefore **multiple genetic means** to produce the **same phenotype**:

1. *Bithorax* by homeotic mutation;
2. assimilated bithorax, which is polygenic with a maternal X-linked component; and
3. extreme bithorax by a single dominant gene.

The environmental treatment exposes subthreshold concentrations of alleles to selection. Selective shifts in gene expression in different environments is the basic genetic mechanism underlying genetic assimilation and the basis for its potential as a mechanism for evolutionary change in morphology. Uniformity of phenotype does not necessarily mean lack of genetic variation when threshold characters are involved and there is variability in the individual genetic components from which the phenotype is produced.[4]

In two critical analyses, Curt Stern (1958, 1959) summarized the essential elements of a genetic mechanism for assimilation:

1. an original population that is polymorphic for the particular phenotype at the outset, but in which genes for the phenotype are not expressed;
2. an environmental stimulus to allow these genes to be expressed; and
3. selection to alter the frequency of these genes in the population, to the point where the phenotype appears in the absence of the environmental stimulus.

As emphasized by Arthur, Grant, and Thomson in reviewing this subject:

'The important point about this and similar experiments is that they provide a conventional selective explanation for a phenomenon which might, in the absence of such an explanation, be considered Lamarckian.'

Arthur, 1984, p. 49

'The new potentialities of phenotypic expression, which can be created inadvertently by selection in one environment, may sometimes have important consequences in evolution, as where a population becomes prepared or "preadapted" in an ancestral environment for entering a new environment.'

Grant, 1963, p. 211

'It (genetic assimilation) is interesting to evolutionists . . . because it shows the capacity of selection to reprogram development in order to alter phenotype expression within a given proscribed range.'

Thomson, 1988, p. 96

That conditions such as crossveinless and bithorax can be induced environmentally and by mutation, indicates that, despite the major differences in the source of the stimulus, environmental and genetic (mutational) factors cannot be distinguished at the level of developmental response and phenotypic change. The heritable component of genetic assimilation lies in the progressive transfer of the origin of the 'morphogenetic signal' from the environmental stimulus to the genotype as alleles favoured by the environmental stimulus are expressed. A progressive decline in the need for the environmental stimulus parallels an increase in independent genotypic control of the response to selection.

19.3 GENETIC FIXATION

A problem which arises is genetic fixation. Why is a character such as bithorax genetically fixed rather than produced in response to the environmental signal in each generation?

One argument is that as the environmental stimulus is an abnormal one, genetic fixation rapidly incorporates the character into the genotype. As assimilated bithorax confers no apparent benefit on the animal, rapid fixation would seem to be a disadvantage. This apparent paradox is discussed below; genetic assimilation may take even adaptive characters beyond their optimal level in the population.

The selection that produces sensitivity to the particular environmental stimulus does not regularly result in corresponding changes in general developmental buffering (i.e. in increased sensitivity to other environmental stimuli) or change the phenotype of other parts of the organism. Genetic assimilation is specific. Waddington used the concept of **canalization** of developmental pathways to explain assimilation of characters such as bithorax with minimal effect on other developmental processes and structures (see s. 12.2). Schmalhausen used the essentially similar concepts of autonomization and stabilizing selection described in section 19.5. Selection of factors that control the capacity to respond to the environment moulds development into a new path and is a means of genetic fixation of the originally environmentally correlated adaptive character. Such a mechanism is in addition to mutation.[5]

19.4 A MECHANISM FOR GENETIC ASSIMILATION

Utilizing experiments establishing genetic assimilation and canalization of developmental pathways, the following general mechanism provides an explanation for the changes involved in the genetic assimilation of a character.

In terms of phenotypic expression, the wild-type genotype has a normal distribution (Figure 19.1). From the point of view of developmental processes, there is canalization about a relatively stable mean (arrow in Figure 19.1A). Canalization means that the phenotype is relatively insensitive to genotypic and environmental influences. By selecting outside the phenotypic range of the wild-type (+) allele (selection for deviation from the canalized mean), a balanced effect of phenotypic expression and environmental influences is obtained. This leads to either of curves A and B in Figure 19.1A, depending on whether selection is for high or low variance from the

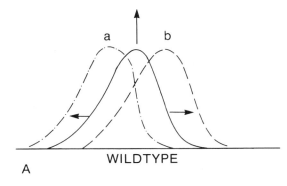

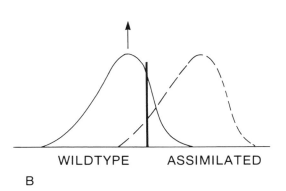

Fig. 19.1 A representation of the mechanism of genetic assimilation. A. Canalization of the wild type genotype and phenotype is around the mean shown by the vertical arrow. Selection away from that mean (horizontal arrows) can shift the population genotype to a new canalized mean (a or b) without altering the phenotype. B. Subjecting the population to an environmental stress and selecting for the new phenotype that arises produces a new canalized mean (assimilated) that only just overlaps the phenotypic variability of the unselected, wild type population. The threshold for the phenotypic switch is shown by the horizontal line. Intrinsic genetic variability, a threshold, an environmental signal and selection combine to produce the new stable phenotype.

+ allele. A new canalized mean for the + allele is obtained.

What Waddington did was to substitute a 'mutant' phenotype for the + allele by subjecting *Drosophila* to an environmental stimulus and selecting for the new phenotype. Where this results in an alteration in the phenotypic range of the thoracic polygenic system, the situation shown in Figure 19.1B is reached. Expression of the new gene complex involving assimilated bithorax is outside the canalized phenotypic range of the + allele. The threshold (horizontal line in Figure 19.1B), which was just exceeded by the bithorax effect to produce slight changes from the normal, is so far exceeded that canalization is now along a new path. Bithorax still lies within the variability of the + allele. In the assimilated bithorax stock, selection for the expression of the subthreshold genes is additive.

19.5 STABILIZING SELECTION AND THE GENETICS OF ASSIMILATION: WADDINGTON AND SCHMALHAUSEN

In Russia, Ivan Schmalhausen independently arrived at mechanisms extraordinarily similar to Waddington's genetic assimilation and canalization. He called his processes **autonomization** and **stabilizing selection** and invoked norms of reaction (see below). Schmalhausen's autonomization was Waddington's genetic assimilation and vice versa. Schmalhausen's views first appeared in English in *Factors of Evolution: the Theory of Stabilizing Selection* (1949), having been published in Russian a decade earlier. Dobzhansky, one of the leading exponents of neo-Darwinism, wrote in the foreword to the English translation:

'The book of I. I. Schmalhausen advances the synthetic treatment of evolution starting from a broad base of comparative embryology, comparative anatomy, and the mechanics of development. It supplies, as it were, an important missing link in the modern view of evolution . . . Ivan Ivanovich Schmalhausen is perhaps the most distinguished among the living biologists in USSR.'

Dobzhansky in Schmalhausen, 1949, pp. xv–xvi

Responding to Waddington's complaint that he, Dobzhansky, always cited Schmalhausen rather than Waddington when referring to 'developmental buffering', Dobzhansky wrote in a letter of August 1959, that he regarded Schmalhausen's "Factors of Evolution as one of the 'basic books' establishing the biological theory of evolution". Schmalhausen and his work are beginning to gain attention; witness David Wake's evaluation of how Schmalhausen's views shaped his approach to research.[6]

Schmalhausen did not discover Waddington's work until after he had developed his own theory. He cites only one of Waddington's works, his book *Organisers and Genes* (1940). Waddington rarely referred to Schmalhausen or his work and when he did, as in his 1961 review, he dismissed it as not based on any original experimental evidence and as not providing any process to explain the concepts. This is not so. Schmalhausen equated autoregulation and autoregulating mechanisms with Waddington's canalization, but considered stabilizing selection a unique contribution.[7] In Schmalhausen's words, albeit in translation:

'The *stabilizing form of selection* is based upon the selective advantage under definite and especially, fluctuating conditions possessed by the normal organization over variations from the norm. It is associated with the elimination of most variations and the establishment of more stable mechanisms of normal morphogenesis.'

Schmalhausen, 1949, p. 73

By acting through the regulating properties of development, stabilizing selection re-establishes stability after an episode of morphological change. Canalization is a consequence of stabilizing selection. The relationship of stabilizing selection to genetic assimilation includes the role of the environment in destabilizing development and revealing previously hidden variation. Günter Wagner and colleagues (1997) have revised

such ideas in drawing a distinction between genetic and environmental canalization, and in their population genetics model of canalization under conditions of stabilizing selection.

The following descriptions of how stabilizing selection and autonomization interact to effect change – 'the creative role of individual selection', as Schmalhausen subtitled a section of Chapter 2 – could just as easily have come from Waddington's pen if we replace autonomization and autonomous with assimilation and assimilated:

'While stabilizing selection does not immediately establish new types, its activity creates a new apparatus of individual development with its regulating mechanisms. There occurs an autonomization of development which involves a subsequent replacement of external developmental factors by internal ones. Dependent developmental processes are either transformed into autoregulative processes or become completely independent – autonomous. This signifies that the genotype with its norm of reaction has changed, since the response to the external stimulus is replaced by a similar response to an internal stimulus. The reactivity of the tissues alters and, apparently, the system of formative stimuli also changes. In every case, the genotype also changes and becomes more stable.'

Schmalhausen, 1949, p. 93

Schmalhausen found the developmental mechanisms for responsiveness to stabilizing selection in changing inductive tissue interactions. Progressive loss of specificity of the environmental stimulus and the inductive interactions accords with Waddington's evidence for genetic assimilation of the bithorax phenotype and with the oft-reported lack of specificity of embryonic inductors, i.e. that multiple inductors can evoke the same differentiative or morphogenetic response. Schmalhausen saw ontogenies evolving through just such modifications of inductive processes.[8]

His concept of the **norm of reaction** is amply demonstrated when different phenotypes can be derived from a single genotype, as occurs in cyclomorphosis (s. 18.3). Pigliucci *et al.* (1996) proposed the developmental reaction norm for the set of multivariate ontogenies than can be produced from a single genotype exposed to environmental variation. They emphasize, as I do:

1. the integration of genetic, environmental and developmental perspectives;
2. an ontogenetic perspective on processes than can alter the phenotype; and
3. that selection, epigenetics and constraint are all major determinants of phenotypes.

Qualls and Shine (1996) devised an explicit and succesful test to reconstruct ancestral reaction norms associated with the evolution of viviparity in lizards, viviparity having evolved some 100 times in 20 or more families of reptiles.

Bateman (1959a,b) demonstrated genetic assimilation of other phenotypes in *Drosophila*, confirming the polygenic basis of genetic assimilation and showing, as Waddington's results had implied, that assimilation should not be expected in genetically invariant inbred populations. Ho and her collaborators (1983) demonstrated that a major aspect of genetic assimilation of bithorax was an extension of the critical time period during which the phenotype was normally induced. Such critical periods are times of heightened susceptibility to genetic and/or environmental perturbations. Thus, the polygenic basis of developmental processes, chance mutations, and shifts in timing of critical periods for organ induction, are genetic/developmental mechanisms that facilitate morphological change.[9]

19.6 GENETIC ASSIMILATION IN NATURE

As far as Waddington was concerned (e.g. 1956, p. 11), genetic assimilation was not merely a phenomenon of artificial selection under laboratory conditions. Matsuda (1987) discusses a substantial number of examples of genetic assimilation, or phenomena that could reasonably only be explained on the basis of genetic assimilation:

1. altitudinal variation in egg size and number in the wood frog, *Rana sylvatica* (see below), and in butterfly colour patterns;
2. temperature-induced asexuality in the turbellarian *Dugesia gonocephala*;
3. shell shape in the mollusc *Limnaea* (see below);
4. winglessness in female embiopterans;
5. plate morphs in the three-spined stickleback, *Gasterosteus aculeatus*; and
6. neoteny in salamanders.

In other situations, such as cyclomorphosis (s. 18.3), genetic assimilation might (would?) be disadvantageous. Plasticity of response to a changing environment rather than genetic assimilation would be optimal when environmental conditions are unpredictable (s. 19.7).

Matsuda (1982, 1987) provided evidence from urodeles and amphipods to argue that heterochronic (neotenic) changes, initially elicited in response to an environmental stimulus, could be genetically assimilated. The variable developmental component proposed by Matsuda is hormonal and only later directly genetic, a neo-Lamarckian view that is not required to accommodate genetic assimilation into mainstream evolutionary theory. Unlike Waddington, who specifically argued against a role of chance mutations in genetic assimilation, Matsuda saw mutation playing a role.

19.6.1 Shell shape in *Limnaea*

An example of genetic assimilation under natural conditions is shell shape in the gastropod *Limnaea*, studied by Jean Piaget, the Swiss embryologist turned psychologist and epistemologist.

In some lakes in Switzerland and Sweden, the normally elongated shell of *Limnaea stagnalis* is replaced by shorter forms adapted to the wind and wave action associated with shallow water and stony lake bottoms, and given species status as *L. lacustris* and *L. bodamica* by Miller (1978).

When individual *L. stagnalis* are raised in turbulent water they develop a shortened shell. This new form, however, continues to produce shortened shells for many generations when bred in the laboratory in still water, or after being transferred to ponds in which there is no wind or wave action. Similar experiments were performed on *L. peregra* and the modified form *L. burnetti*. The interpretation of these experiments is that a morphology that arose in response to an environmental factor was assimilated so that it now forms in the absence of the environmental signal.[10]

Behaviour was modified in the species studied by Piaget: the shortened forms preferentially remain in shallow water. (Indeed, these studies set Piaget on his lifelong pursuit dedicated to behavioural psychology.) Morphology and behaviour are coupled, a morphological and a behavioural trait are assimilated, and, in this case, genetic assimilation has ecological relevance. To quote Waddington's assessment of these experiments:

'One must conclude, therefore, that the contracted phenotype has been definitely assimilated. The form, which has originated as a physiological reaction to the stress of wave action, has been converted, presumably through selection, into a genetically assimilated condition, which develops independently of a particular precipitating stress.'
Waddington, 1975, p. 93

19.6.2 Xerophytic plants

Some xerophytic (drought-adapted) plants permanently lose their leaves as an adaptation to excess water loss. For other xerophytes, leaf loss is seasonal, adaptive and environmentally triggered. Schmalhausen used seasonal leaf loss in xerophytic plants as an example of genetic assimilation (autoregulation) in nature.

With the evolution of permanent leaf loss, shoots are extended and expanded into leaflike branches (phylloclades). In such forms, shedding of leaves and development of phylloclades are independent of environmental cues and initiated by internal, probably hormonal, signals. Leaves are shed even when such plants are maintained under constant environmental conditions, i.e. genetic assimilation has occurred.

19.6.3 Egg size and number in amphibians, sea urchins and lizards

Matsuda essentially argued that any character adapted to a particular environment but produced independently of the environment, arose by genetic assimilation. His summary of Berven's 1982 analysis of egg size and number in the frog *Rana sylvatica* illustrates this thinking.

Egg size affects many aspects of larval life history, including (but not limited to) larval morphology, development rate, growth rate, feeding capacity, size, stage of maturation/ size at metamorphosis, and timing of the metamorphic transition. As altitude increases, *R. sylvatica* lay larger and fewer eggs, larger adult size and delayed reproduction being selectively advantageous at high altitudes. Berven exchanged juvenile frogs between high and low altitudes and found that egg size and number followed parental type and not altitude. Here is a character that develops independently of the environment but where alternate character states are appropriate to one or other environment. Matsuda concluded that egg size and egg number were genetically fixed through genetic assimilation.

Does this analysis mean that every character that is genetically fixed, dimorphic, and environmentally relevant, became fixed through genetic assimilation? We don't know. As with innovations (Chapters 13–16), a methodology is needed to identify genetically assimilated characters in nature. Because such characters are fixed by selection acting on pre-existing genetic variability, as are many other characters (the only difference in genetic assimilation being the stimulus that evoked the response to selection), there may, in fact, be no simple way of detecting genetically assimilated characters other than through experimental induction.

One approach would be to combine experimental manipulation with field studies of natural patterns. Sinervo and McEdward demonstrated that body form and larval patterns in sea urchins can be altered by experimentally manipulating the volume of the egg. Here development plays a pivotal role in integrating evolutionary and life history theory. Reducing the amount of cytoplasm in the egg results in smaller larvae, as has been known since the classic experiments of the 1880s. Reducing egg volume also slows development (which might not be expected), and is partially compensated for by faster growth later in larval life and by regulation of size and shape. Reducing egg volume also produces simpler larvae – certainly an unexpected finding.[11]

The epigenetic effects of smaller egg volume (comparable to the effects of altitude differences in *Rana sylvatica*) preferentially affect larval stages and have consequences beyond the production of smaller larvae. Barry Sinervo and his colleagues have used such an approach to considerable effect with eggs of the lizard *Sceloporus occidentalis*. Southern Californian populations produce large eggs and larger individuals than do populations from the north of Washington state. Mean clutch size increases and egg size decreases from south to north and altitudinally within California. Removal of yolk from eggs at the southerly edge of the range slows development and produces smaller individuals, comparable to the latitudinal and elevational patterns seen across the species range. Sinervo and Huey (1990) christened this approach **allometric engineering** (see Box 19.1 for types of allometry). A combination of such an experimental approach with ecological studies of environmentally related changes holds great promise for the search for evidence of

genetic assimilation in nature. In combination with a phylogenetic approach it should be possible to identify plasticity and norms of reaction in the appropriate outgroup and therefore to map the direction of evolutionary change; plasticity in the outgroup would be consistent with fixation in the group of interest.[12]

19.7 ADAPTATION AND GENETIC ASSIMILATION

Waddington (1962a) discussed another experiment bearing on his model of assimilation acting on canalized pathways of development. Eyeless strains of *Drosophila* show considerable variation in the number of scutellar bristles in comparison with the wild type. Selection for increased number of bristles in the wild type increases the number from generation to generation. If such genes accumulate sufficiently, the wild type eventually shows an increased bristle number; four bristles becomes stabilized or buffered.

BOX 19.1
ALLOMETRY

Allometry comes in three guises:

1. static – individual variation within a population of individuals of known age;
2. ontogenetic, which results from processes of growth and development; and
3. evolutionary, reflecting phylogenetic variation within a group (Klingenberg, 1996a, 1997).

Only ontogenetic allometry deals with processes and mechanisms. Static and evolutionary allometry are statements about patterns, not processes.

Waddington's work was with non-adaptive characters. In terms of adaptation to the general environment, the work of Rendel and Sheldon gives plausibility to an extension of genetic assimilation to adaptive characters in evolution. Their work on the *scute* phenotype in *Drosophila melanogaster* and genetic assimilation in nature sheds light on the problem of variation, adaptation, canalization, and the acquisition of new characters (Rendel and Sheldon, 1960; Rendel, 1967).

Normally, the number of scutellar bristles is canalized at four. This is apparently a non-adaptive character, but it may be correlated with a physiological adaptive character. Selection for low variance of scutellar bristle number resulted in canalization about a mean of two bristles; selection for high variance was ineffective. Waddington's aim in the experiments was to see if selection could make a particular phenotype insensitive to genotypic and environmental influences. As selection proceeded over 28 generations, the low variance line developed reduced phenotypic variance at the temperature of the experiment (25°C) and lowered sensitivity to extreme temperature changes. This should be compared to genetic fixation (s. 19.3). Canalization about a mean of two bristles meant a reduced variance and reduced movement away from the newly canalized mean. This does not appear to be a new adapted mean as the lowered bristle number has no apparent adaptive significance. The fact that phenotypic variance is reduced about the lowered bristle number shows it to be a stable condition; the lowered sensitivity to extreme temperature is one aspect of general environmental buffering. In terms of adaptation to the general environment, the change may be advantageous. Waddington saw the 'tuning' of development as 'perhaps the most important aspect of genetic assimilation' (1957a, p. 188).[13]

Analysis of quantitative trait loci affecting abdominal and sternopleural bristle numbers in *Drosophila* by Trudy Mackay and her colleagues in North Carolina, has identified many loci with small effects. Counting bristles on fruit flies may seem like counting hairs on dogs, but these bristles are important sensory organs, unlike hairs, many of which are dispensed with during life. Selection experiments were used to produce lines of flies with different numbers of bristles. Any association between bristle number and marker loci was then determined by further breeding and backcrossing, allowing quantitative trait loci that affect bristle number to be determined and mapped. Contrary to expectations, a small number of loci with large effects control most of the variation in bristle number. Variable dominance, polymorphism and epistatic and pleiotropic interactions between these loci provide important signposts to what may be a relatively small number of genes regulating most of the variation in the characters studied by Waddington. Rendel and Sheldon's work on *scute* in *Drosophila* may be similarly explained and related to the issue of the adaptiveness of genetically assimilated characters.

There are important lessons to be learned here about how Mendelian inheritance operates, how Mendelian genes vary, how genes such as homeobox genes are conserved over hundreds of millions of years and dozens of phyla, and how selection 'sees' these genes. A major area requiring synthesis and/or differential explanations is the question of how mutation and selection produce such divergent outcomes as conserved, co-linear homeobox genes; quantitative trait loci of large and small effect; and the extensive variability seen in many structural genes.[14]

19.7.1 Adaptation and canalized pathways

The building of a new canalized path involves the selection of genotypes with only restricted responsiveness to the environment. Waddington overcame this problem by proposing that natural selection can act in two ways:

1. to reduce the response of an animal to deleterious environmental modifications by narrowing canalization; and
2. to strengthen an animal's capacities for adaptive responses, leading to loose canalization with respect to advantageous modifications.

Provided that these two selections can operate concurrently, the problem of reduced variability associated with canalization is resolved. Rendel and Sheldon's experiments also could be interpreted in this way. Schmalhausen proposed that stabilizing selection would achieve the same end (s. 19.5).

Reduced variance associated with the canalization of developmental pathways is a block to the assimilation of adaptive characters. Unless there is selection for both 'loose' and 'narrow' canalization, it will take a significant deviation from the norm to enable an adaptive character to be expressed. Once this adaptive shift is made, the reduced variance of the new canalized path is an advantage in the new environment.

The work of Dun, Fraser and colleagues is relevant. Their studies demonstrated that mutant genes have a much greater phenotypic variability than do their normal alleles. In a changed environment they enable adaptation to vary from a previously canalized path. Selection in mice with the sex-linked mutant gene *Tabby*, which affects the teeth, skin, hair and exocrine glands, led to differences in vibrissa number in progeny that lacked the *Tabby* gene. The stress of the mutant gene was employed to expose hidden genetic variability in the same way that Waddington used an environmental stress to evoke bithorax and crossveinless phenotypes. These demonstrations of genetic variability in the absence of phenotypic variability may in part balance the reduced variability initiated by canalized developmental pathways (s. 12.2).[15]

In light of these studies, mutation may provide a way for assimilation to act on a new adaptive character. Once the mutant is present, assimilation increases its frequency. Waddington did not place importance on an initial mutation; the greater expression of assimilated bithorax stems from its polygenic base and selection of sub-threshold genes present.

Many adaptations depend on correlated changes in several characters not expected to be connected with one another in development. Change is often integrated. In studies on the California poppy, Cook (1962), showed that parallel variation occurs when several characters respond concurrently to changes in a single environmental factor, the result being that the species is subdivided into a graded patchwork of distinctive populations, each adapted to particular local conditions, as with assimilation of morphology and behaviour in *Limnaea*. According to Waddington (1957a), canalizing selection for an optimum adaptive response would build up epigenetic systems producing special phenotypes.

Therefore, selection which impinges on the phenotype is selection of the capacity of a genotype to respond to an environment. The epigenetic nature of developmental processes guides the phenotypic effects of available mutations by favouring systems of genes which respond to the local situation by producing well-adapted organisms. In terms of the view of adaptation described in section 18.1, genetic assimilation is a way for organisms to adapt to the environment.

19.7.2 Canalized pathways and evolution

Can adaptation of canalized paths be extended to adaptation in evolution, the time extension of adaptation in ontogeny? On theoretical grounds the answer is 'yes'.

The secondary or adventitious cartilages that develop on avian and mammalian membrane bones in response to biomechanical stresses are adaptive. Once induced, the cartilage is of functional value in resisting the stresses that evoked it. Cartilage formation requires a shift in differentiation from osteogenesis to chondrogenesis in response to mechanical stimuli mediated by N-CAM. Although the character has not been assimilated (possibly because of the continued presence of the environmental stimulus), the ability to respond to the environmental stimulus persists and has, in effect, been assimilated.[16]

The development of pseudoarthroses (false joints) and the developmental lability that allows false joints to form, is also relevant. The lability of bone as a tissue to develop cartilaginous articular surfaces is an adaptation or possibly a pre-adaptation that arose during vertebrate evolution. In terms of canalized pathways of development, there is genetic potential to produce more tissues (cartilage, fibrocartilage) than the one normally produced (bone). Pseudoarthroses, and indeed joints in general, are adaptive consequences of the interacting processes that generate skeletal elements. Assimilation of cartilage formation should show up under the appropriate stimulation and selection, and may explain skeletal changes in cichlid pharyngeal jaws and secondary avian jaw articulations (ss. 16.5, 17.2.2).

Bock's 1959 and 1960 studies on pre-adaptation dealt explicitly with adaptation in evolution and intensification of function (s. 17.2.2). He did not reduce his concept to the genetic level, seeing a structure as pre-adapted for a new function if the original form that enabled it to discharge its original function allowed it to assume a new function whenever need for that function arose. Thus, he saw the acquisition of the new function as not necessarily accompanied by a genetic change. Such flexibility, however, stems from a genotype able to respond to varying environments in an appropriate way. Such flexibility, though advantageous to an animal in a fluctuating environment, is hard to reconcile with the concept of canalized development proposed by Waddington, although the two types of selection he proposed provide a way out of this problem. Selection could act on an assimilated character so that it was pre-adapted for an-

other function. Secondary cartilage may belong to this category.

Direct evidence for adaptation in evolution cannot usually be obtained. Only the adapted structure or function remains as the end product with little or no indication of associated environmental changes (but see the work of Colbert, Vrba and Brain on changes that track climate, in s. 18.1). On *a priori* grounds, if genetic assimilation is operating now, there is no reason for not extending it back in time to account, in part, for past change. Waddington's experiments with *Drosophila* were spread over up to 30 generations, not extensive in terms of evolutionary time, but nevertheless a considerable time extension of ontogeny. Longer-term experiments have been carried out. Populations of *Drosophila pseudoobscura* kept at different temperatures for up to six years show divergence in body size (Anderson, 1966), a divergence that is maintained if the populations are then transferred to the same temperature, i.e. temperature-induced alteration in body size is genetically assimilated.[17]

Brace (1963) studied what in many ways is the antithesis of genetic assimilation, calling it **structural reduction**. His thesis is that as the environment continually changes, so does the selective advantage of a character. If this character is reduced (i.e. loses its adaptive value) and is selected against, and if, as Brace maintains, the probable effect of mutation is toward structural reduction, then the combined trend is to reduce the structure over time. This structural reduction is, in fact, an adaptation to the changed general environment.

19.8 SUMMARY

Genetic assimilation provides a mechanism relating phenotypic change to genotype and environment in terms of specific environmental stimuli, epigenetic interactions, and 'developmental buffering' to the general environment. Changes in gene frequencies vary in different environments. An environmental influence evokes a phenotypic character that is subsequently expressed in the absence of the environmental influence because pre-existing genetic variation is exposed to selection.

Waddington's genetic assimilation experiments reveal that the same phenotype can be achieved by the operation of different sets of genes; different cascades are activated. Development of segmented body regions in *Drosophila* is governed by homeotic genes (s. 9.4). Variation in homeotic genes is expressed as homeotic mutants. Phenotypes that are identical to homeotic mutants, but that arise via genetic assimilation, have a polygenic basis involving genes on many chromosomes. Genetic assimilation therefore has a different genetic basis than homeotic mutations, despite similarities of the phenotypes produced. Canalization and stabilizing selection operate on features such as those responsible for the origins of *Baupläne*, to limit phenotypic variability. Unexpressed but available genetic variability, coupled with canalization of development and an ability to respond to environmental changes, adds genetic assimilation to the list of mechanisms by which evolution can be said to be the control of development by ecology.

19.9 ENDNOTES

1. See Waddington (1943, 1956, 1957a,b) for the basic studies establishing genetic assimilation. Several of Waddington's papers are reprinted in his 1975 'autobiography', *The Evolution of an Evolutionist*. His ideas are discussed by Polikoff (1981), Yoxen (1986), Hall (1992) and Gilbert (1991b, 1994b).
2. See Hadorn (1961) for coincidence of developmental periods, and Goldschmidt (1958) for a summary of different phenocopies evoked by differing intensities of the same environmental stimulus. Heat shock, cold shock, X-rays, and other environmental perturbations produce similar effects.
3. Waddington is clearly coming out against macromutation here; see the discussion in section 13.1.

4. For discussion of variation in threshold characters, see Kindred (1967), Rendel (1967) and Scharloo (1987).
5. For environmental determination of phenotype, see the quantitative genetics models for complex morphological change in development and evolution of Atchley and Hall (1991) outlined in Chapter 20; Boake (1994) in the context of behavioural evolution; and Maze *et al.* (1992) and Maze and Vyse (1993) as applied to Douglas fir and Engelmann spruce.
6. Schmalhausen's two books on autonomization written in Russian were published in 1938 and 1942. His ideas are discussed by Dobzhansky (1949), Adams (1980), Allen (1991), the papers in Adams (1994) and D. B. Wake (1996a). For further discussion of why Dobzhansky favoured Schmalhausen, see Gilbert (1994b). The quote from Dobzhansky is taken from Waddington (1975, p. 98).
7. For stabilizing selection, see the preface in Schmalhausen (1949) and Grant (1963).
8. See ss. 11.3.1, 21.3 and 21.7. for discussions of lens development, multiple ways of eliciting a morphogenetic response, scleral and mandibular membrane induction by heterotypic epithelia and otic capsule cartilage inductions in amphibians.
9. Genetic assimilation has been confirmed in studies by Bateman (1959a,b), Rendel (1968), Capdevila and Garcia-Bellido (1974), Thompson and Thoday (1975) and Ho *et al.* (1983). See Hall (1985b) for a discussion of critical periods. As Ho (1984) emphasized, maternal cytoplasmic changes (s. 7.4.1) are another means of exposing pre-existing genetic variability to selection.
10. Piaget (1974) and Waddington (1975) discuss Piaget's experiments with *Limnaea*. See also the studies by Boycott (1938).
11. For reduction in egg volume of sea urchins, see Sinervo and McEdward (1988) and McEdward (1996). For correlation of such life history traits as timing of metamorphosis, size at metamorphosis and development rate, see McLaren (1965), Raff (1992) and McEdward and Janies (1997).
12. For studies on allometric engineering using lizard eggs, see Sinervo (1990, 1993), Sinervo and Huey (1990), Sinervo and Licht (1991) and Qualls and Shine (1996). For a recent synthesis, see Sinervo and Basolo (1996).

13. Genotype by environment interactions also vary non-linearly with temperature, as demonstrated by Brakefield and Kesbeke (1997) in their studies on pre-adult growth in the tropical butterfly *Bicyclus anynana*.
14. For an introduction to these important studies on quantitative trait loci, see Long *et al.* (1995) and Mackay (1995, 1996). See Leamy *et al.* (1997) for quantitative trait loci and fluctuating asymmetry of murine mandibles. Endler (1995) also invoked epistasis and pleiotropic interactions when considering the developmental basis for correlation and coevolution of traits.
15. For the greater phenotypic variability of mutant genes and exposure of inherent genetic variability to selection, see Dun and Fraser (1959) and Fraser and Kindred (1960). For studies on the *tabby* mutant, see Johnson (1986).
16. For secondary cartilage, see Murray (1957, 1963), Hall (1978) and Fang and Hall (1995, 1996, 1997).
17. Paradoxically, smaller adults arise when growth is limited by temperature than when limited by availability of food. This may reflect a constraint arising from the increase in cell size that is a consequence of temperature affects on cell growth (McLaren, 1965; Atkinson and Sibly, 1997).

20

A QUANTITATIVE GENETICS MODEL FOR MORPHOLOGICAL CHANGE IN DEVELOPMENT AND EVOLUTION

'Developmental biologists and geneticists usually focus on different aspects of genes (translation versus transmission). The geneticist uses a particular view of genes as units of heredity (i.e. transmission to the next generation) and may neglect the role of genes in development. Consequently, the developmental biologist may ask *whether the distinction between genotype and phenotype advances genetics by leaving out development*. Does evolutionary genetics provide a sufficient theory of morphological evolution? The mapping function from genotype to phenotype is not one-to-one. A gene may affect multiple structures (pleiotropy) and traits are often affected by many genes (polygeny). Furthermore, the mapping of gene effects on phenotype may be nonlinear. Because gene action during development is a cyclic series of gene-cell interactions, genes are just one element in the developmental process. Thus the nature of interactions is the primary issue in development.'

Arnold et al., 1989, p. 406; emphasis added

This chapter brings together epigenetic and environmental influences and integrates them with genetic control of development in an approach used to great effect in quantitative genetics. Such an integrative and hierarchical approach is required because there is no one-to-one correspondence between genome and structure, or genotype and phenotype. Indeed, an evolutionary developmental biology is required because (perhaps only because) the genotype does not equal the phenotype.

The integration of evolution with development that existed a century ago came unstuck early this century. After the rediscovery of Mendel's work in 1900 the gene became 'the' particulate basis of evolutionary change, embryos and their development merely the means to the end – necessary but evolutionarily uninteresting links between genotype and adult phenotype. The inadequa-

cies of that separation have become increasingly apparent over the past decade, in part because of differences in the fundamental units of genetics and development, which are the topics of sections 20.1 and 20.2.[1]

Then I briefly discuss some models proposed for morphological change (s. 20.3) before introducing the model Bill Atchley and I developed using the dentary of the mammalian mandible (s. 20.4) and showing its application to butterfly eyespot patterns (s. 20.5).

20.1 FUNDAMENTAL UNITS IN GENETICS

Genetics has a cohesion as a subdiscipline of biology precisely because the fundamental unit of genetics, the gene, has been known since early this century, and because this unit has predictable properties that do not change from organism to organism, place to place, and time to time.

Although the gene is the fundamental genetic unit, to model genetic control of the development of a particular structure is not merely to understand how the genome of the organism is read out. Maternal, paternal and zygotic genomes each have to be taken into account – recall maternal cytoplasmic control of early development (s. 7.4.1) and the inter-related roles of maternal and zygotic genes in *Drosophila* segmentation (s. 9.2.2).

Many genes have pleiotropic actions; for example, selection for increased leg length in chickens produces changes in many other regions of the body. Amplification and inte-gration of gene effects as epigenetic pleiotropy is a fundamental aspect of how development is organized. Because change can occur in epigenetic interactions as a unit, and because pleiotropic changes can occur in genes that affect more than one tissue, Levinson saw epigenetics as best dealt with as a form of pleiotropy, and so distinguished genetic from epigenetic pleiotropy; Bill Atchley and I argue that epigenetics and pleiotropy should be treated separately.[2]

Epigenetic factors and environmental influ-ences that activate the genome, or maintain and/or transmit states of gene activity (such as methylation of DNA, chromatin structure, and peptides that modify transcription; s. 7.5), must be identified and incorporated into any model of morphological change in evolu-tion. The developmental quantitative genetics model required will have to be substantially more sophisticated, multifactorial and hierar-chical than the models used to analyse particulate inheritance.[3]

20.2 FUNDAMENTAL UNITS IN DEVELOPMENT

Development lacks cohesion, in part, because it cannot readily be reduced to fundamental units, and in part because our definitions (perhaps also our knowledge of processes such as epigenetics) are too vague to be incor-porated adequately into models. Evolutionary concepts such as selection, fitness, drift and migration are as complex as developmental processes, but are much more amenable to rigorous analysis because of agreement that they are fundamental units.

Development is hierarchical and epigenetic. Different 'units' are present at different times during ontogeny. The properties of units change through differentiation and morphogenesis. As outlined in Chapters 7–9, these units include cytoplasmic components in the egg, cells in the blastula, germ layers in the gastrula, tissues in the neurula, and so forth.

Stebbins (1968) identified three levels of developmental sequences: biosynthetic path-ways, informational relays and epigenetic sequences. Although fundamental develop-mental units are difficult to identify they must be identified. Atchley and Hall (1991) defined them as the basic structural entities and regu-latory phenomena necessary to assemble a complex morphological structure in ontogeny and alter it in phylogeny, and sought these developmental units for the dentary bone of the mammalian mandible.

20.3 MODELS FOR MORPHOLOGICAL CHANGE

A major thrust of this book is that genetics, development, epigenetics and environment must be integrated because they are causally linked in the production of embryos and adults in both ontogeny and phylogeny. We need to determine how genetic, epigenetic and environmental factors are integrated into a hierarchical set of unified controls. An under-standing of the genome alone, important as that is, will not provide the explanations we need. There is a need for:

- population analyses of developmental regulation and variability;
- quantitative genetic models of develop-ment and evolution;

- incorporation of epigenetic and environmental factors;
- treatment of development as hierarchical and integrative; and
- recognition of the effects of many genes as pleiotropic.

Although epigenetic regulation affects gene expression, epigenetics has both a genetic and an environmental component (ss. 18.2, 20.4.3, and Atchley and Hall, 1991). In proposing models for evolutionary change that incorporate epigenetic control of embryonic development, it is important to distinguish the heritable from the environmental components of epigenetic control, although, as Waddington demonstrated, the ability to respond to an environmental component is often heritable (Chapter 19). In mammals, where embryos develop within the body of the female, genetic, epigenetic and environmental effects that arise maternally have to be considered along with those that flow from the zygotic genome. The biological origin of the epigenetic effect has to be distinguished from its extrinsic mode of action. Phenotypic variability results from intrinsic genetic effects, heritable epigenetic effects and non-genetic environmental effects, some of which act epigenetically. Examples – cyclomorphosis, seasonal polymorphism, and inductive interactions between and among individuals, species and environmental factors – were discussed in sections 18.3–18.7.

In 1988 Sachs developed a model of epigenetic selection based on the premises that:

1. each developmental event is a consequence of previous events; and
2. variability of normal development allows for more than one way to achieve each developmental event.

Selection operates, says Sachs, to initiate one developmental programme over another through competition between cells and tissues for limited developmental signals used by different pathways. An interesting consequence of this model is that signal availability, rather than concentration, effectively becomes the factor for which cells compete. Buss (1987) developed a conceptually similar model, with levels of selection from cell → individual → population, and competition between selection at different levels (or between different cell lines) as the forces that established stable phenotypes as we know them today.

The canalization of development proposed by Waddington (s. 12.2), whereby one pathway of development is preferred over another, would seem to go against epigenetic selection, but under constant conditions, particular epigenetic pathways would be preferentially selected for, even though the potential for other pathways exists. Such a process would result in canalized development. That individual developmental processes need not be associated with the activity of specific genes, is essentially Waddington's concept of changing epigenetic programmes without changing every gene involved in the programme. Epigenetic selection and hierarchical organization both have the economy of minimal information for maximal developmental outcome.

20.4 MODULAR CONSTRUCTION OF THE MAMMALIAN DENTARY

Bill Atchley and I elaborated a model for complex morphological change in 1991, using the dentary of the mandibles in inbred strains of mice. The essential features of that model are presented in sections 20.4.1–20.4.4. [Atchley and Fitch undertook a comprehensive analysis of the genealology of inbred strains of mice; changes in these strains may now be related to robust phylogenies. No longer need inbred mouse strains only represent the model organisms discussed in section 8.1.] The generality of the model is illustrated by its applicability to craniofacial development in general (s. 20.4.4) and to the generation of eyespot patterns in butterfly wings (s. 20.5).[4]

Our model depends critically on two elements:

1. the ability to identify the **fundamental developmental units** which respond to selection to modify mandibular form, or indeed the form of any other structure (s. 20.4.2); and
2. the ability to identify the various genetic, epigenetic and environmental **factors** that interact to generate and modify mandibular morphology or the morphology of any other feature (s. 20.4.3).

First, I need to identify the structural components of the mammalian mandible.

20.4.1 Structural components of the mammalian mandible

The skeleton of the mammalian mandible is comprised of a single cartilage (Meckel's cartilage) and a single bone (the dentary). Our concern is with the dentary.

Although a single bone, the dentary is complex, consisting of four structural components that develop from six morphogenetic units (Figure 20.1). The four components are identified below on the basis of morphology and mode of ossification. The six units are identified in section 20.4.2 on the basis of lineages of cells and epigenetic factors that act upon those cells.

The four components are the ramus or body of the dentary (which develops by intramembranous ossification), and the condylar, coronoid and angular processes that develop by a combination of intramembranous and endochondral ossification and secondary chondrogenesis. The ramus and three processes form four of the six morphogenetic units. The other two are alveolar units of the incisor and molar dentition (Figure 20.1). The number of structural components and hence morphogenetic units, especially alveolar units (canine, premolar) associated with diversification of the dentition, varies across the mammals.

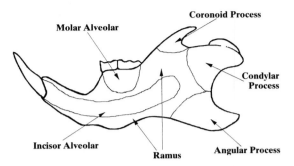

Fig. 20.1 A diagram of the dentary of the mouse mandible to illustrate the four structural components (ramus and coronoid, condylar and angular processes) and six morphogenetic units (ramal, incisor alveolar, molar alveolar, and coronoid, condylar and angular processes). See text for details. Figure courtesy of W. R. Atchley.

The two alveolar units develop from odontogenic cells that produce osteoblasts for alveolar bone and odontoblasts for dentine of the teeth (Figure 20.2). Experimental and evolutionary evidence substantiates alveolar units as separate populations of cells, but formation and maintenance of alveolar bone depends on the presence of the teeth. The ramal unit develops from a separate population of cells that forms only bone (Figure 20.2) and is not influenced by the teeth.[5]

The condylar, angular and coronoid processes develop from a third class of skeletogenic cells, which form membrane bone at the base of the processes and secondary cartilage at their tips (Figure 20.2). These units develop under the influences of muscles that insert onto them. With congenital absence of the appropriate muscle(s), particular processes will fail to form. Once established, continued stimulation is required to maintain the form of the processes, as we have demonstrated with an organ culture approach in which whole joints can be maintained and exposed to mechanical stimulation. Components of cartilage extracellular matrix are also adapted to particular functional requirements of individual cartilage types at different sites. Each unit – alveolar, ramal, bony processes –

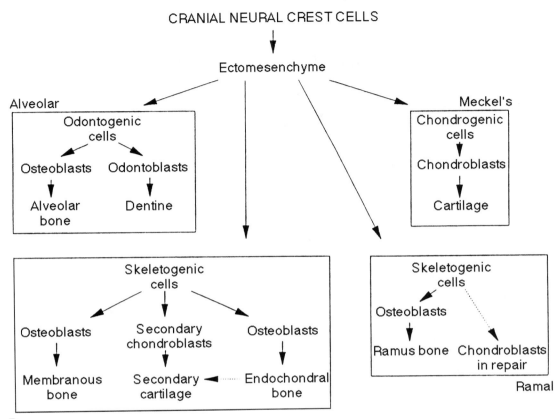

Fig. 20.2 The six morphogenetic units of the mouse dentary identified in Figure 20.1 along with Meckel's cartilage consist of separate cell lines all derived from the cranial neural crest through ectomesenchyme. Meckel's cartilage consists of a chondrogenic population that differentiates as chondroblasts. The alveolar units (incisor and molar; Figure 20.1) consist of an odontogenic population of cells that produces both osteoblasts for the alveolar bone and odontoblasts for dentine of the teeth (see M. M. Smith and Hall, 1990 for details). The three bony processes (coronoid, condylar and angular; Figure 20.1) consist of skeletogenic cells that form osteoblasts of both membrane and endochondral bone and the secondary chondroblasts that form the secondary cartilage of the three bony processes. The ramal unit consists of skeletogenic cells that form osteoblasts of the membrane bone of the ramus and chondroblasts during fracture repair. See text and Atchley and Hall (1991) for details. Figure courtesy of W. R. Atchley.

therefore has a distinctive cellular and epigenetic history.[6]

20.4.2 Fundamental developmental units – the first component of the model

These six morphogenetic units are the fundamental structural units that 'make' the dentary. Furthermore:

- each cell population in these units is distinctive and separate;
- parameters of these populations determine initial dentary morphology; and
- each population arises as an aggregation or **condensation** of cells.

Consequently, condensations of independent cell populations are the **fundamental cellular**

units that 'make' the mandibular skeleton and respond to selection to modify skeletal morphology. Scott Gilbert and his colleagues (1996) placed their emphasis on morphogenetic fields which may be comprised of more than one condensation of cells. For the reasons elaborated below, I believe that condensations form the more natural unit through which selection acts to modify morphology.

An enormous body of evidence demonstrates that condensations represent the initial phase of specific activation of cell populations in the development of many tissues and organs. The condensation process is a threshold because gene activity that is specific for the cell type that differentiates within a condensation, is upregulated at this stage.[7]

Grüneberg (1963) referred to the condensation stage as the 'membranous skeleton', and recognized that many mutants that affect skeletal development disrupt condensations. Differentiation does not commence until condensation has taken place and condensations have reached a critical size; skeletal elements fail to form, or are much smaller than normal, if condensation size is reduced below a critical mass. In mice carrying the gene *congenital hydrocephalus* (*ch*) chondrogenesis is not initiated in the tracheae because tracheal condensations fail to attain a critical mass. Such mass effects of condensations are not restricted to the skeleton or to vertebrates. *Drosophila* embryos with reduced numbers of precursor cells cannot pattern bristles, pigment, or denticle bands. A model with condensations as the fundamental cellular unit is equally applicable to vertebrate skeletal development, *Drosophila* bristles, butterfly eyespots (s. 20.5) and the form of trees.[8]

Having determined the cellular units of mandibular morphology, the challenge is to determine how those units are altered when mandibular morphology changes through artificial selection, through inbreeding, or through evolutionary change.

Parameters of the cellular units are identified as **developmental units**:

1. the number of stem cells (n);
2. the time of initiation of the condensation during development (t);
3. the fraction of cells that is mitotically active (f);
4. their rate of cell division (r); and
5. the rate of cell death (d).

Variation in these five units provides the basis for ontogenetic and phylogenetic modification of mandibular morphology; these are the fundamental development and evolutionary units of morphological change. Other aspects of these cell populations, such as rates of synthesis and deposition of extracellular matrices, affect later aspects of mandibular morphology.

20.4.3 Causal factors: the second component of the model

Mandibular morphology is heritable and modified by natural and artificial selection. For example, just 18 months after house mice were introduced to the Isle of May in Scotland, hybrids between house and feral mice already displayed modified mandibular morphology, 57% of the variation of which was due to genetic differences. Differences in mandibular morphology between inbred strains of mice can also be detected. The greatest differences are often found early in postnatal life, e.g. 15 days postnatal for differences between C3H and C57 mice.[9]

Mandibular morphology is not the simple result of the growth and morphogenetic parameters of a single population of cells developing in isolation. Rather it is the exquisite integration of these different cell populations, their activation by other tissues and developing organs such as muscles and teeth, and the influence of external factors such as litter size and uterine physiology. Uterine maternal influences are not transitory, restricted to early development, or swamped by intrinsic genomic control. Uterine maternal effects extend well into adult life.[10]

Therefore, the second component of the model is the identification of the **causal factors** that impinge on cell populations. Five causal factors contribute to the developmental variability upon which selection acts:

1. intrinsic genetic factors;
2. epigenetic factors;
3. genetic maternal factors;
4. environmental factors; and
5. genotype by environment interactions.

(a) Intrinsic genetic factors

These include genes that act within the skeletal cells and genes that act in cells of other tissues (muscles, nerves, vascular system), which in turn influence skeletal development. Evidence that specific genes control the development of specific cell lineages in the dentary is now available from studies in which individual genes have been deleted from mouse embryos. Six specific examples are mice deficient in *Msx-1*, *goosecoid*, *Dlx-1* and *-2*, TGFβ-2, lymphoid enhancer-binding factor-1 (*Lef-1*) and a zinc finger gene *Gli2* regulated by *hedgehog*.

Teeth fail to develop in *Msx-1*-deficient, *Lef-1* deficient and *Gli2* homozygote mice (Figure 20.3). In *Msx-1*-deficient mice, the tooth germ for the incisor is absent, while the molar teeth are present as germs but do not progress beyond the bud stage. Other aspects of the dentary involving other cell lineages (size and shape of the ramus and bony processes) are normal. These results are consistent with *Msx-1* acting selectively within the alveolar cell lineage. *Lef-1* regulates the epithelial-mesenchymal interaction that initiates transformation of the dental mesenchyme into a dental papilla.

No maxillary molar teeth develop in mice carrying a double mutation of *Dlx-1* and *Dlx-2*; they are replaced by a tissue consisting of a mix of ectopic cartilage nodules, bone and connective tissue. All other teeth (maxillary incisors, mandibular incisors and molars) are

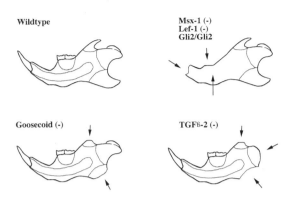

Fig. 20.3 Diagrammatic representations of the dentary bone of wild type mice and from mice in which deletion of specific genes eliminates or reduces specific cell lineages in one or more of the morphogenetic units (angular, coronoid, condylar, incisor, molar). The alveolar units I and M (including the extensive root of the incisor) fail to develop in mice deficient in *Msx*, Lef-1 or Gli2 (arrows in dentary at top right). Coronoid and angular processes are reduced in size in *Goosecoid⁻* mice (arrows, bottom left), while coronoid and condylar processes are reduced and the angular process fails to develop in TGFβ-2⁻ mice (bottom right).

normal in development and size. *Dlx-1* and *-2* therefore act in only one of the alveolar populations of cells, suggesting fine-grained control within classes of fundamental units.

In contrast, *goosecoid*-deficient mice have normal teeth and alveolar units, a normal condylar process, but small coronoid and angular processes, a phenotype consistent with *goosecoid* acting selectively in two of the bony processes or on the environmental signals that act upon those processes (Figure 20.3). TGFβ-2-null mice have normal alveolar units, but the angular process is missing and coronoid and condlyar processes only half their normal size (Figure 20.3).

The anticipation is that other genes that act selectively within cell lineages will be identified and a model of overlapping domains of gene activity targeting specific cell populations will emerge. Indeed the elements of such a model have been presented by

Akimenko *et al.* (1994) and Qiu *et al.* (1997) from their studies on *Dlx* in zebrafish and mice, and by Reecy *et al.* (1997) for members of the NK homeobox transcription factors in chicks.[11]

To model whether selection acts directly on such units or is a correlated response to selection on the phenotype, Nijhout and Paulsen (1997) propose a population genetics model of development that treats each developmental unit as controlled by a single gene. Although in their model (as in Slatkin's 1987 model), selection does not act directly on the developmental units, the authors emphasize that genetic background plays a major role. Whether the same conclusions would emerge from models in which each developmental unit was controlled by multiple genes with pleiotropic or epistatic effects, remains to be determined.

(b) Epigenetic factors

Epigenetic factors act indirectly, often inductively. They include local epigenetic interactions (such as those between the muscular and skeletal systems) and encompass global epigenetic effects (such as the actions of hormones, neural, vascular and metabolic factors). A vast literature documents the roles of such epigenetic factors in ontogeny and phylogeny, much of it discussed in this book.

(c) Maternal genetic effects

Genetic maternal effects come into play because the mammalian embryo develops within the uterus and therefore is influenced by products of the maternal genotype. Both prenatal genetic maternal effects (uterine size, litter size, maternal metabolism and placental physiology) and postnatal genetic maternal effects (nutrition and maternal behaviour) play a role. Although maternal effects are genetic, they involve a genome other than the embryo's, and so are a special type of epigenetic effect, much as in cyclomorphosis

(s. 18.3). In mammals, the genes of the mother essentially condition the expression of the genes of the embryos.

Embryo transfer between females of differing genotypes provides a practical way of determining the relative importance of uterine genetic maternal effects. David Cowley's embryo-transfer experiments with mice of different genetic backgrounds demonstrated substantial prenatal maternal effects on pregnancy rate, embryo survival, parameters of embryonic growth (such as effects on body weight evident as late as ten weeks after birth, or on tail length as late as six weeks after birth), and parameters of maturation (the age at which eyes and ears open). Permanent effects of uterine environment are seen in adult skeletal morphology. Craniofacial size attained is dependent on the strain into which the embryos are transferred, reflecting both female genotype and uterine size.[12]

(d) Environmental factors

These are non-heritable factors that influence morphology. Environmental factors include maternal diet, temperature, and crowding. Nongenetic factors, such as time of weaning, influence skeletal growth and often allow growth rates to be partitioned into early rates under genetic control and later rates subject to epigenetic regulation, as was seen demonstrated in Hallgrimsson's studies of fluctuating asymmetry discussed in section 12.3.1.[13]

(e) Genotype by environment interactions

Genotype by environment interactions come into play because the same environmental factor may have different effects depending on the genome with which it interacts. Cyclomorphosis and genetic assimilation are two examples already discussed. Chromatin structure, genomic imprinting, maternal and zygotic genotypes, and sex chromosomes are examples of different genotypes interacting with environmental factors (s. 7.5). In develop-

Path Analysis Model of Genetic and Epigenetic Effects

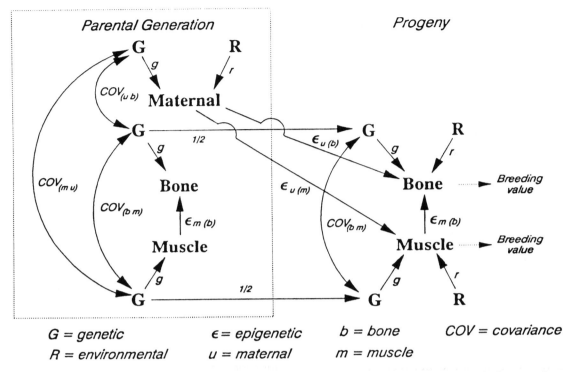

Fig. 20.4 The path analysis model of genetic and epigenetic effects on mandibular morphology to show how zygotic (progeny) and maternal genomes, and environmental and epigenetic factors all influence mandibular morphology. A one-way epigenetic interaction of muscle on bone morphology is shown. See Atchley and Hall (1991) for details. Figure courtesy of W. R. Atchley.

ing their recent population genetics theory of canalization, Wagner, Booth and Bagheri-Chaichian (1997) emphasize that genotype by environment interactions and epistasis are the two major means by which organisms react to environmental and genetic perturbations in ways that are channelled by their genotypes.[14]

20.4.4 Interactions: the third component of the model

Because of epigenetic interactions and pleiotropy, relationships between the five causal factors are neither simple nor linear. Atchley uses path analysis and a multivariate selection model to formalize the interactions between these five factors to show how causal factors act on the developmental units in each of the morphogenetic units to effect change in morphology. Figure 20.4 illustrates the complexity of the interactions between the five causal factors. Zygotic and maternal genomes, the zygotic genome differentially expressed in muscle and bone, maternal, epigenetic and environmental factors, all contribute to the morphology of the dentary as these causal factors operate on the developmental units of time, amount, rate and co-ordination of cellular condensations.[15]

With such models we can place the basis of morphological change as selection acting through integrated genetic, epigenetic and environmental factors to modify ontogenetic processes. Selection results in a change in one or more of the developmental units in one or more of the cell populations that generate the morphological units from which the dentary is constructed. Therefore, it is possible for 'identical' morphological changes to come about through the same selection pressure modifying different developmental units. We saw this for genetic assimilation.

In an insightful analysis of the complexity of developmental control, Miriam Zelditch and her colleagues (1993) addressed selection pressure acting on different developmental units treated as synapomorphies (the shared features that unite taxa but often form using different developmental processes).[16] Using the hierarchical approach outlined in the previous sections, Zelditch *et al.* concluded that different parts of the skull (morphogenetic units), developmental processes that form those parts (fundamental developmental units), and the genetic, epigenetic, environmental, and ecological processes (causal factors) regulating variability in shape, may each represent a synapomorphy at a different cladistic level. Their studies on the ontogeny of skull form in the cotton rat offer a major rationale for the distinctive approach of evolutionary developmental biology. To use their words for how we should proceed to determine how phylogenetic change is shaped by ontogeny and developmental processes:

'Cladistic studies of developmental regulation, ontogenetic processes, and life-history characters in more ecologically diversified genera could examine the congruence and relative lability of these characters.'
Zelditch, Bookstein and Lundrigan, 1993, p. 639

A further example of the applicability of the model is considered below in the exciting work by Paul Brakefield and his colleagues (1996) on development and evolution of butterfly eyespot patterns.

20.5 MODULAR CONSTRUCTION OF BUTTERFLY EYESPOT PATTERNS

The connection between development and evolution of mammalian mandibles, cotton rat skulls and eyespot patterns in butterfly wings is **modularity**.

The colour patterns of butterfly wings in all their infinite variety can be reduced to semi-independent pattern elements, each acting as a signalling source much as a cell condensation or an embryonic inducer acts as a locus. Our understanding of how these patterns develop and evolve is the result of an international and interdisciplinary collaboration that brought molecular genetics, selection, developmental and evolutionary approaches to bear on patterns that are expressed by some 15 000 species of lepidopterans.[17]

The investigators did not attempt to treat all aspects of butterfly wing pattern but rather focused on the eyespot, perhaps the most visually prominent element of butterfly wings and one often duplicated by mimics. The contribution of Brakefield and colleagues is in defining the developmental pathway controlling formation of eyespots, showing how the pathway is modulated in species with divergent eyespot patterns (or in which the eyespot is absent), and demonstrating how seasonal polymorphism for eyespot is regulated.

Eyespots develop through the activity of an inductive centre – the focus – which is the source of a diffusible molecule. Consistent with its patterning role in other systems, the homeobox gene *Dll* is expressed in the focus. A pathway with four stages specifies eyespot position, number, size and colour in the African butterfly *Bicyclus anynana*:

- Stage 1. *Dll* is broadly expressed over the wing imaginal disc late in larval life.
- Stage 2. Expression of *Dll* is restricted to the eyespot focus. The genes *Cyclops* and *Spotty* are required to specify number and position(s) of eyespots.
- Stage 3. Expression of a gene *Bigeye* is activated in the focus early in pupal life. The

gene product diffuses from the focus to control eyespot size.

- Stage 4. Eyespot colour is determined by genes within the eyespot rather than from signals from the focus.

Recognizing that these four stages, with their independent genetic controls, represent points where development could diverge and so produce new patterns (as do the developmental units identified for the dentary in the previous sections), Brakefield and colleagues examined *Dll* expression in species without eyespots, or in which the number of eyespots differed from *B. anynana*. *Precis coenia*, with two eyespots per wing, diverged at stage 2. The monarch butterfly, *Danaus plexippus*, which lacks eyespots, diverged at stage 1, leading to the expectation that a combination of control over these two stages may well regulate presence or absence of eyespots and eyespot number in all butterflies.

Further analysis was of seasonal polymorphism in *B. anynana*, in which eyespots are more prominent and more colourful in the wet than in the dry season, an adaptation to the brown leaf litter on which butterflies are found in the dry season. A norm of reaction relates the temperature at which larvae are reared to wing phenotype. Wet season forms develop at or above 23°C, dry season morphs at 17°C. The sensitive stage at either temperature regime is late-larval, when eyespots are initiated in the wing discs. From artificial selection experiments five or six genes were estimated to regulate eyespot size.

Analysis of *Dll* expression in selected lines demonstrated a common developmental basis for determination of eyespot size, whether under genetic (as in the selection experiments) or environmental control (as in the seasonal morphs). Seasonal polymorphism is mediated by ecdysone hormones and not by a direct effect on pigment precursors of either temperature or photoperiod.[18]

Eyespots vary independently of other aspects of wing patterning. Brakefield *et al.* (1996) therefore concluded that differential regulation of the developmental pathway was mediated by *Dll* acting on the eyespot focus. Developmental modularity is coupled to phenotypic plasticity and the ability to respond rapidly to selection.

Some important consequences for homology of the fact that identical morphological changes can arise through the modification of different developmental processes and units are considered in Chapter 21 in the context of a discussion of the evolution of developmental processes, which is the topic of Part Six.

20.6 ENDNOTES

1. See Mayr (1997a) for a short but pithy analysis of the role of genetics and the separation of genotype from phenotype. See Smocovitis (1996) for a detailed history of the development of the modern evolutionary synthesis. An authoritative scientific biography of Mendel has been produced by Vitezslav Orel, emeritus director of the Mendel Museum in Brno (Orel, 1996).
2. See Lerner (1954), Levinton (1988) and Atchley and Hall (1991) for distinctions between genetics, epigenetics and pleiotropy.
3. See Chapter 7 and Maclean and Hall (1987), Locke (1990), Paro (1993), Jablonka and Lamb (1995) and Hall (1997c) for epigenetic inheritance. For quantitative genetic modelling of interactions over ecological and evolutionary time scales, see Björklund (1996).
4. For how developmental studies relate to evolution, see Hall (1975, 1983a, 1984a,b, 1990b,c, 1994d). For Atchley's development of the mouse mandible as a model for developmental and evolutionary studies, see Atchley *et al.* (1985a,b, 1990, 1992, 1994), Riska *et al.* (1985), Atchley and Newman (1989), Atchley (1990), Cheverud *et al.* (1991), Cowley and Atchley (1992) and Vogl *et al.* (1994). For the genealogy of inbred strains, see Atchley and Fitch (1991, 1993), and see Sage *et al.* (1993) for mice as model organisms.
5. For the separation of odontogenic from osteoblastic cells, see Osborn and Price (1988), Lumsden (1987, 1988) and M. M. Smith and Hall (1990, 1993).

6. For mandibular differentiation and morphogenesis, see Hall (1982a–d, 1984c,d, 1994b,c). See Hall (1978, 1985a, 1986), Tran and Hall (1989), Hall and Herring (1990) and Bertram *et al.* (1997) for epigenetic influences of muscles on the skeleton, and Kantomaa and Hall (1988a,b) and Pirttiniemi and Kantomaa (1996) for an organ culture approach to maintenance of temporomandibular joint form in response to mechanical or electrical stimulation. See Daegling (1996) for documentation of regional differences in growth of mandibular components of the African apes *Pan troglodytes* and *Gorilla gorilla*, and see S. Roth *et al.* (1997) for adaptation of extracellular matrix components to functional demand.

7. For a sample of the vast literature on condensation in various tissues and organs, see Hall (1982b, 1988a, 1991c), Fyfe and Hall (1983), Nathanson (1989), Hall and Miyake (1992, 1995b), Dunlop and Hall (1995) and Miyake *et al.* (1996a,b, 1997a,b). For condensation as a prerequisite for differentiation and up-regulation of gene activity, see Kosher *et al.* (1986a,b), Kulyk *et al.* (1989a,b) and Hurle *et al.* (1989).

8. For the importance of condensation size as detected using vertebrate mutants, see Grüneberg and des Wickramaratne (1974), Grüneberg (1963), Hall (1982d) Hinchliffe and Johnson (1983) and Johnson (1986). For *Drosophila*, see Busturia and Lawrence (1994), and for studies on the form of trees, see Maze *et al.* (1992) and Maze and Vyse (1993).

9. See Scriven and Bauchau (1992) for hybrid mice on the Isle of May, and Vogl *et al.* (1994), Hall and Miyake (1995a) and Miyake *et al.* (1996a,b, 1997b) for differences in mandibular morphology between inbred strains of mice in pre- and postnatal ontogeny. For similar patterns in vertebral morphology and an approach to metameric segmentation as reflected in the vertebral columns of mouse strains and F1 hybrids between strains, see Johnson and O'Higgins (1994).

10. For heritability of mandibular morphology, see Atchley (1993) and Atchley *et al.* (1985a,b). For persistence of uterine maternal effects, see Atchley *et al.* (1991).

11. For the *Msx-1*, *goosecoid*, *Dlx*, TGFβ-2, *Lef-1* and *GLi2* studies, see Satokata and Maas (1994), Qiu *et al.* (1995, 1997), Rivera-Pérez *et al.* (1995), Chen *et al.* (1996), Kratochwil *et al.* (1996), Mo *et al.* (1997) and Sanford *et al.* (1997).

12. For embryo transfer studies, see Roth and Klein (1986), Cowley *et al.* (1989), Shea *et al.* (1990), Atchley *et al.* (1991), Cowley (1991a,b), Cowley and Atchley (1992), Shea (1992), Nomaka *et al.* (1993) and Sasaki *et al.* (1994, 1995).

13. For non-heritable environmental factors, see Hall (1984c), Atchley and Hall (1991), Atchley *et al.* (1994) and Helm and German (1996).

14. Atchley and Hall (1991), Via and Lande (1985), Pigliucci *et al.* (1996) and Wagner *et al.* (1997) contain detailed evaluations of genotype by environment interactions. Thompson (1991) discusses them in relation to phenotypic plasticity and as a buffer to selection.

15. For studies using path analysis or multivariate selection models, see Wright (1934, 1968), Lande (1979), Arnold and Wade (1984), Atchley (1987) and Atchley and Newman (1989).

16. For the position that synapomorphies are homologues, see Patterson (1982), Rieppel (1992a), and others reviewed in the chapters in Hall (1994a).

17. See Brakefield *et al.* (1996) and the accompanying commentary by Nijhout (1996) for these studies in butterfly eyespot patterns. See Nijhout (1991) and French (1997) for an overview of the genetic control of pattern formation in butterfly wings.

18. See Koch (1992) for hormonal rather than temperature mediation of seasonal polymorphism and Brakefield and Kesbeke (1997) for evidence that genotype by environment interactions in the tropical butterfly *Bicyclus anynana* are more pronounced in low- than in high-temperature environments.

PART SIX

DEVELOPMENT EVOLVES

21

DEVELOPMENT EVOLVES: THE DILEMMA FOR HOMOLOGY

'It has become increasingly clear from researchers in embryology that the processes whereby the structures are formed are as important as the structures themselves from the point of view of evolutionary morphology and homology.'

de Beer, 1958, p. 163

'Among evolutionary biologists, homology has a firm reputation as an elusive concept.'

Wagner, 1989b, p. 51

The catchphrase 'development evolves' should be posted over the doorway of every laboratory in which comparative biology is taught or researched.

The fact that selection can evoke different genetic pathways to produce the same phenotype has profound implications for homology. Structures that we confidently regard as homologous in adults can arise from non-homologous developmental processes. A beautiful example is initiation of lens development with and without the inductive involvement of the optic cup (s. 11.3). Another is selection for increased tail length in a highly inbred, genetically homogenous strain of mice. In just seven generations of selection, several lines were produced in which tail length had increased to the same extent. Because the mice were highly inbred and exposed to the same selection regime, the presumption was that the developmental basis for increased tail length would be the same in all lines. In one line, however, the tails had fewer and longer vertebrae than the unselected line. In another, the tails had more and shorter vertebrae. In the latter, increased tail length came about through the addition of new vertebrae; in the former through increased growth of pre-existing vertebrae. As summarized by the authors:

'The responses for replicate one (fewer but longer vertebrae) infer that tail length was increased primarily by elongation of vertebrae, whereas the responses for replicate two (more but shorter vertebrae) infer that tail length was increased primarily by an increase in vertebral number.'

Rutledge, Eisen and Legates, 1974, p. 28

The developmental processes producing increased tail length are profoundly different in these two lines; early re-specification of basic segmentation in the former, expanded growth of elements already present in the latter.[1]

Are the tails and/or vertebrae in these selected lines homologous, either with one another, or with tails and vertebrae in the unselected parental line? I say yes, as does Mary McKitrick (1994), who has also considered this example.

If not regarded as homologous (because the developmental processes that produced them are not the same), then at what level of the biological hierarchy should homology be assigned – to the final structural pattern, or to the developmental processes that produced the pattern? This chapter seeks the answer by concentrating on those aspects of homology that relate to development, essentially the use of developmental criteria as the basis for homology.

Answering such questions is especially difficult when homology can be studied at all levels of organization from molecular → genetic → cellular → tissue → organ → developmental → population → community → behavioural. This is why 'the hierarchical basis of comparative biology' was selected as the subtitle for a recent treatment of homology at all these levels (Hall, 1994a).[2]

Before proceeding, let me recap what was said in the preceding chapters in which homology was an important and recurring theme.

21.1 HOMOLOGY: A RECAP

Geoffroy and other idealistic morphologists used the concept of homology for adult structures in the same relative positions in different organisms and with the same connections to other elements, a typological view of homology that emphasized morphology and unity of types (s. 4.3).

The paper by Meyranx and Laurencet that started the great *Académie* debate sought to homologize each of the organs of vertebrates and cephalopods and so placed homology squarely in the centre of the controversy over animal classification (s. 4.7). Today, homology has resurfaced in systematics, where homology is equated with synapomorphy and where, for some, ontogeny plays a key role (Box 21.1). Primitive shared characters (symplesiomorphies) may be homologous also, which is a complication. McKitrick (1994) has argued from her studies on presence or absence of flexor muscles in avian hindlimbs, that synapomorphies are hypotheses of homology rather than homologues.

Homology as we know it was defined by Richard Owen as the same organ in different animals under every variety of form and function, and by Mayr as a feature in two or more taxa that could be traced back to the same feature in a common ancestor (ss. 5.3, 5.4), reflecting pre- and post-Darwinian views respectively. The post-Darwinian view added an evolutionary inference to homology as morphological equivalence. As Gavin de Beer emphasized, 'Darwin's bombshell of evolution, which burst in 1859, had a profound effect on the concept of the explanation of homology, but without touching the criteria by which it was established' (1971, p. 4). Nevertheless, homology, even if still determined on the basis of equivalent structures, could now be used to identify ancestors.

Owen regarded embryological characters as either irrelevant or, if relevant, subordinate to adult characters. Thus, homology

'. . . is mainly, if not wholly, determined by the relative position and connection of the parts, and may exist independently of . . . similarity of development.'

Owen, 1846, p. 174

Darwin embraced embryology for its relevance to homology and to archetypes, and as a major source of evidence for evolution (s. 5.4). Woodger equated homologous structures with the morphological correspondence in adults that enabled *Baupläne* to be identified (Chapter 6).

The 19th-century discoveries in embryology (notably commonality of germ layers and conservation of embryonic stages), and their expression in the theories of Haeckel and von Baer, brought embryonic development into the mainstream of studies on homology. Now there were three components to homology:

1. equivalence of adult structure;
2. sharing of a common ancestor; and
3. sharing of common developmental origins.

The last was pre-eminent in the latter half of the 19th century. In fact, the coming of the

BOX 21.1
ONTOGENY AND SYSTEMATICS

For some, ontogeny has been linked to phylogeny in cladistic analyses of systematic relationships between organisms since Nelson (1973, 1978) argued that ontogenetic series (character transformations) provide a valid means of ordering organisms. If, as some advocate, more general (more widespread) characters are primitive (ancestral) and less general characters advanced, then we have a modern application of von Baer's Biogenetic Law in phylogenetic systematics.

Ontogeny is used in at least four ways in systematics:

1. to establish the polarity of characters;
2. to determine homology between characters;
3. to order characters; and
4. as a character(s) in establishing cladograms.

Excluded here is any detailed discussion of the use of homology in systematics and classification. The interested reader should consult the works of Nelson, Patterson, Kluge, Strauss, Williams, Rieppel and Bookstein. The context of the paper by Williams, Scotland and Blackmore (1990) is especially interesting; the paper is the introduction to a symposium on 'Developmental Pathways and Evolution' organized by the Linnean Society of London. Williams and colleagues go to some lengths to argue that because the ontogenetic criterion is based on parsimony and not on sequences from ontogeny, claims that it is based on recapitulation are unfounded. Not all agree. De Queiroz, Rieppel and others marshalled arguments against the use of ontogenetic criteria when assessing homologous characters for systematic analysis. In constructing their phylogeny of gastropod molluscs, Ponder and Lindberg (1997) remind us that mistakes in determining homology of ontogenetic characters used to infer relationships can have drastic consequences, including reversing entire phylogenies![3]

'embryological criterion' of homology heralded the birth of evolutionary embryology, as European and British embryologists began their search for ancestors in embryos (s. 5.5).

E. B. Wilson (whose views on epigenetics were discussed in section 7.2), typified the rise and fall of the embryological approach to homology. In 1891 he established two types of homology:

1. **complete homology**, in which homologous adult structures arise from a common embryological origin; and

2. **incomplete homology**, in which homologous adult structures have different embryological origins.

Three years later Wilson did an about-face, perhaps in large part because the embryological criterion was so tightly coupled to recapitulation and germ-layer theory, and because of the paucity of data on comparative embryology. He now maintained that comparative morphology, not embryology, held the key to the identification of homologous structures.[4]

The three components of homology listed above are much too broad a series of topics to cover in this volume. By way of summary and to indicate the compass of homology as viewed today, I paraphrase the questions posed in the introduction to the 1994 volume on homology (the answers are in the endnote):[5]

1. Do we possess a single concept of homology applicable to all elements and all levels in the hierarchy of biological complexity?
2. How should such definitions relate to one another if homology is defined differently at different levels in the biological hierarchy?
3. Should the definition(s) incorporate criteria to recognize homology, and/or provide explanations of mechanisms of homology?
4. Should homology encompass explanations of proximate causation (ontogeny) *and* explanations of ultimate causation (phylogenetic change)?
5. Is homology always tied to evolutionary origins of the cells, organs, organisms, species, populations, communities, under consideration?
6. Is homology always linked to structure, even when behavioural and physiological characters are under investigation?
7. Must homologous features always share a common embryonic development, or can homologous features arise from non-homologous (non-equivalent) developmental processes?

In the remainder of this chapter I concentrate on question seven – can homologous features arise from non-equivalent developmental processes? Equivalent (homologous?) structures such as gametes can arise in different organisms from different germ layers (endoderm and mesoderm) and under different developmental control (s. 9.1.3). The same vertebrate tissues (cartilage and bone) can arise from different germ layers (ectoderm [neural crest] and mesoderm; ss. 10.4, 24.2). These two examples typify the problem of whether to assign homology as common de-

velopmental origin and process, or common final structure.

First I examine serial homology and regeneration as special cases, using as a basis my treatment of these topics in an analysis of homology and embryonic development published in 1995 (Hall, 1995a).

21.2 SERIAL HOMOLOGY

Richard Owen introduced serial (iterative) homology as the repetition of organs and segments within an individual. Segments in vertebrae, insects and annelids, forelimbs and hindlimbs, pectoral and pelvic fins, leaves and petals, have all been regarded as serial homologues. Each either shares common developmental processes with its serial homologues (segments), or represents a homologous transformation of the same fundamental structure (leaves and petals).

Serial homologues that establish the basic parts of animals (segments in insects, neuromeres in vertebrates) share a serial organization and expression of homeotic genes, but even here, examination of additional taxa is beginning to reveal the variability characteristic of biological processes (s. 9.3). Equivalence or lack of equivalence between larval and adult body parts creates some interesting problems in the analysis of serial homology (Cowley, 1991a).[6]

Brian Goodwin has argued for some time that while a set of features can be homologized, individual elements within the set cannot (Goodwin, 1984, 1989). Is it possible to homologize segments, vertebrae, neuromeres, leaves or petals, without accurate identification of individual repeated units? Only if segment number 10 could be identified in different individuals or species of annelids could we homologize segment 10 across those individuals or species. Goodwin would say, homologize the set, not the individual elements. Woodger said the same. With such an approach serial homology becomes a statement about meristic variables that are homologous collectively but not individually. A similar approach applies to paralogous genes that are duplicated within a species. The fact that the genes are duplicated necessitates that all members of the set are homologues.

Serial homologues fail to meet one fundamental criterion of homology. If homology has a historical (evolutionary, phylogenetic) component, then structures cannot be homologous with structures in the same individual, only with a structure in another, related, organism.

21.3 REGENERATION

The question of serial homology raises issues of the homology of structures that are lost and then partially or totally replaced through regeneration or regulation. To quote Richard Goss's analysis of the evolution of regeneration: 'when a given structure can regenerate, its serial homologs can do likewise' (1992, p. 240).[7] Three issues need to be considered:

1. **regeneration** of an adult structure using different developmental pathways from those that formed the structure in the embryo;
2. **regulation** after removal of an embryonic structure by different cells or processes than formed the initial structure; and
3. **replacement** of a structure by a different structure, as occurs when an antenna regenerates in place of an eye in crustaceans or a leg in place of an antenna in homoeotic transformations in *Drosophila*.

Take as an example an annelid worm that loses 20 segments to a predator or zealous researcher:

- If regeneration replaces 18 of these segments, are the 18 homologous to the 20 that were removed, to 18 of the 20, or only to the set of segments removed. If only to 18, then to which 18?
- What if the number of segments that regenerate is variable, some animals regenerating 15, some 16, some 17, and so on? Serial

homology is not helpful at the level of individual segments.

- What if the segments regenerate by different developmental processes, some by budding from a blastema at the posterior extremity of the animal, others from intercalation between existing segments?
- What if some individuals regenerate segments entirely by terminal addition, others entirely by intercalation, and still others by a mix of the two? Are we now unable to homologize segments among regenerates when we could homologize the original segments?

Clearly, in such cases, homology exists neither at the level of individual segments, nor at the level of developmental mechanisms. All that can be said is that segments as a class of repeated structures are homologous.

Perhaps I have unnecessarily biased the relationship between regeneration and homology by choosing an example involving serial homology and serial repetition. Examination of regeneration in lizard and axolotl tails might provide a better example. The developmental processes by which tails regenerate are not those that produce the tail during embryonic development. Regeneration involves the following series of steps:

1. covering the wound with an epithelial outgrowth during the first few days following amputation;
2. innervation of the epithelial cap over the first week or so;
3. dedifferentiation of differentiated cells such as chondrocytes, osteoblasts and fibroblasts at the stump, under the control of factors released by the nerves in the epithelial cap;
4. proliferation of the dedifferentiated cells and formation of a blastema of undifferentiated cells; and
5. redifferentiation of blastemal cells to produce the regenerate.

Differences between tail regeneration and tail development are considerable. Embryonic tails develop from undifferentiated, not from de- and redifferentiated cells, and do not require innervation; other aspects, such as dependence on epithelial–mesenchymal interactions (s. 10.5) are similar in regenerating and embryonic tails. If commonality of developmental processes is to be the criterion for homology of structures, then a regenerated tail is not homologous with the tail it replaces.

Furthermore, various proportions of the tail may be regenerated. Consequently, a population may consist of a mix of individuals, some with their original tails, some with tails that have regenerated completely, others with various proportions of the tail replaced by regeneration. To use process rather than structure as the basis for homology does not serve us well in such cases.

A final example reinforces this point and reintroduces a familiar friend, the lens of the eye, which can regenerate in axolotls. Regeneration is not from residual lens tissue (the lens is removed completely) but from iris cells at the margin of the lens. Iris cells dedifferentiate, cease producing iris proteins, divide, synthesize and accumulate lens proteins, redifferentiate as lens cells, and finally produce a new lens functionally integrated into the eye. This series of regenerative processes differs fundamentally from those processes used to create lenses in embryos. Are regenerated lenses therefore not homologous with the lens they replace?

21.4 LATENT HOMOLOGY

'But no, I am mistaken; from the beginning of all things the Creator knew, that one day the inquisitive children of men would grope about after analogies and homologies, and that Christian naturalists would busy themselves with thinking out his Creative ideas; at any rate, in order to facilitate the discernment by the former that the opercular peduncle of the *Serpulae* is homologous with a branchial filament, He allowed it to make a *détour* in its development, and pass through the form of a barbate branchial filament.'

F. Müller, 1869, p. 114

Because their visceral arch skeletons are homologous, agnathans, although they lack jaws, possess homologous developmental precursors of jaws (s. 16.1). Here the argument for homology is based on developmental characters and knowledge of shared ancestors. Agnathans lack the structures (jaws) that are present in gnathostomes, but possess homologous skeletal elements (the visceral arch skeleton) from which jaws evolved (Figure 16.1). This is what de Beer (1971) called **latent homology**; the developmental programme is in the common ancestor, the structure is not. Other examples discussed or noted earlier are the homology of book-lungs and book-gills in terrestrial and aquatic scorpions, the hypothesized origin of insect wings from gills (s. 16.6.2), and of crustacean mouth parts from locomotory appendages (discussed below).

Over 125 years ago, the pioneer student of crustacean embryology Fritz Müller predicted that developmental change is at the base of structural change in crustaceans, and that homology and analogy would long persist as problem areas; see the epigraph for this section. He was referring to his studies on latent homology between crustacean mouth parts and paired limbs in their ancestors.

Crustacean mouth parts are modified walking legs. Crustacean ancestors show no evidence of such mouth parts but have the homologous paired limbs that are modified in their descendants to form mouth parts. Therefore, although the functions of feeding and locomotion are entirely different (recognition of homology is independent of function; s. 5.3), and even though the ancestral group lacks mouth parts, the appendages are regarded as homologous. Repression of expression of *Ubx* and *AbdA* in anterior trunk segments in association with development of mouthparts in representatives of nine crustacean orders, is consistent with homology of appendages and mouthparts (Averof and Patel, 1997).

Homology of body regions in crustaceans is complicated by studies of gene expression in the brine shrimp, *Artemia franciscana*. Expression boundaries of the homeobox genes *Ubx*, *abd*-A, *Abd*-B and *Antp* do not correspond to boundaries of major body regions in *Artemia* as they do in *Drosophila*. Nevertheless, Averof and Akam (1995) believe that these genes define homologous regions in crustaceans and insects. But taking such a position required that they redefine those homologous regions as head, trunk, genital and post-genital, rather than the traditional head, trunk, and abdomen. It remains to be determined whether we should redefine body regions in such a radical way, search for a deeper homology of body parts within the arthropods, regard homeotic genes as bounding regions of parallel functional adaptation, or await a currently unanticipated, new explanation. See section 21.7.3 for further discussion of these issues.[8]

21.5 ATAVISMS

Atavisms were introduced in section 17.5 as evidence for past evolutionary history. At the individual level, atavistic characters are recognized as the continuation, in some individuals, of a developmental process that is truncated in other individuals of the population or species. At the population level we recognize the re-establishment of an ancestral character as a normal feature in all members of a descendant as phylogenetic character reversal or taxic (phylogenetic) atavism.[9] Darwin was well aware of this:

'I have stated that the most probable hypothesis to account for the reappearance of very ancient characters, is – that there is a tendency in the young of each successive generation to produce the long-lost character, and that this tendency, for unknown causes, sometimes prevails.'

Darwin, 1859, p. 201

As seen in population-level variation in the urodele *Taricha granulosa* (s. 24.1.5), atavisms are backward-looking, reflecting past structural history, but can point to structures likely to be found in more derived taxa. Similarly, the variation in gastrulation in a single species of marine hydroid (*Dynamena pumila*) is identical to the invariant patterns of gastrulation in more derived coelenterates. Stiassny (1992) documented a variety of atavisms in species in a single family of fishes, the Cichlidae, including the relationship between elements of the lateral line system, ancestral character states of muscles and ligaments of the branchial apparatus, and absence of denticles on gill rakers.[10]

A further example, described by Reilly and Lauder (1988), is the use of an atavistic character to establish homology of the branchial arch elements in urodeles. The most medial branchial arch skeletal elements in urodeles are regarded as ceratobranchials and epibranchials. In lungfish, ray-finned and rhipidistian fishes, the most medial elements are regarded as hypobranchials and ceratobranchials, i.e. as not homologous to the urodele elements. The discovery of an atavistic epibranchial as a third branchial arch element in one population of *Notophthalmus viridescens*, is used to argue that urodele branchial arch skeletal elements are homologous with those in other vertebrates.

21.6 THE DILEMMA FOR HOMOLOGY

How can such diversity and variability be accounted for when homology typifies stability? De Beer (1971) accommodated serial homology by dismissing it as a misnomer; it does not trace organs to ancestors. In the crustacean mouth parts example, however, serially repeated structures can be traced to an ancestral condition. Where should the line be drawn? There are at least four ways out of this dilemma:

(1) Regard homology as the pattern expressed in adult structure, not the process by

which that pattern arises during development. This was essentially the position adopted by Geoffroy. It has the advantage that it is operational, no prior assumptions have to be made about ancestors, no knowledge of developmental mechanisms is required, and it is as readily applicable to fossils as to extant forms. Some botanists take this approach (Sattler, 1984). Because ancestry is not taken into account, this approach applies equally to homoplasy and to homology.

(2) Regard homology as pattern expressed in adult structures in organisms sharing a common ancestor – the evolutionary, phylogenetic or historical homology concept.

(3) Regard homology as hierarchical, extending across levels of biological organization (development to adult), through time (evolutionary sequences of common ancestors), and between individuals (equivalent features in different organisms). This was the position adopted by Osche and by Van Valen:

'Homology is resemblance caused by a continuity of information. In biology it is a unified developmental phenomenon.'

Van Valen, 1982, p. 305

'Homologies are non-random similarities of complex structures which are based on common genetic information (in the sense of instruction).'

Osche, 1982, p. 21, as translated by Haszprunar, 1992, p. 14 [11]

Obvious difficulties with these definitions are that 'resemblance' is far too vague and 'information' or 'instruction' far too broad. Homology so defined is, in my view, non-operational. Not all agree.

Haszprunar (1992), building on the definitions of Osche and Van Valen, proposed four levels of comparison and hence four types of homology:

- iterative (serial homology, homonomy), for comparisons within the same individual at the same time. Haszprunar extends this (unreasonably, in my view) to

the hairs of a mammal as homologues of one another;
- ontogenetic homology, for comparisons within the same individual at different times, including larva and adult;
- di- or polymorphic homology, for comparisons between different individuals of the same species; and
- supraspecific homology, for comparisons between different species or higher taxa.

These four types illustrate the difficulty of using continuity of information as the criterion for homology. Every conceivable comparison becomes a homology without regard to the level of the comparison or evolutionary relatedness of the organisms. Are the individual hairs on your head homologues of one another, or is hair as a tissue or organ system homologous across the mammals? Without inclusion of relatedness Haszprunar's fourth definition applies equally to analogy and to homology. Counter to the concept of continuity of information, Gans (1985) provided a thoughtful analysis of the different but overlapping classes of similarities in comparative anatomy. He considered similar structure, similar development and similar function, and their ability to vary independently.

(4) The fourth way out of the dilemma is to regard homology as process, specifically equivalent developmental processes at the tissue, cell and genetic level. This approach is the topic of the remainder of the chapter.

21.7 THE DILEMMA FOR HOMOLOGY AND DEVELOPMENT

In a somewhat testily worded review, Gavin de Beer dealt with the dilemma of homology and embryology, arguing that although later stages of development may be useful in determining homology (when organ rudiments and relationships to other organs are well established), earlier stages are not (de Beer, 1971). In sequence he dismissed common origins, com-

NEURAL PLATE   NEURAL KEEL

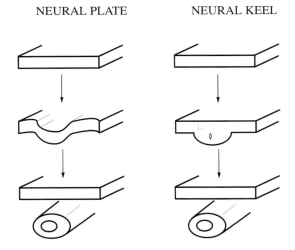

Fig. 21.1 On the left is shown formation of the neural tube by invagination of a neural plate, which is then covered by ectoderm. On the right is shown neural tube formation by cavitation, in which a cavity forms with a solid neural keel that is then covered by ectoderm, a situation seen in many teleosts and lampreys.

mon inductions, and a common genetic basis as criteria for homology.

21.7.1 Common origins

In dismissing common origins, de Beer cited the example of the vertebrate alimentary canal, which is regarded as homologous despite the fact that it forms from the roof of the embryonic gut cavity in sharks and teleosts, the floor of the cavity in lampreys and newts, the roof and floor in frogs, and from the hypoblast or lower layer of the embryonic disc in reptiles and birds. De Beer concluded: 'Therefore, correspondence between homologous structures cannot be pressed back to similarity of position of the cells of the embryo or the parts of the egg out of which these structures are ultimately differentiated' (1971, p. 3).

As an example of homologous structures arising from disparate developmental processes, we may cite the way in which the verte-

brate neural tube develops. The textbook story is of a flat neural plate rolling up at the edges as it sinks beneath the ectoderm to create neural folds, which approximate the midline dorsally and fuse to form the hollow neural tube. The cavity of the tube forms as the neural plate invaginates. This is true for vertebrates other than many teleosts and lampreys, in which a solid rod of cells sinks beneath the ectoderm before the cavity of the hollow neural tube forms (Figure 21.1). These two distinct developmental mechanisms of **invagination** and **cavitation** produce homologous neural tubes. Indeed, as we saw in section 10.5, the caudal end of the neural tube in avian and mammalian embryos develops by secondary neurulation from a tail bud, while the remainder of the neural tube forms by primary neurulation from the primary germ layers. None of us subdivides the neural tube into non-homologous elements because cranial and caudal regions develop differently.

21.7.2 Common inductions

In dismissing common inductive mechanisms, de Beer cites the now familiar example of induction of the lens by the optic cup in *Rana fusca* and lack of induction in the congeneric species *R. esculenta*, and concludes 'that homologous structures can owe their origin and stimulus to differentiate to different organizer-induction processes without forfeiting their homology' (1971, p. 13; and see s. 11.3). If we are prepared to regard lenses in congeneric species as homologous (as surely we must, for they fulfill the first and second criteria of common structure and shared ancestry outlined in section 21.1), then what of the third (developmental) criterion? Is it of lesser and subordinate importance to the other two, or is it irrelevant to homology?

Germ cells remain homologues despite being produced autonomously by transmission of maternal cytoplasmic constituents in the germ plasm, or conditionally following inductive interaction between germ layers (s. 9.1.3).

This is not the only example of homologous features arising through the fundamentally different mechanisms of cytoplasmic segregation and induction. The D quadrant cells in molluscan and annelid early embryogenesis are specified either:

1. by cytoplasmic determinants shunted from the vegetal pole to one blastomere at the 4-cell stage, a consequence of unequal first and second cleavages; or
2. by contact-mediated induction later in embryogenesis between a vegetal pole and an animal pole blastomere.

In an analysis of many species, Freeman and Lundelius (1992) determined that cytoplasmic specification was characteristic of derived taxa with direct development. In ancestral taxa, specification was by induction. Developmental mechanisms producing homologous cells have evolved, a theme that will recur.

21.7.3 Common genetic basis

In dismissing the requirement for a common genetic basis for homologous phenotypes, de Beer cited phenocopies, homeotic mutants and the *Drosophila eyeless* gene. Recent studies on eye development throughout the animal kingdom have revealed some fascinating aspects of homology that bring issues of homology and development into perspective. These, and similar studies on brain development in invertebrates and vertebrates are now discussed.

(a) Development of vertebrate and invertebrate eyes

Drosophila with the *eyeless* (*ey*) mutation lack eyes, but extensive inbreeding of homozygous *eyeless* individuals leads to the re-appearance of eyes. Other genes compensate for the deleterious effects of the *eyeless* mutant to produce eyes that are homologous with those in any other *Drosophila*, just as genetic assimilation evokes homologues using different ge-

netic pathways. Parallel mutations at homologous loci and the independent evolution of features such as industrial melanism in different species using different genes, are further examples of the production of homologous structures from different genetic pathways.[12] De Beer concluded:

'. . . that characters controlled by identical genes are not necessarily homologous . . . Therefore, homologous structures need not be controlled by identical genes, and homology of phenotypes does not imply similarity of genotypes.'

De Beer, 1971, p. 15

There is now more than meets the eye to the story of eye development.

The eyes of invertebrates and of vertebrates have long been considered analogous structures – parallel solutions to the problem of light reception. All that we know about the structure, development and cellular architecture of eyes across the Metazoa is consistent with this interpretation.

A complication arises, however, when we consider the genes that regulate eye development. *Eyeless* in *Drosophila*, its vertebrate homologue *Pax-6*, and human homologue *Aniridia*, function as the master controlling genes for eye development. Similarity or equivalence of the genes (not homology; when the phrases 'homology' and 'percent homology' are used in molecular biology for the amount of overlap between the sequences of bases in DNA or amino acids in proteins, similarity and degree of similarity, not homology, are being compared) is supported by:

1. extensive sequence similarity (94% in the paired domain, 90% in the homeodomain);
2. conserved positions of three intron splice sites;
3. similar patterns of expression (in both developing nervous systems and eyes); and
4. that loss-of-function mutations in *Drosophila* and the mouse reduce or eliminate the eyes.

Small eye (*Sey*) is a mutation of *Pax-6* in mice. Neither lens nor nasal placodes develop in homozygous *Sey* mice, indicating that *Pax-6* acts at early stages in the development of the eye and other sense organs.

Georg Halder and his colleagues (1995) obtained dramatic confirmation that *ey* in *Drosophila* (and by implication its homologues in other organisms) functions as a phylogenetically conserved master control gene for eye development. Targeted expression of *ey* cDNA in *Drosophila* imaginal discs destined to form wings, legs and antennae, evokes ectopic eyes from these discs. Structurally, even down to the level of scanning electron microscopical analysis, these ectopic eyes were surprisingly normal and complete. Halder and colleagues estimate that as many as 2500 of the approximately 17 000 genes in *Drosophila* may be involved in eye development. *Ey* is therefore the master controller of genes representing almost one-seventh of the entire *Drosophila* genome.

The demonstration that mouse *Pax-6* induces ectopic eyes in *Drosophila* leg imaginal discs and that ectopic expression of *Pax-6* in *Xenopus* induces ectopic lenses, further confirms homology of *ey* with *Pax-6* and that *Pax-6* also functions as a master control gene. (Similarly, mouse *Hoxb-6* evokes legs in place of antennae when expressed in *Drosophila*, demonstrating conservation of genetic mechanisms of A-P patterning between the two groups.) Further confirmation of phylogenetic conservatism of *Pax-6* is that the ascidian homologue isolated from *Phallusia mammillata* induces ectopic and supernumerary eyes in *Drosophila*. Two *Pax* genes (*Pax-A* and *Pax-B*) have been isolated from a jellyfish (*Chrysaora quinquecirrha*) and a coelenterate (*Hydra littoralis*) by Sun *et al.* (1997), confirming the ancient origins of these transcription factors and that they may function throughout the Metazoa.[13]

Another gene, *sine oculis*, has now joined *ey* and *Pax-6* in being involved in eye development in *Drosophila*. The mouse homologue, *Six³*, is expressed in the neural plate, optic vesicles and lens primordia. Murine *Six³* has been shown by Oliver *et al.* (1996) to induce an ectopic lens in embryos of the Japanese medaka, *Oryzias latipes*, a teleost fish, and to do so without any involvement of retinal tissue from the optic cup; *Six³* either bypasses the normal induction, or is a component of a normal inductive mechanism.

It would be remarkable if the regulatory cascades flowing from *ey*, *Six³* and *Pax-6* were identical or even very similar between flies, mammals and fish. Fly and vertebrate eyes are certainly neither similarly constructed nor similarly wired to the nervous system. Nor do they function in similar ways.

Do we regard vertebrate and invertebrate eyes as homologous because the same master gene directs their development? Do we shift our concept of homology down to the control genes and away from the structures which have long been regarded as analogous? Do we construct a hierarchical or nested set of definitions for homology as proposed by Abouheif (1997) and others? De Beer's easy summary of genes and homology has been complicated indeed.

Comparisons of patterns of gene expression in developing invertebrate and vertebrate nervous systems are raising similar problems.

(b) Invertebrate and vertebrate brains

Similarity in patterns of gene expression is now being used to define homologous structures, as in comparisons of body regions between insects and crustaceans and between hemichordates and chordates.

Williams and Holland (1996) analysed the expression pattern of *AmphiOtx*, whose vertebrate homologue *Otx* is involved in development of the most rostral portion of the nervous system, head and eyes. *AmphiOtx* is expressed in the most rostral neurectoderm and adjacent neurendoderm, where the single frontal eye develops. This pattern is equivalent to that of *Otx* in vertebrates in association with development of the paired eyes. Williams and Holland argue from the homologous expression patterns that the amphioxus rostral nerve cord is homologous with the vertebrate forebrain and that the frontal eye is an unpaired homologue of the paired vertebrate eyes. P. W. H. Holland and Graham (1995) place these data on gene expression patterns into the historical context of Garstang's theory of the origin of the chordate neural tube.[14]

The use of equivalent patterns of gene expression to determine the homology of structures has prompted some authors to revisit structural homology. For Detlev Arendt and Katharina Nübler-Jung (1996), control of D-V and V-D patterning in *Xenopus* and *Drosophila* by equivalent genes (discussed in section 9.4) has an additional chapter. They argue for homology of specific areas of insect and vertebrate brains from molecular and morphological data to support the conclusion that 'insect and vertebrate brains are built according to a common ground plan, and that specific areas of the insect and vertebrate brains be considered as homologous, meaning that these areas already existed, with their specific functions, in their common ancestor' (Arendt and Nübler-Jung, 1996, p. 255). Three lines of similarity are used:

1. formation of the brains in comparable body regions;
2. early expression patterns of orthologous patterning genes in comparable regions of the brains that have similar functions later in ontogeny; and
3. patterning of early axonal connections.

In these respects they go beyond those who claim homology solely on the basis of equivalence of gene expression patterns.

Others have raised the legitimate concern that to regard equivalent boundaries of gene expression as homologues is to turn structures long regarded as analogous into homologues. Examples used are arthropod/vertebrate eyes (shared *Pax-6* genes) and the vertebrate

brain/notochord and insect hindgut (shared *Brachyury* [*T*] gene).

For most researchers, the expression boundaries of one gene (even of multiple genes) do not provide a sufficient basis to determine homology. Furthermore, there is often no correlation of conservation of patterns of gene expression and conservation of structure. For others, molecular similarities reveal homologies at a deeper level than do structural homologies, but should not cause us to redefine homology; N. D. Holland (1996) argues that expression boundaries can be used, but only for closely related organisms. In Akam's (1995) model of changes in *Hox* gene regulation and the evolution of arthropod body plans, domains of *Hox* gene expression are used to define homologous regions along insect and crustacean bodies. Even while maintaining that *Hox* genes are part of the mechanism that defines homology, Akam is sanguine about this use of gene expression as a surrogate for homology:

'... we would be foolish to think that the expression of one gene, or even one family of genes, will provide a certain guide to homology ... No one part of a developmental mechanism is immutable, so no single gene can define homology.'

Akam, 1995, p. 318[15]

The genes for which most is known (homeotic and *Hox* genes) regulate many downstream genes and are themselves controlled by many upstream genes, such as retinoic acid-response elements in *Drosophila* and mice, the zinc-finger protein *Krox-20* that regulates *Hoxb-2* in the mouse, or genes involved in epithelial-mesenchymal interactions such as BMPs, *Msx* or cell adhesion molecules that up-regulate *Hox* genes. Conservation of *Hox* patterns may reveal nothing about the regulatory changes occurring down- or upstream that specify particular cell, tissue and organ types; *Hox* genes do not determine the identity of characters.[16]

21.8 RESOLVING THE DILEMMA

De Beer's analysis and summary of the situation are as compelling now as they were a quarter of a century ago. Variability of developmental processes on the one hand, and the constancy of homologous structures on the other, render any single concept of homology that attempts to unite the two as an uncomfortable alliance between the constancy of the final pattern and the variability of developmental processes. It does not therefore appear possible to have a single concept or definition of homology that embraces both pattern and process.

Historically, homology encompassed the 'idealistic homology concept' (structures built upon a common plan) and the 'historical homology concept' (derivation from the equivalent structure in a common ancestor). The former implied, and was then explicitly developed as, the embryological criterion of homology – that homologous structures are derived from the same embryonic regions.

Similarity is implicit in the recognition of conservation of plan or basic structure. Yet, particularly over the past 10–15 years, questions have been raised about our ability to recognize individual components within complex structures across phylogenetic groups. The multiple elements of the wrist and ankle, or the numbers of digits, are especially difficult to homologize across the vertebrates. Those who are troubled, emphasize homology as the structural plan, not elements within that plan (s. 21.2).[17]

The tetrapod limb does not pass through an archetypal pattern of condensations. Rather, a common set of developmental processes (in effect a set of archetypal processes), based on branching and segmentation of condensations, provides 'the developmental basis of structural homology' (Hinchliffe, 1989, pp. 172). If conservation of developmental information is at the level of regional patterns rather than individual elements, then homology resides at the level of the developmental processes that produce the overall pattern.

'... the elements themselves may not be readily comparable, but the morphogenetic processes that created them can be compared ... homologies can be readily constructed, and major evolutionary changes can be resolved into iterations of condensation, branching, and segmentation events.'

Oster *et al.*, 1988, p. 877

As is argued below, it is more helpful to distinguish developmental processes as causes of homologous structures than to regard them as homology itself. Development is process, homology is pattern. Obviously it is important to find the 'construction rules' for limb development and evolution (Oster *et al.*, 1988), but a limb built upon one set of rules does not lose its homology with limbs built upon different rules; there may just have been more divergence and evolution of the developmental processes between the two groups than within each group. Change is after all the essence of evolution.

A good example comes from a comparison of digital development in anurans and amniotes with that in urodeles. In the former, digits form from posterior to anterior and are lost in evolution (or following experimental perturbation) in the reverse order. Thus anterior digit 1, the last digit to form, is the first lost. Digits in urodeles form from anterior to posterior, and again digits are lost in the reverse order. Posterior digit 5, the last digit to form, is the first lost. Furthermore, although the normal sequence of skeletal development in vertebrate limbs is from proximal to distal, some distal elements in urodeles form before some proximal elements (Oster *et al.*, 1988). Thus, branching and segmentation sequences are not immutable. Are such differences in developmental pattern sufficient to render urodele limbs non-homologous to all other vertebrate limbs? I argue that they are not. Indeed the straightforward dichotomy of urodeles versus anurans/amniotes unravels as more species are examined. In the developing limbs of the marbled newt (*Triturus marmoratus*) a central axis develops in a distal-to-proximal direction without postaxial branching, while postaxial structures develop independently of the digital arch. Blanco and Alberch (1992) interpreted this as caenogenesis – a cautionary reminder that developmental processes evolve within the urodeles as everywhere else.

21.8.1 Biological homology

The biological homology concept was developed by Louise Roth and elaborated by Günter Wagner as a way out of the dilemma, which continues because some regard homology as process, and common development and developmental processes as axiomatic.

The term 'biological homology' is, however, unfortunate. It implies that definitions and concepts of homology based on adult structure or derivation from a common ancestor are somehow abiological, which they obviously are not. [Having decried the term 'biological homology', I acknowledge that the term 'biological species' was introduced by G. G. Simpson to highlight a mechanism, viz., reproductive isolation as the mechanism of speciation (s. 3.4).] As development is so central to the concept of biological homology, 'developmental homology' might have been a preferable term, although similarity of development is not required to establish homology. Arnold Kluge uses the phrase 'ontogenetic homology' to convey his view that ontogeny is more important than other criteria in deducing homology.[18]

In her comprehensive analysis of homology, Roth took the approach that homology is fundamentally a developmental concept.

'A *necessary* component of homology is *the sharing of a common developmental pathway*. Homologues must, to some extent, follow similar processes of differentiation which, one infers,

depend on the same batteries of (regulatory or structural) genes . . . homology is based on the sharing of pathways of development which are controlled by genealogically related genes.'

Roth, 1984, pp. 17, 27

We have already seen the problem of using common developmental processes. In recognizing such difficulties, Roth (1988) incorporated Van Valen's concept of correspondence caused by continuity of information. Wagner dealt with the problem of diversity of developmental control by modifying the concept of biological homology, restricting it to those aspects of development that constitute developmental constraints. Unique sets of epigenetic interactions are identified by Wagner as the essence of developmental constraints (s. 6.7.2). Feedback between retina and lens is an example of an essential epigenetic interaction, developmental constraint, or epigenetic trap. As homology is concerned with stability of structure, independent of function, one can see the logic of linking homology and constraint. Wagner refined the biological homology concept and formalized it as:

'Structures from two individuals or from the same individual are homologous if they share a set of developmental constraints, caused by locally acting self-regulatory mechanisms of organ differentiation. These structures are thus developmentally individualized parts of the phenotype.'

Wagner, 1989b, p. 62

Individualized parts are those parts of embryos that react to systemic or environmental stimuli with their own specific response. The vertebrate eye and segments in *Drosophila* are two examples. Since we cannot identify all the developmental constraints, this central portion of the definition is not amenable to analysis.

21.8.2 Homology as pattern not process

To limit biological homology to 'developmentally constrained morphological patterns',

Wagner (1989b, p. 66) raises a number of fundamental difficulties, chief of which are:

- our comparative ignorance of developmental constraints; and
- that equivalent adult structures can be produced by different developmental programmes which, although constrained, are neither the same nor homologous.

Indeed, not only are vertebrate lenses produced by different developmental mechanisms, but vertebrates use a variety of unrelated proteins as lens proteins (Piatigorsky and Wistow, 1991). At the level of both common developmental sequences and common gene products, the mechanisms that produce lenses show much greater variation than the structures themselves.

Developmental variability in the production of homologous structures makes it inappropriate to establish a one-to-one or obligatory cause-and-effect relationship between development and homology, although many homologous structures do share common developmental bases and any set of homologues must have been based on common development when they arose. The difficulty arises precisely because developmental mechanisms evolve. Homology is a statement about pattern and should not be conflated with a concept about processes and mechanisms. A conceptually similar situation exists with the use of heterochrony as a term for pattern and process (s. 24.1.1).

Although common developmental processes may aid in the identification of homologous structures, lack of common development speaks neither for nor against homology. This view was expressed by Darwin over 125 years ago:

'Community in embryonic structure reveals community of descent; but dissimilarity in embryonic development does not prove discommunity of descent, for in one of two groups, the developmental stages may have been suppressed, or may have been so greatly modified through adaptation to new habits of life as to be no longer recognizable.'
Darwin, 1872, cited by Kluge, 1988, p. 73

My position is that homology is a statement about pattern which is separate from statements about processes that produce those patterns. Often homologous structures arise through the same developmental programme. But as developmental programmes evolve (genes are substituted, parts of causal sequence are lost or altered, and so forth), many homologous structures no longer share the same set of developmental sequences. Similarly (or conversely?), there are many cases in which the same genes direct the development of different (analogous) structures. Reasoning similar to that used for homologous structures and the developmental processes underlying them, applies to homologous behaviours and the structures underlying them (Box 21.2).

In the first edition, I argued that homologous structures can be produced by **equivalent** and **non-equivalent developmental processes**. Striedter and Northcutt (1991) came to identical conclusions in an analysis published while the first edition of this book was in press. They speak of 'non-homologous developmental precursors and processes'. I used the term 'non-equivalent developmental processes' to avoid another, and potentially confusing, use of homologous and non-homologous. In all other respects I am entirely in agreement with their analysis.

This approach has already been put into effect. Miyake and colleagues (1992b) examined the development, morphology and homology of rostral cartilage in batoid fishes on the basis of equivalent and non-equivalent developmental processes. Striedter (1997) used equivalent and non-equivalent developmental processes to illuminate the long-standing problem of the homology of the telencephalon between reptiles/birds and mammals. Here minor evolutionary changes in development have produced major changes in neuronal

---

BOX 21.2
HOMOLOGY AND BEHAVIOUR

A parallel to homologous structures arising from the same or different developmental processes is the application of homology to behaviour, specifically the question of whether homologous behaviours must be based on homologous structures. Atz (1970) confined homology to shared innate behaviours, or to those behaviours that share fixed actions, while Hodos (1976) equated behavioural homology with homology of the underlying structures.

Lauder (1986, 1994), on the other hand, argued emphatically and persuasively that behaviours stand as homologues in their own right, without reference to commonality of structural basis, a view shared by the few others such as Greene (1994) who have considered these issues recently. Striedter and Northcutt (1991) discussed two interesting examples. The first is that gymnotoid fishes produce electric fields using two different types of organs (one myogenic and one neurogenic). Nevertheless, they argue that generating an electric field should be regarded as a homologous behaviour. The second example is production of song in acridid grasshoppers. The most common pattern is stridulation – rubbing the hindleg against the forewing. Less common is movement of the mandibles, as in *Calliptamus italicus*, which still displays the movements of the hindleg characteristic of other species but in which leg and wing do not come into contact. Here there is evidence for the structural change underlying the homologous behaviour.

---

connections that led to the independent evolution of telencephalic structure.[19]

These are not classes of homology as were E. B. Wilson's complete and incomplete homology. Rather, they are statements that homologous structures arise via developmental processes that may be the same or that may differ, i.e. that are conserved or have changed over the evolutionary history of the homologous feature. In a real sense homology is 'what remains unchanged in the face of change' (Stevens, 1984, p. 394). Equating homology with structural pattern does not mean that homologous structures can only be determined in adults. Homologous structures exist at all stages of the life cycle; recall the agnathan and crustacean examples of latent homology (s. 21.4). Rudy Raff and his colleagues identified homologous cell lineages in direct-developing sea urchin embryos using the same classes of criteria of shared features and shared patterns used to establish homology of adult structures, as well as common embryonic origin and fate. The latter are appropriate criteria for the lineages but may not be appropriate for adult structures that develop from cells in the lineages.[20]

This discussion of the variability of developmental processes that produce homologues, and of the fact that different structures exist at different stages of the life cycle, leads into a discussion of the evolution of ontogeny or life history, the topic of the next chapter. I explore this theme using direct development, in which the larval stage is modified or eliminated altogether from the life cycle.

21.9 ENDNOTES

1. Atchley *et al.* (1997) use the term 'developmental homoplasy for situations in which the same phenotype is produced by different genetic or developmental mechanisms.

2. The literature on homology is enormous, reflecting its fundamental importance in comparative biology. The volume on homology which I edited (Hall, 1994a) and Hall (1994d) contain extensive treatments of all aspects of homology, especially developmental. Other analyses that I have found most useful are Cain (1982), Desmond (1982), Patterson (1982, 1988a), Beer (1984), Roth (1984, 1988, 1991), Wagner (1986, 1989a,b, 1995, 1996), Rieppel (1987, 1992a), Brooks and Wiley (1988), Kluge (1988), Striedter and Northcutt (1991), Donoghue (1992), McKitrick (1994), Sluys (1996), Collazo and Fraser (1996) and Abouheif (1997).

3. Nelson (1978), Patterson (1982), Kluge and Strauss (1985), Kluge (1988), Williams *et al.* (1990) and the chapters by Rieppel, Nelson, Bookstein and others in Hall (1994a) contain discussions of the ontogenetic criterion in systematics and phylogenetics. For evaluations and/or tests of the application of the ontogenetic criterion in phylogenetic analysis, see de Queiroz (1985), Patterson (1988b), Mabee (1989a,b, 1993, 1995), M. H. Wake (1989), Rieppel (1990) and Panchen (1992).

4. See Maienschein (1978) for an analysis of Wilson's changing views on homology. For arguments for the central role of developmental processes in homology, see Spemann (1915), Baltzer (1950), Riedl (1978), Van Valen (1982), Goodwin *et al.* (1983), Roth (1984, 1988), Wagner (1989a,b, 1995, 1996), Minelli and Peruffo (1991) and Minelli (1996b).

5. (1) no; (2) opinions will vary; (3) no; (4) answers will vary; (5) yes, but some would say no; (6) no, but some would say yes; (7) no and yes, but some would say yes and no (see Hall, 1994a, pp. 2–3).

6. For discussions of serial homology in plants, see Sattler (1994) and Sattler and Jeune (1992); in insects, see s. 9.4; and for neuromeres, see Gilland and Baker (1993).

7. Richard Goss's 1969 book *Principles of Regeneration* is still the best source for the fundamentals of regeneration. For current analyses, see Stocum (1995), Tsonis (1996) and Ferretti and Géraudie (1998).

8. See Averof and Akam (1995), Williams and Nagy (1995) and Averof and Patel (1997) for homeobox gene expression in brine shrimp. Minelli (1996a) also proposes that segmental organization be de-emphasized and that 'tagmata' are the basic body units and the units the target of selection. His tagmata are A-P, D-V and P-D axes of appendages and a temporal axis of post-embryonic development.

9. For phylogenetic information in atavisms, see Berg (1969), Hall (1984a, 1995b), D. B. Wake and Larson (1987), Stiassny (1992) and Shubin *et al.* (1995).

10. See Cherdantsev and Krauss (1996) for variant patterns of gastrulation in coelenterates, and Webb (1990) for additional information on relationships between pores and tubes of the lateral line system in fishes.

11. Osche first elaborated his position in 1973 and developed it more fully in 1982, contemporaneously with Van Valen. Thus, 'Homolog sind . . . Strukturen deren nicht zufällige Übereinstimmung auf gemeinsamer Information beruht' (Osche, 1973, p. 156).

12. See Alexander (1976) and Kettlewell (1973) for different genetic programmes resulting in parallel evolution.

13. For homology of master genes for eye development in *Drosophila* and vertebrates, see Quiring *et al.* (1994), Halder *et al.* (1995), Hanson and Van Heyningen (1995), Kaufman *et al.* (1995) and Oliver *et al.* (1996). For ectopic eyes induced in *Drosophila* by mouse and ascidian *Pax-6*, see Halder *et al.* (1955, esp. Figure 5) and Glardon *et al.* (1997). For mouse *Hoxb-6* evoking legs in place of antennae in *Drosophila*, see Malicki *et al.* (1990). For ectopic lens formation in *Xenopus*, see Altmann *et al.* (1997). Specificity of *Pax-6* action in *Xenopus* is indicated by induction of ectopic lenses without concomitant induction of neural tissue. For the role of Pax transcription factors in the development of many organs that arise following epithelial-mesenchymal interactions (including sense organs, kidneys, brain, muscles and skeletal tissues), see Dahl *et al.* (1997).

14. The long history of the search for the forerunner of the vertebrate forebrain is analysed in Northcutt (1995).

15. Akam (1995) and P. W. H. Holland and Garcia-Fernàndez (1996) contain analyses of each of these examples of homologous patterns of gene expression. For methodologies and/or

difficulties associated with using boundaries of gene expression as the primary criterion for homology, see Dickinson (1995), Bolker and Raff (1996), Burke and Nelson (1996), Galis (1996), L. Z. Holland *et al.* (1996), N. D. Holland (1996), Meyer (1996), Müller and Wagner (1996) and Zardoya *et al.* (1996a,b) and Abouheif *et al.* (1997).

16. For changes in genes regulated by or regulating *Hox* genes, see Davidson (1993), Sham *et al.* (1992), Krumlauf (1994) and Frasch *et al.* (1995).

17. There is a considerable literature on our inability to homologize elements within complex structures. Samples from different approaches are Goodwin *et al.* (1983), Goodwin (1984), Alberch and Gale (1985), Shubin and Alberch (1986), Hinchliffe (1985, 1989, 1994), Oster *et al.* (1988), Oster and Murray (1989), Rieppel (1989), Padian (1992), Hecht and Hecht (1994), Shubin (1991, 1995) and Vorobyeva and Hinchliffe (1996).

18. See Roth (1984, 1988) and Wagner (1989a,b, 1995) for the concept of biological homology, and Kluge (1988, p. 74) for ontogenetic homology.

19. For additional examples of variability of developmental bases producing homologous structures, see Striedter and Northcutt (1991), McKitrick (1994), Collazo and Fraser (1996), Meyer (1996) and Müller and Wagner (1996).

20. For their approach to identification of homologous cell lineages in sea urchins, see Raff (1988) and Raff *et al.* (1990).

22

ONTOGENY EVOLVES: THE DILEMMA FOR LARVAE

'There are no questions which are of greater importance for the embryologist than those which concern the nature of the secondary changes likely to occur in the fœtal or in the larval states; since it is on the answer to such questions that our knowledge of the extent to which a record of the ancestral history may be expected to be preserved in development depends.'

Balfour, 1880a, p. 381

Development is not immutable. Development evolves and has evolved. Indeed, Part Four was devoted to the evolution of development and developmental processes. I have already treated the evolution of embryonic development itself when discussing Haeckel's Gastraea theory and the origins of multicellularity, the Metazoa, the chordates and gnathostomes (ss. 5.5.2, 13.5, 14.2, 16.1 and Chapter 15). Ontogeny (life history) also evolves.[1]

Much evolutionary change relates to the production of adult characters, hence the difficulty with homology discussed in the previous chapter. Such evolution is reflected in the developmental plasticity that allows different developmental processes to effect similar morphological changes in response to equivalent selection pressure – the non-equivalent developmental processes responsible for genetic assimilation, changes in tail length and mandibular morphology in mice. Such plasticity of developmental capability reflects the fact that developmental processes have evolved and diversified, and that latent variability in developmental processes brings about morphological change under the appropriate environmental conditions and in response to selection.

22.1 LARVAL ADAPTATION AND EVOLUTION

Other evolutionary changes relate to embryonic and larval adaptations, as for example in extra-embryonic membranes, placentae, the early development of the feeding and respiratory apparatuses in larval amphibians and fishes, or specific genes devoted to establishing the developmental competence of larvae of the red abalone, *Haliotis rufescens*, to settle and initiate metamorphosis (Degnan and Morse, 1995). Recognizing the independence of larval and adult evolution is important: a combined molecular and morphological analysis of 29 extant echinoid species and 13 extinct taxa conducted by A. B. Smith and colleagues in 1995 provided strong support for separate evolution of larvae and adults. Furthermore, larval evolution may be a key that unlocks the evolution of many invertebrate groups (s. 14.2.4).

Much larval evolution is in marine organisms, especially planktonic invertebrate larvae. The three major development patterns of marine larvae – planktotrophy (feeding larvae), lecithotrophy (non-feeding larvae with yolk-rich eggs) and brooding (larvae brooded by their parents) – correlate with nutritional mode (feeding/non-feeding) and ecological environment (benthic/pelagic). Planktotrophic feeding requires rapid development and the elaboration of specialized larval organs. Such larval adaptations illustrate dramatically the independent evolution of life cycle stages, although larval adaptations can be co-opted as features of adult life.

Some larval adaptations, such as those associated with transitions between pelagic and benthic development, are independent of changes in morphology and thus potentially reversible. Because morphological change is not involved, ecological boundaries such as those separating plankotrophic and lecithotrophic larvae can be crossed frequently. On the other hand, adaptations such as those associated with development of non-feeding from feeding larvae involve changes in morphology, are not reversible and therefore limit opportunities to diversify in new environments.[2]

Paradoxically, much of the information on larval adaptation and evolution is derived from studies of animals that have modified or lost the larval stage, and therefore eliminated or modified metamorphic transformation. These direct-developing animals – the topic of the remainder of the chapter – are from taxa that normally develop indirectly by way of a specialized larva with metamorphosis to an adult.

22.2 DIRECT DEVELOPMENT

Direct development and its association with increased egg size (s. 17.1) stands in contrast to indirect development, in which eggs are small and larvae and metamorphosis retained. Direct development, which has evolved repeatedly, has been most fully studied in frogs, sea urchins and ascidians. It is not an esoteric evolutionary oddity but a common evolutionary adaptation to particular environmental (often seasonal) conditions. Through an examination of direct development we see that early embryonic development can vary enormously with dramatic consequences for

life-history stages but without affecting the *Bauplan*, adult structure, or systematic position of the organism.

22.2.1 Direct-developing frogs

Direct development has evolved as many as ten times in the 800 species of nine of the 21 families of anurans.[3]

Direct-developing frogs illustrate that von Baer's 'law' – early embryonic stages of related organisms resemble one another much more than do later stages – is not universal. Under such a law, all frogs would be expected to share a common pattern of early development, especially patterns of cleavage and gastrulation. Egg-brooding hylid frogs of the genus *Gastrotheca* (*G. riobambae* and *G. plumbea*), however, have a pattern of gastrulation from an embryonic disc that is typical of birds, not of amphibians (Figure 22.1). Eggs in these species are enormous, the largest eggs of any amphibians – up to 3.6 mm diam. in *G. riobambae*, 10 mm in *G. ceratophrys*. For comparison, the eggs of *Xenopus laevis* are of the order of 1.3 mm diameter.[4]

During cleavage, these eggs show no sign that anything other than a typical amphibian embryo will develop. Normally, the amphibian primary body axis arises after invagination of future notochord through a small opening to the outside, the blastopore (Figure 22.1 and s. 10.2). During gastrulation in these direct developers however, an embryonic disc develops around the blastopore and gradually expands over the surface of the yolk (Figure 22.1). Thus a superficial disc of embryonic cells (as seen in birds and reptiles), rather than invaginated presumptive layers (as seen in other amphibians) forms the tissues and organs in these direct-developing frogs. Despite this fundamental modification of early development the adults are perfectly respectable frogs. This profound modification in early developmental stages does not affect the basic body plan.

Fig. 22.1 The embryo of *Gastrotheca riobambae* (top) forms from an embryonic disc and a very large egg (10 mm diameter) in a manner that diverges dramatically from amphibians such as *Xenopus laevis* (bottom), in which invagination is through a blastopore shown on the right. Modified from Elinson (1987).

Development of the Puerto Rican species known as coqui (*Eleutherodactylus coqui*) is the subject of studies by Richard Elinson in Toronto and Jim Hanken in Boulder. Coqui are abundant and accessible in Puerto Rico. Although breeding cannot be hormonally induced in the laboratory, mating and breeding do occur. Such serendipitous breeding requires alert researchers especially when, as described below, transplantation of embryonic regions between embryos of coqui and another species are required to answer the questions at hand. The effort is worthwhile; coqui display one of the most extreme forms of direct development, with virtually all tadpole features lost from the life cycle.

Coqui lack tadpole mouth parts and jaw apparatus, lateral line systems and cement glands. Development of adult skeletal and muscle elements is precocious, ossification and myogenesis appearing with what, in species with tadpoles, would be a mid-metamorphic, or even adult configuration. The limbs appear extremely early, appearing to 'float' at the extremity of the embryonic axis in these oversize eggs (Figure 22.2). Preliminary indications from the work of David Jennings are that the thyroid hormone axis is activated precociously with respect to species with tadpoles in the life cycle.[5]

Hanken and Elinson have provided insights into head development in coqui. Patterns of migration of cranial neural crest cells were examined with special attention to the mandibular stream from which tadpole and adult jaw skeletons arise in indirect developers, and from which only the adult skeleton arises in coqui. Using a combination of scanning electron microscopy and labelling with the vital dye DiI, patterns of migrating neural crest cells were shown to be highly conserved in coqui when compared with such indirect-developing species as *Xenopus*, *Bombina* and *Rana*. The only difference in coqui was some enhancement of the mandibular stream. From the data available so far it does not appear that major alterations in patterning of neural crest migration are responsible for the cranial changes in coqui.[6]

In 1996 Fang and Elinson reported the first combined molecular and experimental study of a direct-developing frog with their analysis of *Dll* expression and cement-gland induction in coqui. As discussed in section 10.2.2, *Dll* patterns the vertebrate forebrain and more caudal portions of the developing brain and is expressed in the mandibular stream of migrating neural crest cells in *Xenopus*. Four *Dll* genes were cloned from coqui. Consistent with conservative patterning of the neural crest, patterning of *Dll* is conserved in coqui.[7]

Dll is also expressed in the cement gland in *Xenopus*, but not in the equivalent ectodermal

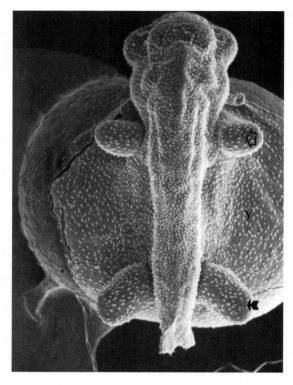
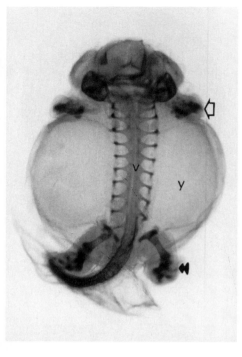

Fig. 22.2 Embryos of *Eleutherodactylus coqui* viewed from the dorsal surface with the head at the top. On the left is a scanning electron micrograph showing the developing regions of the brain and the prominent fore- and hindlimb buds (open and solid arrows). The cartilaginous skeleton of the embryo on the right has been visualized with a type II collagen antibody, Vertebral (v), limb (arrows) and skull skeletal elements are very evident. Y, yolk sac. Photographs courtesy of Dr. James Hanken.

region of coqui, which lacks cement glands. Fang and Elinson used cross-species transplantation and tissue recombinations to investigate the potential developmental mechanisms responsible for loss of the cement glands in coqui; these were essentially the methods used to investigate induction of balancers, described in section 17.5.5. Cement glands, like balancers, arise following a series of inductive interactions. Indeed, balancers in pond-dwelling larval salamanders may be homologous with cement glands.[8]

Cement glands form after inductive interactions between ectoderm and adjacent cranial tissues. Coqui cranial tissues can induce

cement glands from *Xenopus* ectoderm, but coqui ectoderm does not respond to inductive signals from *Xenopus*, demonstrating that competence of coqui ectoderm to respond to induction has been modified without modification of the inductive signal, leading to loss of cement glands in coqui. You will recall that loss of ectodermal competence is also responsible for loss of balancers in some amphibians (s. 17.5.5); loss of limbs in avian mutants such as *limbless* (s. 11.4.5); and loss of teeth in birds.[9]

An important series of messages therefore lies in these four examples of how cell and tissue interactions are modified when structures are lost during evolution:

1. An organ may be lost without loss of the entire developmental system for producing that organ.
2. Loss of organs is often mediated through modification (not loss) of inductive interactions.
3. Modification of competence is the usual means by which inductive interactions are altered.
4. Inductive signalling persists, even when competence to respond is lost.
5. The potential to reform the organ exists, provided that competence can be restored.

22.2.2 Direct-developing sea urchins

The second example comes from the work of Rudy Raff and his colleagues on sea urchins. Typically, echinoderms produce a blastula which transforms into the characteristic pluteus larva with its radiating arms and internal skeletal spicules of calcium carbonate (Figure 22.3). The larva then metamorphoses into an adult.

Direct development has evolved independently a number of times: some 180 out of 900 species of living sea urchins exhibit direct development; all the members of one order, the Echinothurioida, are direct developers.[10]

As with frogs, so with sea urchins. Direct development in sea urchins follows increased egg size and loss or abbreviation of the larval phase, while adult features appear earlier in ontogeny than in indirect developers. Some of the major differences between indirect and direct developers are listed in Table 22.1.

Direct-developing sea urchins are often brooded by the female. There are congeneric species in which one is a direct developer and one develops with a larval stage. Such congeneric species can occur as species pairs, inhabiting the same environment and breeding at the same time, as do the Pacific species *Heliocidaris erythrogramma*, a direct developer, and *H. tuberculata*, an indirect developer.[11]

Raff identified the evolution of large, yolky eggs as the key pre-adaptation in the evolution

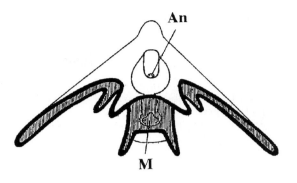

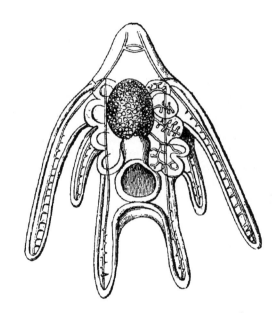

Fig. 22.3 Pluteus larvae from a brittle-star (ophiuroid). An, anus; m, mouth. Modified from Balfour (1980–1, Vol. 1).

of direct development. A large, yolk-filled egg eliminates the need for feeding early in development. Consequently, the feeding larval stage can be modified or eliminated. An intermediate stage would be a larval stage in which feeding was optional. Ecological factors drive these developmental changes associated with adoption of direct development. Time of onset of larval or juvenile structures and larval size

Table 22.1 Major differences in embryonic development between indirect- and direct-developing echinoderms[1]

Feature	Indirect developers	Direct developers
Primary (D-V) embryonic axis	Maternal	Maternal
Secondary (A-P) embryonic axis	Zygotic	Maternal
Cleavage pattern	Radial	Modified radial
Cleavage cell divisions	Equal	±equal
Timing of specification	3rd cleavage	Variable
Dorsal vegetal cells	Endodermal	Ectodermal
Larval mouth	Present	Lost
Primary mesenchyme	>1000 cells	32 cells

[1] Based on data from Wray and Raff (1990b) and Wray (1994). The features for direct development are those of *Heliocidaris erythrogramma*. Patterns vary from species to species.

can also be modified adaptively in response to the amount of food presented. Hart investigated this aspect in the heart urchin *Brisaster latifrons*, in which the larvae can feed but do not have to feed to complete larval development. His studies show that large egg size, lack of dependence on feeding at the larval stage, and/or low rates of larval feeding, are ecological factors that precede the origin of the developmental changes seen in direct development.[12]

Modifications in development are evident even before cleavage, as demonstrated by Henry, Klueg and Raff (1992) with cleavage and axis specification in direct and indirect developers. As early as the 16-cell stage in the direct developer *H. erythrogramma,* the cell lineages of the vegetal cells that form most of the mesoderm in indirect developers are modified to contribute cells to ectoderm and gut. Something as seemingly invariant as the relationship between the first cleavage division and the axis of D-V symmetry is modified in some direct developers. Indeed, as Goldstein and Freeman (1997) demonstrated in their analysis across the animal kingdom, time and mechanism of primary and secondary axis specification vary considerably.

Modification or elimination of larval structures is thus the consequence of changes that

occur very early in development indeed. Maternal determinants are re-allocated between cell lineages, while re-specification of cell lineages modifies the fundamental association between cell division and embryonic axis determination. Consequently, the fate map of direct developers diverges radically from the map of indirect developers. Genetic control over that switch is beginning to be unravelled (s. 24.1.4).

Despite such dramatic changes in early development, the adult morphologies of sea urchins that develop directly or indirectly are typically very similar. This is but one of a growing number of examples of how patterns of early development can be modified substantially without affecting adult morphology. The variant patterns of cleavage seen in cephalopods and derived crustaceans, and of gastrulation in a sea urchin that broods its larvae, are further examples.[13]

22.2.3 Direct-developing ascidians

Ten to 12 species in two ascidian families (the Molgulidae and Styelidae) develop from what are known as anural larvae or anural tadpoles. The chief characteristic of these atypical tadpoles is suppression of development of the tail so familiar in tailed larvae/tadpoles. Anural larvae therefore represent a form of direct development. The small number of tissues (six), cells (2500), and the smallest genome of any chordate (1.8×10^8 base pairs), facilitates analysis of developmental and/or evolutionary changes in ascidian tadpoles.[14] [Tailed larvae are often referred to as urodele larvae or urodele tadpoles. Because of confusion with the term urodele amphibians, I refer to them as tailed larvae, not urodele larvae.]

Direct development accompanies anural development in such species as *Molgula provisionalis* and *M. pacifica*, in which the larval stage is lost completely. Other anural species retain indirect development in that a tail-less larva hatches before metamorphosis begins. As with direct-developing frogs and sea urchins, adult ascidian morphology holds no clues to whether development was direct or indirect and larvae anural or tailed. As in frogs and sea urchins, direct- and indirect-developing ascidians can be closely related; there are congeneric ascidian species with anural and tailed larvae, indirect and direct development. Most of our recent knowledge comes from studies by Jeffery and Swalla on *Molgula occulta*, which has direct development and an anural larva, and *M. oculata*, which has indirect development and a tailed larva.

Early developmental stages of future anural and tailed larvae are morphologically similar, often indistinguishable. Anural larvae diverge after gastrulation, which is when tail development begins in tailed embryos. Numerous larval organs – notochord, muscle, adhesive organ, ocellus, otolith – fail to develop in anural embryos. There is some evidence that failure of tail development is in part due to fewer cells being set aside as notochord or muscle in the early embryo, and an indication that cells that would otherwise form these structures contribute to the endoderm in anural species, a re-specification of cell lineages as seen in direct-developing sea urchins.

Neither alpha-actin nor myosin (markers for myogenesis) appear in *M. occulta* embryos.

This difference was traced back to defective localization of myoplasm in early anural embryos. (Most ascidian cells develop autonomously; myoplasm contains determinants for muscle cells.) The structure of the actin gene is also modified in anural embryos. Deletions, insertions and codon substitutions result in production of a non-functioning gene product. A search by Swalla *et al.* (1993) revealed differential expression of maternal and zygotic products of two genes (*Uro-2* and *Uro-11*). *Uro-11*, which may encode a DNA-binding protein, has maternal and zygotic transcripts in tailed larvae (in early ectoderm and in ectoderm and tail muscle in the gastrula → neurula transition) but only maternal transcripts in the anural species.[15]

In anural larvae, not only are there modifications to the larval stage, but adult development is accelerated and adult tissues differentiate precociously. Swalla and colleagues (1994) refer to this process as adultation, and document it with respect to an adult muscle actin gene expressed precociously in larvae.

The plasticity of anural development and the ease with which it arises were revealed in a fascinating series of experiments in which hybrids were created between an anural and a tailed species; anural eggs can be partially 'rescued' when fertilized with sperm from a tailed species. Hybrids survive past metamorphosis and so features of anural or tailed development can be assessed. Rescued anural larvae form notochordal cells and a short tail. In their 1996 analysis Swalla and Jeffery used hybrids to pinpoint *Manx*, a gene that encodes a zinc finger protein that binds to DNA and may act as a master gene controlling notochord and therefore tail development. *Manx* is down-regulated in the tail-less species but restored in hybrid tailed embryos. It may be that the switches between neural and anural development are few in number.

Manx is also inhibited and notochord and tails fail to develop in hybrid embryos treated with antisense *Manx* DNA. This result implicates *Manx* in the development of such chordate features as dorsal nervous system and tail muscle, whose development requires inductive activity from the notochord. Recalling that the ascidian tadpole was long considered a likely chordate ancestor (s. 5.5.5), *Manx* and/or its upstream regulators (currently unknown) may be key players in the initiation of fundamental caudal chordate features.[16]

Some of the changes in direct developers in the three taxa discussed are based on variation in timing of developmental processes: precocious patterning of cell lineages during cleavage and/or development of adult structures early in embryonic life. Bates (1994) compared the anural *Molgula pacifica* with the tailed *Boltenia villosa*. While developmental time is fixed in *M. pacifica* at some 32–36 hours postfertilization, it is variable and environmentally induced in *B. villosa*.

Do such alterations in timing have evolutionary or ecological significance? I take up the questions of how embryos measure time and how time and place enter into phylogenetic change as heterochrony and heterotopy in the two chapters in the next section.

22.3 ENDNOTES

1. Some classic and more recent treatments of the origin of development are Bonner (1958), and the papers in Grant and Horn (1992) honouring Bonner; Willmer (1960), Buss (1987), Goodwin (1989), Davidson (1990, 1991, 1993), Wolpert (1990, 1992, 1994) and Davidson *et al.* (1995).

2. See Ponder and Lindberg (1997) for a discussion of gastropod planktotrophy and of the relationships between larval and adult evolution in gastropods; and Thorson (1950), Strathmann (1978, 1985, 1988, 1993), Strathmann *et al.* (1992), and McEdward and Janies (1997) for larval development/evolution and planktotrophy/lecithotrophy in echinoderms and other invertebrates.

3. For an entry into the literature on direct development in amphibians, see del Pino and Elinson (1983), G. Roth and Wake (1985), Hanken (1986), Elinson (1987, 1990), D. B. Wake and Roth (1989), Hanken *et al.* (1992), Moury

and Hanken (1995), del Pino (1996), Fang and Elinson (1996), Wake and Hanken (1996) and Hanken *et al.* (1997a,b). For the distribution of direct development among amphibians, see Duellman and Trueb (1986), M. H. Wake (1989) and Duellman (1992).

4. For basic studies on direct development in *Gastrotheca*, see del Pino and Elinson (1983) and Elinson (1987). Del Pino (1996) used *Brachyury* as a marker to compare gastrulation in *Gastrotheca* with other amphibians.

5. For precocious development in coqui, see Hanken *et al.* (1992), Elinson (1994), Jennings (1997), Hanken *et al.* (1997a,b) and Schlosser and Roth (1997). Paradoxically, although the thyroid axis is activated earlier in coqui than in *Xenopus laevis*, responsiveness of coqui tissues to thyroxine is either very low or absent; see Elinson (1994) and Hanken *et al.* (1997a,b). Paradoxically, skeletal elements also arise earlier in ontogeny than normal in laboratory-reared tadpoles whose development is slowed down (Smirnov, 1992). That elements thought to be absent from particular species appear in very old animals (the palatine, preorbital process of the maxillary, postrostrals), or that single elements become paired in very old specimens (Smirnov, 1994, 1995), speaks further to the temporal plasticity of amphibian skeletal development. So too does the decoupling of osteo- from chondrogenesis in amphibians discussed in s. 24.1.5(d).

6. For patterns of neural crest cell migration in coqui and comparisons with other amphibians, see Moury and Hanken (1995), Olsson and Hanken (1996) and Hanken *et al.* (1997a,b).

7. For patterning of the caudal brain by *Dll*, see Akimenko *et al.* (1994).

8. A complex of hierarchical interactions regulates development of the cement gland in *Xenopus*, disruption of any of which could block cement gland formation. For a discussion of positive and negative controls regulating the development of cement glands in *Xenopus*, see Bradley *et al.* (1996).

9. For examples of loss of ectodermal competence, see Maclean and Hall (1987) and Hall (1987a).

10. For an entry into the literature on direct development in echinoderms, see Raff (1987, 1992, 1996), Wray and McClay (1988), Wray and Raff (1989), Henry and Raff (1990) and Wray (1992, 1994).

11. For the distribution of direct development among echinoderms, see Strathmann (1978) and Raff (1987).

12. See Raff (1987) for evolution of large eggs, Hart (1996) for facultative feeding larvae in *Brisaster*, and Strathmann *et al.* (1992) for adaptation of larval features in response to food.

13. For re-specification of cell lineages in direct-developing echinoderms, see Wray and Raff (1989, 1990b) and Raff (1992). For modification of development without affecting adult morphology, see Boletzky (1989), Dohle (1989), Dohle and Scholtz (1988) and Wray and Raff (1991).

14. For an entry into the rapidly growing literature on direct development in ascidians, see Berrill (1931), Jeffery and Swalla (1990), Satoh (1994), Swalla *et al.* (1994), Satoh and Jeffery (1995), Swalla (1996), Swalla and Jeffery (1996) and Jeffrey (1997). See Berrill (1931), Jeffery and Swalla (1990) and Satoh (1994) for descriptions of tailed and anural ascidian larvae.

15. See Jeffery and Swalla (1990, 1991,1992), Swalla (1992, 1996), Nishida (1994), Kusakabe *et al.* (1996), Swalla and Jeffery (1995, 1996) for the studies on cell determinants.

16. For rescue of anural development and the role of *Manx*, see Swalla and Jeffery (1990, 1996), Jeffery and Swalla (1991, 1992) and Jeffrey (1997). For the isolation of cDNA clones of genes expressed in the tail region see Takahashi *et al.* (1997). For tail development in chordates, see s. 10.5. For ascidian tadpoles as chordate ancestors, see Garstang (1928), Hardy (1954), Berrill (1955) and Gee (1996).

PART SEVEN

PATTERNS AND PROCESSES, TIME AND PLACE

23

TIME AND PLACE IN DEVELOPMENT

'Time enters as an essential element into our definition of organism.'

Russell, 1930, p. 171

Timing, duration, rate and sequence of developmental stages and processes vary between individuals, strains and closely related species. Such shifts and changes are usually dichotomized as acceleration, where development is advanced, or retardation, where development is slowed. This statement assumes that embryos measure time and not some other metric such as size, and that we possess stable reference points to compare timing of development between different organisms. Obtaining such reference points, even for closely related organisms, is no easy task.

23.1 STABLE REFERENCE POINTS: TIME, SIZE OR RATES

The most commonly used reference points are:

1. measures of **maturity**, or such major events in ontogeny as morphological stage(s), hatching, birth, attainment of sexual maturity and onset of metamorphosis;
2. measures of **growth,** such as growth rate, size at a particular time during ontogeny and size at maturity; and
3. measures of **chronological age**, either absolute age (days, months, years), or such measures of relative age as percentage of time through ontogeny or percentage of final size.[1]

Although debate over whether time or size should be the metric of choice in analyses of variation in development and growth has raged for decades, there is still no objective way to decide which criterion or combination of criteria to use. Some, such as relative age, are extremely sensitive to temperature, metabolic rate and body size, and are not necessarily correlated with one another during ontogeny, nor under the same control.

Size is influenced by many factors including temperature, nutrition and litter size. For species with external development, temperature is an especially important environmental variable influencing rate of individual development. Because of additive genetic variation, rates of embryonic development can change rapidly in response to selection. Neyfakh and Hartl (1993) demonstrated a response to artificial selection at elevated temperatures in as little as three generations in *Drosophila*. Catch-up growth (and indeterminate growth in organisms whose final size is not fixed) also influences final size.

Maturity and age are less dependent on factors that influence size, but are certainly not independent of them. Attainment of sexual maturity, times of hatching, birth and termination of growth are all subject to evolutionary change. As reference points all must be used with caution.[2]

Table 23.1 The periods of avian ontogeny, their duration, the metrics suggested for their quantification, and estimated mass-specific energy metabolism for each period

Period	Duration	Metric	Metabolized energy (K j/g)
Embryonic	Fertilization to hatching 11–80 days	Incubation time Morphological stages	2 ± 0.8
Postnatal	Hatching to sexual maturity 20 days–years	Growth	20–40
Adult	Sexual maturity to death[1] 8–120 years	Physiological rates such as mass-specific energy metabolism	2400–4300

From data in Starck (1993).

[1] Estimating longevity in birds is notoriously difficult and many records are anecdotal. Accurate estimates from recovery of banded wild birds in N. America yield upper limits of 21 years 9 months for the White-winged dove, *Zenaida asiatica*; 20 years 11 months for the Common grackle, *Quiscalus quiscula*; and 17 years 5 months for Clark's nutcracker, *Nucifraga columbiana* (*J. Field. Ornithol.* [1983], **54**, 123–137, 287–294).

One of the earliest experiments in experimental embryology demonstrated size determination in echinoderms. In 1892, Hans Driesch separated the cells of echinoderm blastulae to produce many small but complete embryos and larvae. The chronology of their development did not differ from their larger siblings. Time, not size, is controlled in species with such regulative development. Conversely, normal, but large, tadpoles develop on time when additional material is added to frog gastrulae.

Another pattern occurs in mice, where deletion of cells at the 2- or 4-cell stages nevertheless produces normally sized mid-gastrula embryos that develop on time. Destruction of as much as 80 percent of a 7-day-old mouse embryo is compensated for in only four to five days. These embryos measure time by cell size, not cell number.[3]

Starck (1993) synthesized an extensive body of literature on the evolution of avian ontogenies, which he divided into the three periods shown in Table 23.1. Duration of these periods varies enormously. As each period is size-dependent, physiological rates rather than chronological time could be used as a time standard when comparing avian ontogenies. Because rates of development vary between embryos, physiological rates are also variable and so not ideal metrics for developmental age. Starck advocated morphological stages, growth, and physiological rates as the most practical metrics for the embryonic, postnatal and adult periods respectively (Table 23.1) and concluded that:

'. . . the constancy of time patterns in birds implies that heterochrony is not a mechanism of changing ontogenies . . . in contrast to other vertebrate groups (e.g., amphibians) where heterochrony . . . plays an important role in the evolution of ontogenies.'

Starck, 1993, p. 292[4]

The postnatal period in birds lasts from 20 days to several years. It has been suggested that growth be related to the time required to increase body weight from 10 to 50% of final size, but growth independent of final size can also be used. Measuring growth over small time intervals in longitudinal studies reveals surprising patterns that show how important it is to make frequent measurement during this period. Humans display saltatory growth during the postnatal period. Indeed, an astonishing 90 to 95% of human development up to 21 months of age is not accompanied by growth but occurs during

Table 23.2   Heartbeat (beat/min.) and wingbeat (beats/sec.) for some representative small and large animals

	Heartbeat		Wingbeat
Shrew	700	Midges	2220
Mice	500	*Chironomus*	650
Daphnia	450	Mosquito	580
Rabbit	200	Honeybee	230
Cat	125	Hummingbird	100
Moths	140	Sparrow	15
Dog	80	Stork	3
Humans	72	Condor	<1
Crayfish	60		
Elephants	40		
Whales	20		
Clams	22		

Source: Prosser and Brown (1961).

periods of stasis (Lampl, Veldhuis and Johnson, 1992).

Energy metabolized per gram of body weight was proposed as a metric to compare ontogenetic phases. Physiological metrics are useful because organisms live on different physiological time scales. In general, small organisms live on shorter time scales ('time' passes more quickly for them) than do larger organisms, for which 'time' passes more slowly. Data for heartbeat and wingbeat, shown in Table 23.2, illustrate this phenomenon, but given the need to normalize for effects of temperature, physiological measures of time should be used with caution.[5]

John Reiss (1989) advocated mass-specific metabolism as the unit of choice for physiological and developmental time. He found a range from 6800 to 61000 calories/g dry weight across selected species of fish, turtles, snakes and birds. He argued for the utility of this measure because congeneric species of albatrosses and terns had closely similar values.

23.2 MORPHOLOGICAL STAGES

The advantage of being able to time development precisely, is to allow temporal and spa-

tial shifts to be recognized when tissues or organs deviate from the sequence of normal stages. Ivan Schmalhausen and Julian Huxley recognized this 60 years ago when they developed growth quotients and growth coefficients – the ratio of the growth of an organ to that of the rest of the body at a specific time, or of the growth of the same organ at different times. Huxley concluded: 'We cannot discover the true growth-coefficient of an organ during its early stage without precise information as to the time-relations of development' (1932, p. 143). This approach has proven useful in other groups. For example, it was used (a) to uncover variability in timing of appearance of skeletal elements and dissociation between cartilage and bone formation in skull development of the Oriental fire-bellied toad, *Bombina orientalis* (Hanken and Hall, 1984); and (b) to demonstrate the appearance of neomorphic skull elements late in adult life in anuran amphibians (Smirnov, 1994 and s. 24.1.5).

Placing embryos and postnatal organisms into morphological stages is a reliable way of measuring the passage of time. Standard staging tables are available for the more common model organisms, and also for a few less-well-studied organisms (Table 23.3). I have used stages explicitly (and implicitly) throughout, without alerting the reader to the nuances of this deceptively simple method of 'ageing' embryos.

While considerable information may be obtained from fresh specimens, more (and different) information can be obtained from fixed and sectioned specimens. Visualization of the onset of skeletal and myogenic condensations comes into this category. Other features may not be revealed without the resolution supplied by scanning or transmission electron microscopy. The latter is especially true for small invertebrates, where surface characters are so critical to systematics, but electron microscopy can augment studies of any organism.[6]

Figure 23.1 illustrates the resolution obtained from 'staging' embryos using several

Table 23.3 Staging tables for some of the species currently used in research

Species	Reference
Xenopus laevis	Stern and Holland (1993)
Laboratory mouse	Theiler (1972), Stern and Holland (1993), Miyake *et al.* (1996a)
Danio rerio (zebrafish)	Stern and Holland (1993), Westerfield (1995)
Oryzias latipes (Japanese Medaka)	Matsui (1949), Yamamoto (1967), Iwamatsu (1994)
Fundulus heteroclitus (killifish)	Armstrong and Child (1965)
Birds	Starck (1993)
Gallus domesticus (common chick)	Hamburger and Hamilton (1951), Eyal-Giladi (1984), Stern and Holland (1993)
Amphibians	Gosner (1960), Rugh (1965); Duellman and Trueb (1986), Burggren and Just (1992)
Loligo pealii (squid)	Arnold (1965)
Oreochromis mossambicus (cichlid fish)	Anken *et al.* (1993)
Scyliorhinus canicula (lesser spotted dogfish)	Ballard *et al.* (1993)
Branchiostoma floridae (lancelet [amphioxus])	Stokes and Holland (1995)
Alosa sapidissima (American Shad)	Shardo (1995)

References cited are not necessarily the original source but may be a recent, accessible source.

levels of analysis of facial development in C57BL/6 mouse embryos. The four embryos are all at stage 21 according to Theiler's (1972) staging table, which assigns each day of gestation to one stage. Closer analysis of facial development reveals aspects of morphology that allow sub-stages to be identified. Such increased precision of staging is important if we are to understand the subtle changes in timing that accompany or elicit alterations in developmental processes (see Miyake, Cameron and Hall, 1996a,b, 1997a).

Precise staging is also required because gestational age is not a reliable metric for mouse embryos; ages of embryos vary within and between litters. The problem is seen in a comparison between gestational age and morphological stage of three inbred strains of mice (Figure 23.2). A given gestational age contains embryos of more than one morphological stage, and embryos of younger gestational ages are represented by a greater range of stages than are older embryos, probably because of the catch-up phenomenon outlined above (Figure 23.2). The relationship between

number of stages and gestational age is best described as two clusters with a sharp transition at 13.5 days gestation (Figure 23.2). Such a time shift could not have been seen in embryos staged on gestational age alone.

These data can be used to calculate the duration of individual stages and to assess differences in development timing, rates, or sequences between individuals, strains or species. We are using this approach to compare inbred strains of mice (Miyake, Cameron and Hall, 1997b), organs in individuals of a single strain, and (in a collaboration with Benedikt Hallgrimsson) fluctuating asymmetry as an index of developmental constraint. Klingenberg addressed the problem in his studies of the water striders *Limnoporus canaliculatus* and *Gerris costae*, using a common principal components model which assumes that similar patterns underlie variation within stages and covariation across stages.[7]

Our approach to developing staging tables for inbred mouse strains is to use as many morphological characters per stage as can be distinguished, either for the entire embryo or

Stage 21.1   Stage 21.2

Stage 21.3   Stage 21.4

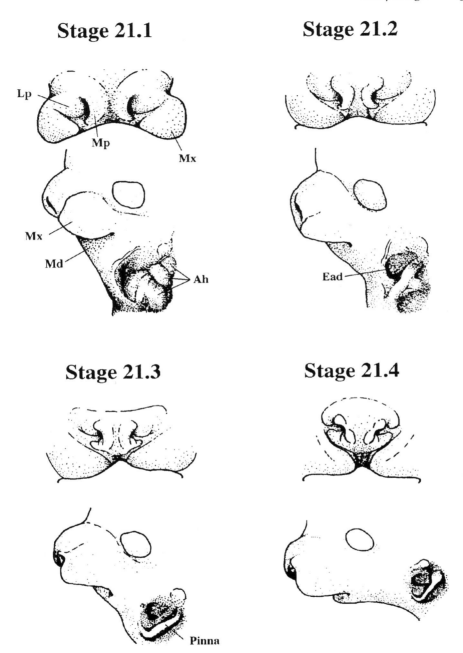

Fig. 23.1 Embryos of C57BL/6 mice designated as Theiler (1972) stage 21 can be divided into sub-stages (21.1–21.4) using detailed analysis of facial development. Each sub-stage is shown in frontal view of the face and in lateral view of the anterior portion of the head. Ah, auricular hillocks from which the pinna arises; Ead, external auditory meatus; Lp, the lateral nasal prominence; Md, mandibular process; Mp, the medial nasal prominence; Mx, the maxillary process. See Miyake *et al.* (1996a) for details.

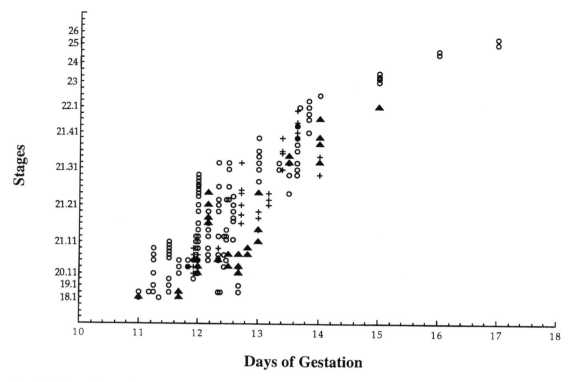

Fig. 23.2 The relationship between days of gestation and Theiler (1972) stages in three inbred strains of mice in inbred C57BL/6 mice. Each day of gestation is represented by embryos of different stages, with more stages/day at earlier than at later ages. Open circles, C57BL/6; +, C3H/He; triangles, CBA/J inbred mice. See Miyake *et al.* (1997b) for details.

for particular embryonic regions. An alternate approach applied to amphibian larvae by Grillitsch (1989) and to teleost fishes by Shardo (1995), defines each of 26 and 35 developmental stages respectively by no more than one or two morphological criteria, each of which is recognized as a fundamental developmental feature of the group of organisms at that stage. Other features present at that stage are treated as variables potentially uncoupled from the 'defining' features. This approach has the drawback of subjective identification of the defining characters and the assumption that all characters not so identified are labile. That being said, the approach holds promise for groups with well-resolved phylogenies and a subset of conserved characters; Grillitsch pro-

vides lists of 'good' and bad' characters. Also, it may lead to identification of fundamental features that characterize *Baupläne* and underlie phylotypic stages.

Confronted with comparing species of centipedes of the genus *Lithobius*, which all metamorphose but which have different numbers of moults, Minelli and his colleagues (1996) divided centipede development into two major periods: hatching to the last larval/first post-larval stage, and the first post-larval to the first mature stage. With this revised scale they were able to reveal previously unreported shifts in timing in developing centipedes. The crucial importance of establishing staging criteria that allow meaningful cross-species comparison is clear.

Table 23.4 Biological time (t_o) for selected phases of embryonic development in selected fishes and amphibians

Period	Fishes	Amphibians
Fertilization to first cleavage furrow	3	3
Fertilization to fall in mitotic index	12–14	
First cleavage furrow to fall in mitotic index	14–20	10–26
Fall in mitotic index to gastrulation	6–16	12–27
Fertilization to onset of RNA synthesis	10–18	
Fertilization to onset of gastrulation	18–29	
Fertilization to 10 pairs of somites	50–61	
Fertilization to end of epiboly	35–66	
Gastrulation to slit-like blastopore	18–19	12–25
Slit-like blastopore to fusion of neural folds	7[1]	12–21
Gastrulation to ten pairs of somites	28–34	
Gastrulation to end of epiboly	17–37	
Fertilization to fusion of neural folds	45[1]	47–67

Based on data in Dettlaff *et al.* (1987) for 7 species of fishes and 8 species of amphibians. See text for details.

[1] Based on single species, *Acipenser güldenstädtii*.

23.3 CELL CYCLES

In a major review including much previously inaccessible Russian work, Dettlaff, Ignatieva and Vassetzky (1987) tackled the problem of using duration of developmental periods and processes as measures of developmental time. By taking temperature into account they were able to develop a 'biological time parameter'. Duration of one cell cycle (t_o), as determined during the synchronized early phase of cleavage divisions, was chosen as the unit of biological time. This approach is not new – it goes back to studies by N. J. Berrill (1935), who used time between the first and second cleavage furrows as the unit of developmental time. The metric used by Dettlaff and colleagues is relative duration of the mitotic phase of the first cleavage division. The relative duration of various early developmental processes as measured by t_o is remarkably stable (Table 23.4).

One biochemical control mechanism for early cleavage stages in *Xenopus* is the protein cyclin E1. Cyclin is required for cells to transit from the pre-replication gap following mitosis (the G_1 phase) to DNA synthesis (the S phase). Howe and Newport (1996) proposed that timing during cleavage can be measured to within some 5 hours by degradation of cyclin. Cyclins and cyclin-dependent protein kinases are candidate molecules regulating cell cycles.

Dettlaff and colleagues discussed evidence relating cycles of DNA replication to onset of RNA synthesis and selective activation of enzyme synthesis. Others emphasize DNA replication and cell cycles as the mechanism timing cell differentiation. Holliday took this to the genetic level by proposing methylation of repeated sequences of DNA as the molecular basis for developmental clocks.

Exciting recent studies indicate that genes for at least some aspects of developmental timing exist and have been conserved phylogenetically. A *Clock* gene that encodes a helix-loop-helix transcription factor regulates circadian rhythm in mice. *Drosophila* share a similar gene, and see section 24.1.4 for a discussion of 'heterochronic genes'. In a study of the evolution of rapid development in

Drosophila, Rothe and colleagues (1992) demonstrated that duration of the cell cycle is coordinated with genome size. Such a coordination constrains the evolution of rapid development, especially during early embryonic development, when cell division is rapid.[8]

23.4 CRITICAL DEVELOPMENTAL EVENTS

In Chapters 8–11, I outlined the major developmental events responsible for transitions from stage to stage during ontogeny, including transitions between morphological stages, between hierarchical inductive events such as sequences of epithelial–mesenchymal interactions, and between such developmental processes as differentiation to growth. Satoh (1982) marshalled evidence that initiation of cell differentiation and morphogenesis are 'timed' by different 'clocks'. Differentiation is set by DNA replication and cell cycles, morphogenesis by cytoplasmic components. **Critical periods** are often emphasized in developmental studies, especially in teratology. The transitions discussed by Satoh are **critical events**.[9]

Cooke and Smith (1990) argued that *Xenopus* embryos measure time elapsing between fertilization and major inductive events such as induction of the mesoderm. The stages that we measure when we assess time since fertilization are, in their view, a true projection of how embryos measure time. Such a metric operating in individual cells (a cell-autonomous timing mechanism) does not measure number of cell divisions. It measures the state of commitment of the cell at a given time during ontogeny; it is an oscillator. Actin genes are turned on at mid-gastrulation irrespective of when mid-gastrulation occurs; they are stage-specific. These authors conclude:

'that all the cells of the embryo monitor the time elapsed since fertilization, in the form of a schedule that lists the activities that cells of the embryo might be directed to undertake and says when these should be undertaken.'
Cooke and Smith, 1990, p. 891

The 'schedule' is not of independent events but a sequence of events that proceeds in a particular order. Servetnick and Grainger (1991b) obtained evidence for autonomous developmental timing in their studies of lens induction in *Xenopus laevis*. The time at which a lens can arise from ectoderm is set by the gradual loss of neural competence and by the appearance and subsequent rapid loss of competence to form a lens. The discovery of an endogenous clock associated with the segmentation of paraxial mesoderm into somites at regular 90 min. intervals in the chick, and that this clock coincides with a repeated 90 min. periodicity in expression of the gene *Ch-1* (*chairy*), opens a window to the discovery of other organ-specific, gene-based clocks in vertebrates (Palmeirim *et al.*, 1997).

The hierarchical integration of inductive events as epigenetic cascades is a sequence of causes and effects that can be quantified. Once epigenetic cascades of embryonic inductions are determined in greater detail it may be possible to use them for comparisons between organisms on the basis of developmental processes. Such comparisons will pave the way for an understanding of mechanisms associated with change in timing – heterochrony, or change in spatial relationships or sequences, and heterotopy, which are the topics of the next chapter.

23.5 ENDNOTES

1. The major points raised in this section are taken from a more detailed analysis by Hall and Miyake (1995a). See Blackstone (1987) and the critique by Strauss (1987), also Blackstone and Yund (1989), German and Meyers (1989), Klingenberg and Spence (1993), Godfrey and Sutherland (1995a,b) and Klingenberg (1997) for discussions of size or time as the metric and for access to the literature; and German *et al.* (1994) for ontogenetic components in sexual dimorphism at maturity. For analyses and discussion of stable reference points, see Dettlaff *et al.* (1987), Jones (1988) and Reiss (1989). Clock (chronological) time is superbly dealt with by Landes (1983) and Whitrow (1990). Stoleson and

Beissinger (1995) discuss hatching asynchrony in birds. Whether size or age controls maturation is discussed with some force by Bernardo (1993a,b).

2. See Cock (1966), Stearns (1976, 1992), Charnov (1982) and Godfrey and Sutherland (1995a,b) for relationships between size and rate of growth.

3. See Driesch (1892) and Waddington (1938) for the echinoderm and amphibian experiments. See Tarkowski (1959), Power and Tam (1993), Snow and Tam (1979) and Henery *et al.* (1992) for studies on regulation and catch-up growth in mammalian embryos.

4. Starck's (1993, 1996) analyses follow the classic studies by Rickleffs (1973, 1983). This work is discussed in more detail in Hall and Miyake (1995a).

5. Calder (1984) contains an extensive discussion of physiological metrics. Metabolic rate is not necessarily correlated with development rate so that such metrics are much less useful in the embryonic or postnatal periods of ontogeny. Although large animals tend to have slower metabolisms and longer lifetimes and generation spans than small animals, 'speed of life' is not necessarily correlated with size (Purvis and Harvey, 1997). Mechanical or hydrodynamic constraints can also confound correlation of metabolism with time.

6. See Dunlop and Hall (1995), Hall and Miyake (1992, 1995b) and Miyake *et al.* (1997a,b) for visualization and timing of condensations, and Tyler (1988) and Snow (1996) for the use of electron microscopy in systematics of small invertebrates and plants respectively.

7. Miyake *et al.* (1993, 1996a,b, 1997a,b) for the mouse studies, Klingenberg (1996a,b, 1997) and Klingenberg *et al.* (1996) for the water striders.

8. See Hall and Miyake (1995a) for the supporting evidence and arguments for a developmental mechanism for time, and Howe and Newport (1996) and M. C. Raff (1996) for molecular bases of such a mechanism. See Satoh (1982), Holliday (1991) and Rothe *et al.* (1992) for possible molecular bases of a cell-cycle clock. See Antoch *et al.* (1997), King *et al.* (1997) and Reppert and Weaver (1997) for positional cloning and functional identification of the circadian *Clock* gene in mice. For a general discussion of factors that affect rate and time of development in *Drosophila*, see Karr and Mittenthal (1992).

9. See MacLean and Hall (1987) and Hall (1992) for examples of transitions between stages in ontogeny, and Hall (1985b) for a discussion of the distinction between critical periods and critical events.

24

TIME AND PLACE IN EVOLUTION: HETEROCHRONY AND HETEROTOPY

'Use time, or time will use you.'

Anon

'From the perspective of the developmental biology of pattern formation, spatial phenomena are just as fundamental as temporal phenomena. In development, time is no more primary or basic than space.'

Zelditch and Fink, 1996, pp. 252–3

Variation in timing of a developmental process in a descendant relative to the timing of the same process in an ancestor is **heterochrony**. Development of a feature in a descendant relative to an ancestor in a different place, from different cells or tissues, or by different mechanisms is **heterotopy**. As a term, process and pattern, heterochrony is much used and much abused. Heterotopy is virtually ignored. In this chapter I address:

1. the separation of heterochronic/heterotopic patterns from heterochronic heterotopic processes;
2. how to recognize whether a shift in timing or position of a developmental event is responsible for a particular heterochronic or heterotopic pattern;
3. recognition that altered timing of developmental processes need not necessarily lead to altered heterochronic patterns;
4. whether heterochronic patterns can arise from non-heterochronic processes; and
5. the ecological context and relevance of heterochrony and heterotopy.

24.1 HETEROCHRONY

The term heterochrony, as coined by Haeckel in 1875, is such a simple word (G. *heteros*, other, different; *chronos*, time), and such a simple concept, that it is not surprising that heterochrony has featured so prominently in developmental explanations of evolutionary change in morphology. In fact, many evolutionary biologists regard heterochrony as the developmental mechanism responsible for evolutionary change in morphology. Others (e.g. Keith Thomson, David Wake, Rudy Raff and myself) have urged caution in the application of heterochrony as the explanation for so much evolutionary change. Thomson neatly summarized the concern:

'Indeed, the first general conclusion that one reaches after reading these three books together is that, while heterochrony is an interesting phenomenon, it is peripheral to the central question and has captured too much attention among evolutionists. The real question involves the way in which fundamental processes of development respond to genetic and other perturbations, and vice versa. That heterochronic mechanisms involving size, shape, and time of sexual maturity can produce powerful effects is not in question, but it seems unlikely that this is a predominant mode of evolutionary change.'

Thomson, 1986, pp. 224–5[1]

Such caution will be apparent in this chapter.

24.1.1 Patterns and processes

Heterochrony as a **pattern** of evolutionary change is brought about by changes in the timing of developmental **processes**. New morphological patterns can (but need not) result from heterochrony. Heterochrony is a consequence of how developmental processes change over evolutionary time and not a developmental mechanism in and of itself. Change in timing cannot be a developmental mechanism. The timing of something always has to change and that something is a developmental process.

We can recognize two categories of heterochrony:

1. **systemic** (global), affecting the whole organism, as in neoteny in salamanders; and
2. **specific** (local), affecting a portion of the organism.

I am primarily concerned with specific heterochrony affecting tissues and organs. Within specific heterochrony, we can recognize:

1. primary specific heterochrony, involving a shift in timing of the organ under investigation; and
2. secondary (regional) specific, involving a shift in timing of development of other tissues and organs or of the hormonal status influencing the organ under investigation.

[I have deliberately not inflicted on the reader the terminology and terminological exactitute/inexactitude that accompanies heterochrony. If interested (desperate?) see Klingenberg (1997) and Reilly, Wiley and Meinhardt (1997) for recent and thoughtful (but not necessarily congruent) attempts to reconcile definitions with meaningful biological processes.]

Hererochronic patterns arise in any of four ways:

1. the **entire development** of an organ or tissue is displaced earlier or later, so that development begins and ends earlier or later;
2. only the **onset** of organ development is shifted. This is seen in achondroplasia (disproportionate dwarfism) in the *creeper* mutant in the chick;
3. development of the organ is **prolonged,** so that organogenesis ceases at a later time or stage; and/or
4. **rate** of development or growth changes without any corresponding shift in either onset or offset times. Sometimes known as **ontogenetic scaling**, this is seen in large and small breeds of dogs, the scaling of skull size in African apes, long-bone size in dogs and marsupials, and in the response to selection for rate of development early but not later in ontogeny in *Drosophila* and mice.[2]

Thus, three characteristics of the development of organs and organisms are subject to heterochronic change: onset, rate, and/or offset. A major reason for invoking heterochrony so frequently as the evolutionary mechanism for change in morphology is that these three parameters embrace the essential components of any change in development, growth, allometry, scaling and functional adaptation. They are fundamental aspects of developmental processes and of the functioning of the developmental units identified in section 20.4.2. It is not surprising that their timing varies in ontogeny.

When invoking heterochrony it is critical to demonstrate that one or other of onset, rate, and offset has been altered in a descendant relative to the condition in an ancestor. Otherwise heterochrony becomes so broad (if not universal) that its resolving power is drastically diminished – a concept that is true but trivial. Furthermore, as Rice (1997) demonstrated so elegantly, there are many instances where changes in size or shape do not produce heterochronic change. There are also those situations of secondary specific heterochrony when altered timing of some other character is responsible for the change in the character of interest.[3]

Changes in morphology that are considered classic examples of heterochrony (those affecting entire organisms, as in neoteny in axolotls) are elicited by alterations in development that do not alter developmental timing, unless all changes in the rate of development of all organ systems are subsumed as heterochrony. Such lumping renders heterochrony so broad as to be meaningless, unoperational, and difficult if not impossible to test. As Brian Shea (1989) emphasized, it is critical to distinguish between rate and timing differences when analysing heterochrony. The difference between disproportionate and proportionate dwarfism (categories 2 and 4 above), and Shea's studies on heterochrony in primates, illustrate his point.[4]

Identification of a change as heterochrony is often based on adult morphology alone. The altered morphological pattern is interpreted as arising from alterations in developmental timing which are neither demonstrated, nor sometimes even sought. Hererochrony as pattern and heterochrony as process are neither necessarily coupled nor causally connected. Reilly, Wiley and Meinhardt (1997) emphasized this in their separation of interspecific from intraspecific phenomena in heterochrony. As heterochrony is a statement about changes from an ancestral condition, only interspecific comparisons can reflect phylogenetic patterns.

24.1.2 Heterochrony as modification of developmental processes

Changes in timing of development encompass all developmental processes (Chapter 23). They can be generalized as:

- timing of onset of development, involving processes such as formation of condensations, initiation of cell differentiation and selective gene activation;
- size at onset of development associated with initiation of growth;
- rate of development affecting growth, allometry, scaling and morphogenesis; and
- time of cessation of development as termination of differentiation, stable morphogenesis, or the triggering of senescence.

Gould (1977) and Alberch *et al.* (1979) provided the clock and ontogenetic trajectory models to identify these parameters as heterochrony. In 1984, I elaborated the developmental mechanisms that underlie heterochrony, using the vertebrate skeleton as a model system. I suggested three integrated mechanisms that:

1. establish (by the number of mitotically active cells in each skeletal condensation and their intrinsic rate of division) the age and size at which a skeletal rudiment began to develop;
2. determine how epigenetic factors (such as muscle action, innervation and vascularity) operating within intrinsic limits, determine the rate of skeletal development; and
3. regulate (by a combination of intrinsic and epigenetic factors) the time, size, and/or shape at which development of an organ stops.[5]

The close similarity of this scheme to the model outlined in Chapter 20 should be evident. This is only to be expected if heterochrony acts by altering the timing of developmental processes, as it does, and if fundamental development units can be identified, which they can.

As emphasized in presenting the model in Chapter 20, the number of cells in a skeletogenic condensation is a prime determinant of the size and shape of the skeletal element that develops. A delay in the time when cells are committed to differentiate results in accumulation of fewer cells to form that skeletal element. Regulation compensates for such delays, but regulative ability is transient and often lost just before formation of the cellular condensation. Thus the condensation, and the developmental units that characterize cells in a condensation, emerge as units of development of organ rudiments especially sensitive to heterochronic changes, especially those that affect onset and rate of development.[6]

24.1.3 Heterochrony as modification of inductive tissue interactions

Inductive tissue interactions provide an important class of epigenetic developmental signalling. They determine where, when and how individual tissues and organs develop, and how much tissue develops. Many inductive interactions initiate a condensation and determine the number of cells in a condensation. The exquisite control of timing by inductions discussed in Chapters 10 and 11 has been demonstrated for many tissues and organs during vertebrate development. It represents one of only a few fundamental characteristics of vertebrate development.[7]

Spatial localization and temporal localization of embryonic inductions means that changes in timing of induction are a developmental basis for heterochrony. When invoking such interactions we must deal only with causal sequences, eliminating temporal sequences that are not causally related. By definition, inductions are causal sequences.[8]

(a) Novelty

The role of timing in the generation of novelty is seen in those situations in which alteration in timing brings about the differentiation of a new tissue type. Perhaps the most well-documented examples are the formation of enameloid rather than enamel during tooth development in the transition from larva to adult in urodele amphibians and lungfishes, and in the teeth of *Polypterus*.[9]

Enamel is deposited by epithelial cells, dentine by mesenchymal cells (s. 11.5). Enameloid is a composite tissue consisting of proteins derived from epithelium and mesenchyme. A delay in the differentiation of epithelial cells relative to mesenchymal cells allows dentine to be deposited before epithelial proteins are secreted, resulting in formation of enamel rather than enameloid. The formation of cosmine in fossil osteichthyans is interpreted as arising from a similar heterochronic shift in inductive interactions. In discussing her work on *Polypterus*, Meinke (1982b) concluded that enamel-enameloid-dentine forms a tissue continuum that has diverged because of changes in developmental timing and matrix production. Shifts in timing of epithelial–mesenchymal interactions therefore can alter structure at the level of tissue type. The demonstration of a timing shift between ancestors and descendants constitutes evidence for heterochrony.

(b) Mutations

Timing of inductive interactions is subject to genetic control, as amply illustrated by mutants that affect timing in one or another of the interacting components. The development of scales on birds' legs, the limb skeleton, and organs that develop under hormonal control, are three representative examples.[10]

The lower portions of birds' legs are covered by scales, but scales fail to develop in birds carrying the *scaleless* (*sc*) mutation. If epidermis from the distal (metatarsal) portion of the lower limb of *sc/sc* embryos is recombined with dermis from wild-type embryos of the same age, a scaleless phenotype develops. Perfectly normal scales form, however, when *scaleless* epidermis is recombined with dermis from older wild-type embryos in which scale development has proceeded further. There is a progressive development of 'scalelessness' in the mutant dermis as a shift in timing prevents epidermis and dermis from interacting. Timing defects can be overcome experimentally, and could also be overcome if selection were to drive a change in developmental timing. It is, however, not known whether sufficient variation exists for natural selection to 'see' such developmental processes as a target for selection, or whether the processes respond because of variation that occurs at a higher level of organization.[11]

24.1.4 'Heterochronic genes'

Richard Goldschmidt (1938, 1940) argued that timing and rate of development are under genetic control, leading some authors to search for 'heterochronic genes'. But given the integration of vertebrate development, heterochrony in those animals is unlikely to reduce simply to the operation of heterochronic genes. In the nematode *Caenorhabditis elegans*, however, where cell fate is much more circumscribed and linked to cytoplasmic determination and cell lineages, a series of genes does regulate timing of the switch from larva to adult.

Identified by Ambros (1989), these *lin* (cell lineage abnormal) genes control timing by regulating the number of stem cell divisions. *Lin* genes are regarded as heterochronic because they alter the timing of cell divisions and decisions concerning cell fate are made during division. *Lin-14*, *lin-28* and *lin-29* act in sequence to modify timing of the specification of cell fate. Subsequently, *lin-4* was added to the list of heterochronic genes involved in what Ambros termed the 'larva-to-adult switch'. This switch is characterized by such changes as formation of the adult cuticle, activation of adult collagen genes and repression of larval genes. Mutants of *lin* genes act as binary switches, switching cells from one lineage to another. Their action is comparable to the homeotic genes presented in section 9.4; in fact, there is considerable sequence similarity between *lin-12* and the *Drosophila* homeotic gene *Notch*.

Duration of development and lifespan in *C. elegans* are regulated by the products of four maternal-effect genes, *Clk-1* to *Clk-3* (*Clk* for Clock) and *gro-1*. As Lakowski and Hekimi (1996) demonstrated so dramatically, mutations in these genes extend the lifespan to five times that of wild-type individuals.

Raff and his colleagues identified genes involved in heterochronic changes in sea urchins. As in *C. elegans*, heterochronic changes in sea urchins come about through alterations in specific cell lineages. The gene *msp130* is restricted to skeletogenic lineages and these lineages are significantly reduced in direct-developing species. Expression of *msp130*, which is transcriptionally regulated, is delayed in the direct developer *Heliocidaris erythrogramma*.[12]

Although genes regulating circadian clocks in *Drosophila* and *Neurospora* were identified 20 years ago, and the *lin* genes in *C. elegans* were identified a decade ago, 'heterochronic genes' are only now beginning to be uncovered in other organisms. *Hoxd-13* acts heterochronically in the mouse by retarding limb development by as much as four days (s. 11.4). Tissues and organs that only appear in adults of other ascidian species develop precociously in larvae of the direct developer *Molgula citrina*. A muscle actin gene that is expressed in adults of other species but in larvae of *M. citrina* (s. 22.2.3) was identified by Swalla *et al.* (1994) as a candidate 'heterochrony gene', but it is unclear whether this gene regulates the time shift or is a consequence of the time shift.[13]

24.1.5 The ecological context of heterochrony

As an evolutionary mechanism, heterochrony couples embryonic development to ecology and *vice versa*. Environmental factors such as diet and population density evoke heterochronic changes in cichlid fishes, ambystomatid salamanders, birds and mammals (see Chapters 16–17). Few studies of embryonic induction or tissue interactions address environmental control over these important aspects of embryonic development. A notable exception is those interspecific inductions involved in cyclomorphosis or seasonal polymorphism discussed in Chapter 18.[14]

(a) To induce or not to induce

Vertebrate lens induction was introduced in section 11.3 (where basic development was discussed) and again in section 21.7.2 in the context of homology and divergent developmental processes. A close, comparative analysis of the timing of lens formation, of the steps in the inductive interactions, and of the influence of temperature on the induction, reveals links between heterochrony, ecology and evolution.

Jacobson and Sater (1988) summarized the data for lens development in 37 species. In the three fish, one bird and one mammal examined, the retina is required for the lens to form. In eight out of 23 species of anurans and four out of nine species of urodeles, however, the lens differentiates without an inductive interaction with the retina, although earlier inductive interactions are required. What should we make of this situation in amphibians?

No obvious phylogenetic pattern is evident. On the contrary, of those genera in which more than one species has been studied:

- of four species of *Bufo,* one (*B. bufo japonicus*) requires retinal induction, three (*B. carens, B. regularis, B. vulgaris*) do not;
- of 12 species of *Rana,* 11 require induction, one (*R. sylvatica*) does not;
- of two species of *Ambystoma,* *A. maculatum* requires induction, *A. mexicanum* does not; and
- of three species of *Triton,* one, *T. pyrrhogaster,* requires induction, *T. cristatus* and *T. taeniatus* do not.

Presence of the interaction is the generalized condition. Parsimony indicates that presence is primitive, loss of the induction being a derived condition. Is this loss due to heterochrony? There are two major possible explanations for this and related examples:

1. a heterochronic pattern, arising via a non-heterochronic process: loss of a causal inductive step (inductive interaction) in development results in the lens being formed earlier or later in its development than in the ancestral condition;
2. a non-heterochronic pattern produced by a heterochronic process: loss of retinal induction as the loss of only the last step in a series of inductive interactions, effectively terminating induction earlier than in other forms, with the lens forming at the same time as in ancestral forms.

Whether we consider these changes as heterochrony depends on whether we are classifying heterochronic patterns or heterochronic processes.

Given the arguments for a link between induction, heterochrony and ecology developed in this section, it may be particularly significant that presence or absence of the induction is temperature-dependent. Early tissue interactions are enhanced at low temperatures. When *Rana esculenta* are reared at 12.5°C before removal of the optic cup, 60% of the operated embryos form normal lenses. When reared at 25°C only 9% form lenses. A similar situation is found in the newt *Taricha torosa.* Lens formation is related to rearing temperature, with the highest percentage of lens formation at 16°C. Jacobson and Sater attributed these differences to the effect of temperature on development rate, which in *T. torosa* ranges from 25 to three days over a temperature range of 5–25°C. Through temperature-dependent variability in embryonic induction, heterochrony can couple development to ecology and ecology to evolution.[15]

(b) Size

Such associations have been verified in a variety of different organisms. Klingenberg's studies of water striders are an especially elegant analysis. Heterochrony is especially evident in

Table 24.1 Three examples of the association of heterochrony with small size

Species displaying heterochrony	*Largest species in taxon*
Salamanders	
Thorius, the smallest at 1.3 cm SVL	*Andrias davidianus*, the largest at 1.5 m SVL
Caecilians	
Idiocranium russeli at 1.1 cm SVL	*Caecilia thompsoni*, the largest at 1.5 m SVL
Elephants	
Elephas falconeri from Sicily at 130 kg	*Loxodonta africana*, African elephant at 10 tonnes

SVL, snout-vent length. Note the order of magnitude difference in sizes from smallest to largest. From M. H. Wake (1986), Roth (1992) and Hanken (1993b).

the smallest member of many taxa, some of which are listed in Table 24.1. Examples are the salamander *Thorius* from Mexico, at 1.3 cm snout-vent length one of the smallest salamanders; *Idiocranium russeli*, one of the smallest caecilians (legless, burrowing amphibians; Table 24.1); and small breeds of dogs. Heterochrony also accompanies increases or decreases in the size of individual organs such as the brain.[16]

(c) Environmental range

Heterochrony can also be the developmental response to climatic or altitudinal variation within the geographic range of a species, often in association with delayed or advanced sexual maturity as the primary temporal shift. Such changes can be genetically assimilated (s. 19.1). An example of heterochrony related to environmental range is the development of nasal bones in individuals within some southern populations of the Olympic salamander, *Rhyacotriton olympicus*, reported by David Wake (1980). Nasal bones are present in all salamanders that metamorphose and in the permanent larvae of other ambystomatid salamanders, appearing late in development. *Rhyacotriton* is unique in lacking nasal bones except in some individuals in the warmest, most southerly portions of the geographical range. The explanation is that all individuals, irrespective of locale, possess the developmen-

tal potential to form nasal bones. With a slowed rate of nasal bone development relative to general body growth, sexual maturity normally occurs before the nasal bones differentiate, except in warmer portions of the species' range. Smirnov (1994, 1995) also demonstrated that bones can appear very late in larval ontogeny; some bones in *Xenopus* are only found in a small number of very old individuals.

Formation of the nasal bones in *Rhyacotriton*, however, need not require a heterochronic event in nasal bone development. Timing of development of the primordium might be the same as in its ancestors, nasal bone development being permitted in some individuals because heterochrony in sexual maturation prolongs development. Developmental change need not directly affect the ossification process of the nasal bones themselves for heterochrony to affect nasal bone development.

(d) Coupling and decoupling

A dilemma emerges. How do we distinguish, developmentally and mechanistically, between altered timing of the organ primordium and a shift in the timing of other organs that secondarily influence the organ of interest? At the level of terminology, we distinguish primary from secondary specific heterochrony (s. 24.1.1). In terms of mechanisms, we look for evidence of coupling or decoupling. Examples will be taken from skeletal development.

Brian Shea has investigated allometry in primates, and analysed giant transgenic mice produced from eggs injected with metallothionein human-growth-hormone fusion genes. He argues for shifts in growth hormone and insulin-like growth factor-I as proximate causes for truncations or extensions of growth trajectories affecting final body size. Scaling of organ to body size also may be based on hormonal regulation which, in turn, is proportional to intrinsic regulation of growth in individual skeletal elements.[17]

Alberch, his colleagues, and most who have worked in this area, have concentrated on patterns of chondrification. Dissociation of the development of two normally integrated tissue or organ systems can be readily identified when hormones are the mediators of change. For example, osteogenesis is dissociated from chondrogenesis following implantation of pellets of thyroxine into amphibian tadpoles. Chondrogenesis is advanced and osteogenesis retarded because chondrogenesis responds earlier, to a greater extent, and to a lower concentration of thyroxine than does osteogenesis. This example demonstrates how two processes whose timing appears to be tightly coupled can be decoupled in ontogeny. Decoupling in a descendant relative to an ancestor would appear as heterochrony.[18]

Decoupling of osteogenesis from chondrogenesis also occurs naturally.

Decoupling has been demonstrated in skeletal formation in such lizards as *Lacerta agilis*, and in *Bufo bufo* and other bufonid amphibians. Uniform states of skull development occur at a range of different stages of limb development in lizards and anurans. (In *Bufo bufo*, ossification of cranial and limb skeletons also occurs at quite different ontogenetic stages.) As measured by the standard Gosner stages used for amphibians, ossification in individual *Bufo bufo* was initiated as early as stage 38 and as late as stage 43. Dunlap and Sanchiz (1996) consider this dissociability an adaptation to rapid larval development in this species. We have similar data on temporal dissociation of ossification and chondrogenesis in inbred strains of mice.[19]

Uncoupling of the sequence in which cartilage and bone develops is also seen in the limbs of primitive diapsid reptiles.

Chondrogenesis of tetrapod limb skeletons proceeds from proximal to distal. This is the same order in which the limb mesenchyme differentiates and in which precartilaginous condensations are deposited. In primitive diapsids, ossification begins proximally down to and including the elements of the zeugopod (radius/ulna or tibia/fibula), skips the carpals and tarsals, beginning again in the more distal metacarpals, metatarsals and digits. Olivier Rieppel has begun to document these sequences in developing stages of living reptiles, while Michael Caldwell has applied this approach to a diversity of fossil reptiles. Caldwell's approach is to determine the skeletogenic sequences in ontogenetic series of Permian diapsid reptiles and to compare them with the extant forms. From his analysis it is clear that patterns of ossification are as constrained as patterns of chondrification. Indeed, there is an amazing correspondence between chondrogenic and ossification sequences, except when the limbs are modified for swimming, as Caldwell documented for ichthyosaur embryos preserved *in utero* and for the paddle-like limbs of both ichthyosaurs and plesiosaurs. With adoption of aquatic locomotion, perichondrial bone (the bone that develops on the surface of long-bone cartilages) is lost progressively phylogenetically (as it is ontogenetically):

1. from the digits in Early Triassic ichthyosaurs;
2. then from the ulna and fibula in Late Triassic forms; and
3. subsequently from the fibula and tibia in Jurassic forms.[20]

Richard Strauss determined ossification sequences for five species of poeciliid fishes by treating each ossification as a developmental event, and evaluated timing differences within and between species. He was then able to use the timing profiles to construct hypothetical phylogenetic trees onto which the character state changes could be mapped. Such an approach gives a hint of future advances as functional morphologists, palaeontologists and developmental biologists integrate their approaches to the evolution of organismal form and function.[21]

24.2 HETEROTOPY

Heterotopy, the displacement of the development of an organ or tissue in space, was used by Haeckel to explain the origin of reproductive organs from different germ layers in different organisms.

Haeckel's reasoning went as follows: diploblastic organisms lack mesoderm but contain reproductive organs and germ cells which arise from ectoderm and endoderm. In many triploblastic organisms these same organs and cells arise from mesoderm. As Haeckel took the diploblastic condition to be primitive and diploblastic organisms to be ancestral to triploblastic, he reasoned that the origin of the reproductive organs shifted from ecto- or endoderm to mesoderm. Hence heterotopy.

Others took up this explanation. In the 1870s, Van Beneden studied separate germ-cell origins in coelenterates and the close links between site of origin/germ-layer theory (sperm from ectoderm, eggs from endoderm), sexual reproduction, and heredity (s. 5.5.3). The dogma that homologous structures must arise from equivalent germ layers (ss. 5.4, 5.5) extended the concept of homology from adults to embryos but placed a straitjacket around homology. The recognition by Wilson, Sedgwick and de Beer of exceptions to a strict relationship between origin and germ layer provided substance to Haeckel's concept of heterotopy.

The origin of diverse cell or tissue types from single germ layers also created difficulties. What to make of ectoderm producing nerve, pigment and epithelial cells, of ectoderm giving rise to somites, of mesoderm giving rise to intestinal or nerve cells, or of secondary neurulation?[22]

E. B. Wilson captured the essence of the dilemma of trying to accommodate heterotopy into germ-layer theory:

'And yet, it must be evident to any candid observer not only that the embryological method [origin from equivalent parts of the embryo] is open to criticism but that the whole fabric of morphology, so far as it rests upon embryological evidence, stands in urgent need of reconstruction . . . it is generally assumed that a safe basis of comparison is to be found in the origin of these parts from particular regions (germ-layers, etc.) of the embryo; and thus the embryological criterion of homology is still, on the whole, accepted. It is just here, however, that some of the most startling contradictions have recently come to light.'

Wilson, 1894, pp. 103, 105

Heterotopy receives little attention these days. Gould (1977) gave it 10 lines out of 500 pages and restricted it to displacement from one germ layer to another during phylogeny – Haeckel's example that started the concept. In discussing why primordial germ cells arise outside the gonads and undergo extensive migration to reach the gonadal primordia, Denis (1994) essentially evoked arguments from heterotopy and the ancestral condition of organisms with a germ line but no gonads.

There are good examples of heterotopy, some of which have been introduced already. Indeed, heterotopy may be about to come into its own as heterochrony wanes and our knowledge of developmental mechanisms increases. Examples of heterotopy include:

1. origin of gametes from endoderm or mesoderm and the initiation of gametic differentiation under maternal cytoplasmic control or via secondary induction (s. 9.1.3);

2. development of gill lamellae and the adenohypophysis of the pituitary gland from endoderm in agnathans but from ectoderm in jawed vertebrates;

3. origin of the enamel organs of teeth from ectoderm and endoderm;

4. origin of vertebrate skeletal tissues from mesoderm and ectoderm (neural crest; s. 10.4);

5. formation of the post-naupliar germ-band stage in cumacean crustaceans such as *Diastylis* from ectoteloblasts and an additional cell layer later in development;

6. formation of the turtle shell following altered pathways of cell migration (s. 16.4);

7. formation of mammalian middle ear ossicles from elements of the reptilian lower jaw (s. 17.3.1);

8. formation of internal and external cheek pouches; and

9. initiation of differentiation of Meckel's cartilage by different epithelia across the vertebrates.[23]

24.2.1 Heterotopy as modification of inductive tissue interactions

The origin of the turtle shell illustrates altered inductive tissue interactions as a basis for heterotopy, as we saw in section 16.4. Two other examples (8 and 9 from the list above) will be discussed in a little more detail for, as with turtle shells, they share altered cell migration and ability to respond to new inductive environments. These three well-studied examples of heterotopy exhibit temporal shifts and flexibility of inductive interactions, and point out the path we must follow to uncover other examples of heterotopy or of heterotopy coupled with, or mediated by, heterochrony.

(a) Internal and external cheek pouches

An analysis of the development and evolution of external cheek pouches in geomyoid rodents was undertaken in my laboratory by Phil Brylski.

Some species of New and Old World mice and squirrels possess cheek pouches that open inside the buccal cavity. Such internal pouches are lined with buccal epithelium and used to store food such as seeds. All members of the geomyoid rodents (pocket gophers, kangaroo rats and allied rodents), have external cheek pouches that open outside the mouth and are lined with fur. Both pouch types function to store food, although external pouches may conserve body water more efficiently than internal pouches.

An internal cheek pouch is primitive ontogenetically and phylogenetically. The shift from an internal to an external cheek pouch in the evolution of the geomyoids has been regarded as a macroevolutionary event in which intermediate states cannot readily be envisaged. Brylski demonstrated that both types of cheek pouch develop as evaginations of the buccal epithelium. In internal pouches the epithelium remains in association with mucous-forming buccal tissues, with which it interacts inductively to produce a typical mucous-lined buccal epithelium lining the pouch.

Brylski examined the development of external cheek pouches in one pocket gopher (*Thomomys bottae*) and three kangaroo rats (*Dipodomys merriami, D. elephantinus* and *D. panamintinus*). An initially internal pouch rudiment is externalized because of an anterior shift of the invagination (Figure 24.1). This shift allows the lip epithelium to be included in the evagination. This shift in location (heterotopy) allows the developing pouch to grow rapidly into the facial mesenchyme lateral to the buccal cavity. Growth into this new location brings the pouch epithelium into contact with hair-forming mesenchyme. An inductive interaction is initiated at this site and hair follicles form to line the pouch rudiment. This is a typical epithelial–mesenchymal interaction as outlined in section 11.2. The novelty (as in formation of the turtle shell) arises because of competence of the pouch epithelium to respond to a newly encountered set of in-

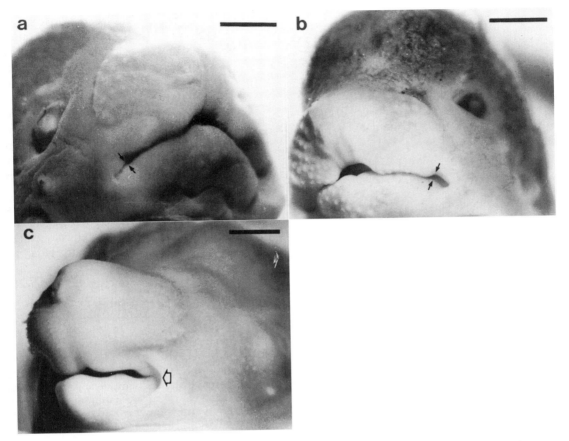

Fig. 24.1 During development of the kangaroo rat, *Dipodomys merriami*, an invagination arises at the corner of the lip (arrows in a). With subsequent development lip epithelium and external pouch invagination are continuous (arrows in b), leading to development of the external pouch rudiment (arrow in c) and its separation from the lip. Scale bars: 1 mm. Modified from Brylski and Hall (1988a).

ductive signals following altered pathways of cell movement.

Experimental studies support such a mechanism. Pouch epithelia and mesenchyme can respond to exogenous signals by modifying their development. Using tissues from hamsters, Covant and Hardy (1990) demonstrated that cheek pouch tissues can modulate their differentiation through either the epithelial or the mesenchymal component; mucous glands form from cheek pouches in response to a retinoid acting through the mesenchyme.

Thus, a simple shift in location coupled with growth into adjacent mesenchyme transforms an internal pouch lined with buccal epithelium into an external, fur-lined pouch. Heterotopy initiates, differential growth facilitates, and the developmental plasticity of buccal epithelium permits the development of a novel organ, the external cheek pouch.[24]

(b) Induction of Meckel's cartilage

Another case involving elements of heterochrony and heterotopy is the tissue interaction

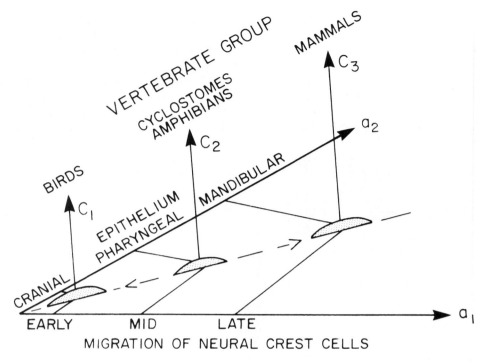

Fig. 24.2 Interactions between embryonic epithelia and neural crest cells in the differentiation of Meckel's cartilage in jawed vertebrates and branchial basket cartilages in cyclostomes (lampreys; jawless vertebrates) illustrate both heterochrony and heterotopy. The time during neural crest cell migration when the epithelial–mesenchymal interaction occurs varies across the vertebrates (shown as early-, mid-, or late-migration; a_1). The epithelium that initiates chondrogenic differentiation also varies (cranial, pharyngeal or mandibular; a_2). In birds the epithelial–mesenchymal interaction is early and involves cranial epithelium. In mammals the interaction is late and involves mandibular epithelium. In urodele and anuran amphibians the interaction occurs during neural crest cell migration and involves pharyngeal epithelium. In cyclostomes, the branchial basket cartilage, which is of neural crest origin, differentiates after interaction with pharyngeal endoderm, a time and place of interaction that is taken as the primitive vertebrate condition. Reproduced from Hall (1984b) with permission of the publisher.

responsible for the production of Meckel's cartilage in vertebrates. In all classes of vertebrates, epithelial–mesenchymal interactions initiate chondrogenesis of Meckel's cartilage but the epithelium involved varies in those representatives of vertebrate classes studied. Mandibular epithelium provides this interaction in the mouse, cranial ectoderm in the chick, and pharyngeal endoderm in numerous anuran and urodele amphibians (Figure 24.2). There has been a heterotopic

change or series of changes during vertebrate evolution.

Lamprey visceral arches are homologous with the visceral arches of jawed vertebrates (see the discussion of the transformation from agnathans to gnathostomes in s. 16.1). Because a similar interaction is found between pharyngeal endoderm and visceral arch (branchial basket) cartilages in lampreys, the pharyngeal endoderm–mesenchyme interaction seen in amphibians is regarded as the generalized vertebrate condition. Given this reasoning, interactions occur earlier in the one bird studied, and later in the one mammal, than in the numerous amphibians investigated.[25]

The interaction in the chick occurs before initiation of migration of neural crest cells. In amphibians the interaction occurs while neural crest cells are migrating toward the lower jaw, while in mice, the interaction occurs with neural-crest-derived mesenchyme in the mandibular arch, i.e. after completion of migration (Figure 24.2). Therefore, these heterotopic changes are linked with heterochrony. Also note that cranial and mandibular **ecto**derm but pharyngeal **endo**derm initiate the interaction in different groups. This transfer between germ layers is also heterotopic.

While neither the morphogenetic nor evolutionary consequences of these differences in timing is completely clear, one can imagine that the extensive growth of the avian beak and the leading role played by Meckel's cartilage in that growth require early specification of Meckelian chondrocytes in birds. Conversely, the extensive repositioning of branchial arch cartilages into the middle ear and their secondment as middle ear ossicles in mammals could only occur in a situation of delayed specification of Meckel's cartilage in the reptile → mammal-like reptile → mammal transitions (s. 17.3.1). Similarly, the development of the new jaw articulation in mammals might be reflected in altered specification of Meckel's cartilage. The evolutionary significance of alterations in time and place is considerable. Indeed, the ancestral pattern of a joint between an ossification around Meckel's cartilage and the temporal bone has been described in a human fetus with Treacher–Collins syndrome, in which altered timing repositions the lower jaw during development. Development of the ancestral pattern follows from the altered positions and the flexibility that vertebrate skeletal systems possess.[26]

Exciting as these studies are, we need information on the timing of such interactions in many more species than is currently available.

Although all urodele and anuran amphibians studied share a pharyngeal endoderm-mesenchyme interaction, there may be subtleties of timing yet to be discovered, especially in direct-developing and paedomorphic species. We would like to know the pattern in reptiles and marsupials to test the thesis of progressively delayed specification of the cartilage of the first arch as a requirement for origin of the middle ear ossicles. We would like to have comparisons between more closely related organisms (strains of mice, conspecifics, monophyletic groups) in order to analyse more subtle heterotopic changes.

We do have some data for heterochronic shifts between three inbred strains of mice. Although the dentary bone and Meckel's cartilage are induced at the same age (10 days) in CBA, C3H and C57 inbred strains, the morphological stages of development at which induction occurs are not constant, being late stage 15, early stage 16 and late stage 16 respectively (MacDonald, 1997). These altered timings are congruent with the relationships between these strains, CBA and C3H being closely related to one another and distant from C57.[27]

24.2.2 Heterochrony tinkers, heterotopy creates

Miriam Zelditch and William Fink focused their attention on the relationships between heterochrony and heterotopy in an important resurrection of the importance of heterotopy as a force for evolutionary change in morphology. They used techniques developed for analysis of skull growth in cotton rats in a study of piranha fishes. Partitioning of the skull into developmentally individualized parts allows epigenetic interactions between parts to be analysed and shifts in growth rates to be identified. The conclusion drawn by Zelditch and Fink is at the head of this chapter.[28]

Zelditch and Fink also drew important conclusions concerning differences between heterochrony (which modifies old structures) and heterotopy (which produces new ones). Heterochrony produces structures constrained by ancestral ontogenies. Changing timing implies that the ancestral developmental programme is unaltered other than in its timing. Consequently, heterochrony results in a parallelism between ontogeny and phylogeny. An especially nice example is the study by Maunz and German (1997) of limb-bone scaling in two species of marsupials which share a common ancestry and are born at the same neonatal size, but in which adult sizes differ by 50 times. The larger species (*Didelphis virginiana*) grows at a higher rate and for a longer period than does the smaller species (*Monodelphis domestica*) but there are few differences in scaling between the two species. Maunz and German conclude that heterochronic patterns affect neither the shape among adults nor shape over evolutionary time. Similarly, in a comparison between three breeds of rabbits that are giant, 'standard', or dwarf in size, Fiorello and German (1997), in looking for variation in ontogenetic trajectories, found that variation in size could be explained by differences in the length of time spent growing at rates that approached maximal growth rates.[29]

On the other hand, heterotopy produces novelties that are based in different developmental pathways than those producing the ancestral character. Turtle shells, external cheek pouches and middle ear ossicles are the best known examples. Rice (1997) draws our attention to the problem that heterochrony is a uniform change in rate of time, not a change in underlying biological processes that contribute to evolutionary novelty. A change in timing could sufficiently alter a developmental process so as to produce an evolutionary novelty but there is greater scope for evolutionary modification with heterotopy than with heterochrony. In general, **heterochrony tinkers, but heterotopy creates**.

This renewed attention to heterotopy is welcome, as it should refocus our attention on mechanisms of evolutionary change in morphology other than temporal. Heterotopic changes entail a reorganization of development that appears more drastic to us but that is wonderfully coordinated as far as embryos are concerned. The potential for self-adjustment that is inherent within developmental processes represents a latent coordination that becomes emergent through such mechanisms as ontogenetic repatterning. Because development is modified to a greater extent with heterotopy, heterotopic changes may be less common than changes based in heterochrony, and their effects more profound. (A heterochronic change early in development, or one involving an organ system upon which many other systems depend, can also have major effects on morphology.) The less frequent discussion of heterotopy as a mechanism may reflect our incomplete knowledge of the potential of embryos to change in coordinated ways, rather than the frequency with which embryos utilize heterotopy to effect that change. The combination of heterochrony and heterotopy provides embryos with the best of both worlds.

24.3 CONCLUSIONS

The examples discussed in section 24.1 illustrate the potential power and the dilemmas of the study of heterochrony, but leave many questions unanswered.

Despite much past work on comparative embryology, identification of a stable metric to measure the passage of developmental time remains a problem (Chapter 23). The basis upon which we should equate and compare developmental processes as mechanisms remains a problem. Embryos may measure time through decisions relating to cell fate. It is now possible to begin to compare the development of different embryos on the basis of when during development decisions are made (s. 16.2) – an exciting prospect. Such a coupling of heterochrony to epigenetic and hierarchical organization of embryonic development sets us on the path to understanding heterochrony as process.

How to identify that a change is a consequence of heterochrony remains unclear. Given that development itself has evolved (see everything in the book preceding this section), it is unclear whether we can compare developmental mechanisms across widely separated groups, or only between closely related (even monophyletic) groups. There is still the paradox that documenting a change in the timing of development does not necessarily mean that the structure evolved through heterochrony. In part, this is:

- the issue of whether altered timing in ontogeny always results in altered morphology in phylogeny;
- that patterns of evolutionary change that appear to be heterochronic, can arise from non-heterochronic processes; and
- the problem of recognizing the organ system in which timing has changed (primary and secondary specific forms of heterochrony).

Heterochrony in adult pattern cannot automatically be equated with a heterochronic change in the development of a descendant relative to its ancestors. Consequently, separating and/or relating heterochronic patterns from and/or to heterochronic processes remain important issues.

Heterochrony couples ecology to development and development to evolution (s. 24.1.5). Developmental programmes such as inductive interactions and tissue differentiation can be regulated by temperature, reinforcing the evolutionary relevance of heterochrony (indeed, reinforcing the evolutionary importance of epigenetic interactions in general) and suggesting a research strategy for analysing the developmental processes subject to heterochronic change.

Heterotopy holds great promise as a mechanism producing novelties in evolution. Heterotopy and heterochrony often go hand in hand. Indeed, a major role for hetero-chrony may be to facilitate heterotopy. If so, heterotopy also couples ecology to development and development to evolution. Alterations in time and place may yet reflect (the same?) unified mechanisms in phylogeny as they do in ontogeny.

24.4 ENDNOTES

1. The literature on heterochrony is enormous. For heterochrony as a mechanism for modification of development during evolution, see Garstang (1922), J. Huxley (1932), de Beer (1940), S. J. Gould (1977, 1992), Alberch *et al.* (1979), Raff and Kaufmann (1983), Hall (1984b, 1998a), Buss (1987), Broadhead (1988), McKinney (1988), Simms (1988), McNamara (1988, 1990, 1993, 1995), McKinney and McNamara (1991), Sessions (1992) and Klingenberg (1997). For the exercise of caution in its application, see Thomson (1986), Hall (1990b), Raff (1996), D. B. Wake (1996c) and the first edition of this book.

2. See Cock (1966) for the *creeper* mutant; Shea (1985, 1992, 1993), Wayne (1986a), Wayne and Ruff (1993), Shea and Bailey (1996) and Maunz and German (1997) for scaling in eutherian and placental mammals; and Atchley (1987, 1993) and Neyfakh and Hartl (1993) for selection early and late in ontogeny.

3. For examples of modification of onset, rate and offset of growth and development, see Huxley (1932), de Beer (1940), Thompson (1942), Gould (1977), Alberch *et al.* (1979) and Hall (1984b).

4. See Raff and Wray (1989). Wray and Raff (1990b) and Shea (1983a) for production of patterns of heterochrony through mechanisms that are not heterochronic. For his work on primates, see Shea (1983a,b, 1985, 1989, 1992, 1993) and Shea and Bailey (1996). Neoteny in salamanders reproduces ancestral patterns, as illustrated in the skull of *Triturus alpestris* (Rocek, 1996).

5. See Hall (1982c,d, 1984b, 1990a,c,d), Atchley (1990) and Atchley and Hall (1991) for the developmental mechanisms that underlie heterochrony. The importance of distinguishing between regulation through growth as a proxy for size, and development as a proxy for shape, has been critically tackled in the context of Gould's clock model by Godfrey and Sutherland (1995a,b, 1996).

6. See Hall (1982b,d, 1984c) and Hall and Miyake (1992, 1995b) for more extensive discussions of condensations. For the correlation of loss of regulative ability and condensation, see Kieny and Pautou (1976).

7. For evidence that epithelial-mesenchymal interactions trigger condensation, see Dunlop and Hall (1995), Hall and Miyake (1995b) and Ruch (1995).

8. For alteration of the timing of causal cell and tissue interactions as a basis for heterochrony, see Maderson (1975, 1983), Hall (1983a, 1984b, 1990a) and Alberch (1985).

9. For transitions in dentition in urodeles, lungfishes and *Polypterus*, see Thomson (1975), Meinke (1982a,b, 1984, 1986), Meinke and Thomson (1983) and Smith and Hall (1990, 1993).

10. See Abbott (1975), McAleese and Sawyer (1982) and Cunha and Young (1992) for supporting evidence for the genetic basis of timing of inductive tissue interactions.

11. See McAleese and Sawyer (1982) and Song and Sawyer (1996) for representative studies on the *scaleless* mutant, and Thomson (1988) and Rollo (1994) for whether such developmental processes are directly subject to selection.

12. For timing genes and specification of cell fate in *C. elegans*, see Ambros and Horvitz (1984), Maclean and Hall (1987), Liu and Ambros (1989), Ambros (1989) and Liu *et al.* (1995). The basic studies on modification of cell lineages in sea urchins are those of Raff *et al.* (1984), Raff (1988, 1992), Wray and Raff (1989) and Wray (1994). Klueg *et al.* (1997) contains the information on *msp130*.

13. See Page (1994) for an overview of genes controlling circadian rhythms. Davidson (1991) organized all metazoan embryos into three types, two of which were interpreted as heterochronic derivatives of the first (see s. 14.2.5).

14. For environmental triggering of heterochrony, see Collins and Cheek (1983), James (1983), Meyer (1987) and Patton and Brylski (1987). For coupling of ecology, development and evolution through heterochrony, see Gould (1977), Alberch *et al.* (1979), McNamara (1982), Calder (1984), McKinney (1986, 1988), McKinney and McNamara (1991) and Hall and Miyake (1995a).

15. See Ten Cate (1953) and Jacobson (1958) for these studies relating temperature to induction.

16. For his studies on water striders, see Klingenberg (1996a,b), Klingenberg and Spence (1993) and Klingenberg and Zimmermann (1992). For the association of heterochronic change with small organs and organisms, see Shea (1983a), Hafner and Hafner (1984), Hanken (1984), M. H. Wake (1986), Wayne and Ruff (1993) and Ponder and Lindberg (1997). Although *Thorius* is the smallest salamander, it is not the smallest vertebrate. That distinction belongs to the Brazilian frog *Psyllophryne didactyla*, which has a snout-vent length of only 0.98 cm.

17. See Palmiter *et al.* (1982), Shea (1992) and Shea *et al.* (1990) for production and analysis of transgenic mice containing the gene for human growth hormone.

18. For dissociation of chondrogenesis from osteogenesis in amphibians, see Hanken and Hall (1984, 1988a,b), Trueb (1985) and Hanken *et al.* (1989).

19. See Rieppel (1992b–d, 1993b, 1994) for the studies on lizards, and Miyake *et al.* (1996a,b, 1997a,b) for those on inbred strains of mice.

20. For analyses of ossification patterns in fossil reptiles, see Caldwell (1994, 1996, 1997) and the discussion in R. L. Carroll (1997).

21. In addition to the study of poeciliid fishes by Strauss (1990), see Huey (1987), Burggren and Bemis (1990), Lauder (1990), Lauder *et al.* (1995) and Schaefer and Lauder (1996) for phylogenetic approaches.

22. For examples and discussions of the demise of the germ-layer theory, see Kolliker (1884), Wilson (1892, 1894), Sedgwick (1910), Oppenheimer (1940), de Beer (1947) and Hall (1997a).

23. For supporting documentation for these examples of heterotopy, see Schaeffer and Thomson (1980), Gorbman (1983), Gorbman and Tamarin (1985a,b), Maclean and Hall (1987), Brylski and Hall (1988a,b), Dohle and Scholtz (1988), Hall and Hörstadius (1988), Northcutt (1990) and Smith and Hall (1990, 1993).

24. See Brylski and Hall (1988a,b) for the studies on development of external cheek pouches and Long (1976) for physiological differences between external and internal pouches.

25. For literature supporting inductive interactions between pharyngeal endoderm and visceral arch mesenchyme in lampreys, see Langille and Hall (1988b, 1989) and Hall and Hörstadius (1988).

26. For the Meckel's cartilage example, see Hall (1983a, 1984b). For the evolutionary significance of such altered timing for production of new organs, see Hall (1975, 1978), Maderson (1975, 1983), Herring *et al.* (1979), Smith and Hall (1990, 1993) and Herring (1993a,b).

27. See MacDonald (1997) for timing of inductions in inbred strains and Atchley and Fitch (1991, 1993) for relationships between inbred strains.

28. See Zelditch *et al.* (1992) and Zelditch and Fink (1995) for the studies on cotton rats, and Fink and Zelditch (1995, 1996) and Zelditch and Fink (1996) for those on piranha.

29. The importance of an awareness of comparative rates of development in *Monodelphis* and *Didelphis* was emphasized by K. K. Smith and van Nievelt (1997) in their analysis of Rowe's (1996a,b) proposition that evolution of mammalian middle ear ossicles tracked heterochronic shifts in growth of the neocortex.

PART EIGHT

PRINCIPLES AND PROCESSES

EVOLUTIONARY DEVELOPMENTAL BIOLOGY: PRINCIPLES AND PROCESSES

'The purpose of the Evolution Society, founded in St. Louis in June 1946, was "the promotion of the study of organic evolution and the integration of the various fields of biology."'

Mayr, 1997a, p. 264, from his address at the 50th anniversary of the founding of the Society for the Study of Evolution, St. Louis, June 19, 1996

This book began with a discussion of the terms and concepts encompassed by evolution and development. I bring it to a close by turning the terms into principles and the concepts into processes. Aside from the symmetry this establishes, which is satisfying, it is important to summarize the status of evolutionary developmental biology and where it is likely to go in the coming years.

25.1 CONNECTIONS

My aim, as set out at the end of section 1.1, was to analyse how development impinges on evolution to effect evolutionary change and how development itself has evolved. A major reason for the historical approach of the early chapters was to demonstrate that evolutionary developmental biology is not a new field of endeavour. Indeed, a search for connections between evolution and development has lain at the heart of much of the scholarship in biology and philosophy for hundreds, in fact, thousands, of years. So close is this connection that evolution, now a term for phylogenetic change, began as a term for ontogenetic change.

The connection between evolution and development, however, is not simply etymological. On the one hand there is a parallel between the stages of embryonic development and the evolution of life on earth and, on the other, a search for causal and mechanistic connections between development and evolution. Although these intellectual inquiries go back at least to the ancient Greek philosophers, this particular intellectual need remains with us. Evolutionary developmental biology is therefore neither new nor dead. As a sub-discipline, however, it has been searching for a cohesion of approach, unity of methodology and corpus of principles. These elements are now in place. Evolutionary developmental biology embraces levels of biological organization and fields of biological endeavour including the following combinations:

- development, genetics and evolution;
- development, ecology and evolution (including behaviour);
- development, evolution and phylogenetics;
- development, palaeontology and evolution;
- development, fitness and adaptation;

- mechanisms of genetic, epigenetic, and environmental control; and
- developmental, genetic, epigenetic and population approaches.

The chief enabling disciplines are phylogenetics, palaeobiology, functional morphology, comparative embryology, and an understanding of the genetic regulation of development. These have been especially brought to bear on basic structural organization, homology, and the origins of body plans and of phyla, as discussed in the following section.

25.2 BASIC STRUCTURAL ORGANIZATION AND HOMOLOGY

The fundamental structural organization that is expressed in the Archetype, body plan or *Bauplan* was elaborated in Chapters 2–6. For two millenia a search for this organization has lain at the foundation of morphology through all its idealistic, transcendental, comparative, functional and evolutionary transformations. It fuelled the great Académie debate over whether form determines function (Geoffroy) or function form (Cuvier). It spurred the search for real, ancestral archetypes in fossils and in the embryos of living descendants. It produced the embryological concept of the Archetype, an embryological basis for homology and an evolutionary embryology. Consequently, morphology, embryology, homology and evolution fused and dominated late-19th-century zoology.

Before the publication of *The Origin*, a typological view of homology prevailed: structural transformation was limited to variations within types. Although the theory of descent with modification from a common ancestor revolutionized approaches to morphology, homology continued to be determined on the basis of equivalent elements and connections.

With von Baer, homology acquired a developmental component. Types were homologous because of shared development rather than shared adult structures. Haeckel took this approach to extremes with his insistence on the primacy of origin from a common germ layer. The conundrum of linking homology of pattern with equivalence of developmental processes was explored in Chapter 21, where it was argued that homology should be reserved for patterns. Processes that produce homologous structures are not the essence of homology. They are the means, not the end. I suggested equivalent and non-equivalent developmental processes as two classes of process that produce homologues. Such a separation can only facilitate the dual search for evidence that developmental processes have themselves evolved, and that variability in developmental processes is available for selection to act upon to effect evolutionary change (Chapter 20). The recognition that development evolves underpins evolutionary developmental biology and will be an important element of any new evolutionary synthesis incorporating development.

25.3 BASIC BODY PLANS

How do the integrated processes of embryonic development maintain basic form yet facilitate change in ontogeny and phylogeny?

An early phase of development is typified by the laying down of the conserved features of the basic body plans of organisms (s. 6.8.1). This phase (and these conserved characters) is differentiated from that which produces aspects of form and function that are more readily modified (plastic), not by different developmental and evolutionary mechanisms, but by a different mix of the same three elements – variation, constraint and selection.

Genetic, developmental, structural and/or functional constraints channel or bound variation (s. 6.7). Consequently, some structures persist (or remain unchanged) for long periods of time, even throughout the history of a group. During the evolution of mammalian brains, for example, allometry and the functional correlation that accompanies allometry constituted a constraint, setting the size of parts by the size of the organism and constraining and conserving the order of neurogenesis over mammalian evolution (Finlay and Darlington, 1995).

Essential features of the development of the body plan (and the conservation of phylotypic stages) are constrained, limited in their variability, and maintained by stabilizing selection, in large measure because of integrated epigenetic processes. Differential survival of early embryos further reduces variability, canalizing those early invariant stages of development in which the basic body plan is specified. Interactions that specify body axes, mesoderm, primary embryonic induction, and segmentation arose early in evolution. With so much of subsequent development dependent upon them, they became highly constrained. So too did structures such as the notochord that are central to these processes. Such evolutionary conservatism provides an explanation for the conservation and constancy of developmental mechanisms used to produce body plans throughout the animal kingdom.

Invariance is reflected in preservation of phylotypic stages such as the germ-band stage in insects and the pharyngula in vertebrates (ss. 8.4.1, 14.2.2). Yet the processes that precede these stages in development can vary considerably and in ways that appear to be fundamental deviations from the norm for the group. Examples include formation of the blastocoele by infiltration or involution, variable mechanisms of segmentation in insects, and diverse patterns of cleavage and gastrulation in vertebrates. Similarly, development following phylotypic stages varies considerably. Examples discussed included formation of the neural cavity by invagination or cavitation, of the alimentary canal from different embryonic regions, and development of homologous structures through equivalent and non-equivalent developmental processes.

One difficulty with endeavouring to document similarities and differences in these epigenetic interactions is the paucity of comparative data, especially for closely related organisms. A large fraction of the work in experimental embryology has been carried out on model organisms – *Drosophila*, *Xenopus*, embryonic chicks and laboratory mice – and extrapolated to all other species (s. 8.1); there is a crying need for comparative studies (ideally on members of monophyletic groups; s. 8.1.2). The gains to be made are amply demonstrated by the fact that the major body axes and basic patterning of the metazoan body are determined by specific and conserved sets of genes, although genes from the same family can substitute for one another (Box 10.1). Waddington envisaged canalized epigenetic pathways as a feedback mechanism(s) that would accommodate such substitutions (s. 12.2).[1]

Current evidence indicates, on the one hand, that such genes are highly conserved evolutionarily, and on the other, that their duplication and modification provide a fundamentally unified basis for specification of animal body plans (ss. 9.2, 15.2). With demonstrations of phylogenetic conservation of *Dll* and *Hox* genes among arthropods, and that *Hox* genes were present in the common arthropod/chordate ancestor, prospects for unravelling the association between phylogenetic changes and homeotic genes are now considerable. A conserved arrangement of homeobox genes defines a zootype for animals, expressed most clearly at the phylotypic stage (s. 14.2.2). This organization is fundamental and may even become diagnostic for phyla and super-phyla as knowledge and understanding increase. Advances in our understanding of the bases for constraint, the origins of body plans and the preservation of phylotypic stages are major steps toward the reintegration of developmental biology with evolutionary theory.

Identifying a zootype, phylotypic stage or phylogenetically conserved suite of characters, does not mean that it (or they) was the key to the evolution of Animalia or of individual phyla. While the A-P axis defined by homeobox genes is fundamental, it is no more fundamental than the D-V axis, L-R symmetry, organization into germ layers, the ability of metazoan cells to create extracellular matrices, and so forth. Just as our inability to assign some Burgess Shale organisms to phyla may reflect our current state of knowledge rather than the existence of unknown phyla, our emphasis on homeobox genes and A-P patterning may reflect our rich state of knowledge concerning these genes and the relative paucity of knowledge of genes acting downstream from them.

One final word of caution. The existence of patterns of similar genes in diverse organisms need not mean that those genes serve the same roles in different organisms. Gene structure can be conserved when gene function changes; differential splicing allows different products to be produced from seemingly identical genes. *Engrailed* exhibits a patterned distribution in arthropods, annelids, amphioxus and chordates, and is associated with segmentation in arthropods and amphioxus. Some genes (*bicoid*, *fushi-tarazu*, *zen-1* and *zen-2*) within the *Drosophila* homeotic complexes are neither highly conserved nor associated with homeotic transformation, but nevertheless are patterning genes. It is not known how such genes have been able to remain within the conserved homeotic complex, retaining some conserved features of homeodomain proteins, yet escaping the strong stabilizing selection operating on the remainder of the genes in the homeobox complex. Closer study of these 'renegade' homeobox genes may reveal the nature of selective constraint on genes in the homeotic complex.[2]

25.4 PLASTIC FORM AND FUNCTION

Cascades of interactions establish major embryonic regions and consolidate basic body plans. Functional interactions between differentiated cells, tissues and/or organs establish and modify embryonic features at a second level of functional and structural integration. Examples in the book include pharyngeal jaws in cichlid fishes, feeding mechanisms in bolitoglossine salamanders, and secondary jaw articulations in birds (Chapters 16 and 17).

These interactions can be more readily modified (or even undone) than those that specify basic plans (although see exceptions in section 2.2.2), in large part because they are functional interactions between already formed structures rather than inductive interactions required to form the structures on which so much subsequent development depends. Selection can change such features through a key innovation, integrated change or ontogenetic repatterning, following intensification of existing function or acquisition of a new function.

Basic and plastic structures do not represent a dichotomy between:

- proximate/ultimate, functional/evolutionary or internal/external causation;
- selection/constraint;
- genetic/epigenetic control;
- typological/populationist thinking;
- invariance/variance; or
- micro-/macroevolution.

Rather, they represent a continuum and the lasting legacy of the Geoffroy–Cuvier debates. Generation of conserved characters of the basic body plan on the one hand and of more plastic characters on the other, represents the fundamentally hierarchical organization of embryonic development and a reconciliation of the Geoffroy–Cuvier debates over whether form determines function or function form, especially if we consider function in relation to conserved embryonic stages. Function follows form in generation of the basic body plan and as a result, fundamental embryonic form is highly constrained. Function determines form in the generation of more plastic féatures and so is less constrained and more variable.

25.5 THE CORE PRINCIPLES OF EVOLUTIONARY DEVELOPMENTAL BIOLOGY

There is no shortage of frameworks for evolutionary developmental biology to choose from: the quantitative approach of the population geneticist; the pattern-generating approach of the morphologist; the reductionist approach of the molecular biologist; the functionalist approach of the functional morphologist; and approaches through life-history strategy, epigenetics, natural selection and comparative anatomy.

An essential message of this book, however, is that evolutionary developmental biology now has its own framework which

- recognizes that development evolves, that evolutionary changes can occur at any stage in ontogeny, and that embryonic, larval and adult stages of the life cycle can evolve independently (s. 8.4, Chapter 22);
- incorporates interacting levels of hierarchical biological organization throughout ontogeny (Chapters 7–11);
- accounts for the operation in embryos of parental and zygotic genomes (Chapter 7);
- integrates actions of the environment with organismal development and ecology (Part Five, Chapter 24);
- recognizes the significance for development and evolution of interactions (including inductive interactions) between organisms in maintaining phenotypic plasticity (Chapter 18);
- considers organisms as more than adults, embryos as more than the means of making adults, and the phenotype as more than the physical expression of the genotype; and
- demonstrates that multiple phenotypes can be produced from the same genotype and that different signals (mutation, phenocopy, environmental trigger, inductive interaction) can elicit those differing phenotypes.

An integration of three elements – epigenetics, genomic and environmental control – underlies and unifies evolutionary developmental biology. Epigenetic mechanisms are essential, and a science of evolutionary developmental biology is required, for one simple reason: there is no one-to-one correspondence between genotype and phenotype. The genotype is the starting point, the phenotype the endpoint of epigenetic control.

As discussed in Chapter 8, inductive control of embryonic development (conditional specification) is the hierarchical mechanism used by embryos of all but the most mosaic and determinative of organisms. But induction is not just an intraspecific phenomenon. Induction

- relates ecology to development as a vehicle for key innovations and integrated change, often acting through the evolutionary mechanisms of heterochrony and heterotopy (Chapters 12–15, 24);
- relates interactions with other organisms or with abiotic environmental factors to embryonic development as interspecific and environment-based induction (ss. 18.3–18.7); and
- relates environmental perturbations to morphological change through interactions between hidden genetic variability and developmental canalization brought to expression through genetic assimilation (Chapter 19).

Inductive interactions thus link development, ecology and evolution, often mediated by heterochrony or heterotopy (Chapter 24). Heterotopy holds great promise as a mechanism producing novelties in evolution. Heterotopy and heterochrony often go hand in hand with a heterochronic change facilitating heterotopy. Developmental programmes can be regulated by such environmental variables as temperature and density of predator, suggesting a research strategy for analysing the developmental processes subject to heterochronic and/or heterotopic change. As summarized in Chapter 18, major future tasks for evolutionary developmental biology are working out the inter- and intraspecific causal links between developmental inductive inter-actions, ecological adaptation and evolution-ary change, for such links will integrate evolutionary developmental biology with life history evolution and life-history theory.

25.6 MODELS AND FUNDAMENTAL UNITS

'The cell is evolution's most brilliant in-vention and development is its triumphant elaboration.'

Wolpert, 1990, p. 109

The unity that evolutionary developmental biology has to offer is exemplified in the models outlined in Chapter 20. Such models incorporate and integrate

- gene regulation;
- heritable and environmental aspects of epigenetic regulation;
- non-heritable environmental factors;
- development as hierarchical and integra-tive; and
- effects of many genes as pleiotropic.

We must identify the causal factors that con-tribute to the developmental variability upon which selection acts, and the fundamental de-velopmental units that are altered in response to selection to effect change. Five fundamental causal factors incorporated into the Atchley–Hall model are intrinsic genetic, epigenetic, maternal and environmental factors, and genotype by environment interactions. Funda-mental developmental units are identified in such cellular properties as stem-cell number and mitotic rate. Variability in developmental processes exists because of variability in these fundamental developmental units. Exposing that variability to selection is a function of the degree of epigenetic and pleiotropic inter-actions among and between components, whether those components are condensations of cells, fields of cells or organ primordia.

These are the units of what Wake and Roth (1989) called the 'combination of internalist, externalist and historical factors' and Kluge and Strauss (1985) referred to as the need to know 'which aspects of ontogeny are histori-cally constrained and which are free to vary'. These are also the units that allow us to deter-mine causal associations between basic phases of development and to understand the compo-nents that vary when homologous structures arise through non-equivalent developmental processes.

A series of fundamental units exist in organismal development and evolution:

- In evolutionary biology, the fundamental unit is the gene, species, or population.
- In developmental biology, the fundamental unit is the embryo.
- In physiology the fundamental unit is the individual organism or organ.
- In evolutionary developmental biology, the fundamental unit is the cell.

The cells are gametes that bridge generations, zygotes that begin new generations, embry-onic cells in lineages whose fate and/or posi-tion is specified, and cells in condensations or fields that have the potential to interact with other cells to evoke yet further cellular, tissue and organ diversity.

Groups of like cells (condensations) are fun-damental units; cellular properties are the fundamental developmental units. Leo Buss (1987) made the cell the target of selection in his analysis of the evolution of individuality. Cells and their immediately adjacent peri- and extracellular matrices carry out the selective responses that allows organisms to develop, adapt to their environment, modify their de-velopment, and translate the effects of gene mutations and genetic assimilation into evolu-tionary change (s. 7.3). An important issue is the extent to which selection reinforces such changes in developmental mechanisms.

The cell is to evolutionary developmental biology as the gene or species is to evolu-tion and as embryos are to development. Epigenetics is to evolutionary developmental biology as natural selection is to evolution and as differentiation, morphogenesis and growth are to development.

The genetic code is embedded in the ordered arrangement of a finite number of base pairs. The epigenetic code, however, potentially encompasses all genetic and non-genetic factors. It is therefore not surprising that there is no single unifying theory explaining development. Epigenetics points us in the right direction, especially with models that distinguish heritable from environmental components of epigenetic regulation, separate the origins of epigenetic control (whether biotic or abiotic, predator or conspecific) from its extrinsic mode of action, and view epigenetics in the context of pleiotropy and genetic correlation. These are overlapping, interlocking and interacting hierarchies of concept and structure, principle and process, cause and effect.

Evolution acts at the three levels of changes in gene frequencies, the appearance of new characters, and the adaptation and radiation of new species. The common denominator of all three is genetic change through time. The common agent of that change is alteration in ontogeny. The common integrator of the three is epigenetic and hierarchical organization. The science of the study of these interactions is evolutionary developmental biology.

25.7 ENDNOTES

1. For interactions between genes that determine body plan, see Cohen and Jürgens (1990) and Hülskamp *et al.* (1990). For gene substitution, see Lumsden and Wilkinson (1990).
2. See Holland (1990), Blau and Baltimore (1991) and Davidson (1991) for discussions of differential regulation and differential methods of regulation; Patel *et al.* (1989) for differential functioning of *engrailed* in different phyla; L. Z. Holland *et al.* (1997) for metameric distribution of *Engrailed* in amphioxus, and Warren and Carroll (1995) for general issues raised by such findings.

ABBREVIATIONS

A list of the more commonly-used abbreviations other than those in figure legends. Members of gene families (e.g. BMP-4, *Hoxa-1*) are listed under the gene family (BMP, *Hox*) and not separately.

abd-A, abdominal-A genes of the bithorax complex

Abd-B, abdominal-B genes of the bithorax complex

AER, apical ectodermal ridge of tetrapod limb buds

AmHNF-3, hepatic nuclear factor-3 (*HNF-3*) gene in *Branchiostoma* (amphioxus)

AmphiEn, the orthologue of the engrailed gene in *Branchiostoma* (amphioxus)

AmphiHox-1–AmphiHox-12, *Hox* genes of *Branchiostoma* (amphioxus)

AmphiOtx, *Branchiostoma* (amphioxus) member of the orthodenticle (*otd*) gene family

ANT-C, the antennapedia complex of *Drosophila* homeotic genes

antp, the antennapedia gene in *Drosophila*

A-P, antero-posterior axis or antero-posterior polarity

ATPase6, adenosine 6'-triphosphatase

bFGF, basic fibroblast growth factor

BMP, bone morphogenetic protein, a member of the TGF-β superfamily

BX-C, the bithorax complex of *Drosophila* homeotic genes

CAM(s), cell adhesion molecule(s)

cDNA, complimentary DNA

ch, congenital hydrocephalus mutation in mice

chn, chinless mutation in zebrafish

CJMs, cell junctional molecules

Clx, clock genes in *Caenorhabditis elegans*

cSnr, chick member of the *snail* family of *Drosophila* genes

cVg-1, Vegetal mRNA or protein in early chick embryos

Dll, distal-less gene in *Drosophila*

Dlx, the vertebrate orthologue of *Dll*

DNA, deoxyribonucleic acid

dpp, the decapentaplegic gene in *Drosophila*, a member of the TGF-β superfamily

D-V, dorso-ventral axis or dorso-ventral polarity

ECM, extracellular matrix

EDB, evolutionary developmental biology

EDTA, ethylendiaminetetraacetic acid, a chelating agent

EGF, epidermal growth factor

EGF-R, epidermal growth factor receptor

En, the engrailed homeobox-containing gene in *Drosophila*

eng, the vertebrate orthologue of *Drosophila en*

Er, *repeated epilation* mutant mice in which both limb and tail ridges are abnormal

eve, even-skipped, a pair-rule gene in *Drosophila*

ey, eyeless mutation in *Drosophila*

FGF, fibroblast growth factor

GDF, growth differentiation factors

Gli2, a zinc finger gene in mice

Her-1, the zebrafish orthologue of the *Drosophila* pair-rule gene *Hairy*

HNF-3, hepatic nuclear factor-3, a DNA-binding protein

Hox, vertebrate genes related to *Drosophila* ANT-C and BX-C gene clusters

HrPax-37, *Pax* gene of the ascidian *Halocynthia roretzi*

IGF, insulin-like growth factor

Krox-20, a zinc finger protein

L-CAM, liver cell adhesion molecule

Lef-1, lymphoid enhancer-binding factor-1

lin, cell lineage abnormal genes in *Caenorhabditis elegans*

L-R, left right axis or left-right polarity

m3, a central tarsal bone in salamanders

mRNA, messenger ribonucleic acid

Msh, muscle segment homeobox gene in *Drosophila*

Msx, orthologues of *Drosophila Msh* in vertebrates

mya, million years ago

MyoD, myoblast determination 1 protein expressed only in muscle cells

NADH1, nicotinamide adenine dinucleotide

N-CAM, neural cell adhesion molecule

Otx, vertebrate members of the *Drosophila* orthodenticle (*otd*) gene family

(*p*), premature death mutation in *Ambystoma mexicanum*

Pax, mouse member of *Drosophila* paired box genes

Pc-G, polycomb response elements of such genes as *Scr*.

P-D, proximo-distal axis or proximo-distal polarity

pdm, a gene involved in initiation of wing primordia in *Drosophila*

RA, retinoic acid

Rop-1, resistance gene to dieldrin

rRNA, proposed ribosomal ribonucleic acid

s, section, referring to another section in the book

SAM(s), surface adhesion molecule(s)

Scl, scalloped wings, modifier of *Rop-1*

Scr, Sex combs reduced, a homeotic gene

Sey, small eye, a mutation of *Pax-6*

ss., sections, referring to other sections in the book

shh, sonic hedgehog gene

Six[6], the mouse orthologue of *Pax-6*

sog, short gastrulation gene in *Drosophila*

TGF-β, transforming growth factor-beta

Trx-G, trithorax response element of such genes as *Scr*

ub, ultrabithorax gene in *Drosophila*

Uro-2, Uro-11, ascidian genes that encode a DNA-binding gene

vestigial tail (vt), mutant mice lacking a tail ridge

Vg-1, vegetal mRNA or protein located in the vegetal pole of *Xenopus* blastulae.

wnt, a family of cysteine-rich glycoproteins in vertebrates, named from wingless (*wn*) and integrated (t), *Drosophila* and vertebrate segment polarity genes

Xbap, bagpipe gene in *Xenopus*

XVent-2, a novel homeobox gene in *Xenopus*

Xwnt, *Xenopus* member of the *Drosophila* wingless gene family

ZPA, zone of polarizing activity in limb buds

REFERENCES

Abbott, U.K. (1975) Genetic approaches to studies of tissue interactions. *Genet. Lect.*, **4**, 69–84.

Abouheif, E. (1997) Developmental genetics and homology: a hierarchical approach. *Trends Ecol. Evol.*, **12**, 405–8.

Abouheif, E., Akam, M., Dickinson, W.J. *et al.* (1997) Homology and developmental genes. *Trends Genet.*, **13**, 432–3.

Acampora, D., Mazan, S., Lallemand, Y. *et al.* (1995) Forebrain and midbrain regions are deleted in $Otx^{-/-}$ mutants due to a defective anterior neurectoderm specification during gastrulation. *Development*, **121**, 3279–90.

Adams, M.B. (1980) Severtzov and Schmalhausen: Russian morphology and the evolutionary synthesis, in *The Evolutionary Synthesis. Perspectives on the Unification of Biology* (eds E. Mayr and W.B. Provine), Harvard University Press, Cambridge, MA., pp. 193–225.

Adams, M.B. (ed.) (1994) *The Evolution of Theodosius Dobzhansky. Essays on His Life and Thought in Russia and America*, Princeton University Press, Princeton.

Adams, M.S. and Niswander, J.D. (1967) Developmental 'noise' and a congenital malformation. *Genet. Res.*, **10**, 313–17.

Adelmann, H.B. (1966) *Marcello Malpighi and the Evolution of Embryology*, 5 volumes, Cornell University Press, Ithaca, N Y.

Agassiz, J.L.R. (1857) Essay on Classification, in *Contributions to the Natural History of the United States*, Vol. 1, Little, Brown & Co., Boston. Reprinted 1962 (ed. E. Luria), Harvard University Press, Cambridge, MA.

Agassiz, J.L.R. (1874) Evolution and permanence of type. *Atlantic Monthly*, **33**, 92–101.

Aguinaldo, A.M.A., Tuberville, J.M., Linford, L.S. *et al.* (1997) Evidence for a clade of nematodes, arthropods and other moulting animals. *Nature*, **387**, 489–93.

Akam, M. (1987) The molecular basis for metameric pattern in the *Drosophila* embryo. *Development*, **101**, 1–22.

Akam, M. (1995) Hox genes and the evolution of diverse body plans. *Phil. Trans. R. Soc. Lond. B*, **349**, 313–19.

Akam, M., Dawson, I. and Tear, G. (1988) Homeotic genes and the control of segment diversity, *Development*, **104**, Suppl., 123–33.

Akam, M., Tear, G. and Kelsh, R. (1991) The evolution of segment patterning mechanisms in insects, in *The Unity of Evolutionary Biology* (ed. E.C. Dudley), Proceedings of the 4th International Congress of Systematics and Evolutionary Biology, Discorides Press, Portland, OR, Vol. 1, pp. 561–7.

Akimenko, M.-A. and Ekker, M. (1995) Anterior duplication of the *Sonic hedgehog* expression pattern in the pectoral fin buds of zebrafish treated with retinoic acid. *Devel. Biol.*, **170**, 243–7.

Akimenko, M.-A., Ekker, M., Wegner, J. *et al.* (1994) Combinatorial expression of three zebrafish genes related to *Distalless*: Part of a homeobox gene code for the head. *J. Neurosci.*, **14**, 3475–86.

Akimenko, M.-A., Ekker, M. and Westerfield, M. (1991) Characterization of three zebrafish genes related to *Hox-7*, in *Developmental Patterns of the Vertebrate Limb* (eds J.R. Hinchliffe, J.M. Hurle and D. Summerbell), Plenum Press, New York, pp. 61–3.

Akimenko, M.-A., Johnson, S.L., Westerfield, M. and Ekker, M. (1995) Differential induction of four *Msx* homeobox genes during fin development and regeneration in zebrafish. *Development*, **121**, 347–57.

Alberch, P. (1980) Ontogenesis and morphological diversification. *Amer. Zool.*, **20**, 653–67.

Alberch, P. (1985) Problems with the interpretation of developmental sequences. *Syst. Zool.*, **34**, 46–58.

Alberch, P. (1989) The logic of monsters: Evidence for internal constraint in development and evolution, in *Ontogenèse et Evolution* (eds B. David, J.L. Dommergues, J. Chaline and B. Laurin), *Geobios, Mém. Spécial* no. 12, 21–57.

Alberch, P. and Alberch, J. (1981) Heterochronic mechanisms of morphological diversification and evolutionary change in the neotropical salamander, *Bolitoglossa occidentalis* (Amphibia: Plethodontidae). *J. Morph.*, **167**, 249–64.

Alberch, P. and Blanco, M.J. (1996) Evolutionary patterns in ontogenetic transformation: From laws to regularities. *Int. J. Devel. Biol.*, **40**, 845–58.

Alberch, P. and Gale, E.A. (1985) A developmental analysis of an evolutionary trend: Digital reduction in amphibians. *Evolution*, **39**, 8–23.

Alberch, P., Gould, S.J., Oster, G.F. and Wake, D.B. (1979) Size and shape in ontogeny and phylogeny. *Paleobiology*, **5**, 296–317.

Alberch, P. and Kollar, E. (1988) Strategies of head development: Workshop report. *Development*, **103**, Suppl., 25–30.

Albers, B. (1987) Competence as the main factor determining the size of the neural plate. *Devel. Growth Differ.*, **29**, 535–45.

Alberts, B., Bray, D., Lewis, J. *et al.* (1983) *Molecular Biology of the Cell*, Garland Publishing Inc., New York.

Aldridge, R.J., Briggs, D.E.G., Clarkson, E.N.K. and Smith, M.P. (1986) The affinities of conodonts – new evidence from the Carboniferous of Edinburgh, Scotland. *Lethaia*, **19**, 279–91.

Aldridge, R.J., Briggs, D.E.G., Smith, M.P. *et al.* (1993) The anatomy of conodonts. *Phil. Trans. R. Soc. Lond.* B, **340**, 405–21.

Aldridge, R.J., Briggs, D.E.G., Sansom, I.J. and Smith, M.P. (1994) The latest vertebrates are the earliest. *Geology Today*, **10**, 141–5.

Aldridge, R.J. and Purnell, M.A. (1996) The conodont controversies. *Trends Ecol. Evol.*, **11**, 463–8.

Aldridge, R.J. and Theron, J.N. (1993) Conodonts with preserved soft tissue from a new Ordovician *Konservat-Lagerstätte*. *J. Micropalaeont.*, **12**, 113–17.

Alexander, M.L. (1976) The genetics of *Drosophila virilis*, in *The Genetics and Biology of* Drosophila (eds M. Ashburner and E. Novitski), Academic Press, London, Volume 1c, pp. 1365–1427.

Alexandre, D., Clark, J.D.W., Oxtoby, E. *et al.* (1996) Ectopic expression of *Hoxa-1* in the zebrafish alters the fate of the mandibular arch neural crest and phenocopies a retinoic acid-induced phenotype. *Development*, **122**, 735–46.

Allen, G.E. (1991) Mechanistic and dialectical materialism in 20th Century Evolutionary Theory: The Work of Ivan I. Schmalhausen, in *New Perspectives on Evolution* (eds L. Warren and H. Koprowski), Wiley-Liss, New York, pp. 15–36.

Alley, K.E. (1989) Myofiber turnover is used to retrofit frog jaw muscles during metamorphosis. *Amer. J. Anat.*, **184**, 1–12.

Alley, K.E. (1990) Retrofitting larval neuromuscular circuits in the metamorphosing frog. *J. Neurobiol.*, **21**, 1092–1107.

Altmann, C.R., Chow, R.L., Lang, R.A. and Hemmati-Brivanlou, A. (1997) Lens induction by Pax-6 in *Xenopus laevis*. *Devel. Biol.*, **185**, 119–23.

Ambros, V. (1989) A hierarchy of regulatory genes controls a larva-to-adult developmental switch in *C. elegans*. *Cell*, **57**, 49–57.

Ambros V. and Horvitz, H.R. (1984) Heterochronic mutants of the nematode, *Caenorhabditis elegans*. *Science*, **226**, 409–16.

Amprino, R., Amprino Bonetti, D. and Ambrosi, G. (1969) Observations on the developmental relations between ectoderm and mesoderm of the chick embryo tail. *Acta Anat.*, **56**, Suppl., 1–26.

Amundson, R. (1994) Two concepts of constraint: Adaptationism and the challenge from developmental biology. *Philos. Sci.*, **61**, 556–78.

Amundson, R. (1996) Historical development of the concept of adaptation, in *Adaptation* (eds M.R. Rose and G.V. Lauder), Academic Press, San Diego, CA, pp. 11–53.

Anderson, D.T. (1973) *Embryology and Phylogeny in Annelids and Arthropods*, Pergamon Press, Oxford.

Anderson, K.V. and Nüsslein-Volhard, C. (1984) Information for the dorso-ventral pattern of the *Drosophila* embryo is stored in maternal mRNA. *Nature*, **311**, 223–7.

Anderson, W.W. (1966) Genetic divergence in M. Vetukliv's experimental populations of *Drosophila pseudobscura*. 3. Divergence in body size. *Genet. Res.*, **7**, 255–66.

Andrews, J.H. (1995) Fungi and the evolution of growth form. *Can. J. Bot.*, **73**, S1206–S1212.

Ang, S.-L., Jin, O., Rhinn, M. *et al.* (1996) A targeted mouse Otx2 mutation leads to severe defects in gastrulation and formation of axial mesoderm and to deletion of rostral brain. *Development*, **122**, 243–52.

Anken, R.H., Kappell, Th., Slenzka, K. and Rahman, H. (1993) The early morphogenetic development of the cichlid fish, *Oreochromis mossambicus* (Perciformes, Teleostei). *Zool. Anz.*, **231**, 1–10.

Anthony, J. and Guibe, J. (1952) Les affinités anatomiques de *Bolyeria* et de *Casarea*. *Mem. Inst. Sci. Madagascar*, Ser. A., **7**, 189–201.

Antoch, M.P., Song, E.-J., Chang, A.-M. *et al.* (1997) Functional identification of the mouse circadian

Clock gene by transgenic BAC rescue. *Cell*, **89**, 655–67.

Antonovics, J. and van Tiederen, P.H. (1991) Ontoecogenophyloconstraints? The chaos of constraint terminology. *Trends Ecol. Evol.*, **6**, 166–7.

Aparicio, S., Hawker, K., Cottage, A. *et al.* (1997) Organization of the *Fugu rubripes Hox* clusters: Evidence for continuing evolution of vertebrate Hox complexes. *Nature Genet.*, **16**, 79–83.

Appel, T.A. (1987) *The Cuvier-Geoffroy Debate. French Biology in the Decades before Darwin*, Oxford University Press, New York.

Arendt, D. and Nübler-Jung, K. (1994) Inversion of dorsoventral axis? *Nature*, **371**, 26.

Arendt, D. and Nübler-Jung, K. (1996) Common ground plans in early brain development in mice and flies. *BioEssays*, **18**, 255–9.

Arendt, D. and Nübler-Jung, K. (1997) Dorsal or ventral: Similarities in fate maps and gastrulation patterns in annelids, arthropods and chordates. *Mech. Devel.*, **61**, 7–21.

Ariizumi, T. and Asashima, M. (1995a) Control of the embryonic body plan by activin during amphibian development. *Zool. Sci.*, **12**, 509–21.

Ariizumi, T. and Asashima, M. (1995b) Head and trunk-tail organizer effects of the gastrula ectoderm of *Cynops pyrrhogaster* after treatment with activin A. *Roux's Arch. Devel. Biol.*, **204**, 427–35.

Arita, H.T. and Fenton, M.B. (1997) Flight and echolocation in the ecology and evolution of bats. *Trends Ecol. Evol.*, **12**, 53–8.

Arkell, R. and Beddington, R.S.P. (1997) BMP-7 influences pattern and growth of the developing hindbrain of mouse embryos. *Development*, **124**, 1–12.

Armstrong, P.B. and Child, J.S. (1965) Stages in the normal development of *Fundulus heteroclitus*. *Biol. Bull.*, **128**, 143–68.

Arnold, A.J., Kelly, D.C. and Parker, W.C. (1995) Causality and Cope's rule: Evidence from the planktonic foraminifera. *J. Paleont.*, **69**, 203–10.

Arnold, J.M. (1965) Normal embryonic stages of the squid, *Loligo pealii*. *Biol. Bull.*, **128**, 24–32.

Arnold, S.J. (1992) Constraints on phenotypic evolution. *Amer. Nat.*, **140**, S85–S107.

Arnold, S.J., Alberch, P., Csányi, V. *et al.* (1989) Group report: How do complex organisms evolve? in *Complex Organismal Functions: Integration and Evolution in Vertebrates* (eds D.B. Wake and G. Roth), John Wiley & Sons, Chichester, pp. 403–33.

Arnold, S.J. and Wade, M.J. (1984) On the measurement of natural and sexual selection. Theory. *Evolution*, **38**, 709–19.

Arnqvist, G. (1994) The cost of male secondary sexual traits: Developmental constraints during ontogeny in a sexually dimorphic water strider. *Amer. Nat.*, **144**, 119–32.

Arthur, W. (1984) *Mechanisms of Morphological Evolution: A Combined Genetic, Developmental and Ecological Approach*. John Wiley & Sons, Chichester.

Arthur, W. (1988) *A Theory of the Evolution of Development*. John Wiley & Sons, Chichester.

Arthur, W. (1997) The Origin of Animal Body Plans: A Study in Evolutionary developmental Biology, Cambridge University Press, Cambridge.

Asashima, M., Nakamo, H., Shimada, K. *et al.* (1990) Mesodermal induction in early amphibian embryos by activin A (erythroid differentiation factor). *Roux's Arch. Devel. Biol.*, **198**, 330–5.

Ashkenas, J., Muschler, J. and Bissell, M.J. (1996) The extracellular matrix in epithelial biology: Shared molecules and common themes in distant phyla. *Devel. Biol.*, **180**, 433–44.

Asma, S.T. (1996) Darwin's causal pluralism. *Biol. Philos.*, **11**, 1–20.

Atchley, W.R. (1987) Developmental quantitative genetics and the evolution of ontogenies. *Evolution*, **41**, 316–30.

Atchley, W.R. (1990) Heterochrony and morphological change. A quantitative genetic approach. *Sem. Devel. Biol.*, **1**, 289–97.

Atchley, W.R. (1993) Genetic aspects of variability in the mammalian mandible, in *The Vertebrate Skull*, Volume 1 (eds J. Hanken and B.K. Hall), The University of Chicago Press, Chicago, IL, pp. 207–47.

Atchley, W.R., Cowley, D.E., Eisen, E.J. *et al.* (1990) Correlated response in mandible form to selection for body composition in the mouse. *Evolution*, **44**, 669–88.

Atchley, W.R., Cowley, D.E., Vogl, C. and McLennan, T. (1992) Evolutionary divergence, shape change and genetic correlation structure in the rodent mandible. *Syst. Biol.*, **41**, 196–221.

Atchley, W.R. and Fitch, W.M. (1991) Gene trees and the origins of inbred strains of mice. *Science*, **154**, 554–8.

Atchley, W.R. and Fitch, W.M. (1993) Genetic affinities of inbred mouse strains of uncertain origin. *Mol. Biol. Evol.*, **10**, 1150–69.

Atchley, W.R., Fitch, W.M. and Bronner Fraser, M. (1994) Molecular evolution of the *MyoD* family of transcription factors. *Proc. Natl Acad. Sci. USA*, **91**, 11522–6.

Atchley, W.R. and Hall, B.K. (1991) A model for development and evolution of complex morphological structures and its application to the mammalian mandible. *Biol. Rev.*, **66**, 101–57.

Atchley, W.R., Logsdon, T. and Cowley, D.E. (1991) Uterine effects, epigenetics and postnatal skeletal development in the mouse. *Evolution*, **45**, 891–909.

Atchley, W.R. and Newman, S. (1989) Developmental quantitative genetic aspects of evolutionary change. *Amer. Nat.*, **134**, 486–512.

Atchley, W.R., Plummer, A.A. and Riska, B. (1985a) Genetics of mandible form in the mouse. *Genetics*, **111**, 555–77.

Atchley, W.R., Plummer, A.A. and Riska, B. (1985b) Genetic analysis of size-scaling patterns in the mouse mandible. *Genetics*, **111**, 579–95.

Atchley, W.R., Xu, S. and Vogl, C. (1994) Developmental quantitative genetic models of evolutionary change. *Devel. Genet.*, **15**, 92–103.

Atchley, W.R., Xu, S., and Cowley, D.E. (1997) Altering developmental trajectories in mice by restricted index selection. *Genetics*, **146**, 629–40.

Atkinson, D. and Sibly, R.M. (1997) Why are organisms usually bigger in colder environments? Making sense of a life history puzzle. *Trends Ecol. Evol.*, **12**, 235–9.

Atkinson, J. (1992) Conceptual issues in the reunion of development and evolution. *Synthese*, **91**, 93–110.

Atz, J.W. (1970) The application of the idea of homology to behavior, in *Development and Evolution of Behavior. Essays in Memory of T.C. Schneirla*, Freeman & Co., San Francisco, CA, pp. 53–74.

Averof, M. (1997) Arthropod evolution: Same *Hox* genes, different body plans. *Curr. Biol.*, **7**, R634–R636.

Averof, M. and Akam, M. (1995) *Hox* genes and the diversification of insect and crustacean body plans. *Nature*, **376**, 420–3.

Averof, M. and Cohen, S. M. (1997) Evolutionary origin of insect wings from ancestral gills. *Nature*, **385**, 627–30.

Averof, M. and Patel, N.H. (1997) Crustacean appendage evolution associated with changes in Hox gene expression. *Nature*, **388**, 682–6.

Ayala, F.J., Rzhetsky, A. and Ayala, F.J. (1998) Origin of the metazoan phyla: molecular clocks confirm paleontological estimates. *Proc. Natl. Acad. USA* **95**, 606–11.

Baer, K.E. von (1828) *Über Entwickelungsgeschichte der Thiere: Beobachtung und Reflexion*, Gebrüder Bornträger, Königsberg. Reprinted 1967 by Culture et Civilisation, Bruxelles.

Baer, K.E. von (1835) *Untersuchungen über die Entwickelungsgeschichte der Fische.* Leipzig.

Balavoine, G. and Telford, M.J. (1995) Identification of planarian homeobox sequences indicates the antiquity of most Hox/homeotic gene subclasses. *Proc. Natl Acad. Sci. USA*, **92**, 7227–31.

Baldwin, I.T. (1996) Inducible defenses and population biology. *Trends Ecol. Evol.*, **11**, 104–5.

Balfour, F.M. (1873a) The development and growth of the layers of the blastoderm. *Quart. J. Microsc. Sci.*, **13**, 266–76; *Studies Physiol. Lab. Camb.*, **1**, 1–10.

Balfour, F.M. (1873b) On the disappearance of the primitive groove in the embryo chick. *Quart. J. Microsc. Sci.*, **13**, 276–80; *Studies Physiol. Lab. Camb.*, **1**, 10–14.

Balfour, F.M. (1874) A preliminary account of the development of the elasmobranch fishes. *Quart. J. Microsc. Sci.*, **14**, 323–64.

Balfour, F.M. (1875) A comparison of the early stages in the development of vertebrates. *Quart. J. Microsc. Sci.*, **15**, 207–26.

Balfour, F.M. (1878) *A Monograph on the Development of Elasmobranch Fishes*, Macmillan and Co., London.

Balfour, F.M. (1880a) Larval forms: Their nature, origin and affinities. *Quart. J. Microsc. Sci.*, **20**, 381–407.

Balfour, F.M. (1880b) Vice-Presidential address before the Anatomical and Physiological section of the British Association at Swansea, August 27 1880. 'Embryology and the Darwinian theory.' *Brit. Assoc. Report*, **1880**, 636–44. Also published in *Nature* (1880), **22**, 417–20.

Balfour, F.M. (1880–1) *A Treatise on Comparative Embryology. Two Volumes.* Macmillan and Co., London.

Balfour, F.M. (1881) On the development of the skeleton of the paired fins of Elasmobranchii, considered in relation to its bearing on the nature of the limbs of the Vertebrata. *Proc. Zool. Soc. Lond.*, **1881**, 656–71.

Balinsky, B.I. (1975) *An Introduction to Embryology*, 4th edn, W.B. Saunders, Co., Philadelphia, PA.

Ballard, J.W.O., Olsen, G.J., Faith, D.P. *et al.* (1992) Evidence from 12S ribosomal RNA sequences that onychophorans are modified arthropods. *Science*, **258**, 1345–8.

Ballard, W.W. (1976) Problems of gastrulation: Real and verbal. *Bioscience*, **26**, 36–9.

Ballard, W.W. (1981) Morphogenetic movements

and fate maps of vertebrates. *Amer. Zool.*, **21**, 391–9.

Ballard, W.W., Melinger, J. and Lechenault, H. (1993) A series of normal stages for development of *Scyliorhinus canicula*, the Lesser Spotted Dogfish (Chondrichthyes: Scyliorhinidae). *J. Exp. Zool.*, **267**, 318–36.

Bally-Cuif, L. and Boncinelli, E. (1997) Transcription factors and head formation in vertebrates. *BioEssays*, **19**, 127–35.

Baltzer, F. (1950) Entwicklungsphysiologische Betrachtungen über Probleme der Homologie und Evolution. *Rev. Suisse Zool.*, **57**, 451–77.

Balzac, H. de (1842) *Père Goriot*. Washington Square Press, New York (1962).

Barlow, A.J. and Francis-West, P.H. (1997) Ectopic application of recombinant BMP-2 and BMP-4 can change patterning of developing chick facial primordia. *Development*, **124**, 391–8.

Barnes, G.L., Alexander, P.G., Hsu, C.W. *et al.* (1997) Cloning and characterization of chicken *Paraxis*: A regulator of paraxial mesoderm development and somite formation. *Devel. Biol.*, **189**, 95–111.

Barrington, E.J.W. (1979) Essential features of lower types, in *Hyman's Comparative Vertebrate Anatomy*, 5th edn (ed. M.H. Wake), pp. 57–63, The University of Chicago Press, Chicago, IL.

Barry, M. (1836–7a) On the unity of structure in the animal kingdom. *Edinburgh New Philos. J.*, **22**, 116–41.

Barry, M. (1836–7b) Further observations on the Unity of Structure in the Animal Kingdom, and on Congenital Anomalies, including 'Hermaphrodites'; with some Remarks on Embryology, as facilitating Animal Nomenclature, Classification, and the Study of Comparative Anatomy. *Edinburgh New Philos. J.*, **22**, 345–64.

Basler, K., Edlund, T., Jessell, T.M. and Yamada, T. (1993) Control of cell pattern in the neural tube: Regulation of cell differentiation by *dorsalin*-1, a novel TGFß family member. *Cell*, **73**, 687–702.

Bateman, K.G. (1959a) The genetic assimilation of the dumpy phenocopy. *J. Genet.*, **56**, 341–51.

Bateman, K.G. (1959b) The genetic assimilation of four venation phenocopies. *J. Genet.*, **56**, 443–74.

Bates, W.R. (1994) Ecological consequences of altering the timing mechanism for metamorphosis in anural ascidians. *Amer. Zool.*, **34**, 333–42.

Bateson, W. (1884a) The early stages in the development of *Balanoglossus* (*sp. incert.*) *Quart. J. Microsc. Sci.*, **24**, 206–36.

Bateson, W. (1884b) On the development of *Balanoglossus*. *Ann. Mag. Nat. Hist.*, **13**, 65.

Bateson, W. (1885a) The later stages in the development of *Balanoglossus Kowalevski*, with a suggestion as to the affinities of the Enteropneusta. *Quart. J. Microsc. Sci.*, **25**, Suppl., 81–122.

Bateson, W. (1885b) Continued account of the later stages in the development of *Balanoglossus Kowalevski*, and of the morphology of the Enteropneusta. *Quart. J. Micros. Sci.*, **26**, 511–33

Bateson, W. (1885c) The Ancestry of the Chordata. *Quart. J. Microsc. Sci.*, **26**, 535–71. Reprinted in *Scientific Papers of William Bateson* (1928), ed. R.C. Punnett, Vol. 1, 1–31, Cambridge at the University Press. Reprinted 1971, Johnson Reprint Company, New York.

Bateson, W. (1894) *Materials for the Study of Variation*, Macmillan and Co., London.

Bateson, W. (1922) Evolutionary faith and modern doubts. *Science*, **55**, 53–61.

Batterham, P., Davies, A.G., Game, A.Y. and McKenzie, J.A. (1996) Asymmetry – where evolutionary and developmental genetics meet. *BioEssays*, **18**, 841–5.

Baumgartner, S. and Noll, M. (1990) Network of interactions among pair-rule genes regulating *paired* expression during primordial segmentation of *Drosophila*. *Genes & Devel.*, **33**, 1–18.

Baur, E. (1922) *Einführung in die experimentelle Vererbungsiehre*. Bornträger, Berlin.

Beer, C.G. (1984) Homology, analogy and ethology. *Human Devel.*, **27**, 297–308.

Beer, G.R. de (1930) *Embryology and Evolution*. Clarendon Press, Oxford.

Beer, G.R. de (1940) *Embryos and Ancestors*, Clarendon Press, Oxford.

Beer, G.R. de (1947) The differentiation of neural crest cells and visceral cartilages and odontoblasts in *Amblystoma*, and a re-examination of the germ-layer theory. *Proc. R. Soc. Lond. B*, **134**, 377–98.

Beer, G.R. de (1958) *Embryos and Ancestors*. Oxford University Press, Oxford.

Beer, G.R. de (1962) *Reflections of a Darwinian*. Thomas Nelson and Sons, Ltd., London.

Beer, G.R. de (1964) *Atlas of Evolution*. Thomas Nelson, & Sons, London.

Beer, G.R. de (1971) *Homology: An unsolved problem*, Oxford Biology Reader No. 11, Oxford University Press, London.

Beeson, R.J. (1978) *Bridging the Gap: The Problems of Vertebrate Ancestry, 1859–1875*, Ph.D. Thesis, 401pp., Oregon State University, Corvallis, OR.

Beetschen, J.-C. (1995) Louis Sébastien Tredern de Lézérec (1780–18?), a forgotten pioneer of chick embryology. *Int. J. Devel. Biol.*, **39**, 299–308.

Beisson, J. and Sonneborn, T.M. (1965) Cytoplasmic inheritance of the organization of the cell cortex in *Paramecium aurelia. Proc. Natl Acad. Sci. USA*, **53**, 275–82.

Bell, G. (1997a) *Selection: The Mechanism of Evolution*, Chapman & Hall, London.

Bell, G. (1997b) *The Basics of Selection*, Chapman & Hall, London.

Bell, G. and Koufopanou, V. (1991) The architecture of the life cycle in small organisms. *Phil. Trans. R. Soc. Lond. B*, **332**, 81–9.

Bell, G. and Mooers, A.O. (1997) Size and complexity among multicellular organisms. *Biol. J. Linn. Soc.*, **60**, 345–63.

Bell, M.A., Ortí, G., Walker, J.A. and Koenings, J.P. (1993) Evolution of pelvic reduction in threespine stickleback fish: A test of competing hypotheses. *Evolution*, **47**, 906–14.

Bellairs, A.d'A. and Gans, C. (1983) A reinterpretation of the Amphisbaenian orbitosphenoid. *Nature*, **302**, 243–4.

Bengtson, S. and Zhao, Y. (1997) Fossilized metazoan embryos from the earliest Cambrian. *Science*, **277**, 1645–8.

Benson, K.R. (1981) Problems of individual development: Descriptive embryology in America at the turn of the Century. *J. Hist. Biol.*, **14**, 115–28.

Béque-Kirn, Smith, A.J., Ruch, J.V. *et al.* (1992) Effect of dentin proteins, transforming growth factor ß1 (TGFß1) and bone morphogenetic protein 2 (BMPs) on the differentiation of odontoblasts *in vitro. Int. J. Devel. Biol.*, **36**, 491–503.

Béque-Kirn, C., Smith, A.J., Loriot, M. *et al.* (1994) Comparative analysis of TGFßs, BMPs, IGF1, M*sx*s, fibronectin, osteonectin and bone sialoprotein gene expression during normal and *in vitro*-induced odontoblast differentiation. *Int. J. Devel. Biol.*, **38**, 405–20.

Berg, L.S. (1969) *Nomogenesis or Evolution determined by Law*, Translated from the Russian by J.N. Rostovtsov, Foreword by T. Dobzhansky, Introduction by D'A.W. Thompson, The MIT Press, Cambridge, MA.

Bernardo, J. (1993a) Determinants of maturation in animals. *Trends Ecol. Evol.*, **8**, 166–73.

Bernardo, J. (1993b) Plasticity versus genetics – reply and letter. *Trends Ecol. Evol.*, **8**, 379–80.

Bernfield, M. and Banerjee, S.D.(1982) The turnover of basal lamina glycosaminoglycan correlates with epithelial morphogenesis. *Devel. Biol.*, **90**, 291–305.

Berrill, N.J. (1931) Studies in tunicate development. II. Abbreviation of development in the Molgulidae. *Phil. Trans. R. Soc. Lond. B*, **219**, 281–346.

Berrill, N.J. (1935) Cell division and differentiation in asexual and sexual development. *J. Morph.*, **57**, 353–427.

Berrill, N.J. (1955) *The Origin of Vertebrates*, Oxford University Press, Oxford.

Berrill, N.J. and Liu, C.K. (1948) Germplasm, Weismann, and Hydrozoa. *Quart. Rev. Biol.*, **23**, 124–32.

Bertalanffy, L. von. (1933) *Modern Theories of Development. An Introduction to Theoretical Biology*, Translated by J.H. Woodger. Oxford University Press, Oxford and London.

Bertram, J.E.A., Greenberg, L., Miyake, T. and Hall, B.K. (1997) Effects of paralysis on long bone growth in the chick: Growth trajectories of the pelvic limb. *Growth, Devel., Ageing* (in press).

Bertram, J.E.A. and Swartz, S.M. (1991) The 'law of bone transformations': A case of crying Wolff? *Biol. Rev.*, **66**, 245–73.

Berven, K.A. (1982) The genetic basis of altitudinal variation in the wood frog *Rana sylvatica*. 1. An experimental basis of life history traits. *Evolution*, **36**, 962–83.

Bijtel, J. (1931) Ueber die Entwicklung des Schwanzes bei Amphibien. *Arch. EntwMech.*, **125**, 448–86.

Bissell, M.J. and Barcellos-Hoff, M.H. (1987) The influence of extracellular matrix in gene expression: Is structure the message? *J. Cell Sci.*, Suppl., **8**, 327–43.

Björklund, M. (1996) The importance of evolutionary constraints in ecological time scales. *Evol. Ecol.*, **10**, 423–31.

Black, R.W.II. and Slobodkin, L.B. (1987) What is cyclomorphosis? *Freshwater Biol.*, **18**, 373–8.

Blackburn, D.G. (1995) Saltationist and punctuated equilibrium models for the evolution of viviparity and placentation. *J. Theor. Biol.*, **174**, 199–216.

Blackstone, N.W. (1987) Allometry and relative growth: Pattern and process in evolutionary studies. *Syst. Zool.*, **36**, 76–8.

Blackstone, N.W. and Yund, P.O. (1989) Morphological variation in a colonial marine hydroid: A comparison of size-based and age-based heterochrony. *Paleobiology*, **15**, 1–10.

Blanco, M.J. and Alberch, P. (1992) Caenogenesis, developmental variability, and evolution in the carpus and tarsus of the marbled newt *Triturus marmoratus. Evolution*, **46**, 677–87.

Blau, H.M., and Baltimore, D. (1991) Differentiation requires continuous regulation. *J. Cell Biol.*, **112**, 781–3.

Boaden, P.J.S. (1975) Anaerobiosis, meiofauna and early metazoan evolution. *Zool. Scripta*, **4**, 21–4.

Boake, C.R.B. (1994) *Quantitative Genetic Studies of Behavioral Evolution*, The University of Chicago Press, Chicago, IL.

Bock, W.J. (1959) Preadaptation and multiple evolutionary pathways. *Evolution*, **13**, 194–211.

Bock, W.J. (1960) Secondary articulations of the avian mandible. *The Auk*, **71**, 19–55.

Bock, W.J. (1991) Explanations in Konstruktionsmorphologie and evolutionary morphology, in *Constructional Morphology and Evolution* (eds N. Schmidt-Kittler and K. Vogel), Springer-Verlag, Berlin, pp. 9–29.

Bock, W.J. and Morioka, H. (1971) Morphology and evolution of the ectethmoid- mandibular articulation in Meliphagidae (Aves) *J. Morph.*, **135**, 13–50.

Bodemer, C.W. (1964) Regeneration and the decline of preformation in eighteenth-century embryology. *Bull. Hist. Med.*, **38**, 20–31.

Boletzky, S.V. (1989) Early ontogeny and evolution: The cephalopod model viewed from the point of developmental morphology. *Geobios, Mém. Spécial*, **12**, 67–78.

Bolker, J.A. (1994) Comparison of gastrulation in frogs and fish. *Amer. Zool.*, **34**, 313–22.

Bolker, J.A. (1995) Model systems in developmental biology. *BioEssays*, **17**, 451–5.

Bolker, J.A. and Raff, R.A. (1996) Developmental genetics and traditional homology. *BioEssays*, **18**, 489–94.

Boncinelli, E., Simeone, A., Acampora, D. and Mavilio, F. (1991) Hox gene activation by retinoic acid. *Trends Genet.*, **7**, 329–34.

Bone, Q. (1957) The problem of the 'amphioxides' larva. *Nature*, **180**, 1462–4.

Bonner, J.T. (1958) *The Evolution of Development. Three Special Lectures Given at University College, London.* Cambridge University Press, Cambridge.

Bonner, J.T. (1974) *On Development. The Biology of Form.* Harvard University Press, Cambridge, MA.

Bonner, J.T. (1988) *The Evolution of Complexity by Means of Natural Selection.* Princeton University Press, Princeton, NJ.

Bookstein, F.L. (1991) *Morphometric Tools for Landmark Data. Geometry and Biology*, Cambridge University Press, Cambridge.

Bookstein, F.L. (1994) Can biometrical shape be a homologous character? in *Homology: The Hierarchical Basis of Comparative Biology* (ed. B.K. Hall), Academic Press, San Diego, CA, pp. 197–227.

Bookstein, F.L., Chernoff, B., Elder, R. *et al.* (1985) *Morphometrics in Evolutionary Biology. The Geometry of Size and Shape Change, with Examples from Fishes*, Special Publication **15**, Academy of Natural Sciences, Philadelphia, PA.

Bourdier, F. (1969) Geoffroy Saint-Hilaire versus Cuvier: The campaign for paleontological evolution (1825–1838), in *Toward a History of Geology* (ed. C.J. Schneer), The MIT Press, Cambridge, MA, pp. 36–61.

Bouwmeester, T., Kim, S.-H., Sasai Y. *et al.* (1996) *Cerberus* is a head-inducing secreted factor expressed in the anterior endoderm of Spemann's organizer. *Nature*, **382**, 595–601.

Bouwmeester, T. and Leyns, L. (1997) Vertebrate head induction by anterior primitive endoderm. *BioEssays*, **19**, 855–63.

Bowler, P.J. (1975) The changing meaning of 'evolution.' *J. History Ideas*, **36**, 95–114.

Bowler, P.J. (1977) Darwinism and the argument from design: Suggestions for a reevaluation. *J. Hist. Biol.*, **10**, 29–43.

Bowler, P.J. (1988) *The Non-Darwinian Revolution. Reinterpreting a Historical Myth.* The Johns Hopkins University Press, Baltimore, MD.

Bowler, P.J. (1989a) *Evolution. The History of an Idea.* Revised edn, University of California Press, Berkeley, CA.

Bowler, P.J. (1989b) *The Mendelian Revolution. The emergence of hereditarian concepts in modern science and society.* The Johns Hopkins University Press, Baltimore, MD.

Bowler, P.J. (1994) Are the Arthropoda a natural group? An episode in the history of evolutionary biology. *J. Hist. Biol.*, **27**, 177–213.

Bowler, P.J. (1996) *Life's Splendid Drama. Evolutionary Biology and the Reconstruction of Life's Ancestry 1860–1940*, The University of Chicago Press, Chicago, IL.

Boycott, A.E. (1938) Experiments on the artificial breeding of *Limnaea involuta*, *Limnaea burnetti* and other forms of *Limnaea peregra*. *Proc. Malac. Soc. Lond.*, **23**, 101–8.

Boycott, A.E., Diver, C., Garstang, S.L. and Turner, F.M. (1930) The inheritance of sinestrality in *Limnaea peregra* (Mollusca: Pulmonata). *Phil. Trans. R. Soc. Lond.* B, **219**, 51–131.

Brace, C.L. (1963) Structural reductions in evolution. *Amer. Nat.*, **97**, 39–49.

Bradley, L., Wainstock, D. and Sive, H. (1996) Positive and negative signals modulate formation of the *Xenopus* cement gland. *Development*, **122**, 2739–50.

Brain, C.K. (1981) Hominid evolution and climate change. *S. Afr. J. Sci.*, **77**, 104–5.

Brain, C.K. (1983) The evolution of man in Africa: Was it a consequence of Caenozoic cooling? *Geol. Soc. S. Afr.*, **84**, 1–19.

Brakefield, P.M., Gates, J., Keys, D. *et al.* (1996) Development, plasticity and evolution of butterfly eyespot patterns. *Nature*, **384**, 236–42.

Brakefield, P.M. and Kesbeke, F. (1997) Genotype-environment interactions for insect growth in constant and fluctuating temperature regimes. *Proc. R. Soc. Lond.* B, **264**, 717–23.

Briggs, D.E.G. (1991) Extraordinary fossils. *Amer. Sci.*, **79** (2), 130–41.

Briggs, D.E.G. (1994) Giant predators from the Cambrian of China. *Science*, **264**, 1283–4.

Briggs, D.E.G., Clarkson, E.N.K. and Aldridge, R.J. (1983) The conodont animal. *Lethaia*, **16**, 1–14.

Briggs, D.E.G., Erwin, D.H. and Collier, F.J., with photographs by Chip Clark (1994) *The Fossils of the Burgess Shale*, Smithsonian Institution Press, Washington and London.

Briggs, D.E.G., Fortey, R.A. and Wills, M.A. (1992a) Morphological disparity in the Cambrian. *Science*, **256**, 1670–3.

Briggs, D.E.G., Fortey, R.A. and Wills, M.A. (1992b) Cambrian and Recent morphological disparity. *Science*, **258**, 1817–18.

Briggs, D.E.G., Fortey, R.A. and Wills, M.A. (1993) How big was the Cambrian evolutionary explosion? A taxonomic and morphological comparison of Cambrian and Recent arthropods, in *Evolutionary Patterns and Processes* (eds D.R. Lees and D. Edwards), Academic Press, London, pp. 33–44.

Briggs, D.E.G., and Kear, A.J. (1994) Decay of *Branchiostoma*: implications for soft-tissue preservation in conodonts and other primitive chordates. *Lethaia*, **26**, 275–87.

Broadhead, T.W. (1988) Heterochrony – a pervasive influence in the evolution of Paleozoic Crinoidea, in *Echinoderm Biology* (eds R.D. Burke, P.V. Mladenov, P. Lambert and R.L. Parsley), A. A. Balkema, Rotterdam, Brookfield, pp. 115–23.

Brockes, J.P. (1997) Amphibian limb regeneration: Rebuilding a complex structure. *Science*, **276**, 81–7.

Brönmark, C. and Miner, J. G. (1992) Predator-induced phenotypical change in body morphology in Crucian carp. *Science*, **258**, 1348–50.

Bronner Fraser, M. (1995a) Origin of the avian neural crest. *Stem Cells*, **13**, 640–6.

Bronner Fraser, M. (1995b) Origins and developmental potential of the neural crest. *Exp. Cell Res.*, **218**, 405–17.

Brooks, D.R. and McLennan, D.A. (eds) (1991) *Phylogeny, Ecology and Behaviour: A Research Program in Comparative Biology*, The University of Chicago Press, Chicago, IL.

Brooks, D.R. and Wiley, E.O. (1988) *Evolution as Entropy. Toward a Unified Theory of Biology*. 2nd edn, University of Chicago Press, Chicago, IL.

Brooks, J.L. (1946) Cyclomorphosis in *Daphnia*. 1. An analysis of *D. retrocurva* and *D. galeata. Ecol. Monographs*, **16**, 409–47.

Brown, S.J., Parrish, J.K. and Denell, R.E. (1994) Genetic control of early embryogenesis in the red flour beetle, *Tribolium castaneum. Amer. Zool.*, **34**, 343–52.

Brown, S.M. (1980) Comparative ossification in tadpoles of the genus *Xenopus* (Anura, Pipidae), Master's Thesis, San Diego State University, San Diego, CA.

Brul, E.L. du (1964) Evolution of the temporomandibular joint, in *The Temporomandibular Joint* (ed. B.G. Sarnat), C.C. Thomas, Springfield, IL, pp. 3–27.

Brun, R.B. (1985) Neural fold and neural crest movement in the Mexican salamander *Ambystoma mexicanum. J. Exp. Zool.*, **234**, 57–61.

Brusca, R.C. and Brusca, G.J. (1990) *Invertebrates*. Sinauer Associates Inc., Sunderland, MA.

Brush, A.H. (1993) The origin of feathers: A novel approach, in *Avian Biology* (eds D.S. Farner, J.R. King and K.C. Parkes), Academic Press, San Diego, CA, Vol. 9, pp. 121–62.

Brush, A.H. (1996) On the origin of feathers. *J. Evol. Biol.*, **9**, 131–42.

Brylski, P. and Hall, B.K. (1988a) Epithelial behaviors and threshold effects in the development and evolution of internal and external cheek pouches in rodents. *Z. zool. Syst. Evolut.-forsch.*, **26**, 144–54.

Brylski, P. and Hall, B.K. (1988b) Ontogeny of a macroevolutionary phenotype: The external cheek pouches of Geomyoid rodents. *Evolution*, **42**, 391–5.

Budd, G.E. (1997) Stem group arthropods from the Lower Cambrian Sirius Passet fauna of North Greenland, in *Arthropod Relationships* (eds R.A. Fortey and R.H. Thomas), Chapman & Hall, London, pp. 125–38.

Bult, C.J., White, O., Olsen, G.J. *et al.* (1996) Complete genome sequence of the methanogenic archaeon, *Methanococcus jannaschii. Science,* **273,** 1058–73.

Burggren, W.W. and Bemis, W.E. (1990) Studying physiological evolution: Paradigms and pitfalls, in *Evolutionary Innovations* (ed. M.H. Nitecki), The University of Chicago Press, Chicago, IL, pp. 191–228.

Burggren, W.W. and Just, J.J. (1992) Developmental changes in physiological systems, in *Environmental Physiology of the Amphibians* (eds M.E. Feder and W.W. Burggren), The University of Chicago Press, Chicago, IL, pp. 467–530.

Burian, R.M. (1992) Adaptation: Historical perspectives, in *Keywords in Evolutionary Biology* (eds E. Fox Keller and E.A. Lloyd), Harvard University Press, Cambridge, MA, pp. 7–12.

Burke, A.C. (1989a) Development of the turtle carapace: Implications for the evolution of a novel Bauplan. *J. Morph.,* **199,** 363–78.

Burke, A.C. (1989b) Epithelial-mesenchymal interactions in the development of the chelonian Bauplan. *Fortsch. Zool.,* **35,** 206–9.

Burke, A.C. (1991) The development and evolution of the turtle body plan: Inferring intrinsic aspects of the evolutionary process from experimental embryology. *Amer. Zool.,* **31,** 616–27.

Burke, A.C. and Feduccia, A. (1997) Developmental patterns and the identification of homologies in the avian hand. *Science,* **278,** 666–8.

Burke, A.C. and Nelson, C.E. (1996) Evolutionary transposition and the vertebrate *Hox* genes: Comparing morphology to gene expression boundaries with *in situ* hybridization, in *Molecular Zoology: Advances, Strategies and Protocols* (eds J.D. Ferraris and S.R. Palumbi), Wiley-Liss Inc., New York, pp. 283–95, 435–40.

Burke, A.C., Nelson, C.E., Morgan, B.A. and Tabin, C. (1995) *Hox* genes and the evolution of vertebrate axial morphology. *Development,* **121,** 333–46.

Buss, L.W. (1987) *The Evolution of Individuality.* Princeton University Press, Princeton.

Buss, L.W. (1988) Diversification and germ-line determination. *Paleobiology,* **14,** 313–21.

Buss, L.W. and Seilacher, A. (1994) The phylum Vendobionta: A sister group of the Eumetazoa? *Paleobiology,* **20,** 1–4.

Busturia, A. and Lawrence, P.A. (1994) Regulation of cell number in *Drosophila. Nature,* **370,** 561–3.

Butler, P.M. (1956) The ontogeny of molar pattern. *Biol. Rev.,* **31,** 30–70.

Butterfield, N.J. (1990a) Organic preservation of non-mineralizing organisms and the taphonomy of the Burgess Shale. *Paleobiology,* **16,** 272–86.

Butterfield, N.J. (1990b) A reassessment of the enigmatic Burgess Shale fossil *Wiwaxia corrugata* (Matthew) and its relationship to the polychaete *Canadia spinosa. Paleobiology,* **16,** 287–303.

Butterfield, N.J. (1994) Burgess Shale-type fossils from a Lower Cambrian shallow-shelf sequence in northwestern Canada. *Nature,* **369,** 477–9.

Buxton, P.G., Kostakopoulou, K., Brickell, P. *et al.* (1997) Expression of the transcription factor *slug* correlates with growth of the limb bud and is regulated by FGF-4 and retinoic acid. *Int. J. Devel. Biol.,* **41,** 559–68.

Cahn, Th. (1962) *La Vie et l'oeuvre d'Etienne Geoffroy Saint-Hilaire.* Presses Universitaires de France, Paris.

Cain, A.J. (1982) On homology and convergence, in *Problems of Phylogenetic Reconstruction* (eds K.A. Joysey and A.E. Friday), Systematics Association Special Volume 21, Academic Press, London, pp. 1–19.

Calder, W.A. (1984) *Size, Function, and Life History,* Harvard University Press, Cambridge, MA.

Caldwell, M.W. (1994) Developmental constraints and limb evolution in Permian and extant lepidosauromorph diapsids. *J. Vert. Paleont.,* **14,** 459–71.

Caldwell, M.W. (1996) Ontogeny and phylogeny of the mesopodial skeleton of mososauroid reptiles. *Zool. J. Linn. Soc.,* **116,** 407–36.

Caldwell, M.W. (1997) Modified perichondrial ossification and the evolution of paddle-like limbs in ichthyosaurs and plesiosaurs. *J. Vert. Paleont.,* **17,** 534–47.

Caldwell, M.W. and Lee, M.S.Y. (1997) A snake with legs from the marine Cretaceous of the Middle East. *Nature,* **386,** 705–9.

Calow, P. (1978) *Life Cycles. An Evolutionary Approach to the Physiology of Reproduction, Development and Ageing.* Chapman & Hall, London.

Calow, P. (1983) *Evolutionary Principles.* Blackie, Glasgow and London.

Campbell, K.S.W. (1987) Evolution evolving. *J. & Proc. R. Soc. N.S.W.,* **120,** 9–19.

Canfield, D.E. and Teske, A. (1996) Late Proterozoic rise in atmospheric oxygen concentration inferred from phylogenetic and sulphur-isotope studies. *Nature,* **382,** 127–32.

Cannadine, D. (1990) *The Pleasures of the Past,* Fontana Press, London.

Cannatella, D.C. and De Sá, R.O. (1993) *Xenopus laevis* as a model organism. *Syst. Biol.*, **42**, 476–507.

Capdevila, M.P. and Garcia-Bellido, A. (1974) Development and genetic analysis of *Bithorax* phenocopies in *Drosophila*. *Nature*, **250**, 500–2.

Carneiro, R.L. (1972) The devolution of evolution. *Social Biol.*, **19**, 248–58.

Carpenter, W.B. (1837) On unity of function in organized beings. *Edinburgh New Philos. J.*, **23**, 92–114.

Carpenter, W.B. (1839) *Principles of General and Comparative Physiology*, John Churchill, London.

Carpenter, W.B. (1851) *Principles of General and Comparative Physiology*, 3rd edn, John Churchill, London.

Carrier, D.R. (1991) Conflict in the hypaxial-musculo-skeletal system: documenting an evolutionary constraint. *Am. Zool.*, **31**, 644–54.

Carroll, R.L. (1987) *Vertebrate Paleontology and Evolution*, W. H. Freeman, New York.

Carroll, R.L. (1997) *Patterns and Processes of Vertebrate Evolution*, Cambridge University Press, Cambridge.

Carroll, S.B. (1995) Homeotic genes and the evolution of arthropods and chordates. *Nature*, **376**, 479–85.

Carroll, S.B., Weatherbee, S.D. and Langeland, J.A. (1995) Homeotic genes and the regulation and evolution of insect wing number. *Nature*, **375**, 58–61.

Carter, C.A. and Wourms, J.P. (1993) Naturally occurring diblastodermic eggs in the annual fish *Cynolebias*: Implications for developmental regulation and determination. *J. Morph.*, **215**, 301–12.

Carter, D.R. and Orr, T.E. (1992) Skeletal development and bone functional adaptation. *J. Bone Min. Res.*, **7**, S389–S395.

Carter, D.R., Van der Meuleun, M.C.H. and Beaupré, G.S. (1996) Mechanical factors in bone growth and development. *Bone*, **18**, 5S–10S.

Carter, D.R., Wong, M. and Orr, T.E. (1991) Musculoskeletal ontogeny, phylogeny and functional adaptation. *J. Biomech.*, **24**, Suppl. 1, 3–16.

Cassin, C. and Capuron, A. (1979) Buccal organogenesis in *Pleurodeles waltlii* Michah (urodele, amphibian). Study by intrablastocelic transplantation and *in vitro* culture. *J. Biol. buccale*, **7**, 61–76.

Castle, W.E. (1911) *Heredity in Relation to Evolution and Animal Breeding*, New York.

Catala, M., Teillet, M.-A., de Robertis, E.M. and Le Douarin, N.M. (1996) A spinal cord fate map in the avian embryo: while regressing Hensen's node lays down the notochord and floor plate thus joining the spinal cord lateral walls. *Development*, **122**, 2599–2610.

Cavalier-Smith, T. (ed.) (1985) *The Evolution of Genome Size*. Wiley, Chichester.

Chambers, R. (1844)*Vestiges of the Natural History of Creation*. John Churchill, London. (10th amended edn issued 1853.)

Chandebois, R. and Faber, J. (1983) *Automation in Animal Development. A New Theory Derived from the Concept of Cell Sociology*, Monographs in Developmental Biology, Volume 16, S. Karger AG, Basle.

Charnov, E.L. (1982) *The Theory of Sex Allocation*, Princeton University Press, Princeton, NJ.

Chen, F. and Capecchi, M.R. (1997) Targeted mutations in *Hoxa-9* and *Hoxb-9* reveal synergistic interactions. *Devel. Biol.*, **181**, 186–96.

Chen, J.-y., Edgecombe, G.D., Ramsköld, L. and Zhou, G.-Q. (1995a) Head segmentation in Early Cambrian *Fuxianhuia*: Implications for arthropod evolution. *Science*, **268**, 1339–43.

Chen, J.-y., Dzik, J., Edgecombe, G.D. *et al.* (1995b) A possible Early Cambrian chordate. *Nature*, **377**, 720–2.

Chen, J.-y., Ramsköld, L. and Zhou, G.-q. (1994) Evidence for monophyly and arthropod affinity of Cambrian giant predators. *Science*, **264**, 1304–8.

Chen, Y., Bei, M., Woo, I. *et al.* (1996) *Msx1* controls inductive signaling in mammalian tooth morphogenesis. *Development*, **122**, 3035–44.

Chen, Y. and Grunz, H. (1997) The final determination of *Xenopus* ectoderm depends on intrinsic and external positional information. *Int. J. Devel. Biol.*, **41**, 525–8.

Cherdantsev, V.G. and Krauss, Y.A. (1996) Gastrulation in the marine hydroid *Dynamena pumila*: An example of evolutionary anticipation based on developmental self-organization. *Evol. Theory*, **11**, 88–98.

Cheverud, J.M. (1984) Quantitative genetics and developmental constraints on evolution by selection. *J. Theor. Biol.*, **110**, 155–71.

Cheverud, J.M. (1996) Developmental integration and the evolution of pleiotropy. *Amer. Zool.*, **36**, 44–50.

Cheverud, J.M., Hartman, S.E., Richtsmeier, J.T. and Atchley, W.R. (1991) A quantitative genetic analysis of localized morphology in mandibles of inbred mice using finite element scaling analysis. *J. Craniofac. Genet. Devel. Biol.*, **11**, 122–37.

Chiang, C., Litingtung, Y., Lee, E. *et al.* (1996) Cyclopia and defective axial patterning in mice lacking *Sonic hedgehog* gene function. *Nature*, **383**, 407–13.

Chisaka, O. and Capecchi, M.R. (1991) Regionally restricted developmental defects resulting from targeted disruption of the mouse homeobox gene *Hox*-1.5. *Nature*, **350**, 473–9.

Cho, K.W.Y. and De Robertis, E.M. (1990) Differential activation of *Xenopus* homeobox genes by mesoderm-inducing growth factors and retinoic acid. *Genes & Devel.*, **4**, 1910–16.

Christian, J.L., McMahon, J.A., McMahon, A.P. and Moon, R.T. (1991) *Xwnt-8*, a *Xenopus Wnt-1/int*-1-related gene responsive to mesoderm-inducing factors, may play a role in ventral mesodermal patterning during embryogenesis. *Development*, **111**, 1045–55.

Christian, J.L. and Moon, R.T. (1993) Interactions between the XWnt and Spemann organizing signalling pathways generate dorsoventral pattern in the embryonic mesoderm of *Xenopus*. *Genes Devel.*, **7**, 13–28.

Chuong, C-M., Oliver, G., Ting, S.A. *et al.* (1990) Gradients of homeoproteins in developing feather buds. *Development*, **110**, 1021–30.

Churchill, F.B. (1986) Weismann, Hydromedusae and the biogenetic imperative: A reconsideration, in *A History of Embryology* (eds T.J. Horder, J.A. Witkowski and C.C. Wylie), British Society for Developmental Biology Symposium 8, Cambridge University Press, Cambridge, pp. 7–33.

Clark, A.G. (1987) Genetic correlations. The quantitative genetics of evolutionary constraint, in *Genetic Constraints on Adaptive Evolution* (ed. V. Loeschcke), Springer-Verlag, Berlin, pp. 25–46.

Clark, C.T. and Smith, K.K. (1993) Cranial osteogenesis in *Monodelphis domestica* (Didelphidae) and *Macropus eugenii* (Macropodidae). *J. Morph.*, **215**, 119–49.

Clayton, R.A., White, O., Ketchum, K.A. and Venter, J.C. (1997) The first genome from the third domain of life. *Nature*, **387**, 459–62.

Clemen, G. (1978) Beziehungen zwischen Gaumenknochen und ihren Zahnleisten bei *Salamandra salamandra* (L.) während der Metamorphose. *Wilhelm Roux Arch. EntwMech. Org.*, **185**, 19–36.

Clemen, G. (1979) Die Bedeutung des *Ramus palatinus* für die Vomerspangenbildung bei *Salamandra salamandra* (L.) *Wilhelm Roux Arch. EntwMech. Org.*, **187**, 219–30.

Clemen, G., Wanninger, A.-C. and Greven, H. (1997) The development of the dentigerous bones and teeth in the hemiramphid fish *Dermogenys pusillus* (Atheriniformes, Teleostei). *Ann, Anat.*, **179**, 165–74.

Coates, M.I. (1991) New palaeontological contributions to limb ontogeny and phylogeny, in *Developmental Patterns of the Vertebrate Limb* (eds J.R. Hinchliffe, J.M. Hurle and D. Summerbell), Plenum Press, New York, pp. 325–37.

Coates, M.I. (1994) The origin of vertebrate limbs. *Development*, 1994 Suppl., 169–80.

Coates, M.I. and Clack, J.A. (1990) Polydactyly in the earliest known tetrapod limbs. *Nature*, **347**, 66–9.

Coates, M.I. and Clack, J.A. (1991) Fish-like gills and breathing in the earliest known tetrapod. *Nature*, **352**, 234–6.

Cock, A.G. (1966) Genetical aspects of metrical growth and form in animals. *Quart. Rev. Biol.*, **41**, 131–90.

Coelho, C.N.D., Krabbenhoft, K.M., Upholt, W.B. *et al.* (1991) Altered expression of the chicken homeobox-containing genes GHox-7 and GHox-8 in the limb buds of *limbless* mutant chick embryos. *Development*, **113**, 1487–93.

Coelho, C.N.D., Upholt, W.B. and Kosher, R.A. (1993a) Ectoderm from various regions of the developing chick limb bud differentially regulates the expression of the chicken homeobox-containing genes GHox-7 and GHox-8 by limb mesenchymal cells. *Devel. Biol.*, **156**, 303–6.

Coelho, C.N.D., Upholt, W.B. and Kosher, R.A. (1993b) The expression pattern of the chicken homeobox-containing gene GHox-7 in developing polydactylous limb buds suggests its involvement in apical ectodermal ridge-directed outgrowth of limb mesoderm and in programmed cell death. *Differentiation*, **52**, 129–37.

Coffin-Collins, P.A. and Hall, B.K. (1989) Chondrogenesis of mandibular mesenchyme from the embryonic chick is inhibited by mandibular epithelium and by epidermal growth factor. *Int. J. Devel. Biol.*, **33**, 297–311.

Cohen, J. (1979) Maternal constraints in development, in *Maternal Effects in Development* (eds D.R. Newth and M. Balls), The Fourth Symposium of the British Society for Developmental Biology, Cambridge University Press, Cambridge, pp. 1–28.

Cohen, J. (1993) Development of the zootype, *Nature*, **363**, 307.

Cohen, J. and Massey, B.D. (1983) Larvae and the origins of major phyla. *Biol. J. Linn. Soc.*, **19**, 321–8.

Cohen, S.M. and Jürgens, G. (1990) Mediation of *Drosophila* head development by gap-like segmentation genes. *Nature*, **346**, 482–5.

Cohn, M.J., Izpisua-Belmonte, J.C., Abud, H. *et al.* (1995) Fibroblast growth factors induce additional limb development from the flank of chick embryos. *Cell*, **80**, 239–46.

Cohn, M.J., Patel, K., Krumlauf, R. *et al.* (1997) *Hox9* genes and vertebrate limb specification. *Nature*, **387**, 97–101.

Coiter, V. (1572–3) *Externarum et internarum Principalium humani corporis partium tabulae*. Nuremberg. (Reprinted in Adelman, H.B. (1933). *Ann. Med. Hist.* n.s., **5**, 327–41, 444–57).

Colbert, E.H. (1949) Progressive adaptation as seen in the fossil record, in *Genetics, Palaeontology and Evolution* (eds G.L. Jepsen, E. Mayr and G.G. Simpson), Princeton University Press, Princeton, NJ., pp. 390–402.

Coleman, W. (1964) *Georges Cuvier, Zoologist*. Harvard University Press, Cambridge, MA.

Coleman, W. (1973) Limits of the recapitulation theory: Carl Friedrich Kielmeyer's critique of the presumed parallelism of earth history, ontogeny, and the present order of organisms. *Isis*, **64**, 341–50.

Collazo, A., Bolker, J.A. and Keller, R. (1994) A phylogenetic perspective on teleost gastrulation. *Amer. Nat.*, **144**, 133–52.

Collazo, A. and Fraser, S.E. (1996) Integrating cellular and molecular approaches into studies of development and evolution: The issue of morphological homology. *Aliso*, **14**, 237–62.

Collazo, A., Fraser, S.E. and Mabee, P.M. (1994) A dual embryonic origin for vertebrate mechanoreceptors. *Science*, **264**, 426–30.

Collazo, A. and Marks, S.B. (1994) Development of *Gyrinophilus porphyriticus*: Identification of the ancestral developmental pattern in the salamander family Plethodontidae. *J. Exp. Zool.*, **268**, 239–58.

Collins, J.P. and Cheek, J.E. (1983) Effect of food and density on development of typical and cannibalistic salamander larvae in *Ambystoma tigrinum nebulosum. Amer. Zool.*, **23**, 77–84.

Collins, J.P., Minton, J.B. and Pierce, B.A. (1980) *Ambystoma tigrinum*: A multispecies conglomerate. *Copeia*, **1980**, 938–41.

Conklin, E.G. (1896) Discussion of the factors of organic evolution from the embryological standpoint. *Proc. Amer. Phil. Soc.*, **1896**, 78–88.

Conway Morris, S. (1979) The Burgess Shale (middle Cambrian) fauna. *Annu. Rev. Ecol. Syst.*, **10**, 327–49.

Conway Morris, S. (1989) Burgess Shale faunas and the Cambrian explosion. *Science*, **246**, 339–46.

Conway Morris, S. (1993a) The fossil record and the early evolution of the Metazoa. *Nature*, **361**, 219–25.

Conway Morris, S. (1993b) Ediacaran-like fossils in Cambrian Burgess Shale-type faunas of North America. *Paleontology*, **36**, 593–635.

Conway Morris, S. (1994a) A palaeontological perspective. *Curr. Opin. Genet. Devel.*, **4**, 802–9.

Conway Morris, S. (1994b) Why molecular biology needs palaeontology. *Development* 1994 Suppl., 1–13.

Conway Morris, S. (1995) A new phylum from the lobster's lips. *Nature*, **378**, 661–2.

Conway Morris, S. (1997) Molecular clocks: Defusing the Cambrian 'explosion'? *Curr. Biol.*, **7**, R71–R74.

Conway Morris, S. and Collins, D.H. (1996) Middle Cambrian ctenophores from the Stephen Formation, British Columbia, Canada. *Phil. Trans. R. Soc. Lond.* B, **351**, 279–308.

Conway Morris, S. and Peel, J.S. (1990) Articulated halkieriids from the Lower Cambrian of north Greenland. *Nature*, **345**, 802–5.

Conway Morris, S. and Whittington, H.B. (1985) Fossils of the Burgess Shale. A national treasure in Yoho National Park, British Columbia. *Geol. Survey Canada Misc. Reports*, **43**, 1–31.

Conway Morris, S., Peel, J.S., Higgins, A.K. *et al.* (1987) A Burgess Shale-like fauna from the Lower Cambrian of north Greenland. *Nature*, **326**, 181–3.

Cook, S.A. (1962) Genetic system, variation and adaptation in *Eschscholzia californica. Evolution*, **16**, 278–99.

Cooke, J. (1990) Proper names for early fingers. *Nature*, **347**, 14–15.

Cooke, J., Nowak, M.A., Boerlijst, M. and Maynard Smith, J. (1997) Evolutionary origins and maintenance of redundant gene expression during metazoan development. *Trends Genet.*, **13**, 360–4.

Cooke, J. and Smith, J.C. (1987) The midblastula cell cycle transition and the character of mesoderm in the u.v.-induced nonaxial *Xenopus* development. *Development*, **99**, 197–210.

Cooke, J. and Smith, J.C. (1990) Measurement of developmental time by cells of early embryos. *Cell*, **60**, 891–4.

Cooke, J. and Wong, A. (1991) Growth-factor-related proteins that are inducers in early amphibian development may mediate similar steps in amniote (bird) embryogenesis. *Development*, **111**, 197–212.

Coope, C.R. (1979) Late Cenozoic fossil Coleoptera: Evolution, biogeography, ecology. *Annu. Rev. Ecol. Syst.*, **10**, 247–67.

Cope, E.D. (1887) *The Origin of the Fittest: Essays in Evolution*, Macmillan & Co., New York.

Cope, E.D. (1889) Synopsis of the families of vertebrata. *Amer. Nat.*, **23**, 1–29.

Cope, E.D. (1896) *The Primary Factors of Organic Evolution*, Open Court Publishing Co., Chicago, IL.

Corbo, J.C., Levine, M. and Zeller, R.W. (1997) Characterization of a notochord-specific enhancer from the *Brachyury* promoter region of the ascidian, *Ciona intestinalis*. *Development*, **124**, 589–602.

Corsi, P. (1988) *The Age of Lamarck. Evolutionary Theories in France 1790–1830*. University of California Press, Berkeley, CA.

Couly, G., Grapin-Botton, A., Coltey, P. and Le Douarin, N.M. (1996) The regeneration of the cephalic neural crest, a problem revisited: The regenerating cells originate from the contralateral or from the anterior and posterior neural fold. *Development*, **122**, 3393–407.

Couly, G. and Le Douarin, N.M. (1990) Head morphogenesis in embryonic avian chimeras: Evidence for a segmental pattern in the ectoderm corresponding to the neuromeres. *Development*, **108**, 543–58.

Covant, H.A. and Hardy, M.H. (1990) Excess retinoid acts through the stroma to produce mucous glands from newborn hamster cheek pouch *in vitro*. *J. Exp. Zool.*, **253**, 271–9.

Cowley, D.E. (1991a) Genetic prenatal maternal effects on organ size in mice and their potential contribution to evolution. *J. Evol. Biol.*, **3**, 363–81.

Cowley, D.E. (1991b) Prenatal effects on mammalian growth: Embryo transfer results, in *The Unity of Evolutionary Biology* (ed. E.C. Dudley), Dioscorides Press, Portland, OR, pp. 762–79.

Cowley, D.E. and Atchley, W.R. (1992) Quantitative genetic models for development, epigenetic selection, and phenotypic evolution. *Evolution*, **46**, 495–518.

Cowley, D.E., Pomp, D., Atchley, W.R. *et al.* (1989) The impact of maternal uterine genotype on postnatal growth and adult body size in mice. *Genetics*, **122**, 193–203.

Cracraft, J. (1990) The origin of evolutionary novelties: Patterns and process at different hierarchical levels, in *Evolutionary Innovations* (ed. M.H.

Nitecki), The University of Chicago Press, Chicago, IL, pp. 21–44.

Crampton, H.E. (1894) Reversal of cleavage in a sinistral gastropod. *Ann. N. Y. Acad. Sci.*, **8**, 167–70.

Crompton, A.W. and Jenkins, F.A. (1979) Origin of mammals, in *Mesozoic Mammals* (eds J.A. Lillegraven, Z. Kielan-Jaworowska and W.A. Clemens), University California Berkeley Press, Berkeley, CA, pp. 59–73.

Cronin, T.M. and Raymo, M.E. (1997) Orbital forcing of deep-sea benthic species diversity. *Nature*, **385**, 624–7.

Crowther, R.J. and Whittaker, J.R. (1992) Structure of the caudal neural tube in an ascidian larva: Vestiges of its possible evolutionary origin from a ciliated band. *J. Neurobiol.*, **23**, 280–92.

Crowther, R.J. and Whittaker, J.R. (1994) Serial repetition of cilia pairs along the tail surface of an ascidian larva. *J. Exp. Zool.*, **268**, 9–16.

Cruz, Y.P., Yousef, A. and Selwood, L. (1996) Fate-map analysis of the epiblast of the dasyurid marsupial *Sminthopsis macroura* (Gould). *Reprod. Fertil. Devel.*, **8**, 779–88.

Cunha, G.R. and Young, P. (1992) Role of stroma in oestrogen-induced epithelial proliferation. *Epith. Cell Biol.*, **1**, 18–31.

Cuvier, G. (1812a) Sur un nouveau rapprochement à établir entre les classes qui composent le règne animal. *Ann. Mus. d'Hist. Nat.* **19**, 73–84.

Cuvier, G. (1812b) *Recherches sur les ossemens fossiles des quadrupèdes*, etc., 4 vols., Déterville, Paris.

Cuvier, G. (1817) Mémoires pour servir à l'histoire naturelle et à l'anatomie des mollusques, Paris, pp. 1–54.

Cuvier, G. (1830) Considérations sur les Mollusques et en particulier sur les Cèphalopodes. *Ann. Sci. Nat.*, **19**, 241–59.

Cuvier, G. and Valenciennes, A. (1828) *Histoire Naturelles des Poissons*. Vol. 1, Levrault, Paris.

Daegling, D.J. (1996) Growth in the mandibles of African Apes. *J. Human Evol.*, **30**, 315–41.

Daeschler, E.B. and Shubin, N. (1998) Fish with fingers? *Nature*, **391**, 133.

Dahl, E., Koseki, H. and Balling, R. (1997) *Pax* genes and organogenesis. *BioEssays*, **19**, 755–65.

Dalcq, A.M. (1968) Form and modern embryology, in *Aspects of Form. A Symposium on Form in Nature and Art* (ed. L.L. Whyte), 2nd edn, Lord Humphries, Publisher, London, pp. 91–120.

Dale, L. and Slack, J.M.W. (1987) Fate map for the 32-cell stage of *Xenopus laevis*. *Development*, **99**, 527–51.

Damuth, J. (1993) Cope's rule, the island rule and the scaling of mammalian population density. *Nature*, **365**, 748–50.

Darlington, P.J.Jr. (1980) *Evolution for Naturalists. The Simple Principles and Complex Reality.* A Wiley-Interscience Publication. John Wiley & Sons, New York.

Darwin, C. (1859) *The Origin of Species by Means of Natural Selection.* John Murray, London.

Darwin, C. (1910) *The Origin of Species by Means of Natural Selection.* John Murray, London. (Popular Impression from which quotations cited herein are taken).

Darwin, F. (ed.) (1887) *The Life and Letters of Charles Darwin, including an Autobiography.* Vols 1 and 2, D. Appleton & Co., New York.

Darwin, F. and Seward, A.C. (eds) (1903) *More Letters of Charles Darwin.* 2 Vols, John Murray, London.

Davidson, D. (1995) The function and evolution of *Msx* genes: Pointers and paradoxes. *Trends Genet.*, **11**, 405–11.

Davidson, E.H. (1968) *Gene Activity in Early Development.* Academic Press, New York.

Davidson, E.H. (1990) How embryos work: A comparative view of diverse modes of cell fate specification. *Development*, **109**, 365–89.

Davidson, E.H. (1991) Spatial mechanisms of gene regulation in metazoan embryos. *Development*, **113**, 1–26.

Davidson, E.H. (1997) Insights from the echinoderms. *Nature*, **389**, 679–80.

Davidson, E.H. (1993) Later embryogenesis: Regulatory circuitry in morphogenetic fields. *Development*, **118**, 665–90.

Davidson, E.H., Peterson, K.J. and Cameron, R.A. (1995) Origin of Bilaterian body plans: Evolution of developmental regulatory mechanisms. *Science*, **270**, 1319–25.

Davies, A.G. Game, A.Y., Chen, Z. *et al.* (1996) *Scalloped wings* is the *Lucilia cuprina Notch* homologue and a candidate for the *Modifier* of fitness and asymmetry of diazinon resistance. *Genetics*, **143**, 1321–37.

Davies, M. (1989) Ontogeny of bone and the role of heterochrony in the Myobatrachine genera *Uperoleia*, *Crinia*, and *Pseudophryne* (Anura: Leptodactylidae: Myobatrachinae). *J. Morph.*, **200**, 269–300.

Davis, A.P. and Capecchi, M.R. (1996) A mutational analysis of the 5' *HoxD* genes: Dissection of genetic interactions during limb development in the mouse. *Development*, **122**, 1175–85.

Davis, A.P., Witte, D.P., Hsieh-Li, M.M. *et al.* (1995) Absence of radius and ulna in mice lacking *hoxa-11* and *hoxd-11*. *Nature*, **375**, 791–5.

Dawkins, R. (1976) *The Selfish Gene*, Oxford University Press, Oxford.

Dawkins, R. (1978) Replicator selection and the extended phenotype. *Z. Tierpsychol.*, **47**, 61–76.

Dean, B. (1899) On the embryology of *Bdellostoma stouti*. A general account of myxinoid development from the egg and segmentation to hatching, in *Festschr. f. C. van Kuppfer*, G. Fischer, Jena, pp 221–77.

Deban, S.M., Wake, D.B. and Roth, G. (1997) Salamander with a ballistic tongue. *Nature*, **389**, 27–8.

De Braga, M. and Carroll, R.L. (1993) The origin of mososaurs as a model of macroevolutionary pattern and processes. *Evol. Biol.*, **27**, 245–322.

De Braga, M. and Rieppel, O. (1997) Reptile phylogeny and the interrelationships of turtles. *Zool. J. Linn. Soc.*, **120**, 281–354.

Debrenne, F. and Wood, R.A. (1990) A new Cambrian sphinctozoan sponge from North America, its relationship to archaeocyaths and the nature of early sphinctozoans. *Geol. Mag.*, **127**, 435–43.

Degnan, N.M. and Morse, D.E. (1993) Identification of eight homeobox-containing transcripts expressed during larval development and at metamorphosis in the gastropod mollusc *Haliotis rufescens*. *Molec. Mar. Biol. Biotech.*, **2**, 1–19.

Degnan, B.M. and Morse, D.E. (1995) Development and morphogenesis gene regulation in *Haliotis rufescens* larvae at metamorphosis. *Amer. Zool.*, **35**, 391–8.

De Loof, A. (1992) All animals develop from a blastula: Consequences of an undervalued definition for thinking on development. *BioEssays*, **14**, 573–5.

del Pino, E.M. (1996) The expression of *Brachyury* (T) during gastrulation in the marsupial frog, *Gastrotheca riobambae*. *Devel. Biol.*, **177**, 64–72.

del Pino, E.M. and Elinson, R.P. (1983) A novel development pattern for frogs: Gastrulation produces an embryonic disk. *Nature*, **306**, 589–91.

Denis, H. (1994) A parallel between development and evolution: Germ cell recruitment by the gonads. *BioEssays*, **16**, 933–8.

de Queiroz, K. (1985) The ontogenetic method for establishing character polarity and the characters of phylogenetic systematics. *Syst. Zool.*, **34**, 280–99.

De Robertis, E.M. (1997) The ancestry of segmentation. *Nature*, **387**, 25–6.

De Robertis, E.M., Fainsod, A., Gont, L.K. and Steinbeisser, H. (1994) The evolution of vertebrate gastrulation. *Development*, 1994 Suppl., 117–24.

De Robertis, E.M., Oliver, G. and Wright, C.V.E. (1990) Homeobox genes and the vertebrate body plan. *Sc. Amer.*, **263** (1), 46–52.

De Robertis, E.M. and Sasai, Y. (1996) A common plan for dorsoventral patterning in Bilateria. *Nature*, **380**, 37–40.

Desmond, A. (1982) *Archetypes and Ancestors. Palaeontology in Victorian London 1850–1875*. The University of Chicago Press, Chicago, IL.

Desmond, A. (1985) Richard Owen's reaction to transmutation in the 1830's. *Brit. J. Hist. Sci.*, **18**, 25–50.

Desmond, A. (1989) *The Politics of Evolution. Morphology, Medicine, and Reform in Radical London*, The University of Chicago Press, Chicago and London.

Dettlaff, T.A., Ignatieva, G.M. and Vassetzky, S.G. (1987) The problem of time in developmental biology: Its study by the use of relative characteristics of developmental duration. *Soviet Sci. Res. Physiol. Genet. Biol.*, **1**, 1–88.

Dhouailly, D., and Maderson, P.F.A. (1984) Ultrastructural observations on the embryonic development of the integument of *Lacerta muralis* (Lacertilia, Reptilia). *J. Morph.*, **179**, 203–28.

Dhouailly, D., and Sengel, P. (1983) Feather-forming properties of the foot integument in avian embryos, in *Epithelial-Mesenchymal Interactions in Development* (eds R.H. Sawyer and J.F. Fallon), Praeger Scientific, New York, pp. 147–61.

Dickinson, M.E., Sellek, M.A.J., McMahon, A.P. and Bronner-Fraser, M. (1995) Dorsalization of the neural tube by the non-neural ectoderm. *Development*, **121**, 2099–106.

Dickinson, W.J. (1995) Molecules and morphology: Where's the homology? *Trends Genet.*, **11**, 119–21.

Dinsmore, C.E. (ed.) (1991) *A History of Regeneration research. Milestones in the Evolution of a Science.* Cambridge University Press, Cambridge.

Dinsmore, C.E. and Hanken, J. (1986) Native variant limb skeletal patterns in the red-backed salamander, *Plethodon cinereus*, are not regenerated. *J. Morph.*, **190**, 191–200.

Dobzhansky, Th. (1937) *Genetics and the Origin of Species*. Columbia University Press, New York.

Dobzhansky, T. (1949) Foreword, in *Factors of Evolution: The Theory of Stabilizing Selection* (I.I. Schmalhausen), The Blakiston Co., Philadelphia,

PA, pp. xv–xvii. (Reprinted 1986 by the University of Chicago Press, Chicago, IL.)

Dobzhansky, T., Ayala, F.J., Stebbins, G.L. and Valentine, J.W. (1977) *Evolution*. W. H. Freeman, San Francisco, CA.

Dodson, S. (1989a) Predator-induced reaction norms: Cyclic changes in shape and size can be protective. *Bioscience*, **39**, 447–52.

Dodson, S. (1989b) The ecological role of chemical stimuli for the zooplankton: Predator-induced morphology in *Daphnia. Oecologia* (Berl.), **78**, 361–7.

Dohle, W. (1989) Differences in cell pattern formation in early embryology and their bearing on evolutionary changes in morphology. *Geobios. Mém. Spécial*, **12**, 145–55.

Dohle, W. and Scholtz, G. (1988) Clonal analysis of the crustacean segment: The discordance between genealogical and segmental borders. *Development*, **104**, Suppl., 147–60.

Dohrn, A. (1875) *Der Ursprung der Wiebelthiere und das Princip des Functionswechsels*. Verlag von Wilhelm Engelmann, Leipzig,

Dollé, P., Dierich, A., LeMeur, M. *et al.* (1993) Disruption of the *Hoxd*-13 genes induces localized heterochrony leading to mice with neotenic limbs. *Cell*, **75**, 431–41.

Dollo, L. (1893) Les lois de l'évolution. *Bull. Soc. Belge Geol.*, **8**, 164–6.

Dollo, L. (1922) Les céphalopodes déroulés et l'irréversibilité de l'évolution. *Bijdr. Dierk. Amsterdam*, **70**, 517–26.

Donoghue, M.J. (1992) Homology, in *Keywords in Evolutionary Biology* (eds E. Fox Keller and E.A. Lloyd), Harvard University Press, Cambridge, MA, pp. 170–9.

Donoghue, M.J. and Sanderson, M.J. (1994) Complexity and homology in plants, in *Homology: The Hierarchical Basis of Comparative Biology* (ed. B.K. Hall), Academic Press, San Diego, CA, pp. 394–421.

Doolittle, W.F. and Sapienza, C. (1980) Selfish genes, the phenotypic paradigm and genomic evolution. *Nature*, **280**, 601–3.

Dover, G.A. (1982) Molecular drive: A cohesive mode of species evolution. *Nature*, **299**, 111–17.

Dover, G.A. (1992) Observing development through evolutionary eyes: A practical approach. *BioEssays*, **14**, 281–7.

Driesch, H. (1892) Entwicklungsmechanische Studien. I. Der Werth der beiden ersten Furchungszellen in der Echinodermentwicklung. Experimentelle Erzeugen von Theil-und

Doppelbildung. *Z. Wissenschaft. Zool.*, **53**, 160–78, 183–4. Tafel VII.

Drieuer, W. and Nüsslein-Volhard, C. (1988a) A gradient of *bicoid* protein in *Drosophila* embryos. *Cell*, **54**, 83–93.

Drieuer, W. and Nüsslein-Volhard, C. (1988b) The *bicoid* protein determines position in the *Drosophila* embryo in a concentration-dependent manner. *Cell*, **54**, 95–114.

Duboule, D. (1992) The vertebrate limb: A model system to study the *Hox*/HOM gene network during development and evolution. *BioEssays*, **14**, 375–84.

Duboule, D. (ed.) (1994a) *Guidebook to the Homeobox Genes*, Oxford University Press, Oxford.

Duboule, D. (1994b) Temporal colinearity and the phylotypic progression: A basis for the stability of a vertebrate Bauplan and the evolution of morphologies through heterochrony. *Development*, 1994 Suppl., 135–42.

Duboule, D. and Morata, G. (1994) Colinearity and functional hierarchy among genes of the homeotic complexes. *Trends Genet.*, **10**, 358–64.

Duellman, W.E. (1992) Reproductive strategies of frogs. *Sci. Amer.*, **267**, 80–7.

Duellman, W.E. and Trueb, L (1986) *Biology of Amphibians*, McGraw-Hill Book, Co., New York.

Dullemeijer, P. (1974) *Concepts and Approaches in Animal Morphology*, Van Gorcum & Comp. B. V., Assen, The Netherlands.

Dullemeijer, P. (1991) Evolution of biological constructions: Concessions, limitations, and pathways, in *Constructional Morphology and Evolution* (eds N. Schmidt-Kittler and K. Vogel), Springer-Verlag, Berlin, pp. 313–29.

Duluc, I., Lorentz, O., Fritsch, C. *et al.* (1997) Changing intestinal connective tissue interactions alters homeobox gene expression in epithelial cells. *J. Cell Sci.*, **110**, 1317–24.

Dulvy, N.K., and Reynolds, J.D. (1997) Evolutionary transitions among egg-laying, live-bearing and maternal inputs in sharks and rays. *Proc. R. Soc. Lond.* B, **264**, 1309–15.

Dun, R.B. and Fraser, A.S. (1959) Selection for an invariant character, vibrissa number, in the house mouse. *Aust. J. Biol. Sci.*, **21**, 506–23.

Dunlap, K.D. and Sanchiz, B. (1996) Temporal dissociation between the development of the cranial and appendicular skeletons in *Bufo bufo* (Amphibia: Bufonidae). *J. Herpetol.*, **30**, 506–13.

Dunlop, L.-L.T. and Hall, B.K. (1995) Relationships between cellular condensation, preosteoblast formation and epithelial-mesenchymal interactions

in initiation of osteogenesis. *Int. J. Devel. Biol.*, **39**, 357–71.

Durston, A.J., Timmermans, J.P.M., Hage, W.J. *et al.* (1989) Retinoic acid causes an anterioposterior transformation in the developing nervous system. *Nature*, **340**, 140–4.

Dziadek, M. and Mitrangas, K. (1989) Differences in the solubility and susceptibility to proteolytic degradation of basement-membrane components in adult and embryonic mouse tissues. *Amer. J. Anat.*, **184**, 298–310.

Ebbesson, S.D. (1984) Evolution and ontogeny of neural circuits. *Behav. Brain Sciences*, **7**, 321–66.

Eddy, E.M. (1996) The germ line and development. *Devel. Genet.*, **19**, 287–9.

Edelman, G.M. (1983) Cell adhesion molecules. *Science*, **219**, 450–7.

Edelman, G.M. (1986) Cell adhesion molecules in the regulation of animal form and pattern. *Ann. Rev. Cell Biol.*, **2**, 81–116.

Edelman, G.M. (1988) *Topobiology: An Introduction to Molecular Embryology*, Basic Books, New York.

Edmonds, H.W. and Sawin, P.B. (1936) Variations of the branches of the aortic arch in rabbits. *Amer. Nat.*, **70**, 65–6.

Eernisse, D.J., Albert, J.S. and Anderson, F.E. (1992) Annelida and Arthropoda are not sister taxa: A phylogenetic analysis of spiralian metazoan morphology. *Syst. Biol.*, **41**, 305–30.

Eernisse, D.J. and Kluge, A.G. (1993) Taxonomic congruence versus total evidence, and amniote phylogeny inferred from fossils, molecules, and morphology. *Mol. Biol. Evol.*, **10**, 1170–95.

Ekanayake, S. and Hall, B.K. (1997) The *in vivo* and *in vitro* effects of bone morphogenetic protein-2 on the development of the chick mandible. *Int. J. Devel. Biol.*, **41**, 67–81.

Eldredge, N. (1985) *Unfinished Synthesis. Biological Hierarchies and Modern Evolutionary Thought*, Oxford University Press, New York.

Eldredge, N. (1989) *Macroevolutionary Dynamics: Species, Niches and Adaptive Peaks*, McGraw-Hill Co., New York.

Eldredge, N. and Gould, S.J. (1972) Punctuated equilibria: An alternative to phyletic gradualism, in *Models of Paleobiology* (ed. T.J.M. Schopf), Freeman, San Francisco, CA, pp. 82–115.

Elinson, R.P. (1987) Change in developmental patterns: Embryos of amphibians with large eggs, in *Development as an Evolutionary Process* (eds R.A. Raff and E.C. Raff), Alan R. Liss, Inc., New York, pp. 1–21.

Elinson, R.P. (1990) Direct development in frogs:

Wiping the recapitulationist slate clean. *Sem. Devel. Biol.*, **1**, 263–70.

Elinson, R.P. (1994) Leg development in a frog without a tadpole (*Eleutherodactylus coqui*). *J. Exp. Zool.*, **270**, 202–10.

Elinson, R.P. and Kao, K.R. (1993) Axis specification and head induction in vertebrate embryos, in *The Skull, Volume 1, Development* (eds. J. Hanken and B.K. Hall), The University of Chicago Press, Chicago, IL, pp. 1–41.

Elinson, R.P., King, M.L. and Forristall, C. (1993) Isolated vegetal cortex from *Xenopus* oocytes selectively retains localized mRNAs. *Devel. Biol.*, **160**, 554–62.

Elinson, R.P. and Pasceri, P. (1989) Two UV-sensitive targets in dorsoanterior specification of frog embryos. *Development*, **16**, 511–18.

Elinson, R.P. and Rowning, B. (1988) A transient array of parallel microtubules in frog eggs: Potential tracks for a cytoplasmic rotation that specifies the dorso-anterior axis. *Devel. Biol.*, **128**, 185–97.

Ellies, D.L., Langille, R.M., Martin, C.C. *et al.* (1997) Specific craniofacial cartilage dysmorphogenesis coincides with a loss of *Dlx* gene expression in retinoic acid-treated zebrafish embryos. *Mech. Devel.*, **61**, 23–36.

Emerald, B.S. and Roy, J.K. (1997) Homeotic transformations in *Drosophila*. *Nature*, **389**, 684.

Emerson, S.B. (1987) The effect of chemically produced shifts in developmental timing on postmetamorphic morphology in *Bombina orientalis*. *Exp. Biol.*, **47**, 105–9.

Emerson, S.B. (1988a) Convergence and morphological constraint in frogs: Variation in postcranial morphology. *Fieldiana Zoology*, n.s. **43** (1386), 1–19.

Emerson, S.B. (1988b) Testing for historical patterns of change: A case study with frog pectoral girdles. *Paleobiology*, **14**, 174–86.

Emerson, S.B., Greene, H.W. and Charnov, E.L. (1994) Allometric aspects of predator-prey interactions, in *Ecological Morphology: Integrative Organismal Biology* (eds P.C. Wainwright and S.M. Reilly), The University of Chicago Press, Chicago, IL, pp. 123–39.

Emerson, S.B. and Koehl, M.A.R. (1990) The interaction of behavioral and morphological change in the evolution of a novel locomotor type: 'flying' frogs. *Evolution*, **44**, 1931–46.

Emerson, S.B., Travis, J. and Koehl, M.A.R. (1990) Functional complexes and additivity in performance: A test case with 'flying' frogs. *Evolution*, **44**, 2153–7.

Endler, J.A. (1995) Multiple-trait coevolution and environmental gradients in guppies. *Trends Ecol. Evol.*, **10**, 22–9.

Ephrussi, A. and Lehmann, R. (1992) Induction of germ cell formation by *oskar*. *Nature*, **358**, 387–92.

Erdelyi, M., Michon, A.M., Guichet, A. *et al.* (1995) Requirement for *Drosophila* cytoplasmic tropomyosin in *oskar* mRNA localization. *Nature*, **377**, 524–7.

Erwin, D.H. (1992) A preliminary classification of evolutionary radiations. *Hist. Biol.*, **6**, 133–47.

Erwin, D.H. (1993) The origin of metazoan development: A palaeobiological perspective. *Biol. J. Linn. Soc.*, **50**, 255–74.

Erwin, D.H., Valentine, J.W. and Jablonski, D. (1997) The origin of animal body plans. *Amer. Sci.*, **85**, 126–37.

Etheridge, A.L. (1968) Determination of the mesonephric kidney. *J. Exp. Zool.*, **169**, 357–70.

Etheridge, A.L. (1972) Suppression of kidney formation by neural crest cells. *Roux's Arch. Devel. Biol.*, **169**, 268–70.

Ettinger, L. and Doljanski, F. (1992) On the generation of form by the continuous interactions between cells and their extracellular matrix. *Biol. Rev.*, **67**, 459–89.

Eyal-Giladi, H. (1984) The gradual establishment of cell commitment during the early stages of chick development. *Cell Differ.*, **14**, 245–55.

Fallon, J. F., Lopez, A., Ros, M. A. *et al.* (1994) FGF-2: Apical ectodermal ridge growth signal for chick limb development. *Science*, **264**, 104–7.

Fan, C.-M., Lee, C.S. and Tessier-Lavigne, M. (1997) A role for WNT proteins in induction of dermomyotome. *Devel. Biol.*, **191**, 160–5.

Fang, J. and Hall, B.K. (1995) Differential expression of neural cell adhesion molecule (NCAM) during osteogenesis and secondary chondrogenesis in the embryonic chick. *Int. J. Devel. Biol.*, **39**, 519–28.

Fang, J. and Hall, B.K. (1996) *In vitro* differentiation potential of the periosteal cells from a membrane bone, the quadratojugal of the embryonic chick. *Devel. Biol.*, **180**, 701–12.

Fang, J. and Hall, B.K. (1997) Chondrogenic cell differentiation from membrane bone periostea. *Anat. Embryol.*, **196**, 349–62.

Fang, H. and Elinson, R.P. (1996) Patterns of *Distalless* gene expression and inductive interactions in the head of the direct developing frog *Eleutherodactylus coqui*. *Devel. Biol.*, **179**, 160–72.

Farber, P.L. (1976) The type-concept in zoology during the first half of the nineteenth century. *J. Hist. Biol.*, **9**, 93–119.

Farley, J. (1977) *The Spontaneous Generation Controversy from Descartes to Oparin*, The Johns Hopkins University Press, Baltimore, MD.

Farley, F. (1982) *Gametes & Spores: Ideas about Sexual Reproduction, 1750–1914*, The Johns Hopkins University Press, Baltimore, MD.

Favier, B., Rijli, F. M., Fromental-Ramain, C. *et al.* (1996) Functional cooperation between the non-paralogous genes *Hoxa*-10 and *Hoxd*-11 in the developing forelimb and axial skeleton. *Development*, **122**, 449–60.

Fedonkin, M.A. and Waggoner, B.M. (1997) The Late Precambrian fossil *Kimberella* is a mollusc-like bilateral organism. *Nature*, **388**, 868–71.

Feduccia, A. (1996) *The Origin and Evolution of Birds*, Yale University Press, New Haven, CT.

Ferguson, E.L. (1996) Conservation of dorsal-ventral patterning in arthropods and chordates. *Curr. Opin. Genet. Devel.*, **6**, 424–31.

Ferguson, E.L. and Anderson, K.V. (1992) Localized enhancement and repression of the TGF-ß family member decapentaplegic, is necessary for dorso-ventral pattern formation in the *Drosophila* embryo. *Development*, **114**, 583–97.

Ferguson, M.M. (1986) Developmental stability of rainbow trout hybrids: Genomic coadaptation or heterozygosity? *Evolution*, **40**, 323–30.

Fernholm, B., Bremer, K. and Jornvall, H. (eds) (1989) *The Hierarchy of Life: Molecules and Morphology in Phylogenetic Analysis*, Elsevier, Amsterdam.

Ferretti, P. and Géraudie, J. (eds) (1998) *Cellular and Molecular Basis of regeneration: From Invertebrates to Humans*, John Wiley & Sons, Ltd., London. (in press).

Filan, S.L. (1991) Development of the middle ear region in *Monodelphis domestica* (Marsupialia, Didelphidae): Marsupial solutions to an early birth. *J. Zool. Lond.*, **225**, 577–88.

Filatow, D. (1925) Über die unabhängige Entstehung (Selbstdifferenzierung) der Linse bei *Rana esculenta*. *Wilhelm Roux Arch. EntwMech. Org.*, **104**, 50–71.

Fink, W.L. and Zelditch, M.L. (1995) Phylogenetic analysis of ontogenetic shape transformations: A reassessment of the piranha genus *Pygocentrus* (Teleostei). *Syst. Biol.*, **44**, 343–60.

Fink, W.L. and Zelditch, M.L. (1996) Historical patterns of developmental integration in Piranhas. *Amer. Zool.*, **36**, 61–9.

Finkelstein, R. and Perrimon, N. (1990) The *orthodenticle* gene is regulated by *bicoid* and *torso* and specifies *Drosophila* head development. *Nature*, **346**, 485–8.

Finlay, B.L. and Darlington, R.B. (1995) Linked regularities in the development and evolution of mammalian brains. *Science*, **268**, 1578–84.

Fiorello, C.V. and German, R.Z. (1997) Heterochrony within species: Craniofacial growth in giant, standard, and dwarf rabbits. *Evolution*, **51**, 250–61.

Fisher, R.A. (1930) *Genetical Theory of Natural Selection*, Clarendon Press, Oxford.

Fitch, W.M. (1970) Distinguishing homologues form analogous proteins. *Syst. Zool.*, **19**, 99–106.

Flick, R.W. (1992) *Effects of Hybridization between the Salamanders Ambystoma tigrinum and A. texanum on Skull Development and Morphology*, M. A. Thesis, South Illinois University, Carbondale, IL.

Fong, D.W., Kane, T.C. and Culver, D.C. (1995) Vestigialization and loss of nonfunctional characters. *Annu. Rev. Ecol. Syst.*, **26**, 249–68.

Foote, M. (1996) Ecological controls on the evolutionary recovery of Post-Paleozoic crinoids. *Science*, **274**, 1492–5.

Foote, M. and Gould, S.J. (1992) Cambrian and Recent morphological disparity. *Science*, **258**, 1816.

Forbes, E. (1854) On the manifestation of Polarity in the distribution of Organized beings in Time. *Notices Meet. Royal Inst.*, April, 1854, 428–33.

Ford, V.S. and Gottlieb, L.D. (1992) *Bicalyx* is a natural homeotic floral variant. *Nature*, **358**, 671–3.

Forey, P. (1995) Agnathans recent and fossil, and the origin of jawed vertebrates. *Rev. Fish Biol. & Fisheries*, **5**, 267–303.

Forey, P. and Janvier, P. (1993) Agnathans and the origin of jawed vertebrates. *Nature*, **361**, 129–34.

Forey, P. and Janvier, P. (1994) Evolution of the early vertebrates. *Amer. Sci.*, **82**, 554–65.

Fortelius, M. (1985) Ungulate cheek teeth: Developmental, functional and evolutionary interrelations. *Acta Zool. Fennica*, **180**, 1–76.

Fortey, R.A., Briggs, D.E.G. and Wills, M.A. (1996) The Cambrian evolutionary 'explosion': Decoupling cladogenesis from morphological disparity. *Biol. J. Linn. Soc.*, **57**, 13–33.

Fortey, R.A., Briggs, D.E.G. and Wills, M.A. (1997) The Cambrian evolutionary 'explosion' recalibrated. *BioEssays*, **19**, 429–34.

Franchina, C.R. and Hopkins, C.D. (1996) The dorsal filament of the weakly electric Apteronotidae (Gymnotiformes; Teleostei) is specialized for electroreception. *Brain Behav. Evol.*, **47**, 165–78.

Francis, P.H., Richardson, M.K., Brickell, P.M. and Tickle, C. (1994) Bone morphogenetic proteins and a signalling pathway that controls patterning in the developing chick limb. *Development*, **120**, 209–18.

Francis-West, P.H., Robertson, K.E., Ede, D.A. *et al.* (1995) Expression of genes encoding bone morphogenetic proteins and Sonic hedgehog in Talpid (*ta³*) limb buds: Their relationships in the signalling cascade involved in limb patterning. *Devel. Dyn.*, **203**, 187–97.

Frasch, M., Chen, X. and Lufkin, T. (1995) Evolutionary-conserved enhancers direct region-specific expression of the murine *Hoxa-1* and *Hoxa-2* loci in both mice and *Drosophila*. *Development*, **121**, 957–74.

Fraser, A.S. and Kindred, D.M. (1960) Selection for an invariant character, vibrissa number, in the house mouse. ii. Limits to variability. *Aust. J. Biol. Sci.*, **13**, 48–58.

Fraser, N. (1991) The true turtle's story . . . *Nature*, **349**, 278–279.

Fraser, S., Keynes, R. and Lumsden, A. (1990) Segmentation in the chick embryo hind brain is defined by cell lineage restrictions. *Nature*, **344**, 431–5.

Frazzetta, T.H. (1970) From hopeful monsters to bolyerine snakes. *Amer. Nat.*, **104**, 55–72.

Freebairn, K., Yen, J.L. and McKenzie, J.A. (1996) Environmental and genetic effects on the asymmetry phenotype: Diazinon resistance in the Australian sheep blowfly, *Lucilia cuprina*. *Genetics*, **144**, 229–39.

Freeman, D. (1974) The evolutionary theories of Charles Darwin and Herbert Spencer. *Curr. Anthropol.*, **15**, 211–37.

Freeman, G. and Lundelius, J.W. (1982) The developmental genetics of dextrality and sinistrality in the gastropod *Limnaea peregra*. *Roux's Arch. Devel. Biol.*, **191**, 69–83.

Freeman, G. and Lundelius, J.W. (1992) Evolutionary implications of the mode of D quadrant specification in coelomates with spiral cleavage. *J. Evol. Biol.*, **5**, 205–47.

French, V. (1997) Pattern formation in colour on butterfly wings. *Curr. Opin. Genet. Devel.*, **7**, 524–9.

Frenz, D.A., Galinovic-Schwartz, V., Liu, W. *et al.* (1992) Transforming growth Factor ß1 is an epithelial-derived signal peptide that influences otic capsule formation. *Devel. Biol.*, **153**, 324–36.

Frenz, D.A., Liu, W., Williams, J.D. *et al.* (1994) Induction of chondrogenesis: Requirement for synergistic interaction of basic fibroblast growth factor and transforming growth Factor-beta. *Development*, **120**, 415–24.

Frolich, L.M. (1991) Osteological conservatism and developmental constraint in the polymorphic 'ring species' *Ensatina eschscholtzii* (Amphibia: Plethodontidae). *Biol. J. Linn. Soc.*, **43**, 81–100.

Funch, P. and Kristensen, R.M. (1995) Cycliophora is a new phylum with affinities to Entoprocta and Ectoprocta. *Nature*, **378**, 711–14.

Futuyma, D.J. (1986) *Evolutionary Biology*, 2nd edn, Sinauer Associates, Sunderland, MA.

Futuyma, D.J., Keese, M.C. and Funk, D.J. (1995) Genetic constraints on macroevolution: The evolution of host affiliation in the leaf beetle genus *Ophraella*. *Evolution*, **49**, 797–809.

Fyfe, D.M. and Hall, B.K. (1983) The origin of ectomesenchyme condensations which precede development of the bony scleral ossicles in the eyes of embryonic chicks. *J. Embryol. exp. Morph.*, **73**, 69–83.

Gabbott, S.E., Aldridge, R.J. and Theron, J.N. (1995) A giant conodont with preserved muscle tissue from the Upper Ordovician of South Africa. *Nature*, **374**, 800–3.

Gaffney, E.S. and Kitching, J.W. (1994) The most ancient African turtle. *Nature*, **369**, 55–8.

Gaffney, E.S., Meylan, P.A. and Wyss, A.R. (1991) A computer assisted analysis of the relationships of the higher categories of turtles. *Cladistics*, **7**, 313–35.

Gajovic, S. and Kostovic-Knezevic, L. (1995) Ventral ectodermal ridge and ventral ectodermal groove: Two distinct morphological features in the developing rat embryo tail, *Anat. Embryol.*, **192**, 181–7.

Galis, F. (1993a) Morphological constraints on behaviour through ontogeny: The importance of developmental constraints. *Mar. Behav. Physiol.*, **23**, 119–35.

Galis, F. (1993b) Interactions between the pharyngeal jaw apparatus, feeding behaviour, and ontogeny in the cichlid fish, *Haplochromis piceatus*: a study of morphological constraints in evolutionary ecology. *J. Exp. Zool.*, **267**, 135–54.

Galis, F. (1996) The evolution of insects and vertebrates: Homeobox genes and homology. *Trends Ecol. Evol.*, **11**, 402–3.

Galis, F. and Drucker, E.G. (1996) Pharyngeal biting mechanisms in centrarchid and cichlid fishes: Insights into a key evolutionary innovation. *J. Evol. Biol.*, **9**, 641–70.

Ganan, Y., Macias, D., Duterque-Coquillaud, M. *et al.* (1996) Role of TGFßs and BMPs as signals controlling the position of the digits and the areas of interdigital cell death in the developing chick limb autopod. *Development*, **122**, 2349–57.

Gans, C. (1985) Differences and similarities: Comparative methods in mastication. *Amer. Zool.*, **25**, 291–301.

Gans, C. (1987) The neural crest: A spectacular invention, in *Developmental and Evolutionary Aspects of the Neural Crest* (ed. P.F.A. Maderson), John Wiley & Sons, New York, pp. 361–79.

Gans, C. (1989) Stages in the origin of vertebrates. Analysis by means of selection. *Biol. Rev.*, **64**, 221–68.

Gans, C. (1993) Evolutionary origins of the vertebrate skull, in *The Vertebrate Skull, Volume 2. Patterns of Structural and Systematic Diversity* (eds J. Hanken and B.K. Hall), The University of Chicago Press, Chicago, IL, pp. 1–35.

Gans, C., Kemp, N. and Poss, S. (eds) (1996) *The lancelets: A new look at some old beasts. Israel J. Zool.*, **42**, Suppl., S1–S446.

Gans, C. and Northcutt, R.G. (1983) Neural crest and the origin of vertebrates: A new head. *Science*, **220**, 268–74.

Gans, C. and Northcutt, R.G. (1985) Neural crest: The implications for comparative anatomy. *Fortschr. Zool.*, **30**, 507–14.

García-Castro, M.I., Anderson, R., Heasman, J. and Wylie, C. (1997) Interactions between germ cells and extracellular matrix glycoproteins during migration and gonad assembly in the mouse embryo. *J. Cell Biol.*, **138**, 471–80.

Garcia-Fernàndez, J., Bagunà, J. and Saló, E. (1991) Planarian homeobox genes: Cloning, sequence analysis, and expression. *Proc. Natl Acad. Sci. USA*, **88**, 7338–42.

Garcia-Fernàndez, J. and Holland, P.W.H. (1994) Archetypal organization of the amphioxus *Hox* gene cluster. *Nature*, **370**, 563–6.

Garcia-Martinez, V., Darnell, D.K., Lopez-Sanchez, C. *et al.* (1997) State of commitment of prospective neural plate and prospective mesoderm in late gastrula/early neurula stages of avian embryos. *Devel. Biol.*, **181**, 102–15.

Garland, T. and Adolph, S.C. (1994) Why not do 2-species comparisons: Limitations on inferring adaptation. *Physiol. Zool.*, **67**, 797–828.

Garrone, R. (1978) *Phylogenesis of Connective Tissues*, S. Karger, Basel.

Garstang, W. (1894) Preliminary note on a new theory of the phylogeny of the Chordata. *Zool. Anz.*, **17**, 122–5; *Quart. J. Microsc. Sci.*, **72**, 51–4.

Garstang, W. (1896) The origin of vertebrates. *Proc. Camb. Phil. Soc.*, **9**, 19–47.

Garstang, W. (1922) The theory of recapitulation. A critical restatement of the biogenetic law. *J. Linn. Soc. Lond. (Zool.)*, **35**, 81–101.

Garstang, W. (1928) The morphology of the Tunicata and its bearing on the phylogeny of the Chordata. *Quart. J. Microsc. Sci.*, **72**, 51–187.

Garstang, W. (1929) The origin and evolution of larval forms. *Report of the 96th meeting of the B.A.A.S.*, 77–98.

Gaunt, S.J. (1994) Conservation in the *Hox* code during morphological evolution, *Int. J. Devel. Biol.*, **38**, 549–52.

Gaunt, S.J. and Singh, P.B. (1990) Homeogene expression patterns and chromosomal imprinting. *Trends Genet.*, **6**, 208–12.

Geddes, P. and Mitchell, P.C. (1910–11) Morphology. *Encyclopaedia Britannica*, 11th edn, **18**, 863–9.

Gee, H. (1996) *Before the Backbone: Views on the Origin of the Vertebrates*, Chapman & Hall, London.

Gegenbaur, C. (1870) *Grundzüge der vergleichenden Anatomie*. 2nd edn, Wilhelm Engelmann, Leipzig.

Gehlbach, F.R. (1969) Determination of the relationship of tiger salamander larval populations to different stages of pond succession at the Grand Canyon, Arizona. *Year Book Amer. Phil. Soc.*, **1969**, 299–302.

Gendron-Maguire, M., Mallo, M., Zhang, M. and Gridley, T. (1993) *Hoxa*-2 mutant mice exhibit homeotic transformation of skeletal elements derived from cranial neural crest. *Cell*, **75**, 1317–31.

Geoffroy Saint-Hilaire, E. (1802) Jean Baptiste Lamarck, Georges Cuvier, and Bernard de Lacépède, 'Rapport des professeurs du Muséum, sur les collections d'histoire naturelle rapportés d'Egypte, par E. Geoffroy.' *Ann. Mus. d'Hist. Nat.*, **1**, 234–41.

Geoffroy Saint-Hilaire, E. (1807a) Premier mémoire sur les poissons, où l'on compare les pièces osseuses de leurs nageoires pectorales avec les os de l'extrémité antérieur des autres animaux à vertèbres. *Ann. Mus. Natl Hist. nat.* **9**, 357–72.

Geoffroy Saint-Hilaire, E. (1807b) Second mémoire sur les poissons. Considérations sur l'os furculaire, une des pièces de la Nageoire pectorale. *Ann. Mus. Natl Hist. nat.*, **9**, 413–27.

Geoffroy Saint-Hilaire, E. (1807c) Troisième mémoire sur les poissons, où l'on traite de leur sternum sous le point de vue de sa détermination et de ses formes générales. *Ann. Mus. Natl Hist. nat.*, **10**, 87–104.

Geoffroy Saint-Hilaire, E. (1807d) Considérations sur les pièces de la tête osseuse des animaux vertébrés, et particulièrement sur celles du crâne des oiseaux. *Ann. Mus. Natl Hist. nat.*, **10**, 342–65.

Geoffroy Saint-Hilaire, E. (1818–1822) *Philosophie Anatomique.* Vol.1, *Des Organes respiratoires sous le rapport de la détermination et de l'identité de leurs pièces osseuses.* Vol. 2, *Des monstruosités humaines.* J.-B. Baillière, Paris.

Geoffroy Saint-Hilaire, E. (1825a) Recherches sur l'organisation des Gavials; sur leurs affinités naturelles, desquelles résulte la nécessité d'une autre distribution générique, Gavials, Teleosaurus, et Steneosaurus; et sur cette question, si les Gavials (*Gavialis*), aujourd'hui répandus dans les parties orientales de l'Asie, descendent, par voie non interrompue de génération, des Gavials antédiluviens, soit des Gavials fossiles, dits Crocodiles de Caen (Teleosaurus), soit des Gavials fossiles du Havre et de Horfleur (Steneosaurus). *Mem. Mus. Hist. nat., Paris,* **12**, 97–115, 155.

Geoffroy Saint-Hilaire, E. (1825b) Rapport sur l' 'Anatomie comparée des monstruosités animales' par M. Serres. *Mem. Mus. Hist. nat. Paris,* **13**, 82–92.

Geoffroy Saint-Hilaire, E. (1830) *Principes de Philosophie Zoologique, discutés en Mars 1830, au sein de l'Académie Royale des Sciences.* Paris.

Geoffroy Saint-Hilaire, E. (1835) *Etudes Progressives d'un Naturaliste pendant les années 1834 et 1835, faisant suite à ses Publications dans les 42 volumes des Mémoires et Annales du Muséum d'Histoire Naturelle.* Paris

Geoffroy Saint-Hilaire, E. (1838) *Notions Synthétiques, Historiques et Physiologiques de Philosophie Naturelle.* Paris.

Geoffroy Saint-Hilaire, I. (1847) *Vie Travaux et Doctrine scientifique d'Etienne Geoffroy St. Hilaire,* Paris.

Géraudie, J. (1978) The fine structure of the early pelvic fin bud of the trout *Salmo gairdneri* and *S. trutta fario. Acta Zool.,* **59**, 85–96.

Géraudie, J. and Ferretti, P. (1997) Correlation between RA-induced apoptosis and patterning defects in regenerating fins and limbs. *Int. J. Devel. Biol.,* **41**, 529–32.

Gerhart, J. (1995) Summing up: Conservation and diversification in metazoan eukaryotic cells. *Phil. Trans. R. Soc. Lond. B,* **349**, 333–6.

Gerhart, J. (1997) In Memoriam: Pieter D. Nieuwkoop (1917–1996). *Devel. Biol.,* **182**, 1–4.

Gerhart, J., Danilchik, M., Doniach, T. *et al.* (1989) Cortical rotation of the *Xenopus* egg: Consequences for the anterioposterior pattern of embryonic dorsal development. *Development,* 1989 Suppl., 37–51.

Gerhart, J., Doniach, T. and Stewart, R. (1991) Organizing the *Xenopus* organizer, in *Gastrulation* (eds R. Keller, W.H. Clark. Jr. and F. Griffin), Plenum Press, New York, pp. 57–76.

German, R.Z., Hertweck, D.W., Sirianni, J.E. and Swindler, D.R. (1994) Heterochrony and sexual dimorphism in the pigtailed macaque (*Macaca nemestrina*). *Amer. J. Phys. Anthrop.,* **93**, 373–80.

German, R.Z. and Meyers, L.L. (1989) The role of time and size in ontogenetic allometry: I. Review. *Growth Differ. Aging,* **53**, 101–6.

Ghiselin, M.T. (1994) The origin of vertebrates and the principle of succession of functions. Genealogical sketches by Anton Dohrn 1875. An English translation from the German, introduction and bibliography. *Hist. Philos. Life Sci.,* **16**, 3–96.

Ghiselin, M.T. (1996) Rediscovering the science of the history of life. *Hist. Phil. Life. Sci.,* **18**, 123–8.

Ghiselin, M.T. and Groeben, C. (1997) Elias Metschnikoff, Anton Dohrn, and the metazoan common ancestor. *J. Hist. Biol.,* **30**, 211–28.

Gibbs, P.E. and Wickstead, J.H. (1996) The myotome formula of the lancelet *Epigonichthys lucayanum* (Acrania): Can variations be related to larval dispersal patterns? *J. Nat. Hist.,* **30**, 615–27.

Gibson, G. and Hogness, D.S. (1996) Effect of polymorphism in the *Drosophila* regulatory gene *Ultrabithorax* on homeotic stability. *Science,* **271**, 200–3.

Gilbert, J.J. (1966) Rotifer ecology and embryological induction. *Science,* **151**, 1234–7.

Gilbert, J.J. (1980) Female polymorphism and sexual reproduction in the rotifer *Asplanchna.* Evolution of their relationship and control by dietary tocopherol. *Amer. Nat.,* **116**, 409–31.

Gilbert, J.-M., Mouchel-Viehl, E. and Deutsch, J.S. (1997) *Engrailed* duplication events during the evolution of barnacles. *J. Mol. Evol.,* **44**, 585–94.

Gilbert, S.F. (1988) *Developmental Biology,* 2nd edn, Sinauer Associates, Sunderland, MA.

Gilbert, S.F. (ed.) (1991a) *Developmental Biology: A Comprehensive Synthesis. Volume 7, A Conceptual History of Modern Embryology,* Plenum Press, New York.

Gilbert, S.F. (1991b) Epigenetic landscaping: Waddington's use of cell fate bifurcation diagrams. *Biol. Philos.,* **6**, 135–54.

Gilbert, S.F. (1992) Cells in search of community: Critiques of Weismannism and selectable units in ontogeny. *Biol. Philos.,* **7**, 473–87.

Gilbert, S.F. (1994a) *Developmental Biology.* 4th edn, Sinauer Inc., New York.

Gilbert, S.F. (1994b) Dobzhansky, Waddington and Schmalhausen: Embryology and the modern synthesis, in *The Evolution of Theodosius Dobzhansky* (ed. M.B. Adams), Princeton University Press, Princeton, NJ, pp. 143–54.

Gilbert, S.F., Opitz, J.M. and Raff, R.A. (1996) Resynthesizing evolutionary and developmental biology. *Devel. Biol.*, **173**, 357–72.

Gilbert, S.F. and Raunio, A.M. (eds) (1997) *Embryology: Constructing the Organism*, Sinauer Associates, Inc., Sunderland, MA.

Gilland, E. and Baker, R. (1993) Conservation of neuroepithelial and mesodermal segments in the embryonic vertebrate head. *Acta Anat.*, **148**, 110–23.

Gingerich, P.D., Raza, S.M., Arif, M. *et al.* (1994) New whale from the Eocene of Pakistan and the origin of cetacean swimming. *Nature*, **368**, 844–7.

Gingerich, P.D., Smith, B.H. and Simons, E.L. (1990) Hind limbs of Eocene *Basilosaurus*: Evidence of feet in whales. *Science*, **249**, 154–7.

Glaessner, M.F. (1984) *The Dawn of Animal Life*, Cambridge University Press, Cambridge

Glardon, S., Callaerts, P., Halder, G., and Gehring, W.J. (1997) Conservation of *Pax-6* in a lower chordate, the ascidian *Phallusia mammillata*. *Development*, **124**, 817–25.

Glenner, H. and Høeg, J.T. (1995) A new motile, multicellular stage involved in host invasion by parasitic barnacles (Rhizocephala). *Nature*, **377**, 147–150.

Glinka, A., Wu, W., Onichtchouk, D. *et al.* (1997) Head induction by simultaneous respression of Bmp and Wnt signalking in *Xenopus*. *Nature*, **389**, 517–19.

Gluckman, P.D. and Liggins, G.C. (1984) Regulation of fetal growth, in *Fetal Physiology and Medicine* (eds R.W. Beard and P.W. Nathaniels), Marcel Dekker, Inc., New York, pp. 511–57.

Goday, C. and Pimpinelli, S. (1993) The occurrence, role and evolution of chromatin diminution in nematodes. *Parasitol. Today*, **9**, 319–22.

Godfrey, L.R. (ed.) (1985) *What Darwin Begat. Modern Darwinism and Non-Darwinian Perspectives on Evolution*. Allyn & Bacon, Boston.

Godfrey, L.R. and Sutherland, M.R. (1995a) Flawed inference: Why size-based tests of heterochronic processes do not work. *J. Theor. Biol.*, **172**, 43–61.

Godfrey, L.R. and Sutherland, M.R. (1995b) What's growth got to do with it? Process and product in the evolution of ontogeny. *J. Human Evol.*, **29**, 405–31.

Godfrey, L.R. and Sutherland, M.R. (1996) Paradox of peramorphic paedomorphosis: Heterochrony and human evolution. *Amer. J. Phys. Anthrop.*, **99**, 17–42.

Godsave, S.F., Isaacs, H.V. and Slack, J.M.W. (1988) Mesoderm-inducing factors: A small class of molecules. *Development*, **102**, 555–66.

Godsave, S.F. and Slack, J.M.W. (1991) Single cell analysis of mesoderm formation in the zebrafish embryo. *Development*, **111**, 523–30.

Goertzel, B. (1992) What is hierarchical selection? *Biol. Philos.*, **7**, 27–33.

Goethe, J.W. (1807) *Bildung und Umbildung organischer Naturen, in Goethes Morphologische Schriften* (ed. W. Troll), Eugen Diederichsverlag, Jena [1932], pp. 111–23.

Goethe, J.W. (1831) Reflexions de Goethe sur les débats scientifiques de mars 1830 dans le sein de l'Académie des Sciences, publiées à Berlin dans les *Ann. Critique Sci. Ann. Sci. Nat.*, **22**, 179–88.

Goethe, J.W. (1832) Derniers pages de Goethe expliquant à l'Allemagne les sujets de philosophie naturelle controversées au sein de l'Académie des Sciences de Paris. *Revue Encyclopédique*, **53**, 563–573; **54**, 554–68.

Goff, D.J. and Tabin, C.J. (1997) Analysis of *Hoxd-13* and *Hoxd-11* misexpression in chick limb buds reveals that *Hox* genes affect both bone condensation and growth. *Development*, **124**, 627–36.

Goldschmidt, R.B. (1933) Some aspects of evolution. *Science*, **78**, 539–47.

Goldschmidt, R.B. (1938) *Physiological Genetics*, McGraw Hill, New York.

Goldschmidt, R.B. (1940) *The Material Basis of Evolution*, Yale University Press, New Haven, CT.

Goldschmidt, R.B. (1958) *Theoretical Genetics.* University of California Press, Berkeley, CA.

Goldstein, B. and Freeman, G. (1997) Axis specification in animal development. BioEssays, **19**, 105–16.

Gollmann, G. (1991) Osteological variation in *Geocrinia laevis, Geocrinia victoriana*, and their hybrid populations (Amphibia, Anura, Myobatrachinae). *Z. zool. Syst. Evolut.-Forsch.*, **29**, 289–303.

Gombrich, E.H. (1989) Evolution in the arts: The altar painting, its ancestry and progeny, in *Evolution and its Influence* (ed. A. Grafen), Clarendon Press, Oxford, pp. 107–25.

Gont, L.K., Steinbesser, H., Blumberg, B. and De Robertis, E.M. (1993) Tail formation as a continuation of gastrulation: The multiple cell populations of the *Xenopus* tailbud derive from the late blastopore lip. *Development*, **119**, 991–1004.

Goodrich, J., Puangsomiee, P., Martin, M. *et al.* (1997) A Polycomb-group gene regulates homeotic gene expression in *Arabidopsis. Nature*, **386**, 44–51.

Goodwin, B.C. (1984) Changing from an evolutionary to a generative paradigm in biology, in *Evolutionary Theory. Paths into the Future* (ed. J.W. Pollard), John Wiley & Sons, Chichester, pp. 99–120.

Goodwin, B.C. (1989) Evolution and the generative order, in *Theoretical Biology* (eds B.C. Goodwin and P. Saunders), Edinburgh University Press, Edinburgh, pp. 89–100.

Goodwin, B.C. (1994) *How the Leopard Changed its Spots: The Evolution of Complexity*, Charles Scribner's Sons, New York.

Goodwin, B.C., Holder, N. and Wylie, C.C. (eds) (1983) *Development and Evolution.* The Sixth Symposium of the British Society for Developmental Biology, Cambridge University Press, Cambridge.

Goodwin, B.C., Sibatani, A. and Webster, G. (eds) (1989) *Dynamic Structures in Biology.* Edinburgh University Press, Edinburgh.

Goodwin, B.C. and Trainor, L.E.H. (1983) The ontogeny and phylogeny of the pentadactyl limb, in *Development and Evolution* (eds B.C. Goodwin, N. Holder. and C.C. Wylie), The Sixth Symposium of the British Society for Developmental Biology. Cambridge University Press, Cambridge, pp. 75–98.

Gorbman, A. (1983) Early development of the hagfish pituitary gland: Evidence for the endodermal origin of the adenohypophysis. *Amer. Zool.*, **23**, 639–54.

Gorbman, A. and Tamarin, A. (1985a) Early development of oral, olfactory and adenohypophyseal structures of agnathans and its evolutionary implications, in *Evolutionary Biology of Primitive Fishes* (eds R.E. Foreman, A. Gorbman, J.M. Dodd. and R. Olsson), Plenum Press, New York., pp. 165–85.

Gorbman, A. and Tamarin, A. (1985b) Head development in relation to hypophysial development in a myxinoid, *Eptatretus* and a petromyzontid, *Petromyzon*, in *The Pars Distalis: Structure, Function and Regulation* (eds F. Yoshimura. and A. Gorbman), Elsevier, Amsterdam, pp. 34–43.

Gordon, D.M. (1992) Phenotypic plasticity, in *Keywords in Evolutionary Biology* (eds E. Fox Keller and E.A. Lloyd), Harvard University Press, Cambridge, MA, pp. 255–62.

Gosner, K.L. (1960) A simplified table for staging anuran embryos and larvae with notes on identification. *Herpetologica*, **16**, 183–190.

Goss, R.J. (1969) *Principles of Regeneration*. Academic Press, New York.

Goss, R.J. (1992) Evolution of regeneration. *J. Theor. Biol.*, **159**, 241–60.

Gould, A. (1997) Functions of mammalian Polycomb group and trithorax group related genes. *Curr. Opin. Genet. Devel.*, **7**, 488–94.

Gould, S.J. (1970) Dollo on Dollo's law: Irreversibility and the status of evolutionary laws. *J. Hist. Biol.*, **3**, 189–212.

Gould, S.J. (1977) *Ontogeny and Phylogeny*. The Belknap Press of Harvard University Press, Cambridge, MA.

Gould, S.J. (ed.) (1980) *Louis Dollo's Papers on Paleontology and Evolution*, Arno Press, New York.

Gould, S.J. (1986) Geoffroy and the homeobox, in *Progress in Developmental Biology. Part A* (ed. H.C. Slavkin), Proceedings of the Tenth International Congress of the International Society of Developmental Biologists, held in Los Angeles, CA, August 4–9, 1985. Alan R. Liss, New York, pp. 205–18.

Gould, S.J. (1989) *Wonderful Life. The Burgess Shale and the Nature of History*. W. W. Norton & Co., New York.

Gould, S.J. (1991) The disparity of the Burgess Shale arthropod fauna and the limits of cladistic analysis: Why we must strive to quantify morphospace. *Paleobiology*, **17**, 411–23.

Gould, S.J. (1992) Heterochrony, in *Keywords in Evolutionary Biology* (eds E. Fox Keller and E.A. Lloyd), Harvard University Press, Cambridge, MA, pp. 158–65.

Gould, S.J. (1993) How to analyze Burgess Shale disparity – a reply to Ridley. *Paleobiology*, **19**, 522–23.

Gould, S.J. and Lewontin, R.C. (1979) The spandrels of San Marco and the Panglossian paradigm: A critique of the adaptationist programme. *Proc. R. Soc. Lond.* B, **205**, 581–98.

Gould, S.J. and Vrba, E.S. (1982) Exaptation – a missing term in the science of form. *Paleobiology*, **8**, 4–15.

Graba, Y., Aragnol, D. and Pradel, J. (1997) *Drosophila* Hox complex downstream targets and the function of homeotic genes. *BioEssays*, **19**, 379–88.

Graham, A., Francis-West, P.H., Brickell, P. and Lumsden, A. (1994) The signalling molecule

BMP-4 mediates apoptosis in the rhomben-cephalic neural crest. *Nature, 372*, 684–6.

Grainger, R.M. (1992) Embryonic lens induction: Shedding light on vertebrate tissue determination. *Trends Genet., 8*, 349–55.

Grande, L. and Rieppel, O. (eds) (1994) *Interpreting the Hierarchy of Nature: From Systematic Patterns to Evolutionary Process Theories*, Academic Press, San Diego, CA.

Grant, P.R. and Horn, H.S. (eds) (1992) *Molds, Molecules, and Metazoa. Growing Points in Evolutionary Biology*, Princeton University Press, Princeton, NJ.

Grant, R.E. (1826) Observations on the nature and importance of Geology. *Edinburgh New Philos. J., 1*, 293–302.

Grant, R.E. (1835) *Outlines of Comparative Anatomy*. Bailliere, London.

Grant, V. (1963) *The Origin of Adaptations*. Columbia University Press, New York.

Graveson, A.C. (1993) Neural crest: Contributions to the development of the vertebrate head. *Amer. Zool., 33*, 424–33.

Graveson, A.C. and Armstrong, J.B. (1990) The premature death (*p*) mutation of *Ambystoma mexicanum* affects a subpopulation of neural crest cells. *Differ. Ontog. Neoplasia, 45*, 71–5.

Graveson, A.C. and Armstrong, J.B. (1996) Premature death (*p*) mutation of *Ambystoma mexicanum* affects the ability of ectoderm to respond to neural induction. *J. Exp. Zool., 274*, 248–54.

Graveson, A.C., Smith, M.M. and Hall, B.K. (1997) Neural crest potential for tooth development in a urodele amphibian: Developmental and evolutionary significance. *Devel. Biol., 188*, 34–42.

Graw, J. (1996) Genetic aspects of embryonic eye development in vertebrates. *Devel. Genet., 18*, 181–97.

Green, J.B.A. (1990) Retinoic acid: The morphogen of the main body axis? *BioEssays, 12*, 437–9.

Green, J.B.A., Howes, G., Symes, K. and Smith, J.C. (1990) The biological effects of XTC-MIF: Quantitative comparison with *Xenopus* bFGF. *Development, 108*, 173–83.

Green, J.B.A. and Smith, J.C. (1990) Graded changes in dose of a *Xenopus* activin A homologue elicits stepwise transitions in embryonic cell fate. *Nature, 347*, 391–4.

Greenburg, G. and Hay, E.D. (1982) Epithelia suspended in collagen gels can lose polarity and express characteristics of migrating mesenchymal cells. *J. Cell Biol., 95*, 333–9.

Greene, E. (1989) A diet-induced developmental polymorphism in a caterpillar. *Science, 243*, 643–6.

Greene, H.W. (1994) Homology and behavioral repertoires, in *Homology: The Hierarchical Basis of Comparative Biology* (ed. B.K. Hall), Academic Press, San Diego, CA, pp. 370–91.

Greene, J.C. (1981) *Science, Ideology, and World Views*. University of California Press, Berkeley, CA.

Gregg, J.R. and Harris, F.T.C. (eds) (1964) *Form and Strategy in Science. Studies dedicated to Joseph Henry Woodger on the Occasion of his Seventieth Birthday*. D. Reidel Publishing Co., Dordrecht-Holland.

Gregory, W.K. (1936) On the meanings and limits of irreversibility of evolution. *Amer. Nat., 70*, 517–28.

Gregory, W.K. (1951) *Evolution Emerging. A Survey of Changing Patterns from Primeval Life to Man*, Macmillan, New York.

Grenier, J.K., Garber, T.L., Warren, R. *et al.* (1997) Evolution of the entire arthropod *Hox* gene set predated the origin and radiation of the onychophoran/arthropod clade. *Curr. Biol., 7*, 547–53.

Griesemer, J.R. (1996) Periodization and models in evolutionary history, in *New Perspectives on the History of Life: Essays on Systematic Biology as Historical Narrative* (eds M.T. Ghiselin and G. Pinna), pp. 19–30, *Mem. Calif. Acad. Sciences* Number 20, California Academy of Sciences, San Francisco, CA.

Griffith, C.M., Wiley, M.J. and Sanders, E.J. (1992) The vertebrate tail bud: Three germ layers from one tissue, *Anat Embryol., 185*, 101–13.

Grillitsch, B. (1989) On the relevance of certain external features in staging ranoid larvae (*Amphibia: Anura*), in *Trends in Vertebrate Morphology* (eds H. Splechtna and H. Hilgers), Gustav Fischer Verlag, Stuttgart, pp. 269–75.

Grindhart, J.G.J. and Kaufman, T.C. (1995) Identification of *polycomb* and *trithorax* group responsive elements in the regulatory region of the *Drosophila* homeotic gene sex comb reduced. *Genetics, 139*, 797–814.

Gruber, J.W., and Thackray, J.L. (eds) (1992). *Richard Owen: Commemoration*, London.

Gruenert, D.C. and Cozens, A.L. (1991) Inheritance of phenotype in mammalian cells – genetic vs. epigenetic mechanisms. *Amer. J. Physiol., 260*, L386-L394.

Grüneberg, H. (1956) A ventral ectodermal ridge of the tail in mouse embryos, *Nature, 177*, 787–8.

Grüneberg, H. (1963) *The Pathology of Development*. Blackwell Scientific Publications, Oxford.

Grüneberg, H. and des Wickramaratne, G.A. (1974) A re-examination of two skeletal mutants of the mouse, vestigial tail (*vt*) and congenital hydrocephalus (*ch*). *J. Embryol. exp. Morph.*, **31**, 207–22.

Gurdon, J.B. (1987) Embryonic induction – molecular prospects. *Development*, **99**, 285–306.

Gurdon, J.B. (1988) A community effect in animal development. *Nature*, **336**, 772–4.

Gurdon, J.B. (1989) The localization of an inductive response. *Development*, **105**, 27–33.

Gurdon, J.B. (1992) The generation of diversity and pattern in animal development. *Cell*, **68**, 185–99.

Hadley, E.A. (1997) Evolutionary and ecological response of pocket gophers (Thomomys talpoides) to late-Holocene climatic change. *Biol. J. Linn. Soc.*, **60**, 277–96.

Hadorn, E. (1961) *Developmental Physiology and Lethal Factors*. John Wiley & Sons, New York.

Haeckel, E. (1866) *Generelle Morphologie der Organismen: Allgemeine Grundzüge der organischen Formen-Wissenschaft, mechanisch begründet durch die von* Charles Darwin *reformite Descendenz-Theorie*. 2 Vols., Georg Reimer, Berlin.

Haeckel, E. (1868) *Natürliche Schöpfungsgeschichte: Gemeinverstanliche wissenschaftliche Vorträge über die Entwickelungslehre in Allgemeinen und die jenige von Darwin, Goethe, und Lamarck in Besonderen*. Georg Reimer, Berlin.

Haeckel, E. (1872) *Die Kalkschwämme: Eine Monographie*. Three Volumes, Berlin.

Haeckel, E. (1874) Die Gastraea-Theorie, die phylogenetische Klassification des Tierreiches und Homologie der Keimblätter. *Jena Z. Naturwiss.*, **8**, 1–55.

Haeckel, E. (1875) Die Gastraea und die Eifurchung der Thiere. *Jena Z. Naturwiss.*, **9**, 402–508.

Haeckel, E. (1876a) Die Physemarien, Gastræaden der Gegenwart und Nachträge zur Gastræa-Theorie. *Jena Z. Naturwiss.*, **10**, 55–98.

Haeckel, E. (1876b) *The History of Creation: or, the Development of the Earth and its Inhabitants by the Action of Natural Causes: A popular Exposition of the Doctrine of Evolution in General and of that of Darwin, Goethe, and Lamarck in Particular*. 2 Volumes, Appleton, New York.

Haeckel, E. (1891) *Anthropogenie oder Entwickelungsgeschichte des Menschen. Keimes-und Stammes-Geschichte* 4th edn, Wilhelm Engelmann, Leipzig.

Hafner, M.S. and Hafner, J.C. (1984) Brain size, adaptation and heterochrony in Geomyoid rodents. *Evolution*, **38**, 1088–98.

Haigh, L.S., Owens, B.B., Hellewell, S. and Ingram, V.M. (1982) DNA methylation in chicken alpha-globin gene expression. *Proc. Natl Acad. Sci. USA*, **79**, 5332–36.

Hainski, A.M. and Moody, S.A. (1996) Activin-like signal activates dorsal-specific maternal RNA between 8- and 16-cell stages of *Xenopus*. *Devel. Genet.*, **19**, 210–21.

Halder, G., Callaerts, P. and Gehring, W.J. (1995) Induction of ectopic eyes by targeted expression of the eyeless gene in *Drosophila*. *Science*, **267**, 1788–92.

Hall, B.K. (1968) The annual interrenal tissue cycle within the adrenal gland of the eastern rosella, *Platycercus eximius* (Aves:Psittaciformes). *Aust. J. Zool.*, **16**, 609–17.

Hall, B.K. (1975) Evolutionary consequences of skeletal differentiation. *Amer. Zool.*, **15**, 329–50.

Hall, B.K. (1978) *Developmental and Cellular Skeletal Biology*. Academic Press, New York.

Hall, B.K. (1981) Specificity in the differentiation and morphogenesis of neural crest-derived scleral ossicles and of epithelial scleral papillae in the eye of the embryonic chick. *J. Embryol. exp. Morph.*, **66**, 175–90.

Hall, B.K. (1982a) Tissue interactions and chondrogenesis, in *Cartilage, Volume 2. Development, Differentiation and Growth* (ed. B.K. Hall), Academic Press, New York, pp. 187–222.

Hall, B.K. (1982b) Distribution of osteo- and chondrogenic neural crest cells and of osteogenically inductive epithelia in mandibular arches of embryonic chicks. *J. Embryol. exp. Morph.*, **68**, 127–36.

Hall, B.K. (1982c) How is mandibular growth controlled during development and evolution? *J. Craniofac. Genet. Devel. Biol.*, **2**, 45–9.

Hall, B.K. (1982d) Mandibular morphogenesis and craniofacial malformations. *J. Craniofac. Genet. Devel. Biol.*, **2**, 309–22.

Hall, B.K. (1983a) Epigenetic control in development and evolution, in *Development and Evolution* (eds B.C. Goodwin, N. Holder and C.C. Wylie), The Sixth Symposium of the British Society for Developmental Biology, Cambridge University Press, Cambridge, pp. 353–79.

Hall, B.K. (1983b) Tissue interactions and chondrogenesis, in *Cartilage, Volume 2, Development, Differentiation and Growth* (ed. B.K. Hall), Academic Press, Orlando, pp. 187–222.

Hall, B.K. (1984a) Developmental mechanisms underlying the formation of atavisms. *Biol. Rev.*, **59**, 89–124.

Hall, B.K. (1984b) Developmental processes underlying heterochrony as an evolutionary mechanism. *Can. J. Zool.*, **62**, 1–7.

Hall, B.K. (1984c) Genetic and epigenetic control of connective tissues in the craniofacial structures. *Birth Defects, Orig. Article Ser.*, **20**, 1–17.

Hall, B.K. (1984d) Matrices control the differentiation of cartilage and bone, in *Matrices and Cell Differentiation* (eds R.B. Kemp and J.R. Hinchliffe), Alan R. Liss, New York, pp. 147–69.

Hall, B.K. (1985a) Research in the development and structure of the skeleton since the publication of Murray's *Bones*. Introduction to reissue of *Bones: A study of the Development and Structure of the Vertebrate Skeleton* by P.D.F. Murray, Cambridge University Press, Cambridge, pp. xi–xlix.

Hall, B.K. (1985b) Critical periods during development as assessed by thallium-induced inhibition of growth of embryonic chick tibiae *in vitro*. *Teratology*, **31**, 353–61.

Hall, B.K. (1986) The role of movement and tissue interactions in the development and growth of bone and secondary cartilage in the clavicle of the embryonic chick. *J. Embryol. exp. Morph.*, **93**, l33–52.

Hall, B.K. (1987a) Tissue interactions in the development and evolution of the vertebrate head, in *Developmental and Evolutionary Aspects of the Neural Crest* (ed. P.F.A. Maderson), John Wiley & Sons., New York, pp. 215–60.

Hall, B.K. (1987b) Development of the mandibular skeleton in the embryonic chick as evaluated using the DNA-inhibiting agent 5-fluoro-2'-deoxyuridine (FUDR). *J. Craniofac. Genet. Devel. Biol.*, **7**, 145–59.

Hall, B.K. (1988a) The embryonic development of bone. *Amer. Sci.*, **76** (2), 174–81.

Hall, B.K. (1988b) Patterning of connective tissues in the head: Discussion report. *Development*, **103**, Suppl., 171–4.

Hall, B.K. (1989) Morphogenesis of the skeleton: Epithelial or mesenchymal control? in *Trends in Vertebrate Morphology* (eds H. Splechtna and H. Hilgers), Gustav Fischer Verlag, Stuttgart, pp. 198–201.

Hall, B.K. (1990a) Genetic and epigenetic control of vertebrate development. *Neth. J. Zool.*, **40**, 352–61.

Hall, B.K. (1990b) Heterochronic change in vertebrate development. *Sem. Devel. Biol.*, **1**, 237–43.

Hall, B.K. (1990c) Evolutionary issues in craniofacial biology. Proceedings of the Symposium on Advances in Craniofacial Developmental Biology and Clinical Implications held at the American Cleft Palate-Craniofacial Association meeting, April 23–24, 1989. *Cleft Palate J.*, **27**, 95–100.

Hall, B.K. (1991a) Cellular interactions during cartilage and bone development. *J. Craniofac. Genet. Devel. Biol.*, **11**, 238–50.

Hall, B.K. (1991b) Evolution of connective and skeletal tissues, in *Developmental Patterning of the Vertebrate Limb* (eds J.R. Hinchliffe, J.M. Hurle and D. Summerbell), NATO ASI Series A: Life Sciences Vol. 205, Plenum Press, New York, pp. 303–11.

Hall, B.K. (1991c) What is bone growth? in *Fundamentals of Bone Growth* (eds B.G. Sarnat, and A.D. Dixon), CRC Press, Boca Raton, FL., pp. 605–12.

Hall, B.K. (1992) Waddington's legacy to development and evolution. *Amer. Zool.*, **32**, 113–22.

Hall, B.K. (ed.) (1994a) *Homology: The Hierarchical Basis of Comparative Biology.* Academic Press, San Diego, CA.

Hall, B.K. (1994b) Biology and mechanisms of tissue interactions in developing systems, in *Developmental Biology and Cancer* (eds G.M. Hodges and C. Rowlatt), CRC Press, Boca Raton, FL., pp. 161–85

Hall, B.K. (1994c) Embryonic bone formation with special reference to epithelial-mesenchymal interactions and growth factors, in *Bone, Volume 8: Mechanisms of Bone Development and Growth* (ed. B.K. Hall), CRC Press, Boca Raton, FL., pp. 137–92.

Hall, B.K. (1994d) Introduction, in *Homology: The Hierarchical Basis of Comparative Biology*, pp. 1–19. Academic Press, San Diego, CA.

Hall, B.K. (ed.) (1994e) *Bone. Volume 8: Mechanisms of Bone Development and Growth*, CRC Press, Boca Raton, FL.

Hall, B.K. (ed.) (1994f) *Bone, Volume 9: Differentiation and Morphogenesis of Bone.* CRC Press, Boca Raton, FL.

Hall, B.K. (1995a) Homology and embryonic development. *Evol. Biol.*, **28**, 1–37.

Hall, B.K. (1995b) Atavisms and atavistic mutations. *Nature Genet.*, **10**, 126–7.

Hall, B.K. (1996a) Evolutionary developmental biology, in *McGraw-Hill Yearbook of Science & Technology*, 1996, McGraw-Hill, New York, pp. 110–12.

Hall, B.K. (1996b) *Baupläne*, phylotypic stages, and constraint: Why are there so few types of animals? *Evol. Biol.*, **29**, 215–61.

Hall, B.K. (1997a) Germ layers and the germ-layer theory revisited: Primary and secondary germ layers, neural crest as a fourth germ layer, homology, demise of the germ-layer theory. *Evol. Biol.*, **30**, (in press).

Hall, B.K. (1997b) Bone, Embryonic development. *Encyclop. Human Biol.*, 2nd ed., **2**, 105–14, Academic Press, San Diego, CA.

Hall, B.K. (1997c) Epigenetics: Regulation not replication. *J. Evol. Biol.*, **10**, (in press).

Hall, B.K. (1997d) Phylotypic stage or phantom: is there a highly conserved embryonic stage in vertebrates? *Trends Ecol. Evol.*, **12**, 461–3.

Hall, B.K. (1998a) Francis Maitland Balfour. *New Dict. Natl Biog.* (in press).

Hall, B.K. (1998b) Developmental and cellular origins of the amphibian skeleton, in *Amphibian Biology, Volume 6. Osteology* (eds H. Heatwole and M. Davies), Surrey Beatty & Sons, Chipping Norton, NSW. (in press).

Hall, B.K. (1998c) The Bone, in *Cellular and Molecular Basis of regeneration: From Invertebrates to Humans* (eds P. Ferretti and J. Géraudie), John Wiley & Sons, Ltd., London, pp. 289–307.

Hall, B.K. (1998d) Evolutionary developmental biology: Where embryos and fossils meet, in *Human Evolution Through Developmental Change* (eds N. Minugh-Purves and M. McKinney), John Wiley & Sons, New York. (in press.)

Hall, B.K. and Coffin-Collins, P.A. (1990) Reciprocal interactions between epithelium, mesenchyme, and epidermal growth factor (EGF) in the regulation of mandibular mitotic activity in the embryonic chick. *J. Craniofac. Genet. Devel. Biol.*, **10**, 241–61.

Hall, B.K. and Ekanayake, S. (1991) Effects of growth factors on the differentiation of neural crest cells and neural crest cell-derivatives. *Int. J. Devel. Biol.*, **35**, 367–87.

Hall, B.K. and Hanken, J. (1985a) Foreword to reissue of *The Development of the Vertebrate Skull* by G.R. de Beer, The University of Chicago Press, Chicago, IL, pp. vii–xxviii.

Hall, B.K. and Hanken, J. (1985b) Repair of fractured lower jaws in the spotted salamander: Do amphibians form secondary cartilage? *J. Exp. Zool.*, **233**, 359–68.

Hall, B.K. and Herring, S.W. (1990) Paralysis and growth of the musculoskeletal system in the embryonic chick. *J. Morph.*, **206**, 45–56.

Hall, B.K. and Hörstadius, S. (1988) *The Neural Crest*. Oxford University Press, Oxford.

Hall, B.K. and MacSween, M.C. (1984) An SEM analysis of the epithelial- mesenchymal interface in the mandible of the embryonic chick. *J. Craniofac. Genet. Devel. Biol.*, **4**, 59–76.

Hall, B.K. and Miyake, T. (1992) The membranous skeleton: The role of cell condensations in vertebrate skeletogenesis. *Anat. Embryol.*, **186**, 107–24.

Hall, B.K. and Miyake, T. (1995a) How do embryos measure time? in *Evolutionary Change and heterochrony* (ed. K.J. McNamara), John Wiley & Sons, New York, pp. 3–20.

Hall, B.K. and Miyake, T. (1995b) Divide, accumulate, differentiate: Cell condensation in skeletal development revisited. *Int. J. Devel. Biol.*, **39**, 881–93.

Hall, B.K. and Van Exan, R.J. (1982) Induction of bone by epithelial cell products. *J. Embryol. exp. Morph.*, **69**, 37–46.

Hall, B.K. and Wake, M.H. (eds) (1998) *Larval Development and Evolution*, Academic Press, San Diego, CA (in press).

Hallam, A. (1990) Biotic and abiotic factors in the evolution of early Mesozoic marine molluscs, in *Causes of Evolution, A Paleontological Perspective* (eds R.M. Ross and W.D. Allmon), University of Chicago Press, Chicago, IL, pp. 249–69.

Hallgrimsson, B. (1993) Fluctuating asymmetry in *Macaca fascicularis*: A study of the etiology of developmental noise. *Int. J. Primatol.* **14**, 421–43.

Hallgrimsson, B. (1995) *Fluctuating asymmetry and maturational spans in mammals: Implications for the evolution of prolonged development in primates*, Ph.D. Thesis, University of Chicago, IL.

Hamburger, V. (1980) Embryology and the modern synthesis in evolutionary theory. In *The Evolutionary Synthesis. Perspectives on the Unification of Biology* (eds E. Mayr and W.B. Provine), Harvard University Press, Cambridge, MA, pp. 97–112.

Hamburger, V. (1988) *The Heritage of Experimental Embryology. Hans Spemann and the Organizer*, Oxford University Press, Oxford.

Hamburger, V. (1997) Wilhelm Roux: Visionary with a blind spot. *J. Hist. Biol.*, **30**, 229–38.

Hamburger, V. and Hamilton, H.L. (1951) A series of normal stages in development of the chick embryo. *J. Morph.*, **88**, 49–92.

Hammerschmidt, M., Serbedzija, G.N. and McMahon, A.P. (1996) Genetic analysis of dorsoventral pattern formation in the zebrafish: Requirement of a BMP-like ventralizing activity and its dorsal repressor. *Genes Devel.*, **10**, 2452–61.

Hanazato, T. (1991) Pesticides as chemical agents inducing helmet formation in *Daphnia ambigua*. *Freshwater Biol.*, **26**, 419–24.

Hanken, J. (1983) High incidence of limb skeletal variants in a peripheral population of the red-backed salamander, *Plethodon cinereus*

(Amphibia: Plethodontidae) from Nova Scotia. *Can. J. Zool.*, **61**, 1925–31.

Hanken, J. (1984) Miniaturization and its effects on cranial morphology in plethodontid salamanders, genus *Thorius* (Amphibia: Plethodontidae). I. Osteological variations. *Biol. J. Linn. Soc.*, **23**, 55–75.

Hanken, J. (1986) Developmental evidence for amphibian origins. *Evol. Biol.*, **20**, 389–417.

Hanken, J. (1993a) Model systems versus outgroups: Alternative approaches to the study of head development and evolution. *Amer. Zool.*, 448–56.

Hanken, J. (1993b) Adaptation of bone growth to miniaturization of body size, in *Bone Volume 7: Bone Growth-B.* (ed. B.K. Hall), CRC Press, Boca Raton, FL., 79–104.

Hanken, J. and Dinsmore, C.E. (1986) Geographic variation in the limb skeleton of the red-backed salamander, *Plethodon cinereus*. *J. Herpetol.*, **20**, 97–101.

Hanken, J. and Hall, B.K. (1984) Variation and timing of the cranial ossification sequence of the Oriental fire-bellied toad, *Bombina orientalis* (Amphibia, Discoglossidae). *J. Morphol.*, **182**, 245–55.

Hanken, J. and Hall, B.K. (1988a) Skull development during anuran metamorphosis. II. Role of thyroid hormone in osteogenesis. *Anat. Embryol.*, **178**, 219–27.

Hanken, J. and Hall, B.K. (1988b) Skull development during anuran metamorphosis. III. Role of thyroid hormone in chondrogenesis. *J. Exp. Zool.*, **246**, 156–70.

Hanken, J. and Hall, B.K. (eds) (1993a) *The Vertebrate Skull*, Volumes 1–3, The University of Chicago Press, Chicago, IL.

Hanken, J. and Hall, B.K. (1993b) Mechanisms of skull diversity and evolution, in *The Vertebrate Skull. Volume 3, Function and Evolutionary Mechanisms* (eds J. Hanken and B.K. Hall), The University of Chicago Press, Chicago, IL, pp. 1–36.

Hanken, K., Jennings, D.H. and Olsson, L. (1997a) Mechanistic basis of life-history evolution in anuran amphibians; Direct development. *Amer. Zool.*, **37**, 160–71.

Hanken, J., Klymkowsky, M.W., Alley, K.E., and Jennings, D.H. (1997b) Jaw muscle development as evidence for embryonic repatterning in direct-developing frogs. *Proc. R. Soc. Lond.* B, **264**, 1349–54.

Hanken, J., Klymkowsky, M.W., Summers, C.H. *et al.* (1992) Cranial ontogeny in the direct-developing frog, *Eleutherodactylus coqui* (Anura: Leptodactylidae), analyzed using whole-mount immunohistochemistry. *J. Morph.*, **211**, 95–118.

Hanken, J., Summers, C.H. and Hall, B.K. (1989) Morphological integration in the cranium during anuran metamorphosis. *Experientia*, **45**, 872–5.

Hanken, J. and Thorogood, P. (1993) Evolution and development of the vertebrate skull: The role of pattern formation. *Trends Ecol. Evol.*, **8**, 9–14.

Hanken, J. and Wake, D.B. (1993) Miniaturization of body size: Organismal consequences and evolutionary significance. *Annu. Rev. Ecol. Syst.*, **24**, 501–21.

Hanson, I. and Van Heyningen, V. (1995) Pax6: More than meets the eye. *Trends Genet.*, **11**, 268–72.

Harada, Y., Yasuo, H. and Satoh, N. (1995) A sea urchin homologue of the chordate *Brachyury* (*T*) gene is expressed in the secondary mesenchyme founder cells. *Development*, **121**, 2747–54.

Haraway, D. (1976) *Crystals, Fabrics and Fields.* Yale University Press, New Haven and London.

Hardy, A.C. (1954) Escape from specialization, in *Evolution as a Process* (eds J. Huxley, A.C. Hardy and E.B. Ford), Allen and Unwin, London, pp. 122–42.

Harrison, R.G. (1918) Experiments on the development of the forelimb of *Amblystoma*, a self-differentiating, equipotential system. *J. Exp. Zool.*, **25**, 413–62.

Hart, M.W. (1996) Evolutionary loss of larval feeding: Development, form and function in a facultatively feeding larva, *Brisaster latifrons. Evolution*, **50**, 174–87.

Hartl, D.L. (1989) *Principles of Population Genetics* 2nd edn, Sinauer Associates Inc., Sunderland, MA.

Hartsoeker, N. (1694) *Essai de Dioptrique*, Anisson, Paris.

Harvell, C.D. (1990) The ecology and evolution of inducible defenses. *Quart. Rev. Biol.*, **65**, 323–40.

Harvey, P.H., Brown, A.J.L., Maynard Smith, J. and Nee, S. (eds) (1996) *New Uses for New Phylogenies.* Oxford University Press, Oxford.

Harvey, P.H. and Pagel, M.D. (eds) (1991) *The Comparative Method in Evolutionary Biology.* Oxford University Press, Oxford.

Harvey, W. (1651) *Disputations Touching the Generation of Animals.* Translated with introductions and notes by G. Whitteridge (1981), Blackwells, London.

Haszprunar, G. (1992) The types of homology and their significance for evolutionary biology and phylogenetics. *J. Evol. Biol.*, **5**, 13–24.

Hatta, K., Schilling, T.F., BreMiller, R.A. and Kimmel, C.B. (1990) Specification of jaw muscle identity in zebrafish: Correlation with *engrailed*-homeoprotein expression. *Science*, **250**, 802–5.

Hawley, S.H.B., Wunnerberg-Stapleton, K., Hashimoto, C. *et al.* (1995) Disruption of BMP signals in embryonic *Xenopus* ectoderm leads to direct neural induction. *Genes Devel.*, **9**, 2923–35.

Hay, E.D. (1982) Interaction of embryonic cell surface and cytoskeleton with extracellular matrix. *Amer. J. Anat.*, **165**, 1–12.

Hay, E.D. (1990) Role of cell-matrix contacts in cell migration and epithelial-mesenchymal transformation. *Cell Differ. Devel.*, **32**, 367–76.

Hecht, M.K. and Hecht, B.M. (1994) Conflicting developmental and paleontological data: The case of the bird manus. *Acta Palaeont. Polonica*, **38**, 329–38.

Hecht, M.K., Ostrom, J.H., Viohl, G. and Wellnhofer, P. (eds) (1985) *The Beginning of Birds*, Proceedings of the International Archaeopteryx Conference, Freunde des Jura Museums, Eichstätt.

Hedrick, P.W. (1983) *Genetics of Populations*. Science Books International, Boston; Van Norstrand Reinhold Co., New York.

Helff, O.M. (1927) Influence of annular tympanic cartilage on development of tympanic membrane (*Rana pipiens*). *Proc. Soc. Exp. Biol. Med.*, **25**, 158–9.

Helff, O.M. (1934) Studies on amphibian metamorphosis. XII. Potential influences of the quadrate and supra-scapular on tympanic membrane formation in the anuran of the columella on the formation of the lamina propria of the tympanic membrane. *J. Exp. Zool.*, **68**, 305–18.

Helff, O.M. (1940) Studies on amphibian metamorphosis. XVII. Influence of non-living annular tympanic cartilage on tympanic membrane formation. *J. Exp. Biol.*, **17**, 45–60.

Helm, J.W. and German, R.Z. (1996) The epigenetic impact of weaning on craniofacial morphology during growth. *J. Exp. Zool.*, **276**, 243–53.

Hemmati-Brivanlou, A., Stewart, R.M. and Harland, R.M. (1990) Region-specific neural induction of an *engrailed* protein by anterior notochord in *Xenopus Science*, **250**, 800–2.

Henery, C.C., Bard, J.B.L., and Kaufman, M.H. (1992) Tetraploidy in mice, embryonic cell number, and the grain of the developmental map. *Devel. Biol.*, **152**, 233–41.

Henery, C.C., Miranda, M., Wiekowski, M. *et al.* (1995) Repression of gene expression at the beginning of mouse development. *Devel. Biol.*, **169**, 448–60.

Hennig, W. (1966) *Phylogenetic Systematics*, University of Illinois Press, Urbana, IL.

Henningsmoen, G. (1964) Zig-zag evolution. *Norsk. Geologisk Tidsskrift*, **44**, 341–52.

Henery, C.C., Bard, J.B.L. and Kaufman, M.H. (1992) Tetraploidy in mice, embryonic cell number, and the grain of the developmental map. *Devel. Biol.*, **152**, 233–41.

Henry, J.J. and Grainger, R.M. (1990) Early tissue interactions leading to embryonic lens formation in *Xenopus laevis*. *Devel. Biol.*, **141**, 149–63.

Henry, J.J., Klueg, K.M. and Raff, R.A. (1992) Evolutionary dissociation between cleavage, cell lineage and embryonic axes in sea urchin embryos. *Development*, **114**, 931–8.

Henry, J.J. and Mittelman, J.M. (1995) The matured eye of *Xenopus laevis* tadpoles produces factors that elicit a lens-forming response in embryonic ectoderm. *Devel. Biol.*, **171**, 39–50.

Henry, J.J. and Raff, R.A. (1990) Evolutionary change in the process of dorsoventral axis determination in the direct developing sea urchin, *Heliocidaris erythrogramma*. *Devel. Biol.*, **141**, 55–69.

Herring, S.W. (1990) Concluding remarks: Trends in vertebrate morphology. *Neth. J. Zool.*, **40**, 403–8.

Herring, S.W. (1993a) Formation of the vertebrate face: Epigenetic and functional influences. *Amer. Zool.*, **33**, 472–83.

Herring, S.W. (1993b) Epigenetic and functional influences on skull growth, in *The Skull, Volume 1, Development* (eds J. Hanken and B.K. Hall), The University of Chicago Press, Chicago, IL, pp. 153–206.

Herring, S.W., Rowlatt, U.F. and Pruzansky, S. (1979) Anatomical abnormalities in mandibulofacial dysostosis. *Amer. J. Med. Genet.*, **3**, 225–59.

Herrmann, B.G. (1991) Expression pattern of the *Brachyury* gene in whole-mount T^{Wis}/T^{Wis} mutant embryos. *Development*, **113**, 913–17.

Hertwig, O. and Hertwig, R. (1879) *Studien zur Blättertheorie. Heft I. Die Aktinien anatomisch und histologisch mit besonderer Berücksichtigung des Nervenmuskelsystems untersucht*, Jena.

Hervé, P., Chenuil, A. and Adoutte, A. (1994) Can the Cambrian explosion be inferred through molecular phylogeny? *Development* 1994, Suppl., pp. 15–25.

Hilderbrand, M., Bramble, D.M., Liem, K.F. and Wake, D.B. (1985) *Functional Vertebrate Morphology*. Belknap Press of Harvard University, Cambridge, MA.

Hill, R., Jones, P.F., Ress, A.R. *et al.* (1989) A new family of mouse homeobox-containing genes: Molecular structure, chromosomal location, and developmental expression of Hox-7.1. *Genes & Devel.*, **3**, 26–37.

Hinchliffe, J.R. (1985) 'One, two, three' or 'two, three, four': An embryologist's view of the homologies of the digits and carpus of modern birds, in *The Beginnings of Birds*. Proceedings of the International *Archaeopteryx* Conference, Eichstätt, 1984 (eds M.K. Hecht, J.H. Ostrom, G. Viohl and P. Wellnhofer), Eichstätt, pp. 141–7.

Hinchliffe, J.R. (1989) Reconstructing the Archetype: Innovation and conservatism in the evolution and development of the pentadactyl limb, in *Complex Organismal Functions: Integration and Evolution in Vertebrates* (eds D.B. Wake and G. Roth), John Wiley & Sons, New York, pp. 171–89.

Hinchliffe, J.R. (1991) Developmental approaches to the problem of transformation of limb structure in evolution, in *Developmental Patterns of the Vertebrate Limb* (eds J.R. Hinchliffe, J.M. Hurle and D. Summerbell), Plenum Press, New York, pp. 313–23.

Hinchliffe, J.R. (1994) Evolutionary developmental biology of the tetrapod limb. *Development*, 1994 Suppl., 163–8.

Hinchliffe, J.R. (1997) The forward march of the bird-dinosaurs halted? *Science*, **278**, 596–7.

Hinchliffe, J.R. and Johnson, D.R. (1980) *The Development of the Vertebrate Limb*. Clarendon Press, Oxford.

Hinchliffe, J.R. and Johnson, D.R. (1983) Growth of cartilage, in *Cartilage, Volume 2. Development, Differentiation and Growth* (ed. B.K. Hall), Academic Press, New York, pp. 255–96.

Hinchliffe, J.R., Hurle, J.M. and Summerbell, D. (eds) (1991) *Developmental Patterning of the Vertebrate Limb*, NATO ASI Series A: Life Sciences Vol. 205, Plenum Press, New York.

Hinegardner, R. (1976) Evolution of genome size, in *Molecular Evolution* (ed. F.J. Ayala), Sinauer Associates, Sunderland, MA, pp. 179–99.

Hinegardner, R. and Engleberg, J. (1983) Biological complexity. *J. Theor. Biol.*, **104**, 7–20.

His, W. (1868) *Untersuchungen über die erste Anlage des Wirbeltierleibes. Die erste Entwicklung des Hühnchens im Ei*, F.C.W. Vogel, Leipzig.

Ho, M.-W. (1984) Environment and heredity in development and evolution, in *Beyond Neo-Darwinism. An Introduction to the New Evolutionary Paradigm* (eds M.-W. Ho and P.T. Saunders), Academic Press, London, pp. 267–89.

Ho, M.-W., Bolton, E. and Saunders, P.T. (1983) The bithorax phenocopy and pattern formation. I. Spatiotemporal characteristics of the phenocopy response. *Exp. Cell Biol.*, **51**, 282–90.

Hodge, M.J.S. (1972) The universal gestation of Nature: Chambers' Vestiges and explanations. *J. Hist. Biol.*, **5**, 127–52.

Hodges, G.M. and Rowlatt, C. (1994) *Developmental Biology and Cancer*, CRC Press, Boca Raton, FL.

Hodos, W. (1976) The concept of homology and the evolution of behavior, in *Evolution, Brain and Behavior: Persistent Problems* (eds R.B. Masterton, W. Hodos, and H. Jerison), Lawrence Erlbaum Assoc., Hillsdale, NJ, pp. 152–67.

Hogan, B.L.M. (1996a) Bone morphogenetic proteins: Multifunctional regulators of vertebrate development. *Genes Devel.*, **10**, 1580–94.

Hogan, B.L.M. (1996b) Bone morphogenetic proteins in development. *Curr. Opin. Genet. Devel.*, **6**, 432–8.

Holland, L.Z. and Holland, N.D. (1996) Expression of *AmphiHox*-1 and *AmphiPax*-1 in amphioxus embryos treated with retinoic acid: Insights into evolution and patterning of the chordate nerve cord and pharynx. *Development*, **122**, 1829–38.

Holland, L.Z., Holland, P.W.H. and Holland, N.D. (1996) Revealing homologies between body parts of distantly related animals by *in situ* hybridization to developmental genes: Amphioxus versus vertebrates, in *Molecular Zoology: Advances, Strategies and Protocols* (eds J.D. Ferraris and S.R. Palumbi), Wiley-Liss Inc., New York, pp. 267–82, 473–83.

Holland, L.Z., Kene, M., Williams, N.A., and Holland, N.D. (1997) Sequence and embryonic expresion of the amphioxus *engrailed* gene (*AmphiEn*): The metameric pattern of transcription resembles that of its segment-olarity homolog in *Drosophila*. *Development*, **124**, 1723–32.

Holland, N.D. (1996) Homology, homeobox genes, and the early evolution of the vertebrates, in *New Perspectives on the History of Life: Essays in Systematic Biology as Historical Narrative* (eds M.T. Ghiselin and G. Pinna), Mem. Cal. Acad. Sciences Number 20, California Academy of Sciences, San Francisco, CA, pp. 63–70.

Holland, N.D., Holland, L.Z., Honma, Y. and Fujii, T. (1993) *Engrailed* expression during development of a lamprey, *Lampetra japonica*: A possible clue to homologies between agnathan and gnathostome muscles of the mandibular arch. *Devel. Growth Differ.*, **35**, 153–60.

Holland, N.D., Panganiban, G., Henyey, E.L. and Holland, L.Z. (1996) Sequence and developmental expression of *AmphiDll*, an amphioxus *Distalless* gene transcribed in the ectoderm, epidermis and nervous system: Insights into evolution of craniate forebrain and neural crest. *Development*, **122**, 2911–20.

Holland, P.W.H. (1990) Homeobox genes and segmentation: Co-option, co-evolution, and convergence. *Sem. Devel. Biol.*, **1**, 135–45.

Holland, P.W.H. (1991) Cloning and evolutionary analysis of *Msh*-like homeobox genes from mouse, zebrafish and ascidian. *Gene* (Amst.), **98**, 253–7.

Holland, P.W.H. (1996) Molecular biology of lancelets: Insights into development and evolution. *Israel J. Zool.*, **42**, S247–S272.

Holland, P.W.H. (1997) Vertebrate evolution: Something fishy about *Hox* genes. *Curr. Biol.*, **7**, R570–R572.

Holland, P.W.H. and Garcia-Fernàndez, J. (1996) Hox genes and chordate evolution. *Devel. Biol.*, **173**, 382–95.

Holland, P.W.H., Garcia-Fernàndez, J., Williams, N.A. and Sidow, A. (1994) Gene duplications and the origins of vertebrate development. *Development* 1994 Suppl., 125–33.

Holland, P.W.H. and Graham, A. (1995) Evolution of regional identity in the vertebrate nervous system. *Persp. Devel. Neurobiol.*, **3**, 17–27.

Holland, P.W.H., Hacker, A. and Williams, N.A. (1991) A molecular analysis of the phylogenetic affinities of *Saccoglossus cambrensis* Brambell & Cole (Hemichordata). *Phil. Trans. R. Soc. Lond. B*, **332**, 185–89.

Holland, P.W.H. and Hogan, B.L.M. (1986) Phylogenetic distribution of *Antennapedia*- like homeoboxes. *Nature*, **321**, 251–3.

Holland, P.W.H. and Hogan, B.L.M. (1988) Expression of homeobox genes during mouse development: A review. *Genes & Devel.*, **2**, 773–82.

Holland, P.W.H., Holland, L.Z., Williams, N.A. and Holland, N.D. (1992) An amphioxus homeobox gene: Sequence conservation, spatial expression during development and insights into vertebrate evolution. *Development*, **116**, 653–61.

Holland, P.W.H., Koschorz, B., Holland, L.Z. and Herrmann, B.G. (1995) Conservation of *Brachyury* (*T*) genes in amphioxus and vertebrates: Developmental and evolutionary implications. *Development*, **121**, 4283–91.

Holland, P.W.H. and Williams, N.A. (1990) Conservation of engrailed-like homeobox sequences during vertebrate evolution. *FEBS Letters*, **277**, 250–2.

Holley, S.A. and Ferguson, E.L. (1997) Fish are like flies are like frogs: Conservation of dorsal-ventral patterning mechanisms. *BioEssays*, **19**, 281–4.

Holley, S.A., Jackson, P.D., Sasai, Y. *et al.* (1995) A conserved system for dorso-ventral patterning in insects and vertebrates involving *sog* and *chordin*. *Nature*, **376**, 249–53.

Holliday, R. (1987) The inheritance of epigenetic defects. *Science*, **238**, 163–70.

Holliday, R. (1990) Mechanisms for the control of gene activity during development. *Biol. Rev.*, **65**, 431–72.

Holliday, R. (1991) Quantitative genetic variation and developmental clocks. *J. Theor. Biol.*, **151**, 351–8.

Holliday, R. (1994) Epigenetics: An overview. *Devel. Genet.*, **15**, 453–7.

Holman, E.W. (1989) Some evolutionary correlates of higher taxa. *Paleobiology*, **15**, 357–63.

Holmdahl, D.E. (1925a) Experimentelle Untersuchungen über die Lage der Grenze zwischen primärer und sekundärer Körperentwicklung beim Huhn, *Anat. Anz.*, **59**, 393–6.

Holmdahl, D.E. (1925b) Die erst Entwicklung des Körpels bei den Vögeln und Säugetoeren inkl. dem Menschen, besonders mit Rücksicht auf die Bildung des Rückenmarks, des Zoloms und der entodermalen Kloake nebst einem Exkürs über die Entstehung der Spina bifida in der Lumbosakralregion. I, *Gegenbaurs Morphol. Jahrb.*, **54**, 333–84.

Holmdahl, D.E. (1925c) Die erst Entwicklung des Körpels bei den Vögeln und Säugetoeren inkl. dem Menschen, besonders mit Rücksicht auf die Bildung des Rückenmarks, des Zoloms und der entodermalen Kloake nebst einem Exkürs über die Entstehung der Spina bifida in der Lumbosakralregion. II-V, *Gegenbaurs Morphol. Jahrb.*, **55**, 112–208.

Holowacz, T. and Elinson, R.P. (1993) Cortical cytoplasm, which induces dorsal axis formation in *Xenopus*, is inactivated by UV irradiation of the oocyte. *Development*, **119**, 277–85.

Holtfreter, J. (1933) Der Einfluss von Wirtsalter und verschiedenen Organbezirken aud die Differenzierung von angelagerten Gastrulaektoderm. *Wilhelm Roux Arch. EntwMech. Org.*, **127**, 619–775.

Holtfreter, J., and Hamburger, V. (1955) Amphibians, in *Analysis of Development* (eds B.H. Willier, P.A. Weiss. and V. Hamburger), W.B. Saunders Co., Philadelphia, PA, pp. 230–96. (Reprinted 1971, Hafner Publishing Co., New York).

Honeycutt, R.L. and Adkins, R.M. (1993) Higher level systematics of eutherian mammals: An assessment of molecular characters and phylogenetic hypotheses. *Annu. Rev. Ecol. Syst.*, **24**, 279–305.

Horan, G.S.B., Kovàcs, E.N., Behringer, R.R. and Featherstone, M.S. (1995a) Mutations in paralogous *Hox* genes result in overlapping homeotic transformations of the axial skeleton: Evidence for unique and redundant function. *Devel. Biol.*, **169**, 359–72.

Horan, G.S.B., Ramírez-Solis, R., Featherstone, M.S. *et al.* (1995b) Compound mutants for the paralogous *hoxa*-4, *hoxb*-4, and *hoxd*-4 genes show more complete homeotic transformations and a dose-dependent increase in the number of vertebrae transformed. *Genes Devel.*, **9**, 1667–77.

Horder, T.J., Witkowski, J.A. and Wylie, C.C. (eds), (1986) *A History of Embryology*. The Eighth Symposium of The British Society for Developmental Biology. Cambridge University Press, Cambridge.

Hörstadius, S. (1950) *The Neural Crest. Its Properties and Derivatives in the Light of Experimental Research*, Oxford, Oxford University Press. [reprinted 1969 by Hafner Publishing Co., New York, and as a facsimile reprint in Hall and Hörstadius (1988)].

Hou, L., Martin, L.D., Zhou, Z. and Feduccia, A. (1996) Early adaptive radiations of birds: Evidence from fossils from Northeastern China. *Science*, **274**, 1164–7.

Hou, X. and Bergström, J. (1995) Cambrian lobopodians – ancestors of extant onychophorans? *Zool. J. Linn. Soc.*, **114**, 3–19.

Hou, X-g. and Sun, W-g. (1988) Discovery of Chengjiang fauna at Meishucun, Jining, Yunnan. *Acta Palaeontol. Sinica*, **27**, 1–12.

Howe, J.A. and Newport, J.W. (1996) A developmental timer regulates degradation of cyclin E1 at the midblastula transition during *Xenopus* embryogenesis. *Proc. Natl Acad. Sci. USA*, **93**, 2060–4.

Hu, C-C., Sakakura, Y., Sasano, Y. *et al.* (1992) Endogenous epidermal growth factor regulates the timing and pattern of embryonic mouse molar tooth morphogenesis. *Int. J. Devel. Biol.*, **36**, 505–16.

Huelsenbeck, J.P. and Rannala, B. (1997) Phylogenetic methods come of age: Testing hypotheses in an evolutionary context. *Science*, **276**, 227–32.

Huey, R.B. (1987) Phylogeny, history, and the comparative method, in *New Directions in Ecological Physiology* (eds M. Feder, A.F. Bennett, W.W. Burggren and R.B. Huey), Cambridge University Press, Cambridge, pp. 76–101.

Hull, D.L. (1980) Individuality and selection. *Ann. Rev. Ecol. Syst.*, **11**, 311–32.

Hull, D.L. (1981) Units of evolution: A metaphysical essay, in *The Philosophy of Evolution* (eds U.L. Jensen and R. Harré), Harvester Press, Brighton, pp. 23–44.

Hull, D.L. (1988) A mechanism and its metaphysics: An evolutionary account of the social and conceptual development of science. *Biol. Philos.*, **3**, 123–55.

Hülskamp, M., Pfeifle, C. and Tautz, D. (1990) A morphogenetic gradient of *hunchback* protein organizes the expression of the gap genes *Krüppel* and *knirps* in the early *Drosophila* embryo. *Nature*, **346**, 577–80.

Hunt, P., Gulisano, M., Cook, M. *et al.* (1991a) A distinct Hox code for the branchial region of the vertebrate head. *Nature*, **353**, 861–4.

Hunt, P. and Krumlauf, R. (1992) Hox codes and positional specification in vertebrate embryonic axes. *Ann. Rev. Cell*, **8**, 227–56.

Hunt, P., Whiting, J., Muchamore, I. *et al.* (1991b) Homeobox genes and models for patterning the hindbrain and branchial arches. *Development* Suppl. **1**, 187–96.

Hunt, P., Whiting, J., Nonchev, S. *et al.* (1991c) The branchial *Hox* code and its implications for gene regulation, patterning of the nervous system and head evolution. *Development* Suppl. **2**, 63–77.

Hunt, P., Wilkinson, D. and Krumlauf, R. (1991d) Patterning the vertebrate head: Murine *Hox*-2 genes mark distinct subpopulations of premigrating and migrating cranial neural crest. *Development*, **112**, 43–50.

Hunt, P., Ferretti, P., Krumlauf, R. and Thorogood, P. (1995) Restoration of normal Hox code and branchial arch morphogenesis after extensive deletion of hindbrain neural crest. *Devel. Biol.*, **168**, 584–97.

Hunter, J.P. and Jernvall, J. (1995) The hypocone as a key innovation in mammalian evolution. *Proc. Natl Acad. Sci. USA*, **92**, 10718–22.

Hunt von Herbing, I., Miyake, T., Hall, B.K. and Boutilier, R.G. (1996a) The ontogeny of feeding and respiration in larval Atlantic cod, *Gadus morhua* (Teleostei; Gadiformes): (1) Morphology. *J. Morphol.*, **227**, 15–35.

Hunt von Herbing, I., Miyake, T., Hall, B.K. and Boutilier, R.G. (1996b) The ontogeny of feeding and respiration in larval Atlantic cod, *Gadus morhua* (Teleostei; Gadiformes): (2) Function. *J. Morphol.*, **227**, 37–50.

Hurle, J.M., Gana, Y. and Marcias, D. (1989) Experimental analysis of the *in vivo* chondrogenic potential of the interdigital mesenchyme of the chick leg bud subjected to local ectodermal removal. *Devel. Biol.*, **132**, 368–74.

Hutchinson, G.E. (1967) *A Treatise on Limnology, Volume II, Introduction to Lake Biology and the Limnoplankton.* John Wiley & Sons, New York.

Huxley, J.S. (1932) *Problems of Relative Growth*, MacVeagh, London. (2nd edn, 1972, Dover Publications, New York).

Huxley, J.S. (1942) *Evolution: The Modern Synthesis*, Allen and Unwin, London.

Huxley, T.H. (1849) On the anatomy and the affinities of the family of the Medusæ. *Phil. Trans. R. Soc.*, **139**, 413–34.

Huxley, T.H. (1853) On the morphology of the Cephalous Mollusca, as illustrated by the anatomy of certain Heteropoda and Pteropoda collected during the voyage of H.M.S. 'Rattlesnake' in 1846–50. *Phil. Trans. R. Soc. Lond.*, **143**, 29–65.

Huxley, T.H. (1864) *Lectures on the Elements of Comparative Anatomy.* London.

Huxley, T.H. (1868) On the animals which are most nearly intermediate between the birds and reptiles. *Ann. Mag. Nat. Hist.*, **4**, 66–75.

Huxley, T.H. (1894) Owen's position in the history of anatomical science, in *The Life of Richard Owen by his Grandson the Rev. Richard Owen, M.A*, John Murray, London, Volume 2, pp. 273–332.

Huxley, T.H. (1910) *Lectures and Lay Sermons.* With an Introduction by Sir Oliver Lodge, J. M. Dent & Sons, London.

Huysseune, A. (1983) Observations on tooth development and implantation in the upper pharyngeal jaws in *Astatotilapia elegans* (Teleostei, Cichlidae). *J. Morph.*, **175**, 217–34.

Huysseune, A. (1989) Morphogenetic aspects of the pharyngeal jaws and neurocranial apophysis in postembryonic *Astatotilapia elegans* (Trewavas, 1933) (Teleostei, Cichlidae). *Med. Kon. Acad. Wetenschappen, Letteren en Schone Kunsten van België*, **51**, 11–35.

Huysseune, A., Sire, J-Y. and Meunier, F.J. (1994) Comparative study of lower pharyngeal jaw structure in two phenotypes of *Astatoreochromis alluaudi* (Teleostei: Cichlidae). *J. Morphol.*, **221**, 25–43.

Huysseune, A. and Verraes, W. (1987) Relationship between cartilage and bone growth in pharyngeal jaw development in a cichlid fish. *Biol. Jb. Dodonaea*, **55**, 121–35.

Hyatt, B.A., Lohr, J.L. and Yost, H.J. (1996) Initiation of vertebrate left-right axis formation by maternal *Vg1*. *Nature*, **384**, 62–5.

Illmensee, K. and Mahowald, A.P. (1974) Transplantation of posterior polar plasm in *Drosophila*. Induction of germ cells at the anterior pole of the egg. *Proc. Natl Acad. Sci. USA*, **71**, 1016–20.

Imai, H., Osumi-Yamashita, N., Ninomiya, Y. and Eto, K. (1996) Contribution of early-migrating midbrain crest cells to the dental mesenchyme of mandibular molar teeth in rat embryos. *Devel. Biol.*, **176**, 151–65.

Irish, F.J. (1989) The role of heterochrony in the origin of a novel bauplan: evolution of the ophidian skull. *Geobios. Mém. Spécial*, **12**, 227–33.

Isaac, A., Sargent, M.G. and Cooke, J. (1997) Control of vertebrate left-right asymmetry by a Snail-related zinc finger gene. *Science*, **275**, 1301–4.

Ismail, M.H., Verraes, W. and Huysseune, A. (1982) Developmental aspects of the pharyngeal jaws in *Astatotilapia elegans* (Trewavas, 1933) (Teleostei: Cichlidae). *Neth. J. Zool.*, **32**, 513–43.

Ivanov, A.V. (1963) *Pogonophora.* Consultants Bureau, New York.

Ivanov, A.V. (1975) Embryonalentwicklung der Pogonophora und ihre Systematische Stellung. *Z. Zool. Syst. Evolutionforsch.* (Special Issue), The phylogeny and systematic position of the Pogonophora, 10–44.

Iwamatsu, T. (1994) Stages of normal development in the medaka *Oryzias latipes*. *Zool. Sci.*, **11**, 825–39.

Jaanusson, V. (1981) Functional thresholds in evolutionary progress. *Lethaia*, **14**, 251–60.

Jablonka, E. and Lamb. M.J. (1989) The inheritance of acquired epigenetic variations. *J. Theor. Biol.*, **139**, 69–83.

Jablonka, E. and Lamb. M.J. (1995) *Epigenetic Inheritance and Evolution: The Lamarckian Dimension*, Oxford University Press, Oxford.

Jablonka, E. and Lamb. M.J. (1997) Epigenetic Inheritance in Evolution. *J. Evol. Biol.*, **10**, (in press).

Jablonski, D. (1996) Body size and macroevolution, in *Evolutionary Paleobiology* (eds D. Jablonski,

D.H. Erwin and J.H. Lipps), University of Chicago Press, Chicago, IL, pp. 256–89.

Jablonski, D. (1997) Body-size evolution in Cretaceous molluscs and the status of Cope's rule. *Nature*, **385**, 250–2.

Jackson, J.B.C. and Cheetham, A.H. (1990) Evolutionary significance of morphospecies: A test with cheilostome Bryozoa. *Science*, **248**, 579–83.

Jacob, F. (1977) Evolution and tinkering. *Science*, **196**, 1161–6.

Jacobs, D.K. (1990) Selector genes and the Cambrian radiation of Bilateria. *Proc. Natl Acad. Sci. USA*, **87**, 4406–10.

Jacobson, A.G. (1958) The roles of neural and non-neural tissues in lens induction. *J. Exp Zool.*, **139**, 525–57.

Jacobson, A.G. (1966) Inductive processes in embryonic development. *Science*, **152**, 25–34.

Jacobson, A.G. (1987) Determination and morphogenesis of axial structures: Mesodermal metamerism, shaping of the neural plate and tube, and segregation and functions of the neural crest, in *Developmental and Evolutionary Aspects of the Neural Crest* (ed. P.F.A. Maderson), John Wiley & Sons, New York, pp. 147–80.

Jacobson, A.G. and Sater, A.K. (1988) Features of embryonic induction. *Development*, **104**, 341–59.

Jacobson, M.D., Weil, M. and Raff, M.C. (1997) Programmed cell death in animal development. *Cell*, **88**, 347–54.

James, F.C. (1983) Environmental components of morphological differentiation in birds. *Science*, **221**, 184–6.

James, R.R. (1984) *Prince Albert. A Biography*, Alfred A. Knopf, New York.

Janvier, P. (1995) Conodonts join the club. *Nature*, **374**, 761–2.

Janvier, P. (1996a) *Early Vertebrates*, Clarendon Press, Oxford.

Janvier, P. (1996b) Fishy fragments tip the scales. *Nature*, **383**, 757–8.

Jaskoll, T.F. and Maderson, P.F.A. (1978) A histological study of the development of the avian middle ear and tympanum. *Anat. Rec.*, **190**, 177–200.

Jeffery, W.R. and Swalla, B.J. (1990) Anural development in ascidians: Evolutionary modifications and elimination of the tadpole larva. *Sem. Devel. Biol.*, **1**, 253–61.

Jeffery, W.R. and Swalla, B.J. (1991) An evolutionary change in the muscle lineage of an anural ascidian embryo is restored by interspecific hybridization with a urodele ascidian. *Devel. Biol.*, **145**, 328–37.

Jeffery, W.R. and Swalla, B.J. (1992) Factors necessary for restoring an evolutionary change in an anural ascidian embryo. *Devel. Biol.*, **153**, 194–205.

Jeffery, W.R. (1997) Evolution of ascidian development. *BioScience*, **47**, 417–25.

Jenkinson, J.W. (1906) Remarks on the germinal layers of vertebrates and on the significance of germinal layers in general. *Mem. Proc. Manchester Lit. Phil. Soc.*, **50**, 1–89.

Jennings, D.H. (1997) Evolution of endocrine control in amphibians with derived life-history strategies, Ph.D. Thesis, University of Colorado, Boulder, CO.

Jepsen, G.L., Mayr, E. and Simpson, G.G. (1949) *Genetics, Paleontology and Evolution*, Princeton University Press, Princeton.

Jeram, A.J. (1990) Book-lungs in a lower Carboniferous scorpion. *Nature*, **343**, 360–1.

Jernvall, J. (1995) Mammalian molar cusp patterns: Developmental mechanisms of diversity. *Acta Zool. Fennica*, **198**, 1–61.

Jernvall, J, Hunter, J.P. and Fortelius, M. (1996) Molar tooth diversity, disparity, and ecology in Caenozoic ungulate radiations. *Science*, **274**, 1489–92.

Jernvall, J., Kettunen, P., Karavanova, I. *et al.* (1994) Evidence for the role of the enamel knot as a control center in mammalian tooth cusp formation: Non-dividing cells express growth stimulating *fgf*-4 gene. *Int. J. Devel. Biol.*, **38**, 463–9.

Jockusch, E.L. and Nagy, L.M. (1997) Insect evolution: How did insect wings originate? *Curr. Biol.*, **7**, R358–R361.

Johnson, D.R. (1986) *The Genetics of the Skeleton. Animal Models of Skeletal Development*. The Clarendon Press, Oxford.

Johnson, D.R. and O'Higgins, P. (1994) The inheritance of pattern of metameric segmentation in the mouse (*Mus musculus*) vertebral column. *J. Zool. Lond.*, **233**, 473–92.

Johnson, D.R. and O'Higgins, P. (1996) Is there a link between changes in the vertebral 'hox code' and the shape of vertebrae? A quantitative study of shape change in the cervical vertebral column of mice. *J. Theor. Biol.*, **183**, 89–93.

Johnson, R.L., Riddle, R.D. and Tabin, C. (1994) *Sonic hedgehog*: A key mediator of anterior-posterior patterning of the limb and dorso-ventral patterning of axial embryonic structures. *Biochem. Soc. Trans.*, **22**, 569–74.

Johnston, I.A., Cole, N.J., Vieira, V.L.A. and Davidson, I. (1997) Temperature and developmental plasticity of muscle phenotype in herring larvae. *J. Exp. Biol.*, **200**, 849–68.

Johnston, M.C. (1966) A radioautographic study of the migration and fate of cranial neural crest cells in the chick embryo. *Anat. Rec.*, **156**, 143–56.

Jones, C.M., Armes, N. and Smith, J.C. (1996) Signalling by TGF-β family members: Short-range effects of *Xnr*-2 and BMP-4 contrast with the long-range effects of activin. *Curr. Biol.*, **6**, 1468–75.

Jones, C.M., Lyons, K.M. and Hogan, B.L.M. (1991) Involvement of bone morphogenetic protein-4 (BMP-4) and Vgr-1 in morphogenesis and neurogenesis in the mouse. *Development*, **111**, 531–42.

Jones, C.M. and Smith, J.C. (1995) Revolving vertebrates. *Curr. Biol.*, **5**, 574–6.

Jones, C.J. and Price, R.A. (1996) Diversity and evolution of seedling *Baupläne* in *Pelargonium* (Geraniaceae). *Aliso*, **14**, 281–95.

Jones, D.S. (1988) Sclerochronology and the size versus age problem, in *Heterochrony in Evolution. A multidisciplinary approach* (ed. M.L. McKinney), Plenum Press, New York and London, pp. 93–108.

Jones, M.L. (1985) On the Vestimentifera, a new phylum. Six new species and other taxa, from hydrothermal vents and elsewhere. *Bull. Biol. Soc. Wash.*, **6**, 117–58.

Jones, M.L. and Gardiner, S.L. (1989) On the early development of the Vestimentiferan tube worm *Ridgeia* sp. and observations on the nervous system and trophosome of *Ridgeia* sp. and *Riftia pachyptila*. *Biol. Bull.*, **177**, 254–76.

Jowett, A.K., Vainio, S., Ferguson, M.W.J. *et al.* (1993) Epithelial-mesenchymal interactions are required for *msx* 1 and *msx* 2 gene expression in the developing murine molar tooth. *Development*, **117**, 461–70.

Joyner, A.L. (1996) *Engrailed, Wnt* and *Pax* genes regulate midbrain-hindbrain development. *Trends Genet.* **12**, 15–20.

Kafri, T., Ariel, M., Brandeis, M. *et al.* (1992) Developmental pattern of gene-specific DNA methylation in the mouse embryo and germ line. *Genes Devel.*, **6**, 705–14.

Kantomaa, T. and Hall, B.K. (1988a) Mechanism of adaptation in the mandibular condyle of the mouse. An organ culture study. *Acta Anat.*, **132**, 114–19.

Kantomaa, T. and Hall, B.K. (1988b) Organ culture providing an articulating function for the temporomandibular joint. *J. Anat.*, **161**, 195–201.

Kao, K.R. and Elinson, R.P. (1988) The entire mesodermal mantle behaves as Spemann's organizer in dorsoanterior enhanced *Xenopus laevis* embryos. *Devel. Biol.*, **127**, 64–77.

Kao, K.R., Masui, Y. and Elinson, R.P. (1986) Lithium-induced respecification of pattern in *Xenopus laevis* embryos. *Nature*, **322**, 371–3.

Kappen, C., Schughart, K. and Ruddle, F.H. (1989a) Organization and expression of homeobox genes in mouse and man. *Annals N. Y. Acad. Sci.*, **567**, 243–52.

Kappen, C., Schughart, K. and Ruddle, F.H. (1989b) Two steps in the evolution of Antennapedia-class vertebrate homeobox genes. *Proc. Nat. Acad. Sci. USA*, **86**, 5459–63.

Kappen, C., Schughart, K. and Ruddle, F.H. (1993) Early evolutionary origin of major homeodomain sequence classes. *Genomics*, **18**, 54–70.

Karr, T.L. and Mittenthal, J.E. (1992) Adaptive mechanisms that accelerate embryonic development in *Drosophila*, in *Principles of Organization in Organisms* (eds J. Mittenthal, and A. Baskin), SFI Studies in the Sciences of Complexity. Proceedings Volume XIII, Addison-Wesley, Menlo Park, CA, pp. 95–108.

Katsuyama, Y., Wada, S., Yasugi, S. and Saiga, H. (1995) Expression of the *labial* group Hox gene *HrHox*-1 and its alteration by retinoic acid in development of the ascidian *Halocynthia roretzi*. *Development*, **121**, 3197–205.

Katz, M.J. and Goffman, W. (1981) Preformation of ontogenetic patterns. *Phil. Sci.*, **48**, 438–53.

Kaufman, D.M. (1995) Diversity of new world mammals: Universality of the latitudinal gradients of species and Bauplans. *J. Mammal.*, **76**, 322–34.

Kaufman, M.H., Chang, H.-H. and Shaw, J.P. (1995) Craniofacial abnormalities in homozygous *Small eye (Sey/Sey)* embryos and newborn mice. *J. Anat.*, **186**, 607–17.

Kawakami, Y., Ishikawa, T., Shimabara, M. *et al.*, (1996) BMP signalling during bone pattern determination in the developing limb. *Development*, **122**, 3557–66.

Kellogg, E.A. and Shaffer, H.B. (1993) Model organisms in evolutionary studies. *Syst. Biol.*, **42**, 409–14.

Kemp, T.S. (1982) *Mammal-like Reptiles and the Origin of Mammals*. Academic Press, London.

Kengaku, M. and Okamoto, H. (1993) Basic fibroblast growth factor induces differentiation of neural tube and neural crest lineages of cultured ectodermal cells from *Xenopus* gastrulae. *Development*, **119**, 1067–78.

Kenyon, C. (1985) Cell lineage and the control of *Caenorhabditis elegans* development. *Phil. Trans. R. Soc. B*, **312**, 21–38.

Kenyon, C. and Wang, B. (1991) A cluster of *Antennapedia*-class homeobox genes in a non-segmented animal. *Science*, **253**, pp. 516–17.

Kessel, M. (1991) Molecular coding of axial positions by *Hox* genes. *Sem. Devel. Biol.*, **2**, 367–73.

Kessel, M. (1992) Respecification of vertebral identities by retinoic acid. *Development*, **115**, 487–501.

Kessel, M., Balling, R. and Gruss, P. (1990) Variations of cervical vertebrae after expression of a Hox-1.1 transgene in mice. *Cell*, **61**, 301–8.

Kessel, M. and Gruss, P. (1990) Murine developmental control genes. *Science*, **249**, 374–9.

Kessel, M. and Gruss, P. (1991) Homeotic transformations of murine vertebrae and concomitant alteration of *Hox* codes induced by retinoic acid. *Cell*, **67**, 89–104.

Kettlewell, H.B.D. (1973) *The Evolution of Melanism.* Clarendon Press, Oxford.

Kielmeyer, C.F. (1793) Ideen einer Entwicklungsgeschichte der Erde und ihrer Organisation, Schreiben an Windischmann, in *Carl Friedrich Kielmeyer, Gesammelte* (ed. F.H. Holler), F. Keiper, Berlin, 1938, pp. 203–10,

Kieny, M. and Pautou, M.P. (1976) Experimental analysis of excendentary regulation in xenoplastic quail-chick limb bud recombinants. *Roux's Arch. Devel. Biol.*, **179**, 327–38.

Kimelman, D., Christian, J.L. and Moon, R.T. (1992) Synergistic principles of development: Overlapping patterning system in *Xenopus* mesoderm induction. *Development*, **116**, 1–9.

Kimmel, C.B. (1996) Was *Urbilateria* segmented? *Trends Genet.*, **12**, 329–31.

Kimmel, C.B., Warga, R.M. and Schilling, T.F. (1990) Origin and organization of the zebrafish fate map. *Development*, **108**, 581–94.

Kindred, B. (1967) Selection for an invariant character, vibrissa number in the house mouse. V. Selection on non-*Tabby* segregants from *Tabby* selection lines. *Genetics*, **55**, 365–73.

King, D.P., Zhao, Y., Sangoram, A.M. *et al.* (1997) Positional cloning of the mouse circadian Clock gene. *Cell*, **89**, 641–53.

King, J.A., Storm, E.E., Marker, P.C. *et al.* (1996) The role of BMPs and GDFs in development of region-specific skeletal structures. *Ann. N. Y. Acad. Sci.*, **785**, 70–9.

King, M.L. (1996) Molecular basis for cytoplasmic localization. *Devel. Genet.*, **19**, 183–89.

King, R.C. and Stansfield, W.D. (1985) *A Dictionary of Genetics*. 3rd end, Oxford University Press, New York and Oxford.

Kingdon, J. (1990) *Island Africa. The Evolution of Africa's Rare Animals and Plants*. Collins, London.

Kingsley, C. (1885) *The Water Babies: A Fairy Tale for a Land-baby*, Macmillan, London. (Facsimile edition 1984, Chancellor Press, London.)

Kingsley, D.M. (1994a) What do BMPs do in mammals? Clues from the mouse short-ear mutation. *Trends Genet.*, **10**, 16–21.

Kingsley, D.M. (1994b) The TGF-β superfamily: New members, new receptors, and new genetic tests of function in different organisms. *Genes Devel.*, **8**, 133–46.

Klingenberg, C.P. (1996a) Multivariate allometry, in *Advances in Morphometrics* (eds L.F. Marcus, M. Corti, A. Loy *et al.*), Plenum Press, New York, pp. 23–49.

Klingenberg, C.P. (1996b) Individual variation of ontogenies: A longitudinal study of growth and timing. *Evolution*, **50**, 2412–28.

Klingenberg, C.P. (1997) Heterochrony and allometry: The analysis of evolutionary change in ontogeny. *Biol. Rev.*, (in press).

Klingenberg, C.P., Neuenschwander, B.E. and Flury, B.D. (1996) Ontogeny and individual variation: Analysis of patterned covariance matrices with common principal components. *Syst. Biol.*, **45**, 135–50.

Klingenberg, C.P. and Spence, J.R. (1993) Heterochrony and allometry: Lessons from the water strider genus *Limnoporus*. *Evolution*, **47**, 1834–53.

Klingenberg, C.P. and Zimmermann, M. (1992) Static, ontogenetic, and evolutionary allometry: A multivariate comparison in nine species of water striders. *Amer. Nat.*, **140**, 601–20.

Klueg, K.M., Harkey, M.A. and Raff, R.A. (1997) Mechanisms of evolutionary changes in timing, spatial expression, and mRNA processing in the *msp130* gene in a direct-developing sea urchin, *Heliocidaris erythrogramma*. *Devel. Biol.*, **182**, 121–33.

Kluge, A.G. (1988) The characterization of ontogeny, in *Ontogeny and Systematics* (ed. C.J. Humphries), Columbia University Press, New York, pp. 57–81.

Kluge, A.G. and Strauss, R.E. (1985) Ontogeny and systematics. *Annu. Rev. Ecol. Syst.*, **16**, 247–68.

Knoll, A.H. (1996) Daughter of time. *Paleobiology*, **22**, 1–7.

Kobel, H.R. and Du Pasquier, L. (1986) Genetics of polyploid *Xenopus*. *Trends Genet.*, **2**, 310–15.

Koch, P.B. (1992) Seasonal polyphenism in butterflies: A hormonally controlled phenomenon of pattern formation. *Zool. Jb. Physiol.*, **96**, 227–40.

Kölliker, A. von (1879) *Entwickelungsgeschichte des Menschen und der Höheren Thiere. Zweite ganz umgearbeitere Auflage*. Leipzig.

Kölliker, A. von (1884) Die Embryonalen Keimblätter und die Gewebe. *Zeit. wiss. Zool.*, **40**, 179–213.

Kölliker, A. von (1889) *Handbuch der Gewebelehre des Menschen. 6. Umgearbeitere Auflage. Erster Band: Die allgemeine Gewebelehre und die Systems der Haut, Knochen und Muskeln*. Leipzig.

Kollros, J.J. (1940) The disappearance of the balancer in *Amblystoma* larvae. *J. Exp. Zool.*, **85**, 33–52.

Kolodziejczyk, S.M. and Hall, B.K. (1996) Signal transduction and TGF-β superfamily receptors. *Biochem. Cell Biol.*, **74**, 299–314.

Kono, T., Obata, Y., Yoshimzu, T. *et al.* (1996) Epigenetic modifications during oocyte growth correlates with extended parthenogenetic development in the mouse. *Nature Genet.*, **13**, 91–4.

Köntges, G. and Lumsden, A. (1996) Rhombencephalic neural crest segmentation is preserved throughout craniofacial ontogeny. *Development*, **122**, 3229–42.

Kosher, R.A., Gay, S.W., Kamanitz, J.R. *et al.* (1986a) Cartilage proteoglycan core protein gene expression during limb cartilage differentiation. *Devel. Biol.*, **118**, 112–17.

Kosher, R.A., Kulyk, W.M. and Gay, S.W. (1986b) Collagen gene expression during limb cartilage differentiation. *J. Cell Biol.*, **102**, 1151–6.

Kostakopoulou, K., Vogel, A., Brickell, P. and Tickle, C. (1996) 'Regeneration' of wing bud stumps of chick embryos and reactivation of *Msx-1* and *Shh* expression in response to FGF-4 and ridge signals. *Mech. Devel.*, **55**, 119–31.

Kostovic-Knezevic, L., Gajovic, S. and Svajger, A. (1991) Morphogenetic features in the tail region of the rat embryo, *Int. J. Devel. Biol.*, **35**, 191–5.

Kovalevsky, A.O. (1866a) Entwickelungsgeschichte der einfachen Ascidien. *Mém. Acad. Sci. St. Petersbourg*, **7** (15), 19pp.

Kovalevsky, A.O. (1866b) Anatomie des Balanoglossus. *Mém. Acad. Sci. St. Petersbourg*, **7**, 1–10.

Kovalevsky, A.O. (1867) Entwickelungsgeschichte des *Amphioxus lanceolatus*. *Mém. Acad. Sci. St. Petersbourg* **11** (4), 1–17.

Kovalevsky, A.O. (1871) Weitere Studien über die Entwicklung der einfachen Ascidien. *Arch. Mikrosk. Anat.*, **7**, 101–30.

Kovalevsky, A.O. (1877) Weitere Studien über die Entwickelungsgeschichte des *Amphioxus lanceolatus*. *Arch. Mikrosk. Anat.* **13**, 181–204.

Kraak, S.B.M. (1997) Fluctuating around directional asymmetry? *Trends Ecol. Evol.*, **12**, 230.

Kratochwil, K., Dull, M., Farinas, I. *et al.* (1996) Lef1 expression is activated by BMP-4 and regulates inductive tissue interactions in tooth and hair development. *Genes Devel.*, **10**, 1382–94.

Kratochwil, K. and Schwartz, P. (1976) Tissue interactions in androgen response of embryonic mammary rudiment of mouse: Identification of target tissue for testosterone. *Proc. Natl Acad. Sci. USA*, **73**, 4041–4.

Kristensen, R.M. (1983) Loricifera, a new phylum with Aschelminthes characters from the meiobenthos. *Z. Zool. Syst. Evolut.-Forsch.*, **21**, 163–80.

Kronmiller, J.E. (1995) Spatial distribution of epidermal growth-factor transcripts and effects of exogenous epidermal growth factor on the pattern of the mouse dental lamina. *Archs Oral Biol.*, **40**, 137–43.

Kronmiller, J.E., Nguyen, T. and Berndt, W. (1995) Instruction by retinoic acid of incisor morphology in the mouse embryonic mandible. *Archs Oral Biol.*, **40**, 589–95.

Krumlauf, R. (1994) *Hox* genes in vertebrate development. *Cell*, **78**, 191–201.

Kulyk, W.M., Rodgers, B.J., Greer, K. and Kosher, R.A. (1989a) Promotion of embryonic chick limb cartilage differentiation by transforming growth factor β. *Devel. Biol.*, **135**, 424–30.

Kulyk, W.M., Upholt, W.B. and Kosher, R.A. (1989b) Fibronectin gene expression during limb cartilage differentiation. *Development*, **106**, 449–56.

Kuratani, S. (1997) Spatial distribution of postotic crest cells defines the head/trunk interface of the vertebrate body: Embryological interpretation of peripheral nerve morphology and evolution of the vertebrate head. *Anat. Embryol.*, **195**, 1–13.

Kurtén, B. (1963) Return of a lost structure in the evolution of the felid dentition. *Soc. Sci. Fennica Comm. Biol.*, **26** (4), 1–12

Kusakabe, T., Swalla, B.J., Satoh, N. and Jeffery, W.R. (1996) Mechanism of an evolutionary change in muscle cell differentiation in ascidians with different modes of development. *Devel. Biol.*, **174**, 379–92.

Laale, H.W. (1984) Naturally occurring diblastodermic eggs in the zebrafish, *Brachydanio rerio. Can. J. Zool.*, **62**, 386–90.

Lacalli, T.C. (1996a) Dorsoventral axis inversion: A phylogenetic perspective. *BioEssays*, **18**, 251–4.

Lacalli, T.C. (1996b) Landmarks and subdomains in the larval brain of *Branchiostoma*: Vertebrate homologs and invertebrate antecedents. *Israel J. Zool.*, **42**, S131–S146.

Lacalli, T.C., Holland, N.D. and West, J.E. (1994) Landmarks in the anterior central nervous system of amphioxus larvae. *Phil. Trans. R. Soc. Lond.* B, **344**, 165–85.

Lacy, R.C. and Horner, B.E. (1996) Effects of inbreeding on skeletal development of *Rattus villosissimus. J. Hered.*, **87**, 277–87.

Lake, J.A. (1989) Origin of the multicellular animals, in *The Hierarchy of Life: Molecules and Morphology in Phylogenetic Analysis* (eds B. Fernholm, K. Bremer and H. Jornvall), Elsevier, Amsterdam, pp. 273–8.

Lakowski, B. and Hekimi, S. (1996) Determination of life-span in *Caenorhabditis elegans* by four clock genes. *Science*, **272**, 1010–13.

Lambert, D.M. and Hughes, A.J. (1988) Keywords and concepts in structuralist and functionalist biology. *J. Theor. Biol.*, **133**, 133–45.

Lampert, W. and Wolf, H.G. (1986) Cyclomorphosis in *Daphnia cucullata*: Morphometric and population genetics analysis. *J. Plankton Res.*, **8**, 289–303.

Lampl, M., Veldhuis, J.D. and Johnson, M.L. (1992) Saltation and stasis: A model of human growth. *Science*, **258**, 801–3.

Landauer, W. (1958) On phenocopies, their developmental, physiological and genetic meaning. *Amer. Nat.*, **92**, 201–13.

Lande, R. (1979) Quantitative genetic analysis of multivariate evolution, applied to brain:body size allometry. *Evolution*, **33**, 402–16.

Landes, D.S. (1983) *Revolution in Time. Clocks and the Making of the Modern World*, The Belknap Press of Harvard University Press, Cambridge, MA.

Langecker, T.G., Schmale, H. and H. Wilkins (1993) Transcription of the opsin gene in degenerate eyes of cave-dwelling *Astyanax fasciatus* (Teleostei, Characidae) and of its conspecific epigean ancestor during early ontogeny. *Cell Tiss. Res.*, **273**, 183–92.

Langille, R.M. and Hall, B.K. (1988a) The role of the neural crest in the development of the cartilaginous cranial and visceral skeleton of the medaka, *Oryzias latipes. Anat. Embryol.*, **177**, 297–305.

Langille, R.M. and Hall, B.K. (1988b) Role of the neural crest in development of the trabecular and branchial arches in embryonic sea lamprey, *Petromyzon marinus* (L.). *Development*, **102**, 302–10.

Langille, R.M. and Hall, B.K. (1989) Developmental processes, developmental sequences and early vertebrate phylogeny. *Biol. Rev.*, **64**, 73–91.

Lankester, E. Ray (1870) On the use of the term homology in modern Zoology, and the distinction between homogenetic and homoplastic agreements. *Ann. Mag. Nat. Hist.*, **6**, 34–43.

Lankester, E. Ray (1873) On the primitive cell layers of the embryo as the basis of genealogical classification of animals, and on the origin of vascular and lymph systems. *Ann. Mag. nat. Hist.* ser. 4., **11**, 321–8.

Lankester, E. Ray (1877) Notes on the embryology and classification of the animal kingdom: Comprising a revision of speculations relative to the origin and significance of germ layers. *Quart. J. Microsc. Soc.*, **17**, 399–454.

Lankester, E. Ray (1878) Balfour on Elasmobranch Fishes. *Nature*, **18**, 113–15.

Lankester, E. Ray (1880) *Degeneration. A Chapter in Darwinism*, Macmillan, London.

Lankester, E. Ray (1888) Zoology. *Encyclopaedia Britannica* 9th edn, Vol. 24, 799–820.

Lankester, E. Ray (1909) *Extinct Animals*, Archibald Constable & Co., London.

Larsen, E.W. (1992) Tissue strategies as developmental constraints: Implications for animal evolution. *Trends Ecol. Evol.*, **7**, 414–17.

Larson, A., Wake, D.B., Maxson, L.R. and Highton, R. (1981) A molecular phylogenetic perspective on the origins of morphological novelties in the salamanders of the Tribe Plethodontini (Amphibia, Plethodontidae). *Evolution*, **35**, 405–22.

Lauder, G.V. (1981) Form and function: Structural analysis in evolutionary morphology. *Paleobiology*, **7**, 430–42.

Lauder, G.V. (1982) Historical biology and the problem of design. *J. Theor. Biol.*, **97**, 57–67.

Lauder, G.V. (1986) Homology, analogy, and the evolution of behavior, in *Evolution of Animal Behavior: Paleontological and Field Approaches* (eds M.H. Nitecki and J.A. Kitchell), Oxford University Press, New York, pp. 9–40.

Lauder, G.V. (1990) Functional morphology and systematics: Studying functional patterns in an historical context. *Annu. Rev. Ecol. Syst.*, **21**, 317–40.

Lauder, G.V. (1994) Homology, Form and Function, in *Homology: The Hierarchical Basis of Comparative Biology*, Academic Press, San Diego, CA, pp. 151–196.

Lauder, G.V., Huey, R.B., Monson, R.K. and Jensen, R.J. (1995) Systematics and the study of organismal form and function. *BioScience*, **45**, 696–704.

Lauder, G.V. and Liem, K.F. (1989) The role of historical factors in the evolution of complex organismal functions, in *Complex Organismal Functions: Integration and Evolution in Vertebrates* (eds D.B. Wake and G. Roth), John Wiley & Sons, Chichester, pp. 63–78.

Laufer, E., Dahn, R., Orozco, O.E. *et al.* (1997) Expression of *Radical fringe* in limb-bud ectoderm regulates apical ectodermal ridge formation. *Nature*, **386**, 366–73.

Laufer, E., Nelson, C.E., Johnson, R.L. *et al.* (1994) *Sonic hedgehog* and *FGF*-4 act through a signalling cascade and feedback loop to integrate growth and patterning of the developing limb bud. *Cell*, **79**, 993–1003.

Lauterborn, R. (1904) Die cyklische oder temporale Variation von *Anuraea cochlearis*. *Verh. Naturhistorisch-Medizinischer verein Heidelberg* n.f., **7**, 529–621.

Laverack, M.S., and Dando, J. (1987) *Lecture Notes on invertebrate Zoology* (3rd edn), Blackwell Scientific London, Oxford.

Lawrence, E. (1990) *Henderson's Dictionary of Biological Terms* 10th edn, Longman Scientific and Technical, London.

Leamy, L. (1993) Morphological integration of fluctuating asymmetry in the mouse mandible. *Genetica*, **89**, 139–53.

Leamy, L. and Atchley, W.R. (1985) Directional selection and developmental stability: Evidence from fluctuating asymmetry of morphometric characters in rats. *Growth*, **49**, 8–18.

Leamy, L.J., Routman, E.J. and Cheverud, J.M. (1997) A search for quantitative trait loci affecting asymmetry of mandibular characters in mice. *Evolution*, **51**, 957–69.

Leblond, C.P. and Inoue, S. (1989) Structure, composition and assembly of basement membrane. *Amer. J. Anat.*, **185**, 367–90.

Leclerc, R.F. and Regier, J.C. (1990) Heterochrony in insect development and evolution. *Sem. Devel. Biol.*, **1**, 271–9.

Lecointre, G. (1994) Aspects historiques et heuristiques de l'ichthyologie systématique. *Cybium*, **18**, 339–430.

Lecointre, G. (1996) Methodological aspects of molecular phylogeny of fishes. *Zool. Studies*, **35**, 161–77.

Le Douarin, N.M. (1969) Particularités du noyau interphasique chez la Caille Japonaise (*Coturnix coturnix japonica*). Utilisation de ces particularites comme 'marquage biologique' dans les recherches sur les interactions tissulaires et les migrations cellulaires au cours de l'ontogenèse. *Bull. Biol. Fr. Belg.*, **103**, 435–52.

Le Douarin, N.M. (1974) Cell recognition based on natural morphological nuclear markers. *Med. Biol.*, **52**, 281–319.

Le Douarin, N.M. (1982) *The Neural Crest*, Cambridge University Press, Cambridge and London.

Le Douarin, N.M., Dupin, E. and Ziller, C. (1994) Genetic and epigenetic control in neural crest development. *Curr. Opin. Genet. Devel.*, **4**, 685–95.

Lee, M.S.Y. (1997) Reptile relationships turn turtle . . . *Nature*, **389**, 245–6.

Lee, M.S.Y. (1992) Cambrian and Recent morphological disparity. *Science*, **258**, 1816–17.

Lee, M.S.Y. (1993) The origin of the turtle body plan: Bridging a famous morphological gap. *Science*, **261**, 1716–20.

Lee, M.S.Y. (1996) Correlated progression and the origin of turtles. *Nature*, **379**, 812–15.

Lee, S.M.K., Danielian, P.S., Fritzsch, B. and McMahon, A.P. (1997) Evidence that FGF8 signalling from the midbrain-hindbrain junction regulates growth and polarity in the developing midbrain. *Development*, **124**, 959–69.

Lehmann, R. and Ephrussi, A. (1994) Germ plasm formation and germ cell determination in *Drosophila*. *Ciba Foundation Symp.*, **182**, 282–96.

Leloup, J. (1955) Part of ammocoete endostyle becoming thyroid. *J. Physiol.*, **47**, 671–7.

Lemaire, P. (1996) The coming of age of ventralizing homeobox genes in amphibian development. *BioEssays*, **18**, 701–4.

Lemus, D. (1995) Contributions of heterospecific tissue recombinations to odontogenesis. *Int. J. Devel. Biol.*, **39**, 291–7.

Lerner, I.M. (1954) *Genetic Homeostasis*. Oliver and Boyd, Edinburgh.

Leroi, A.M., Rose, M.R. and Lauder, G.V. (1994) What does the comparative method reveal about adaptation? *Amer. Nat.*, **143**, 381–402.

Lessa, E.P. and Stein, B.R. (1992) Morphological constraints in the digging apparatus of pocket gophers (Mammalia: Geomyidae). *Biol. J. Linn. Soc.*, **47**, 439–53.

Lester, J. and Bowler, P.J. (1995) *E. Ray Lankester and the Making of Modern British Biology*, British Society for the History of Science, Stanford in the Vale, UK.

Levi, G., Crossin, K.L. and Edelman, G.M. (1987) Expression, sequences and distribution of two primary cell adhesion molecules during embryonic development of *Xenopus laevis*. *J. Cell Biol.*, **105**, 2359–72.

Levin, M. (1997) Left-right asymmetry in vertebrate embryogenesis. *BioEssays*, **19**, 287–96.

Levinton, J. (1988) *Genetics, Paleontology and Macroevolution*. Cambridge University Press, Cambridge.

Levinton, J.S., Bandel, K., Charlesworth, B. *et al.* (1986) Organismic evolution: The interaction of microevolutionary and macroevolutionary processes, in *Patterns and Processes in the History of Life* (eds D.M. Raup and D. Jablonski), Dahlem Conference 1986, Springer-Verlag, Berlin, pp. 167–82.

Levy, B.M., Detwiler, S.R. and Copenhaver, W.M. (1956) The production of developmental abnormalities of the oral structures in *Amblystoma punctatum*. *J. Dental Res.*, **36**, 659–62.

Lewis, E.B. (1978) A gene complex controlling segmentation in *Drosophila*. *Nature*, **276**, 565–70.

Lewis, E.B. (1985) Regulation of the genes of the *bithorax* complex in *Drosophila*. *Cold Spring Harb. Symp. Quant. Biol.*, **50**, 155–64.

Lewontin, R.C. (1970) The units of selection. *Annu. Rev. Ecol. Syst.*, **1**, 1–18.

Lewontin, R.C. (1974) *The Genetic Basis of Evolutionary Change*. Columbia University Press, New York.

Lewontin, R.C. (1992) Genotype and phenotype, in *Keywords in Evolutionary Biology* (eds E. Fox Keller and E.A. Lloyd), Harvard University Press, Cambridge, MA, pp. 137–44.

Li, S., Anderson, R., Reginelli, A.D. and Muneoka, K. (1996) FGF-2 influences cell movements and gene expression during limb development. *J. Exp. Zool.*, **274**, 234–47.

Libbin, R.M., Singh, I.J., Hirschman, A. and Mitchell, O.G. (1988) A prolonged cartilaginous phase in newt forelimb skeletal regeneration. *J. Exp. Zool.*, **248**, 238–42.

Libbin, R.M., Mitchell, O.G., Guerra, L. and Person, P. (1989) Delayed carpal ossification in *N. viridescens* efts: Relation to the progress of mesopodial completion in newt forelimb regenerates. *J. Exp. Zool.*, **252**, 207–11.

Liem, K.F. (1974) Evolutionary strategies and morphological innovations: Cichlid pharyngeal jaws. *Syst. Zool.*, **22**, 425–41.

Liem, K.F. (1980) Adaptive significance of intra- and interspecific differences in the feeding repertoire of cichlid fishes. *Amer. Zool.*, **20**, 295–314.

Liem, K.F. (1991a) Toward a new morphology: Pluralism in research and education. *Amer. Zool.*, **31**, 759–67.

Liem, K.F. (1991b) A functional approach to the development of the head of teleosts: Implications on constructional morphology and constraints, in *Constructional Morphology and Evolution* (eds N. Schmidt-Kittler and K. Vogel), Springer-Verlag, Berlin, pp. 231–49.

Liem, K.F. and Greenwood, P.H. (1981) A functional approach to the phylogeny of the pharyngognath teleosts. *Amer. Zool.*, **21**, 83–101.

Liem, K.F. and Wake, D.B. (1985) Morphology: Current approaches and concepts, in *Functional Vertebrate Morphology* (eds M. Hilderbrand, D.M. Bramble, K.F. Liem and D.B. Wake), Harvard University Press, Cambridge, MA, pp. 366–77.

Liem, K.F., Jr., Tremml, G., Roelink, H. and Jessell, T.M. (1995) Dorsal differentiation of neural plate cells induced by BMP-mediated signals from epidermal ectoderm. *Cell*, **82**, 969–79.

Lindberg, D.R. and Ponder, W.F. (1996) An evolutionary tree for the Mollusca: Branches or roots? in *Origin and Evolutionary Radiations of the Mollusca* (ed. J. Taylor), Oxford University Press, Oxford, pp. 67–75.

Linnaeus, C. von (1758) *Systema Naturae*, 10th edn, Stockholm.

Liu, Z. and Ambros, V. (1989) Heterochronic genes control the stage-specific initiation and expression of the dauer larva developmental program in *Caenorhabditis elegans*. *Genes Devel.*, **3**, 2039–49.

Liu, Z., Kirch, S. and Ambros, V. (1995) The *Caenorhabditis elegans* heterochronic gene pathway controls stage-specific transcription of collagen genes. *Development*, **121**, 2471–78.

Lloyd, E.A. (1992) Units of selection, in *Keywords in Evolutionary Biology* (eds E. Fox Keller and E.A. Lloyd), Harvard University Press, Cambridge, MA, pp. 334–40.

Locke, M. (1990) Is there somatic inheritance of intracellular patterns? *J. Cell Sci.*, **96**, 563–7.

Loeschcke, V. (1987) *Genetic Constraints on Adaptive Evolution*. Springer-Verlag, Berlin.

Lombard, R.E. (1991) Experiment and comprehending the evolution of function. *Amer. Zool.*, **31**, 743–56.

Long, A.D., Mullaney, S.L., Reid, L.A. *et al.* (1995) High resolution mapping of genetic factors affecting abdominal bristle number in *Drosophila melanogaster*. *Genetics*, **139**, 1273–91.

Long, C.A. (1976) Evolution of mammalian cheek pouches and a possibly discontinuous origin of a higher taxon (Geomyoidea). *Amer. Nat.*, **110**, 1093–1111.

Lovejoy, A.O. (1936) *The Great Chain of Being. A Study of the History of an Idea*. Harvard University Press, Cambridge, MA. (Reprinted 1964 *et seq.*)

Løvtrup, S. (1974) *Epigenetics – A Treatise on Theoretical Biology*. John Wiley, London.

Løvtrup, S. (1977) *The Phylogeny of the Vertebrata*, John Wiley, London.

Løvtrup, S. (1982) The four theories of evolution. *Riv. Biol.*, **75**, 53–9, 231–55, 385–98.

Løvtrup, S. (1983) Reduction and emergence. *Riv. Biol.*, **76**, 437–61.

Løvtrup, S. (1984a) Ontogeny and phylogeny, in *Beyond Neo-Darwinism* (eds M.-W. Ho. and P.T. Saunders), Academic Press, London, pp. 159–90.

Løvtrup, S. (1984b) Ontogeny and phylogeny from an epigenetic point of view. *Human Devel.*, **27**, 249–61.

Løvtrup, S. (1986) Evolution, morphogenesis, and recapitulation: An essay on metazoan evolution. *Cladistics*, **2**, 68–82.

Løvtrup, S. (1987) *Darwinism: the Refutation of a Myth*. Croom Helm, London.

Løvtrup, S. (1988) Epigenetics, in *Ontogeny and Systematics* (ed. C.J. Humphries), Columbia University Press, New York, pp. 189–227.

Lowe, C.J. and Wray, G.A. (1997) Radical alterations in the roles of homeobox genes during echinoderm evolution. *Nature*, **389**, 718–21.

Ludwig, W. (1932) *Das Rechts-Links Problem im Tierreich und beim Menschen*, Springer, Berlin.

Lufkin, T., Dierich, A., LeMeur, M. *et al.* (1991) Disruption of the *Hox*-1.6 homeobox gene results in defects in a region corresponding to its rostral domain of expression. *Cell*, **66**, 1105–19.

Lufkin, T., Mark, M., Hart, C.P. *et al.* (1992) Homeotic transformation of the occipital bones of the skull by ectopic expression of a homeobox gene. *Nature*, **359**, 839–41.

Lumsden, A. (1987) Neural crest contribution to tooth development in the mammalian embryo, in *Developmental and Evolutionary Aspects of the Neural Crest* (ed. P.F.A. Maderson), John Wiley & Sons, New York, pp. 261–300.

Lumsden, A. (1988) Spatial organization of the epithelium and the role of neural crest cells in the initiation of the mammalian tooth germ. *Development*, **103**, Suppl., 155–70.

Lumsden, A. (1991) Motorizing the spinal cord. *Cell*, **64**, 471–3.

Lumsden, A. and Graham, A. (1995) A forward role for Hedgehog. *Curr. Biol.*, **5**, 1347–50.

Lumsden, A. and Krumlauf, R. (1996) Patterning the vertebrate neuraxis. *Science*, **274**, 1109–15.

Lumsden, A. and Wilkinson, D. (1990) Developmental biology – The promise of gene ablation. *Nature*, **347**, 335–6.

Luo, G., Hofmann, C., Bronckers, A.L.J.J. *et al.* (1995) BMP-7 is an inducer of nephrogenesis and is also required for eye development and skeletal patterning. *Genes Devel.* **9**, 2808–20.

Lurie, E. (1960) *Louis Agassiz: A Life in Science*. The University of Chicago Press, Chicago, IL.

Lyell, C. (1832) *Principles of Geology*. 2 Vols., John Murray, London.

Lyon, M.F. (1993) Epigenetic inheritance in mammals. *Trends Genet.*, **9**, 123–8.

Lyons, K.M., Hogan, B.L.M, and Robertson, E.J. (1995) Colocalization of BMP 7 and BMP 2 RNAs suggest that these factors cooperatively mediate tissue interactions during murine development. *Mech. Devel.*, **50**, 71–83.

Lyons, S.L. (1995) The origins of T.H. Huxley's saltationism: History in Darwin's shadow. *J. Hist. Biol.*, **28**, 463–94.

Mabee, P.M. (1989a) Assumptions underlying the use of ontogenetic sequences for determining character state order. *Trans. Amer. Fish. Soc.*, **118**, 151–8.

Mabee, P.M. (1989b) An empirical rejection of the ontogenetic polarity criterion. *Cladistics*, **5**, 409–16.

Mabee, P.M. (1993) Phylogenetic interpretation of ontogenetic change: Sorting out the actual and artifactual in an empirical case study of centrarchid fishes. *Zool. J. Linn. Soc.*, **107**, 175–291.

Mabee, P.M. (1995) Evolution of pigment pattern development in centrarchid fishes. *Copeia*, **1995** (3), 586–607.

MacDonald, M.E. (1997) *Epithelial-mesenchymal interactions leading to mandibular bone and cartilage formation in three mouse strains*, M.Sc. Thesis, Dalhousie University, Halifax, NS.

MacDonald, P.M. and Struhl, G. (1986) A molecular gradient in early *Drosophila* embryos and its role in specifying body pattern. *Nature*, **324**, 537–45.

Macdonald, R., Xu, Q., Barth, K.A. *et al.* (1994) Regulatory gene expression boundaries demarcate sites of neuronal differentiation in the embryonic zebrafish forebrain. *Neuron*, **13**, 1039–53.

Macdonald, R., Barth, K.A., Xu, Q. *et al.* (1995) Midline signalling is required for *Pax* gene regulation and patterning of the eyes. *Development*, **121**, 3267–78.

Macias, D., Ganan, Y., Sampath, T.K. *et al.* (1997) Role of BMP-2 and OP-1 (BMP-7) in programmed cell death and skeletogenesis during chick limb development. *Development*, **124**, 1109–17.

Mackay, T.F.C. (1995) The genetic basis of quantitative variation: Numbers of sensory bristles of *Drosophila melanogaster* as a model system. *Trends Genet.*, **11**, 464–70.

Mackay, T.F.C. (1996) The nature of quantitative genetic variation revisited: Lessons from *Drosophila* bristles. *BioEssays*, **18**, 113–21.

MacKenzie, A., Ferguson, M.W.J. and Sharpe, P.T. (1992) Expression patterns of the homeobox gene, Hox-8, in the mouse embryo suggest a role in specifying tooth initiation and shape. *Development*, **115**, 403–20.

Maclean, N. and Hall, B.K. (1987) *Cell Commitment and Differentiation*. Cambridge University Press, Cambridge.

Macleay, W.S. (1821) *Horae Entomologicae: or Essays on the Annulose Animals. Volume I, Part II. Containing an Attempt to Ascertain the Rank and Situation which the Celebrated Insect, Scarabaeus sacer, Holds Among Organised Beings*, S. Bagster, London.

Maclise, J. (1846) On the nomenclature of anatomy. *Lancet*, March 14, 298–301.

Maden, M. (1993) The homeotic transformation of tails into limbs in *Rana temporaria* by retinoids. *Devel. Biol.*, **159**, 379–91.

Maderson, P.F.A. (1972) On how an archosaurian scale might have given rise to an avian feather. *Amer. Nat.*, **106**, 424–8.

Maderson, P.F.A. (1975) Embryonic tissue interactions as the basis for morphological change in evolution. *Amer. Zool.*, **15**, 315–27.

Maderson, P.F.A. (1983) An evolutionary view of epithelial-mesenchymal interactions, in *Epithelial-Mesenchymal Interactions in Development* (eds R.H. Sawyer and J.F. Fallon), Praeger Press, New York, pp. 215–42.

Maderson, P.F.A. (1987) *Developmental and Evolutionary Aspects of the Neural Crest*, John Wiley & Sons, New York.

Madigan, M.T. and Marrs, B.L. (1997) Extremophiles. *Sci. Amer.*, **276** (4), 82–7.

Maienschein, J. (1978) Cell lineage, ancestral reminiscence, and the Biogenetic Law. *J. Hist. Biol.*, **11**, 129–58.

Maienschein, J. (1981) Shifting assumptions in American biology: Embryology. *J. Hist. Biol.*, **14**, 89–113.

Maienschein, J. (1994) 'It's a long way from Amphioxus.' Anton Dohrn and late nineteenth century debates about vertebrate origins. *Hist. Phil. Life Sci.*, **16**, 465–78.

Maier, W. (1990) Phylogeny and ontogeny of mammalian middle ear structures. *Neth. J. Zool.*, **40**, 55–74.

Maisey, J.G. (1986) Heads and tails: A chordate phylogeny. *Cladistics*, **2**, 201–56.

Maisey, J.G. (1988) Phylogeny of early vertebrate skeletal induction and ossification patterns. *Evolutionary Biology*, **22**, 1–36.

Malicki, J., Schugart, K. and McGinnis, W. (1990) Mouse *Hox-2.2* specifies thoracic segmental identity in *Drosophila* embryos and larvae. *Cell*, **63**, 961–7.

Mallatt, J. (1996) Ventilation and the origin of jawed vertebrates: A new mouth. *Zool. J. Linn. Soc.*, **117**, 329–404.

Mancilla, A. and Mayor, R. (1996) Neural crest formation in *Xenopus laevis*: Mechanisms of *Xslug* induction. *Devel. Biol.*, **177**, 580–9.

Maniotis, A.J., Chen, C.S. and Ingber, D.E. (1997) Demonstration of mechanical connections between integrins, cytoskeletal filaments, and nucleoplasm that stabilize nuclear structure. *Proc. Natl Acad. Sci. USA*, **94**, 849–54.

Maret, T.J. and Collins, J.P. (1997) Ecological origin of morphological diversity: a study of alternative trophic phenotypes in larval salamanders. *Evolution*, **51**, 898–905.

Margulis, L. and Schwartz, K.V. (1988) *Five Kingdoms. An Illustrated Guide to the Phyla of Life on Earth* 2nd edn, W.H. Freeman and Co., New York.

Markow, T.A. (1994) *Developmental Instability: Its Origins and Evolutionary Implications*, Kluwer Academic Publishers, Dordrecht.

Marsh, O.C. (1892) Recent polydactyle horses. *Amer. J. Sci.*, **43**, 23–355.

Marshall, C.R., Raff, E.C. and Raff, R.A. (1994) Dollo's law and the death and resurrection of genes. *Proc. Natl Acad. Sci. USA*, **91**, 12283–87.

Martinez, I., Alvarez, R., Herraes, I. and Herraes, P. (1992) Skeletal malformations in hatchery reared *Rana pereri* tadpoles. *Anat. Rec.*, **233**, 314–20.

Mathews, W.W. (1986) *Atlas of Descriptive Embryology* 4th edn, Macmillan Publishing Co., New York.

Matsuda, R. (1982) The evolutionary process in Talitrid amphipods and salamanders in changing environments, with a discussion of "genetic assimilation" and some other evolutionary concepts. *Can. J. Zool.*, **60**, 733–49.

Matsuda, R. (1987) *Animal Evolution in Changing Environments with Special Reference to Abnormal Metamorphosis*. John Wiley & Sons, New York.

Matsui, K. (1949) Illustration of the normal course of development in the fish *Oryzias latipes*. *Zikken-keitaigaku*, **5**, 32–42.

Matzuk, M.M., Kumar, T.R., Vassali, A. *et al.* (1995) Functional analysis of activins during mammalian development. *Nature*, **375**, 354–6.

Maunz, M. and German, R.Z. (1997) Ontogeny and limb bone scaling in two New World marsupials, *Monodelphis domestica* and *Didelphis virginiana*. *J. Morph.*, **231**, 117–30.

Maynard Smith, J. (1958) *The Theory of Evolution*, Penguin Books, Harmondsworth, Middlesex.

Maynard Smith, J. (1983) Evolution and development, in *Development and Evolution* (eds B.C. Goodwin, N. Holder and C.C. Wylie), British Society for Developmental Biology Symposium 6, Cambridge University Press, Cambridge, pp. 33–45.

Maynard Smith, J. (1989a) *Evolutionary Genetics*. Oxford University Press, Oxford.

Maynard Smith, J. (1989b) Weismann and modern biology, in *Oxford Surveys in Evolutionary Biology*, Volume 6 (eds P.H. Harvey and L. Partridge), Oxford University Press, Oxford, pp. 1–12.

Maynard Smith, J. (1990) Models of a dual inheritance system. *J. Theor. Biol.*, **143**, 41–53.

Maynard Smith, J., Burian, R., Kauffman, S. *et al.* (1985) Developmental constraints and evolution. *Quart. Rev. Biol.*, **60**, 265–87.

Maynard Smith J. and Szathmáry, E. (1995) The *Major Transitions in Evolution*. Freeman Press, Oxford.

Mayo, O. (1983) *Natural Selection and Its Constraints*, Academic Press, London.

Mayor, R., Guerrero, N. and Martínez, C. (1997) Role of FGF and *Noggin* in neural crest induction. *Devel. Biol.*, **189**, 1–12.

Mayor, R., Morgan, R. and Sargent, M.G. (1995) Induction of the prospective neural crest of *Xenopus*. *Development*, **121**, 767–77.

Mayr, E. (1942) *Systematics and the Origin of Species*, Columbia University Press, New York.

Mayr, E. (1954) Change of genetic environment and evolution, in *Evolution as a Process* (eds J. Huxley, A.C. Hardy and E.B. Ford), George Allen & Unwin Ltd., London, pp. 157–80.

Mayr, E. (1959) Agassiz, Darwin, and evolution. *Harvard Libr. Bull.*, **13**, 165–94.

Mayr, E. (1960) The emergence of evolutionary novelties, in *Evolution after Darwin. 1. The Evolution of Life: Its Origin, History, and Future* (ed. S. Tax), University of Chicago Press, Chicago, IL, pp. 349–80,

Mayr, E. (1961) Cause and effect in biology. *Science*, **134**, 1501–6.

Mayr, E. (1963) *Animal Species and Evolution*, Belknap Press of Harvard University, Cambridge, MA.

Mayr, E. (1980) Prologue: Some Thoughts on the History of the Evolutionary Synthesis, in *The Evolutionary Synthesis. Perspectives on the Unification of Biology* (eds E. Mayr and W.B. Provine), Harvard University Press, Cambridge and London, pp. 1–48.

Mayr, E. (1982) *The Growth of Biological Thought. Diversity, Evolution, and Inheritance.* The Belknap Press of Harvard University Press, Cambridge, MA.

Mayr, E. (1994) Recapitulation reinterpreted: The somatic program. *Quart. Rev. Biol.*, **69**, 223–32.

Mayr, E. (1997a) The establishment of evolutionary biology as a discrete biological discipline. *BioEssays*, **19**, 263–6.

Mayr, E. (1997b) Review of *The Origin of Birds* by A. Feduccia. *Amer. Zool.*, **37**, 210–11.

Mayr, E., and Provine, W.B. (eds) (1980) *The Evolutionary Synthesis: Perspectives on the Unification of Biology*, Harvard University Press, Cambridge, MA.

Maze, J., Banerjee, S., Elkassab, Ya. and Bohm, L.R. (1992) A quantitative genetic analysis of morphological integration in Douglas-fir. *Int. J. Plant Sci.*, **153**, 333–40.

Maze, J. and Vyse, A. (1993) An analysis of growth, growth increments, and the integration of growth increments in a fertilizer test of *Picea engelmannii* in south-central British Columbia. *Can. J. Bot.*, **71**, 1449–57.

McAleese, S.R. and Sawyer, R.H. (1982) Avian scale development. IX. Scale formation by scaleless (*sc/sc*) epidermis under the influence of normal scale dermis. *Devel. Biol.*, **89**, 493–502.

McDowell, N., Zorn, A.M., Crease, D.J., and Gurdon, J.B. (1997) Activin has direct long-range signalling activity and can form a concentration gradient by diffusion. *Curr. Biol.*, **7**, 671–81.

McEdward, L.R. (1996) Experimental manipulation of parental investment in echinoid echinoderms. *Amer. Zool.*, **36**, 169–79.

McEdward, L.R. and Janies, D.A. (1997) Relationships among development, ecology, and morphology in the evolution of echinoderm larvae and life cycles. *Biol. J. Linn. Soc.*, **60**, 381–400.

McGhee, G.R.Jr. (1980) Shell form in the biconvex articulate Brachiopoda: A geometric analysis. *Paleobiology*, **6**, 57–76.

McKenzie, J.A. and Batterham, P. (1994) The genetic, molecular and phenotypic consequences of selection for insecticide resistance. *Trends Ecol. Evol.*, **9**, 166–9.

McKenzie, J.A. and Yen, J.L. (1995) Genotype, environment and the asymmetry phenotype. Dieldrin-resistance in *Lucilia cuprina* (the Australian sheep blowfly). *Heredity*, **75**, 181–7.

McKinney, M.L. (1986) Ecological causation of heterochrony: A test and implications for evolutionary theory. *Paleobiology*, **12**, 282–9.

McKinney, M.L. (1988) *Heterochrony in Evolution. A multidisciplinary approach*. Plenum Press, New York and London.

McKinney, M.L. and McNamara, K.J. (1991) *Heterochrony: The Evolution of Ontogeny*, Plenum Press, New York.

McKitrick, M.C. (1986) Individual variation in the flexor cruris lateralis muscle of the Tyrannidae (Aves: Passeriformes) and its possible significance. *J. Zool. Lond.*, **209**, 251–70.

McKitrick, M.C. (1991) Phylogenetic analysis of avian hindlimb musculature. *Misc. Publ. Mus. Zool. Univ. Michigan* #**179**, 1–85.

McKitrick, M.C. (1993) Phylogenetic constraint in evolutionary theory: Has it any explanatory power? *Annu. Rev. Ecol. Syst.*, **24**, 307–30.

McKitrick, M.C. (1994) On homology and the ontological relationship of parts. *Syst. Biol.*, **43**, 1–10.

McLaren, I.A. (1965) Temperature and frogs eggs. A reconsideration of metabolic control. *J. Genet. Physiol.*, **48**, 1071–9.

McMenamin, M.A.S. (1996) Ediacaran biota from Sonora, Mexico. *Proc. Natl Acad. Sci. USA*, **93**, 4990–3.

McNamara, K.J. (1982) Heterochrony and phylogenetic trends. *Paleobiology*, **8**, 130–42.

McNamara, K.J. (1988) The abundance of heterochrony in the fossil record, in *Heterochrony in Evolution. A multidisciplinary approach* (ed. M.L. McKinney), Plenum Press, New York and London, pp. 287–325.

McNamara, K.J. (1990) *Evolutionary Trends*, The University of Arizona Press, Tucson, AZ.

McNamara, K.J. (1993) Inside evolution: 1992 presidential address. *J. Roy. Soc. W. Aust.*, **76**, 3–12.

McNamara, K.J. (ed.) (1995) *Evolutionary Change and Heterochrony*, John Wiley & Sons, Chichester.

McShea, D.W. (1991) Complexity and evolution: What everybody knows. *Biol. Philos.*, **6**, 303–24.

McShea, D.W. (1993) Arguments, tests, and the Burgess Shale – a commentary on the debate. *Paleobiology*, **19**, 399–402.

McShea, D.W. (1996) Metazoan complexity and evolution: Is there a trend? *Evolution*, **50**, 477–92.

Meckel, J.F. (1811) *Entwurf einer Darstellung der zwischen dem Embryozustande der höheren Tiere und dem permanenten der niederen stattfindenen Parallele: Beyträge zur vergleichenden Anatomie*, Volume 2, Carl Heinrich Reclam, Leipzig.

Meckel, J.F. (1821) *System der Vergleichenden Anatomie*. 7 volumes, Rengerschen Buchhandlung, Halle.

Medawar, P.B. (1954) The significance of inductive relationships in the development of vertebrates. *J. Embryol. exp. Morph.*, **2**, 172–4.

Medawar, P.B. (1986) *Memoirs of a Thinking Radish. An Autobiography*. Oxford University Press, Oxford.

Medawar, P.B. and Medawar, J.S. (1983) *Aristotle to Zoos. A Philosophical Dictionary of Biology*, Harvard University Press, Cambridge, MA.

Medvedeva, I.M. (1986a) On the origin of nasolacrimal duct in Tetrapoda, in Studies in Herpetology (ed. Z. Rocek), Charles University, Prague, pp. 37–40.

Medvedeva, I.M. (1986b) Nasolachrymal canal of Ambystomatidae in the light of its origin in other terrestrial vertebrates, in *Morphology and Evolution of Animals* (eds E.I. Vorobyeva and N.S. Lebedkina), Nauka, Moscow, pp. 138–56.

Meier, S. and Packard, D.S.Jr. (1984) Morphogenesis of the cranial segments and distribution of neural crest in the embryo of the snapping turtle, *Chelydra serpentina*. *Devel. Biol.*, **102**, 309–23.

Meinke, D.K. (1982a) A light and scanning electron microscopic study of microstructure, growth and development of the dermal skeleton of *Polypterus* (Pisces: Actinopterygii). *J. Zool. Lond.*, **187**, 355–82.

Meinke, D.K. (1982b) A histological and histochemical study of developing teeth in *Polypterus* (Pisces: Actinopterygii). *Arch. Oral Biol.*, **27**, 197–206.

Meinke, D.K. (1984) A review of cosmine: Its structure, development and relationship to other forms of the dermal skeleton in osteichthyans. *J. Vert. Paleont.*, **4**, 457–70.

Meinke, D.K. (1986) Morphology and evolution of the dermal skeleton in lungfishes. *J. Morph.*, Suppl. 1, 113–49.

Meinke, D.K. and Thomson, K.S. (1983) The distribution and significance of enamel and enameloid in the dermal skeleton of osteolepiform rhipidistian fishes. *Paleobiology*, **9**, 138–49.

Mes-Hartree, M. and Armstrong, J.B. (1980) Evidence that the premature death mutation (*p*) in the Mexican axolotl (*Ambystoma mexicanum*) is not an autonomous cell lethal. *J. Embryol. exp. Morph.*, **60**, 295–302.

Metschnikoff, E. (1869) Uber ein Larvenstadium von *Euphasia*. *Z. Wiss. Zool.*, **19**, 479–81.

Metschnikoff, E. (1876) Beiträge zur Morphologie der Spongien. *Z. Wiss. Zool.*, **27**, 275–86.

Meyer, A. (1987) Phenotypic plasticity and heterochrony in *Cichlasoma managuense* (Pisces, Cichlidae) and their implications for speciation in cichlid fishes. *Evolution*, **41**, 1357–9.

Meyer, A. (1990) Ecological and evolutionary consequences of the trophic polymorphism in *Cichlasoma citrinellum* (Pisces: Cichlidae). *Biol. J. Linn. Soc.*, **39**, 279–99.

Meyer, A. (1996) The evolution of body plans: HOM/Hox cluster evolution, model systems, and the importance of phylogeny, in *New Uses for New Phylogenies* (eds P.H. Harvey, A.J.L. Brown, J. Maynard Smith and S. Nee), Oxford University Press, Oxford, pp. 322–40.

Meyer, A., Biermann, C.H. and Ortí, G. (1993) The phylogenetic position of the zebrafish (*Danio rerio*), a model system in developmental biology: An invitation to the comparative method. *Proc. R. Soc. Lond. B*, **252**, 231–6.

Meyer, A., Kocher, T.D., Basasibwaki, P. and Wilson, A.C. (1990) Monophyletic origin of Lake Victoria cichlid fishes suggested by mitochondrial DNA sequences. *Nature*, **347**, 550–3.

Meyer, A., Ritchie, P.A. and Witte, K.-E. (1995) Predicting developmental processes from evolutionary patterns: A molecular phylogeny of the zebrafish (*Danio rerio*) and its relatives. *Phil. Trans. R. Soc. Lond. B*, **349**, 103–11.

Meyer, A.W. (1939) *The Rise of Embryology*. Oxford University Press, London and Oxford.

Michaud, J.L., Lapointe, F. and Le Douarin, N.M. (1997) The dorsoventral polarity of the presumptive limb is determined by signals produced by the somites and by the lateral somatopleure. *Development*, **124**, 1453–63.

Michod, R.E. (1997) Cooperation and conflict in the evolution of individuality. I. Multilevel selection of the organism. *Amer. Nat.*, **149**, 607–45.

Mikhailov, A.T. and Gorgolyuk, N.A. (1988) Concanavalin A induces neural tissue and cartilage in amphibian early gastrula ectoderm. *Cell Differ.*, **22**, 145–54.

Milinkovitch, M.C. (1995) Molecular phylogeny of cetaceans prompts revision of morphological transformations. *Trends Ecol. Evol.*, **10**, 328–34.

Milinkovitch, M.C., Orti, G. and Meyer, A. (1993) Revised phylogeny of whales suggested by mitochondrial ribosomal DNA sequences. *Nature*, **361**, 346–8.

Milinkovitch, M.C. and Thewissen, J.G.M. (1997) Even-toed fingerprints on whale ancestry. *Nature*, **388**, 622–4.

Milkman, R.D. (1962) The genetic basis of natural variation. IV. On the natural distribution of *eve* polygenes of *Drosophila melanogaster*. *Genetics*, **47**, 261–72.

Miller, A.H. (1949) Some ecological and morphological considerations in the evolution of higher taxonomic categories, in *Ornithologie als Biologische Wissenschaft* (eds E. Mayr and E. Schuzs), Carl Winter, Heidelberg, pp. 84–8.

Miller, B.B. (1978) Non-marine molluscs in quaternary paleoecology. *Malacol. Rev.*, **11**, 27–38.

Miller, D.J. and Miles, A. (1993) Homeobox genes and the zootype. *Nature*, **365**, 215–16.

Millhauser, M. (1959) *Just before Darwin: Robert Chambers and 'Vestiges'*. Wesleyan University Press, Middletown, CT.

Milne-Edwards, H. (1853) *Introduction à la zoologie générale, ou considérations sur les tendances de la nature dans la constitution du règne animal*. Paris.

Minelli, A. (1996a) Segments, body regions, and the control of development through time. *Mem. Calif. Acad. Sci.*, number 20, pp. 55–61.

Minelli, A. (1996b) Some thoughts on homology 150 years after Owen's definition. *Mem. Soc. Ital. Sci. Nat. Mus. Civ. Storia Nat. Milan*, **27**, 71–79.

Minelli, A., Negrisolo, E. and Fusco, G. (1996) Developmental trends in the post-embryonic development of lithobiomorph centipedes. *Mém. Mus. Natn. Hist. nat.*, **169**, 351–8.

Minelli, A. and Peruffo, B. (1991) Developmental pathways, homology and homonomy in metameric animals. *J. Evol. Biol.*, **3**, 429–45.

Minelli, A. and Schram, F.R. (1994) Owen revisited: A reappraisal of morphology in evolutionary biology. *Bijdragen Dierkunde*, **64**, 65–74.

Misof, B.Y., Blanco, M.J. and Wagner, G.P. (1996) PCR-survey of Hox-genes of the zebrafish: New sequence information and evolutionary implications. *J. Exp. Zool.*, **274**, 193–206.

Misof, B.Y. and Wagner, G.P. (1996) Evidence for four *Hox* clusters in the killifish *Fundulus heteroclitus* (Teleostei). *Mol. Phylog. Evol.*, **5**, 309–22.

Mitari, S. and Okamoto, H. (1991) Inductive differentiation of two neural lineages reconstituted in a microculture system from *Xenopus* early gastrula cells. *Development*, **112**, 21–31.

Mitrani, E., Ziv, T., Thomsen, G. *et al.* (1990) Activin can induce the formation of axial structures and is expressed in the hypoblast of the chick. *Cell*, **63**, 495–501.

Mitsiadis, T.A., Henrique, D., Thesleff, I. and Lendahl, U. (1997) Mouse *Serrate-1* (*Jagged-1*): Expression in the developing tooth is regulated by epithelial-mesenchymal interactions and fibroblast growth factor-4. *Development*, **124**, 1473–83.

Mitsiadis, T.A., Lardelli, M., Lendahl, U. and Thesleff, I. (1995) Expression of *Notch 1, 2,* and *3* is regulated by epithelial-mesenchymal interactions and retinoic acid in the developing mouse tooth and associated with determination of ameloblast cell fate. *J. Cell. Biol.*, **130**, 407–18.

Mivart, St.G. (1879) On fins of elasmobranchs. *Trans. Zool. Soc. Lond.*, **10**, 1–76.

Miya, T., Morita, K., Ueno, N. and Satoh, N. (1996) An ascidian homologue of vertebrate BMPs-5–8 is expressed in the midline of the anterior neuroectoderm and in the midline of the ventral epidermis of the embryo. *Mech. Devel.*, **57**, 181–90.

Miyake, T., Cameron, A.C. and Hall, B.K. (1993) Detailed timing of onset of the mandibular skeleton in inbred C57BL/6 mice. *Molec. Biol. Cell*, **4**, 145a.

Miyake, T., Cameron, A.M. and Hall, B.K. (1996a) Detailed staging of inbred C57BL/6 mice between Theiler's [1972] stages 18 and 21 (11–13 days of gestation) based on craniofacial development. *J. Craniofac. Genet. Devel. Biol.*, **16**, 1–31.

Miyake, T., Cameron, A.M. and Hall, B.K. (1996b) Stage-specific onset of condensation and matrix deposition for Meckel's and other first arch cartilages in inbred C57BL/6 mice. *J. Craniofac. Genet. Devel. Biol.*, **16**, 32–47.

Miyake, T., Cameron, A.M. and Hall, B.K. (1997a) Stage-specific expression patterns of alkaline phosphatase during development of the first arch skeleton in inbred C57BL/6 mouse embryos. *J. Anat.*, **190**, 239–260.

Miyake, T., Cameron, A.M. and Hall, B.K. (1997b) Variability and constancy of embryonic development in and among three inbred strains of mice. *Growth, Devel. Aging* (in press).

Miyake, T., McEachran, J.D. and Hall, B.K. (1992a) Edgeworth's legacy of cranial muscle development with an analysis of muscles in the ventral gill arch region of batoid fishes (Chondrichthyes: Batoidea). *J. Morph.*, **212**, 213–56.

Miyake, T., McEachran, J.D., Walton, P.J. and Hall, B.K. (1992b) Development and morphology of rostral cartilages in batoid fishes (Chondrichthyes: Batoidea), with comments on homology within vertebrates. *Biol. J. Linn. Soc.*, **46**, 259–96.

Mo, R., Freer, A.M., Zinyk, D.L. *et al.* (1997) Specific and redundant functions of *Gli2* and *Gli3* zinc finger genes in skeletal patterning and development. *Development*, **124**, 113–23.

Model, P.G., Jarrett, L.S. and Bonazzoli, R. (1981) Cellular contacts between hind-brain and prospective ear during inductive interactions in the axolotl embryo. *J. Embryol. exp. Morph.*, **66**, 27–41.

Mohanty-Hejmadi, P., Dutta, S.K. and Mahapatra, P. (1992) Limbs generated at site of tail amputation in marbled balloon frogs after vitamin A treatment. *Nature*, **355**, 352–3.

Molven, A., Wright, C.V.E., Bremiller, R. *et al.* (1990) Expression of a homeobox gene product in normal and mutant zebrafish embryos: Evolution of the tetrapod body plan. *Development*, **109**, 279–88.

Monge-Najera, J. (1995) Phylogeny, biogeography and reproductive trends in the Onychophora. *Zool. J. Linn. Soc.*, **114**, 21–60.

Monk, M. and Surani, A. (1990) Genomic imprinting. *Development*, 1990 Suppl., 1–155.

Monsoro-Burq, A.-H., Duprez, D., Watanabe, Y. *et al.* (1996) The role of bone morphogenetic proteins in vertebral development. *Development*, **122**, 3607–16.

Montgomery, M.K. and McFall-Ngai, M.J. (1995) The inductive role of bacterial symbionts in the morphogenesis of a squid light organ. *Amer. Zool.*, **35**, 372–80.

Moore, J.A. (1993) *Science as a Way of Knowing: The Foundations of Modern Biology*, Harvard University Press, Cambridge, MA.

Moore, J. and Willmer, P. (1997) Convergent evolution in invertebrates. *Biol. Rev.*, **72**, 1–60.

Morell, V. (1997a) The origin of birds: The dinosaur debate. *Audubon*, **99** (2), 36–45.

Morell, V. (1997b) Dino debate rages anew. *Audubon*, **99** (5), 23.

Morris, P.J. (1993) The developmental role of the extracellular matrix suggests a monophyletic origin of the kingdom Animalia. *Evolution*, **47**, 152–65.

Morton, J.E. (1958) *Molluscs*. Hutchinson & Co., London.

Moury, J.D. and Hanken, J. (1995) Early cranial neural crest migration in the direct-developing frog, *Eleutherodactylus coqui. Acta Anat.*, **153**, 243–53.

Moury, J.D. and Jacobson, A. (1990) The origins of neural crest cells in the axolotl. *Devel. Biol.*, **141**, 243–53.

Müller, F. (1864) *Für Darwin*, Engelmann, Leipzig.

Müller, F. (1869) *Facts and Arguments for Darwin*, with additions by the author, translated from the German by W.S. Dallas, John Murray, London.

Müller, G.B. (1986) Effects of skeletal change on muscle pattern formation. *Bibliotheca Anat.*, **29**, 91–108.

Müller, G.B. (1989) Ancestral patterns in bird limb development: a new look at Hampé's experiment. *J. Evol. Biol.*, **2**, 31–48.

Müller, G.B. (1990) Developmental mechanisms at the origin of morphological novelty: A side-effect hypothesis, in *Evolutionary Innovations* (ed. M.H. Nitecki), The University of Chicago Press, Chicago, IL, pp. 99–130.

Müller, G.B. (1991) Experimental strategies in evolutionary embryology. *Amer. Zool.*, **31**, 605–15.

Müller, G.B. (1994) Evolutionäre Entwicklungsbiologie: Grundlagen einer neuen Synthese, in *Die Evolution der Evolutionstheorie: Von Darwin zur DNA* (ed. W. Wieser), Spektrum Akadaemischer Verlag, Heidelberg, pp. 155–93.

Müller, G.B. and Streicher, J. (1989) Ontogeny of the syndesmosis tibiofibularis and the evolution of the bird hindlimb: A caenogenetic feature triggers phenotypic novelty. *Anat. Embryol.*, **179**, 327–39.

Müller, G.B., Streicher, J. and Müller, R.J. (1996) Homeotic duplication of the pelvic body segment in regenerating tadpole tails induced by retinoic acid. *Devel. Genet. Evol.*, **206**, 344–8.

Müller, G.B. and Wagner, G.P. (1991) Novelty in evolution: Restructuring the concept. *Annu. Rev. Ecol. Syst.*, **22**, 229–56.

Müller, G.B. and Wagner, G.P. (1996) Homology, *Hox* genes, and developmental integration. *Amer. Zool.*, **36**, 4–13.

Müller, G.B., Wagner, G.P. and Hall, B.K. (1989) Experimental vertebrate embryology and the study of evolution: Report of a Workshop, in *Trends in Vertebrate Morphology* (eds H. Splechna and H. Hilgers) Gustav Fischer Verlag, Stuttgart, pp. 299–303.

Müller, M.M., Carrasco, A.E. and de Robertis, E.M. (1984) A homeobox-containing gene expressed during oogenesis in *Xenopus. Cell*, **39**, 157–62.

Müller, M., Weizsäcker, E.V. and Campos-Ortega, J.A. (1996) Expression domains of a zebrafish homologue of the *Drosophila* pair-rule gene *hairy* correspond to primordia of alternating somites. *Development*, **122**, 2071–8.

Murray, J.D., Deeming, D.C. and Ferguson, M.W.J. (1990) Size-dependent pigmentation-pattern formation in embryos of *Alligator mississippiensis*: Time of initiation of pattern generation mechanisms. *Proc. R. Soc. Lond.* B, **239**, 279–93.

Murray, P.D.F. (1936) *Bones. A study of the development and structure of the vertebrate skeleton.* Cambridge University Press, Cambridge. (Reissued 1985 with a new introduction by B.K. Hall).

Murray, P.D.F. (1957) Cartilage and bone: A problem in tissue differentiation. *Aust. J. Sci.*, **19**, 65–73.

Murray, P.D.F. (1963) Adventitious (secondary) cartilage in the chick embryo, and the development of certain bones and articulations in the chick skull. *Aust. J. Zool.*, **11**, 368–430.

Murtha, M.T., Leckman, J.F. and Ruddle, F.H. (1991) Detection of homeobox genes in development and evolution. *Proc. Natl Acad. Sci. USA*, **88**, 10711–15.

Nagy, L.M. and Carroll, S. (1994) Conservation of *wingless* patterning functions in the short-germ embryos of *Tribolium castaneum. Nature*, **367**, 460–4.

Nakai, Y., Kubota, S. and Kohno, S. (1991) Chromatin diminution and chromosome elimination in four Japanese hagfish species. *Cytogenet. Cell Genet.*, **56**, 196–8.

Nagata, S. (1997) Apoptosis by death factor. *Cell*, **88**, 355–65.

Nakatani, Y. and Nishida, H. (1994) Induction of notochord during ascidian embryogenesis. *Devel. Biol.*, **166**, 289–99.

Nakatani, Y., Yasuo, H., Satoh, N. and Nishida, H. (1996) Basic fibroblast growth factor induces notochord formation and the expression of *As-T*, a *Brachyury* homolog, during ascidian embryogenesis. *Development*, **122**, 2023–31.

Nardonne, G.M., Saylor, B.Z. and Grotzinger, J.P. (1997) The youngest Ediacaran fossils from Southern Africa. *J. Paleont.*, **71**, 953–67.

Nathanson, M.A. (1989) Differentiation of musculoskeletal tissues. *Int. Rev. Cytol.*, **116**, 89–164.

Naylor, G.J.P. and Brown, W. M. (1997) Structural biology and phylogenetic estimation. *Nature*, **388**, 527–8.

Needham, J. (1933) On the dissociability of the fundamental processes in ontogenesis. *Biol. Rev.*, **8**, 180–223.

Needham, J. (1942) *Biochemistry and Morphogenesis*, Cambridge University Press, Cambridge.

Needham, J. (1959) *A History of Embryology*. (2nd edn), Cambridge University Press, Cambridge.

Neill, W.E. (1992) Population variation in the ontogeny of predator-induced vertical migration of copepods. *Nature*, **356**, 54–7.

Nelson, C.E., Morgan, B.A., Burke, A.C. *et al.* (1996) Analysis of *Hox* gene expression in the chick limb bud. *Development*, **122**, 1449–66.

Nelson, G.J. (1973) The higher-level phylogeny of vertebrates. *Syst. Zool.*, **22**, 87–91.

Nelson, G.J. (1978) Ontogeny, phylogeny, paleontology, and the biogenetic law. *Syst. Zool.*, **27**, 324–45.

Neufeld, D.A. (1985) Bone healing after amputation of mouse digits and newt limbs: Implications for induced regeneration in mammals. *Anat. Rec.*, **211**, 156–65.

Neuhass, S.C.F., Solnica-Krezel, L., Schier, A.F. *et al.* (1996) Mutations affecting craniofacial development in zebrafish. *Development*, **123**, 357–67.

Newell, N.D. (1949) Phyletic size increase – an important trend illustrated by fossil invertebrates. *Evolution*, **3**, 103–24.

Newgreen, D. (ed.) (1995) Epithelial-mesenchymal transitions, Part I. *Acta Anat.*, **154**, 1–97.

Newgreen, D.F. (1985) Control of the timing of commencement of migration of embryonic neural crest cells. *Exp. Biol. Med.*, **10**, 209–21.

Newman, C.S., Grow, M.W., Cleaver, O. *et al.* (1997) *Xbap*, a vertebrate gene related to *bagpipe*, is expressed in developing craniofacial structures and in anterior gut muscle. *Devel. Biol.*, **181**, 223–33.

Newman, S.A. (1996) Sticky fingers: *Hox* genes and cell adhesion in vertebrate limb development. *BioEssays*, **18**, 171–4.

Neyfakh, A.A. and Hartl, D.L. (1993) Genetic control of the rate of embryonic development: Selection for faster development at elevated temperatures. *Evolution*, **47**, 1625–31.

Nielsen, C. (1995) *Animal Evolution: Interrelationships of the Living Phyla*. Oxford University Press, Oxford.

Nieuwkoop, P.D. (1977) Origin and establishment of embryonic polar axes in amphibian development. *Curr. Top. Devel. Biol.*, **11**, 115–32.

Nieuwkoop, P.D. (1987) Gerhart, J.C. The epigenetic nature of vertebrate development: An interview of Pieter D. Nieuwkoop on the occasion of his 70th birthday. *Development*, **101**, 653–7.

Nieuwkoop, P.D. (1992) The formation of the mesoderm in urodele amphibians. VI. The self-organizing capacity of the induced meso-endoderm. *Roux's Arch. Devel. Biol.*, **201**, 18–29.

Nieuwkoop, P.D. and Albers, B. (1990) The role of competence in the cranio-caudal segregation of the central nervous system. *Devel. Growth Differ.*, **32**, 23–31.

Nieuwkoop, P.D., Johnen, A.G. and Albers, B. (1985) *The Epigenetic Nature of Early Chordate Development. Inductive Interaction and Competence.* Cambridge University Press, Cambridge.

Nieuwkoop, P.D. and Sutasurya, L.A. (1979) *Primordial Germ Cells in the Chordates: Embryogenesis and Phylogenesis*. Cambridge University Press, Cambridge.

Nijhout, H.F. (1990) Metaphors and the role of genes in development. *BioEssays*, **12**, 441–6.

Nijhout, H.F. (1991) *The Development and Evolution of Butterfly Wing Patterns*, Smithsonian Institution Press, Washington.

Nijhout, H.F. (1996) Focus on butterfly eyespot development. *Nature*, **384**, 209–10.

Nijhout, H.F. (1998) Hormonal control of larval development and evolution – Insects, in *Larval Development and Evolution* (eds B.K. Hall and M.H. Wake), Academic Press, San Diego, CA (in press).

Nijhout, H.F. and Paulsen, S.M. (1997) Developmental models and polygenic characters. *Amer. Nat.*, **149**, 394–405.

Nijhout, H.F. and Wheeler, D.E. (1996) Growth models of complex allometries in holometabolous insects. *Amer. Nat.*, **148**, 40–56.

Nijhout, H.F., Wray, G.A., Kremen, C. and Teragawa, C.K. (1986) Ontogeny, phylogeny and evolution of form: An algorithmic approach. *Syst. Zool.*, **35**, 445–57.

Niklas, K.J. (1997) *The Evolutionary Biology of Plants*, The University of Chicago Press, Chicago, IL.

Niklas, K.J. and Kaplan, D.R. (1991) Biomechanics and the adaptive significance of multicellularity in plants, in *The Unity of Evolutionary Biology* (ed. E.C. Dudley), Dioscorides Press, Portland, OR, pp. 489–502.

Nilsson, D.-E. and Pelger, S. (1994) A pessimistic estimate of the time required for an eye to evolve. *Proc. R. Soc. Lond. B*, **256**, 53–8.

Nishida, H. (1991) Induction of brain and sensory pigment cells in the ascidian embryo analyzed by experiments with isolated blastomeres. *Development*, **112**, 389–95.

Nishida, H. (1994) Localization of determinants for formation of the anterior-posterior axis in eggs of the ascidian *Halocynthia roretzi*. *Development*, **120**, 3093–104.

Nishida, H. and Satoh, N. (1983) Cell lineage analysis in ascidian embryos by intracellular injection of a tracer enzyme. I. Up to the eight-cell stage. *Devel. Biol.*, **99**, 382–94.

Niswander, L. (1997) Limb mutants: what can they tell us about normal limb development? *Curr. Opin. Genet. Devel.*, **7**, 530–6.

Niswander, L., Jeffrey, S., Martin, G.R. and Tickle, C. (1994) A positive feedback loop coordinates growth and patterning in the vertebrate limb. *Nature*, **371**, 609–12.

Nitecki, M.H. (1990) *Evolutionary Innovations*. The University of Chicago Press, Chicago and London.

Noack, K., Zardoya, R. and Meyer, A. (1996) The complete mitochondrial DNA sequence of the Bichir (*Polypterus ornatipinnis*), a basal ray-finned fish: Ancient establishment of the consensus vertebrate gene order. *Genetics*, **144**, 1165–80.

Noden, D.M. (1983) The role of the neural crest in patterning of avian cranial skeletal, connective and muscle tissues. *Devel. Biol.*, **96**, 144–65.

Nomaka, K., Sasaki, Y., Yamagita, K.-i. *et al.* (1993) Intrauterine effect of dam on prenatal development of craniofacial complex of mouse embryo. *J. Craniofac. Genet. Devel. Biol.*, **13**, 206–12.

Nørrevang, A. (1970) On the embryology of *Siboglinum* and its implications for the systematic position of the Pogonophora. *Sarsia*, **42**, 7–16.

Nørrevang, A. (ed.) (1975) *The Phylogeny and Systematic Position of Pogonophora*. Proceedings of a Symposium held at Zoological Central Institute, University of Copenhagen, November 1st–3rd, 1973. Special issue of *Z. Zool. Syst. Evolutionforsch*. Springer-Verlag, Berlin.

Northcutt, R.G. (1990) Ontogeny and phylogeny: A re-evaluation of conceptual relationships and some applications. *Brain Behav. Evol.*, **36**, 116–40.

Northcutt, R.G. (1992) The phylogeny of octavolateralis ontogenies: A reaffirmation of Garstang's phylogenetic hypothesis, in *The Evolutionary Biology of Hearing* (eds A. Popper, D. Webster and R. Fay), Springer-Verlag, New York, pp. 21–47.

Northcutt, R.G. (1995) The forebrain of Gnathostomes – In search of a morphotype. *Brain Behav. Evol.*, **46**, 275–318.

Northcutt, R.G. (1996) The origin of craniates – neural crest, neurogenic placodes, and homeobox genes. *Israel J. Zool.*, **42**, S273–S313.

Northcutt, R.G. and Brändle, K. (1995) Development of branchiomeric and lateral line nerves in the axolotl. *J. Comp. Neurol.*, **355**, 427–54.

Northcutt, R.G., Brändle, K. and Fritzsch, B. (1995) Electroreceptors and mechanosensory lateral line organs arise from single placodes in axolotls. *Devel. Biol.*, **168**, 358–73.

Northcutt, R.G., Catania, K.C. and Criley, B.B. (1994) Development of lateral line organs in the axolotl. *J. Comp. Neurol.*, **340**, 480–514.

Northcutt, R.G. and Gans, C. (1983) The genesis of neural crest and epidermal placodes: A reinterpretation of vertebrate origins. *Quart. Rev. Biol.*, **58**, 1–28.

Novacek, M.J. (1996) Paleontological data and the study of adaptation, in *Adaptation* (eds M.R. Rose and G.V. Lauder), Academic Press, San Diego, CA, pp. 311–59.

Novas, F.E. and Puerta, P.F. (1997) New evidence concerning avian origins from the Late Cretaceous of Patagonia. *Nature*, **387**, 390–2.

Nübler-Jung, K. and Arendt, D. (1994) Is ventral in insects dorsal in vertebrates? A history of embryological arguments favouring axis inversion in chordate ancestors. *Roux's Arch. Devel. Biol.*, **203**, 357–66.

Nüsslein-Volhard, C. and Wieschaus, E. (1980) Mutations affecting segment number and polarity in *Drosophila*. *Nature*, **287**, 795–801.

Nyhart, L.K. (1995) *Biology Takes Form. Animal Morphology and the German Universities, 1800–1900*, The University of Chicago Press, Chicago, IL.

Ohno, S. (1970) *Evolution by Gene Duplication*, Springer, New York.

Ohno, S. (1996) The notion of the Cambrian pananimalia genome. *Proc. Natl Acad. Sci. USA*, **93**, 8475–8.

Okulitch, V.J. (1955) Archaeocyatha, in *Treatise of Invertebrate Paleontology, Part E, Archaeocyatha and Porifera*. Geological Society of America and University of Kansas Press, Lawrence, KS, pp. E1–E20.

Oliver, G., Loosli, F., Köster, R. *et al.* (1996) Ectopic lens induction in fish in response to the murine homeobox gene Six[3]. *Mech. Devel.* **60**, 233–9.

Olson, E.C. (1959) The evolution of mammalian characters. *Evolution*, **13**, 344–53.

Olsson, L. and Hanken, J. (1996) Cranial neural-crest migration and chondrogenic fate in the Oriental Fire-Bellied toad, *Bombina orientalis*: Defining the ancestral pattern of head development in anuran amphibians. *J. Morph.*, **229**, 105–20.

Onichtchouk, D., Gawantka, V., Dosch, R. *et al.* (1996) The *XVent-2* homeobox gene is part of the BMP-4 signalling pathway controlling dorsoventral patterning of *Xenopus* mesoderm. *Development*, **122**, 3045–53.

Opitz, J.M., Gorlin, R.J., Reynolds, J.F. and Spano, L.M. (1988) *Neural Crest and Craniofacial Disorders. Genetic Aspects.* Proceedings of the March of Dimes Clinical Genetic Conference, Minneapolis, 1987, Alan. R. Liss, Inc., New York.

Oppenheimer, J. (1940) The non-specificity of the germ layers. *Quart. Rev. Biol.*, **15**, 1–27.

Oppenheimer, J. (1951) Problems, concepts and their history, in *Analysis of Development* (eds B.J. Willier, P.A. Weiss and V. Hamburger), W.B. Saunders, Philadelphia, PA, pp. 1–24. (Reprinted in 1971 by Hafner Publ. House, New York.)

Oppenheimer, J. (1959) An embryological enigma in the *Origin of Species*, in *Forerunners of Darwin, 1745–1859* (eds B. Glass, O. Temkin and W.L. Strauss, Jr.), Johns Hopkins University Press, Baltimore, MD, pp. 292–322.

Oppenheimer, J. (1967) *Essays in the History of Embryology and Biology.* The MIT Press, Cambridge, MA.

Orel, V. (1996) *Gregor Mendel: The First Geneticist*, Translated by S. Finn, Oxford University Press, New York.

Orgel, L.E. and Crick, F.H.C. (1980) Selfish DNA: The ultimate parasite. *Nature*, **284**, 604–7.

Orlando, V. and Paro, R. (1995) Chromatin multi-protein complexes involved in the maintenance of transcription patterns. *Curr. Opin. Genet. Devel.*, **5**, 174–9.

Osborn, J.W. and Price, D.G. (1988) An auto-radiographic study of periodontal development in the mouse. *J. Dental Res.*, **67**, 455–61.

Osche, G. (1973) Das homologisieren als eine grundlegende Methode der Phylogenetik. *Auf. Reden Senck. Nat. Ges.*, **24**, 155–66.

Osche, G. (1982) Rekapitulationsentwicklung und ihre Bedeutung für die Phylogenetik – Wann gilt die "Biogenetische Grundregel"? *Verh. Naturw. Ver. Hamburg* (N. F.), **25**, 5–31.

Ospovat, D. (1976) The influence of Karl Ernst von Baer's embryology, 1825–1859: A reappraisal in light of Richard Owen's and William B. Carpenter's "palaeontological application of 'Von Baer's Law'". *J. Hist. Biol.*, **9**, 1–28.

Ospovat, D. (1978) Perfect adaptation and teleological explanation: Approaches to the problem of the history of life in the mid-nineteenth century. *Stud. Hist. Biol.*, **2**, 33–56.

Ospovat, D. (1981) *The Development of Darwin's Theory. Natural History, Natural Theology, and Natural Selection, 1838–1859.* Cambridge University Press, Cambridge.

Oster, G.F. and Murray, J.D. (1989) Pattern formation models and developmental constraints. *J. Exp. Zool.*, **251**, 186–202.

Oster, G.F., Shubin, N., Murray, J.D. and Alberch, P. (1988) Evolution and morphogenetic rules: The shape of the vertebrate limb in ontogeny and phylogeny. *Evolution*, **42**, 862–84.

Ostrom, J.H. (1974) *Archaeopteryx* and the origin of flight. *Quart. Rev. Biol.*, **49**, 27–47.

Ostrom, J.H. (1976) *Archaeopteryx* and the origin of birds. *Biol. J. Linn. Soc.*, **8**, 91–182.

Ostrom, J.H. (1994) On the origin of birds and of avian flight, in *Major Features of Vertebrate Evolution* (eds D.R. Prothero and R.M. Schoch), University of Tennessee Paleontological Society, Knoxville, TN, pp. 160–77.

Outram, D. (1984) *Georges Cuvier. Vocation, Science and Authority in post-revolutionary France.* Manchester University Press, Manchester.

Owen, R. (1836) Cephalopoda, in *Cyclopaedia of Anatomy and Physiology* (ed. Todd, R.B), Vol. 1, pp. 517–52, London.

Owen, R. (1843) *Lectures on Comparative Anatomy and Physiology of the Invertebrate Animals., Delivered at the Royal College of Surgeons* in 1843, London.

Owen, R. (1846) Report on the Archetype and Homologies of the Vertebrate Skeleton. *Report of the sixteenth meeting of the British Association for the Advancement of Science*, Murray, London, pp. 169–340.

Owen, R. (1848) *On the Archetype and Homologies of the Vertebrate Skeleton.* John van Voorst, London.

Owen, R. (1849) *On the Nature of Limbs.* John van Voorst, London.

Owen, R. (1861) *Palaeontology* 2nd edn, Black, Edinburgh.

Owen, R. (1866–1868) *On the Anatomy of Vertebrates*, Three Volumes, Longman, London.

Owen, R.B., Crossley, R., Johnson, T.C. *et al.* (1990) Major low levels of Lake Malawi and their implications for speciation rates in cichlid fishes. *Proc. R. Soc. Lond. B*, **240**, 519–53.

Oyama, S. (1993) Constraints and development. *Neth. J. Zool.*, **43**, 6–16.

Pacces Zaffaroni, N., Arias, E., Lombardi, S. and Zavanella, T. (1996) Natural variation in the appendicular skeleton of *Triturus carnifex* (Amphibia: Salamandridae). *J. Morphol.*, **230**, 167–75.

Padgett, R.W., Wozney, J.M. and Gelbart, W.M. (1993) Human BMP sequences can confer normal dorsal-ventral patterning in the *Drosophila* embryo. *Proc. Natl Acad. Sci. USA*, **90**, 2905–9.

Padian, K. (1992) A proposal to standardize phalangeal formula designations. *J. Vert. Paleont.*, **12**, 260–2.

Padian, K. (1995a) Form versus function: The evolution of a dialectic, in *Functional Morphology in Vertebrate Paleontology* (ed. J.J. Thomason), Cambridge University Press, Cambridge, pp. 264–77.

Padian, K. (1995b) Pterosaurs and typology: Archetypal physiology in the Owen-Seeley dispute of 1870, in *Vertebrate Fossils and the Evolution of Scientific Concepts* (ed. W.A.S. Sargeant), Gordon and Breach, Australia, pp. 285–98.

Padian, K. (1995c) A missing Hunterian lecture on vertebrae by Richard Owen, 1837. *J. Hist. Biol.*, **28**, 333–68.

Padian, K. (1997) The rehabilitation of Sir Richard Owen. *BioScience*, **47**, 446–53.

Page, L.R. (1997) Ontogenetic torsion and protoconch form in the Archaeogastropod *Haliotis kamtschatkama*: Evolutionary implications. *Acta Zool.* (Stock.), **78**, 227–45.

Page, T.L. (1994) Time is the essence: Molecular analysis of the biological clock. *Science*, **263**, 1570–2.

Palmeirim, I., Henrique, D. and Pourquié, O. (1997) Evidence for a molecular clock underlying avian somitogenesis. *Int. J. Devel. Biol.*, **41**, 26S.

Palmer, A.R. and Strobeck, C. (1992) Fluctuating asymmetry as a measure of developmental stability: Implications of non-normal distributions and power of statistical tests. *Acta Zool. Fennica*, **191**, 57–72.

Palmer, A.R. and Strobeck, C. (1997) Fluctuating asymmetry and developmental stability: Heritability of observable variation vs. hereditability of inferred cause. *J. Evol. Biol.*, **10**, 39–49.

Palmiter, R.D., Brinster, R.L., Hammer, R.E. *et al.* (1982) Dramatic growth in mice that develop from eggs micro-injected with metallothionein-growth hormone fusion genes. *Nature*, **300**, 611–15.

Palopoli, M.F. and Patel, N.H. (1996) Neo-Darwinian developmental evolution: Can we bridge the gap between pattern and process? *Curr. Opin. Genet. Devel.*, **6**, 502–8.

Panchen, A.L. (1992) *Classification, Evolution, and the Nature of Biology*, Cambridge University Press, Cambridge.

Panchen, A.L. (1994) Richard Owen and the concept of homology, in *Homology: The Hierarchical Basis of Comparative Biology* (ed. B.K. Hall), Academic Press, San Diego, CA, pp. 21–62.

Pander, C.H. (1817) *Dissertatio inauhuralis, sistens historiam metamorphoseos quam ovum incubatum prioribus quinque diebus subit.* Würzburg.

Panganiban, G., Irvine, S.M., Lowe, C. *et al.* (1997) The origin and evolution of animal appendages. *Proc. Natl Acad. Sci. USA*, **94**, 5162–6.

Panganiban, G., Sebring, A., Nagy, L. and Carroll, S. (1995) The development of crustacean limbs and the evolution of arthropods. *Science*, **270**, 1363–6.

Parker, T.J. and Haswell, W.A. (1960) *A Text-Book of Zoology*, Volumes 1 and 2, Macmillan and Co., New York.

Paro, R. (1990) Imprinting a determined state into the chromatin of *Drosophila*. *Trends Genet.*, **6**, 416–21.

Paro, R. (1993) Mechanisms of heritable gene expression during development of *Drosophila*. *Curr. Opin. Cell Biol.*, **5**, 999–1005.

Paro, R. and Hognes, D.S. (1991) The Polycomb protein shares a homologous domain with a heterochromatin-associated protein of *Drosophila*. *Proc. Natl Acad. Sci. USA*, **88**, 263–7.

Parr, B.A. and McMahon, A.P. (1995) Dorsalizing signal *Wnt*-7a required for normal polarity of D-V and A-P axes of mouse limbs. *Nature*, **374**, 350–3.

Parsons, P.A. (1990) Fluctuating asymmetry: An epigenetic measure of stress. *Biol. Rev.*, **65**, 131–45.

Pasteels, J. (1943) Proliférations et croissance dans la gastrulation et la formation de la queue des Vertébrés. *Arch. Biol.* (Liège), **54**, 1–51.

Patel, N.H. (1994) Developmental evolution: Insights from studies of insect segmentation. *Science*, **266**, 581–9.

Patel, N.H., Ball, E.E. and Goodman, C.S. (1992) Changing role of *even-skipped* during the evolution of insect pattern formation. *Nature*, **357**, 339–42.

Patel, N.H., Condron, B.G. and Zinn, K. (1994) Pair-rule expression patterns of *even-skipped* are found in both short- and long-germ beetles. *Nature*, **367**, 429–34.

Patel, N.H., Martin-Blanco, E., Coleman, K.G. *et al.* (1989) Expression of *engrailed* proteins in arthropods, annelids, and chordates. *Cell*, **58**, 955–68.

Patterson, B. (1949) Rates of evolution in Taeniodonts, in *Genetics, Palaeontology, and Evolution* (eds G.L. Jepsen, E. Mayr and G.G. Simpson), Princeton University Press, Princeton, NJ, pp. 243–78.

Patterson, C. (1978) *Evolution*, Routledge & Kegan Paul, London and Henley.

Patterson, C. (1982) Morphological characters and homology, in *Problems of Phylogenetic Reconstruction* (eds K.A. Joysey and A.E. Friday), Systematics Association Special Volume 21, Academic Press, London, pp. 21–74.

Patterson, C. (1983) How does phylogeny differ from ontogeny? in *Development and Evolution* (eds B.C. Goodwin, N. Holder and C.C. Wylie), British Society for Developmental Biology Symposium 6, Cambridge University Press, Cambridge, pp. 1–31.

Patterson, C. (1987) *Molecules and Morphology in Evolution: Conflict or Compromise*, Cambridge University Press, Cambridge.

Patterson, C. (1988a) Homology in classical and molecular biology. *Mol. Biol. Evol.*, **5**, 603–25.

Patterson, C. (1988b) The impact of evolutionary theories on systematics, in *Prospects in Systematics* (ed. D.L. Hawksworth), Systematics Association Special Volume 36, The Clarendon Press, Oxford, pp. 59–91.

Patterson, C. (1990) Metazoan phylogeny: Reassessing relationships. *Nature*, **344**, 199–200.

Patterson, C., Williams, D.M. and Humphries, C.J. (1993) Congruence between molecular and morphological phylogenies. *Annu. Rev. Ecol. Syst.*, **24**, 153–88.

Patton, J.L. and Brylski, P.V. (1987) Pocket gophers in alfalfa fields: Causes and consequences of habitat-related body size variation. *Amer. Nat.*, **130**, 493–506.

Peel, J.D.Y. (1971) *Herbert Spencer: The Evolution of a Sociologist*. Heinemann, London.

Perrin, N. and Travis, J. (1992) On the use of constraints in evolutionary biology and some allergic reactions to them. *Funct. Ecol.*, **6**, 361–3.

Person, P. (1983) Invertebrate Cartilages, in *Cartilage, Volume 1. Structure, Function, and Biochemistry* (ed. B.K. Hall), Academic Press, New York, pp. 31–57.

Peterson, K.J. (1995) A phylogenetic test of the calcichordate scenario. *Lethaia*, **28**, 25–38.

Pettersson, M. (1996) *Complexity and Evolution*, Cambridge University Press, Cambridge.

Pettigrew, J.D. (1991) Wing or brain? Convergent evolution in the origin of bats. *Syst. Zool.*, **40**, 199–216.

Pfennig, D.W. (1990) The adaptive significance of an environmentally cued developmental switch in an anural tadpole. *Oecologia*, **85**, 101–7.

Pfennig, D.W. (1992) Polyphenism in spadefoot toad tadpoles as a locally adjusted evolutionary stable strategy. *Evolution*, **46**, 1408–20.

Phillips, J.A.S. (ed.) (1981) *Prince Albert and the Victorian Age: A Seminar held in May 1980 in Coburg under the Auspices of the University of Bayreuth and the City of Coburg*, Cambridge University Press, Cambridge.

Phippard, D.J., Weber-Hall, S.J., Sharpe, P.T. *et al.* (1996) Regulation of *Msx*-1, *Msx*-2, *Bmp*-2 and *Bmp*-4 during foetal and postnatal mammary gland development. *Development*, **122**, 2729–37.

Piaget, J. (1968) *Structuralism*. Routledge and Kegan Paul, London.

Piaget, J. (1974) *Adaptation and Intelligence: Organic Selection and Phenocopy*. University of Chicago Press, Chicago, IL.

Piatigorsky, J. and Wistow, G. (1991) The recruitment of crystallins: New functions precede gene duplication. *Science*, **252**, 1078–9.

Piersma, T. and Lindström, Å. (1997) Rapid reversible changes in organ size as a component of adaptive behaviour. *Trends Ecol. Evol.*, **12**, 134–8.

Pigliucci, M. (1996) How organisms respond to environmental changes: From phenotypes to molecules (and *vice versa*). *Trends Ecol. Evol.*, **11**, 168–73.

Pigliucci, M., Schlichting, C.D., Jones, C.J. and Schwenk, K. (1996) Developmental reaction norms: The interactions among allometry, ontogeny and plasticity. *Plant Species Biol.*, **11**, 69–85.

Pinto, C. and Hall, B.K. (1991) Toward an understanding of the epithelial requirement for osteogenesis in scleral mesenchyme in the embryonic chick. *J. Exp. Zool.*, **259**, 92–108.

Piotrowski, T., Schilling, T.F., Brand, M. *et al.* (1996) Jaw and branchial arch mutants in zebrafish. II: Anterior arches and cartilage differentiation. *Development*, **123**, 345–56.

Pirttiniemi, P. and Kantomaa, T. (1996) Electrical stimulation of masseter muscles maintains condylar cartilage in long-term organ culture. *J. Dent. Res.*, **75**, 1365–71.

Polikoff, D. (1981) C. H. Waddington and modern evolutionary theory. *Evol. Theory*, **5**, 143–68.

Ponder, W.F. and Lindberg, D.R. (1997) Towards a phylogeny of gastropod molluscs – An analysis using morphological characters. *Biol. J. Linn. Soc.* (in press).

Popadic, A., Rusch, D., Peterson, M. *et al.* (1996) Origin of the arthropod mandible. *Nature*, **380**, 395.

Pourquié, O., Fan, C.-M., Coltey, M. *et al.* (1996) Lateral and axial signals involved in avian somite patterning: A role for BMP4. *Cell*, **84**, 461–71.

Power, M.-A. and Tam, P.P.L. (1993) Onset of gastrulation, morphogenesis and somitogenesis in mouse embryos displaying compensatory growth. *Anat. Embryol.*, **187**, 493–504.

Prentiss, C.W. (1903) Polydactylism in man and the domestic animals with especial reference to digital variations in swine. *Bull. Mus. Comp. Zool. Harvard.*, **40**, 245–313.

Presley, R., Horder, T.J. and Slipka, J. (1996) Lancelet development as evidence of ancestral chordate structure. *Israel J. Zool.*, **42**, S97–S116.

Prosser, C.L. and Brown, F.A.Jr. (1961) *Comparative Animal Physiology* 2nd edn, W.B. Saunders, Philadelphia, PA.

Provine, W.B. (1986) *Sewall Wright and Evolutionary Biology*, University of Chicago Press, Chicago, IL.

Purnell, M.A. (1995) Microwear on conodont elements and macrophagy in the first vertebrate. *Nature*, **374**, 798–800.

Purnell, M.A. and von Bitter, P.H. (1992) Blade-shaped conodont elements functioned as cutting teeth. *Nature*, **359**, 629–31.

Purvis, A. and Harvey, P.H. (1997) The right size for a mammal. *Nature*, **386**, 332–3.

Qualls, C.P. Andrews, R.M. and Mathies, T. (1997) The evolution of viviparity and placentation revisited. *J. Theor. Biol.*, **185**, 129–35.

Qualls, C.P. and Shine, R. (1996) Reconstructing ancestral reaction norms: An example using the evolution of reptilian viviparity. *Funct. Ecol.*, **10**, 688–97.

Qiu, M., Bulfone, A., Martinez, S. *et al.* (1995) Role of *Dlx-2* in head development and evolution: Null mutation of *Dlx-2* results in abnormal morphogenesis of proximal first and second branchial arch derivatives and abnormal differentiation in the forebrain. *Genes Devel.*, **9**, 2523–38.

Qiu, M., Bulfone, A., Ghattas, I. *et al.* (1997) Role of the Dlx homeobox genes in proximodistal patterning of the branchial arches: Mutations of *Dlx-1*, *Dlx-2*, and *Dlx-1* and *-2* alter morphogenesis of proximal skeletal and soft tissue structures derived from the first and second arches. *Devel. Biol.*, **185**, 165–84.

Quiring, R., Walldorf, U., Kloter, U. and Gehring, W.J. (1994) Homology of the *eyeless* gene of *Drosophila* to the *Smalleye* gene in mice and *Aniridia* in humans. *Science*, **265**, 785–9.

Raff, M.C. (1996) Size control: The regulation of cell numbers in animal development. *Cell*, **86**, 173–5.

Raff, R.A. (1987) Constraint, flexibility and phylogenetic history in the evolution of direct development in sea urchins. *Devel. Biol.*, **119**, 6–19.

Raff, R.A. (1988) Direct-developing sea urchins: A system for the study of developmental processes in evolution, in *Echinoderm Biology* (eds R.D. Burke, P.V. Mladenov, P. Lambert and R.L. Parsley), Proceedings of the Sixth International Echinoderm Conference, Victoria, 23–28 August, 1989. A. A. Balkema, Rotterdam, pp. 63–9.

Raff, R.A. (1992) Direct-developing sea urchins and the evolutionary reorganization of early development. *BioEssays*, **14**, 211–18.

Raff, R.A. (1996) *The Shape of Life: Genes: Development, and the Evolution of Animal Form*, University of Chicago Press, Chicago, IL.

Raff, R.A., Anstrom, J.A., Huffman, C.J. *et al.* (1984) Origin of a gene regulatory mechanism in the evolution of echinoderms. *Nature*, **310**, 312–14.

Raff, R.A. and Kaufmann, T.C. (1983) *Embryos, Genes and Evolution*. Macmillan, New York.

Raff, R.A., Parr, B.A., Parks, A.L. and Wray, G.A. (1990) Heterochrony and other mechanisms of radical evolutionary change in early development, in *Evolutionary Innovations* (ed. M.H. Nitecki), University of Chicago Press, Chicago, IL, pp. 71–98.

Raff, R.A. and Wray, G.A. (1989) Heterochrony: Developmental mechanisms and evolutionary results. *J. Evol. Biol.*, **2**, 409–34.

Raikow, R.J. (1975) The evolutionary reappearance of ancestral muscles as developmental anomalies in two species of birds. *Condor*, **77**, 514–17.

Raikow, R.J., Bledsoe, A.H., Myers, B.A. and Welsh, C.J. (1990) Individual variation in avian muscles and its significance for the reconstruction of phylogeny. *Syst. Zool.*, **39**, 362–70.

Raikow, R.J., Borecky, S.R. and Berman, S.L. (1979) The evolutionary re-establishment of a lost ancestral muscle in the bower bird assemblage. *Condor*, **81**, 203–6.

Rainger, R. (1985) Paleontology and philosophy: A critique. *J. Hist. Biol.*, **18**, 267–87.

Rainger, R. (1989) What's the use: William King Gregory and the functional morphology of fossil vertebrates. *J. Hist. Biol.*, **22**, 103–39.

Ramsköld, L. (1992) The second leg row of *Hallucigenia* discovered. *Lethaia*, **25**, 221–4.

Ramsköld, L. and Hou, X.-g. (1991) New early Cambrian animal and onychophoran affinities of enigmatic metazoans. *Nature*, **351**, 225–8.

Ransick, A., Cameron, R.A. and Davidson, E.H. (1996) Postembryonic segregation of the germ line in sea urchins in relation to indirect development. *Proc. Natl Acad. Sci. USA*, **93**, 6759–63.

Rappaport, R. (1974) Cleavage, in *Concepts of Development* (eds J. Lash and J.R. Whittaker), Sinauer Associates, Stamford, CT, pp. 76–98.

Rathke, H. (1839) *Bemerkungen über die Entwickelung des Schädels der Wirbelthiere.* Königsberg.

Raup, D.M. (1966) Geometric analysis of shell coiling: General problems. *J. Paleontol.*, **40**, 1178–90.

Raup, D.M. and Stanley, S.M. (1971) *Principles of Paleontology*, W. H. Freeman and Co., San Francisco, CA.

Raven, C.P. (1935) Zur Entwicklung der Ganglienleiste. IV. Untersuchungen über Zeitpunkt und Verlauf der 'materiellen Determination' des präsumptiven Kopfganglienleistenmaterials der Urodelen. *Wilhelm Roux. Arch. EntwMech. Org.*, **132**, 509–75.

Raven, C.P. and Kloos, J. (1945) Induction by medial and lateral pieces of the archenteron roof with special reference to the determination of the neural crest. *Acta néerl. Morph.*, **5**, 348–62.

Ray, J. (1693) *Synopsis animalium quadripedum*, London.

Raynaud, A. (1985) Development of limbs and embryonic limb reduction, in *Biology of the Reptilia, Volume 15. Development B* (eds C. Gans and F. Billett), John Wiley & Sons, New York, pp. 59–148.

Reecy, J.M., Yamada, M., Cummings, K. *et al.* (1997) Chicken Nkx-2.8: a novel homeobox gene expressed in early heart progenitor cells and pharyngeal pouch-2 and -3 endoderm. *Devel. Biol.*, **188**, 295–311.

Regal, P.J. (1975) The evolutionary origin of feathers. *Quart. Rev. Biol.*, **50**, 35–66.

Regan, C.T. and Trewavas, E. (1932) Deep-sea angler fishes (Ceratioidea). *Reports Carlsberg Ocean. Exped.*, **2**, 1928–30.

Reik, W., Howlett, S.K. and Surani, M.A. (1990) Imprinting by DNA methylation: From transgenes to endogenous gene sequences. *Development*, 1990, Suppl., 99–106.

Reik, W., Romer, I., Barton, S.C. *et al.* (1993) Adult phenotype in the mouse can be affected by epigenetic events in the early embryo. *Development*, **119**, 933–42.

Reilly, S.M. and Lauder, G.V. (1988) Atavisms and the homology of hyobranchial elements in lower vertebrates. *J. Morph.*, **195**, 237–46.

Reilly, S.M. and Lauder, G.V. (1990) Metamorphosis of cranial design in Tiger Salamanders (*Ambystoma tigrinum*): a morphometric analysis of ontogenetic change. *J. Morph.*, **204**, 121–37.

Reilly, S.M., Wiley, E.O. and Meinhardt, D.J. (1997) An integrative approach to heterochrony: The distinction between interspecific and intraspecific phenomena. *Biol. J. Linn. Soc.*, **60**, 119–43.

Reiss, J.O (1989) The meaning of developmental time: A metric for comparative embryology. *Amer. Nat.*, **134**, 170–89.

Reisz, R.R. and Laurin, M. (1991) *Owenetta* and the origin of turtles. *Nature*, **349**, 324–6.

Rendel, J.M. (1967) *Canalization and Gene Control.* Logos Press, London.

Rendel, J.M. (1968) Genetic control of developmental processes, in *Population Biology and Evolution* (ed. R.C. Lewontin), Syracuse University Press, Syracuse, NY, pp. 47–68.

Rendel, J.M. and Sheldon, B.L. (1960) Selection for canalization of the *scute* phenotype in *D. melanogaster. Aust. J. Biol. Sci.*, **13**, 36–47.

Reppert, S. M. and Weaver, D. R. (1997) Forward genetic approach strikes gold: Cloning of a mammalian *Clock* gene. *Cell*, **89**, 487–90.

Reverberi, G., Ortolani, G. and Farinella-Ferruzza, N. (1960) The causal formation of the brain in the ascidian larval. *Acta Embryol. Morph. Exp.*, **3**, 296–336.

Reznick, D.N., Bryga, H. and Endler, J.A. (1990) Experimentally induced life-history evolution in a natural population. *Nature*, **346**, 357–9.

Reznick, D.N., Shaw, F.H., Rod, F.H. and Shaw, R.G. (1997) Evaluations of the rate of evolution in natural populations of guppies (*Poecilia reticulata*). *Science*, **275**, 1934–7.

Rice, S.H. (1997) The analysis of ontogenetic trajectories: When a change in size or shape is not heterochrony. *Proc. Natl Acad. Sci. USA*, **94**, 907–12.

Richards, E. (1987) A question of property rights: Richard Owen's evolutionism reassessed. *Brit. J. Hist. Sci.*, **20**, 129–71.

Richards, R.J. (1992) *The Meaning of Evolution: The Morphological Construction and Ideological Reconstruction of Darwin's Theory*, The University of Chicago Press, Chicago, IL.

Richardson, M.K. (1995) Heterochrony and the phylotypic period. *Devel. Biol.*, **172**, 412–21.

Richardson, M.K., Hanken, J., Gooneratne, M.L. *et al.* (1997) There is no highly conserved embryonic stage in vertebrates: Implications for current theories of evolution and development. *Anat. Embryol.*, **196**, 91–106.

Rickleffs, R.E. (1973) Patterns of growth in birds. II. Growth rate and mode of development. *Ibis*, **115**, 177–201.

Rickleffs, R.E. (1983) Avian postnatal development. *Avian Biol.*, **7**, 2–83.

Riddle, R.D., Johnson, R.L., Laufer, E. and Tabin, C. (1993) *Sonic hedgehog* mediates the polarizing activity of the ZPA. *Cell*, **75**, 1401–16.

Ridley, M. (1986) Embryology and classical zoology in Great Britain, in *A History of Embryology* (eds T.J. Horder, J.A. Witkowski, and C.C. Wylie), The Eighth Symposium of the British Society for Developmental Biology. Cambridge University Press, Cambridge, pp. 35–68.

Ridley, M. (1993) Analysis of the Burgess Shale. *Paleobiology*, **19**, 519–21.

Riedl, R. (1969) Gnathostomulida from America. *Science*, **163**, 445–62.

Riedl, R. (1978) *Order in Living Organisms.* John Wiley & Sons, New York.

Rieger, R., Michaelis, A. and Green, M.M. (1976) *Glossary of Genetics and Cytogenetics. Classical and Molecular* 4th edn, Springer-Verlag, Berlin.

Rienesl, J. and Wagner, G.P. (1992) Constancy and change of basipodial variation patterns: A comparative study of crested and marbled newts – *Triturus cristatus, Triturus marmoratus* – and their natural hybrids. *J. Evol. Biol.*, **5**, 307–24.

Rieppel, O.C. (1986) Atomism, epigenesis, preformation and pre-existence: A clarification of terms and consequences. *Biol. J. Linn. Soc.*, **28**, 331–41.

Rieppel, O.C. (1987) Pattern and process: The early classification of snakes. *Biol. J. Linn. Soc.*, **31**, 405–20.

Rieppel, O.C. (1988a) *Fundamentals of Comparative Biology.* Birkhäuser Verlag, Basel.

Rieppel, O.C. (1988b) A review of the origin of snakes. *Evol. Biol.*, **22**, 37–130.

Rieppel, O.C. (1989) The hind limb of *Macrocnemus bassanii* (Nopcsa) (Reptilia, Diapsida): Development and functional anatomy. *J. Vert. Paleont.*, **9**, 373–87.

Rieppel, O.C. (1990) Ontogeny – a way forward for systematics, a way backwards for phylogeny. *Biol. J. Linn. Soc.*, **39**, 177–91.

Rieppel, O.C. (1991) Progress in evolution: Snakes as an example. *Z. Zool. Syst. Evolut. Forsch.*, **29**, 208–12.

Rieppel, O.C. (1992a) Homology and logical fallacy. *J. Evol. Biol.*, **5**, 701–15.

Rieppel, O.C. (1992b) Studies on skeleton formation in reptiles. I. The postembryonic development of the skeleton in *Cyrtodactylus pubisulcus* (Reptilia, Gekkonidae). *J. Zool. Lond.*, **227**, 87–100.

Rieppel, O.C. (1992c) Studies on skeleton formation in reptiles. III. Patterns of ossification in the skeleton of *Lacerta vivipara* (Reptilia, Squamata). *Fieldiana Zool.* n.s., **68**, 1–25.

Rieppel, O.C. (1992d) The skeleton of a juvenile *Lanthanotus* (Varanoidea). *Amph.-Rept.*, **13**, 27–34.

Rieppel, O.C. (1993a) The conceptual relationship of ontogeny, phylogeny and classification: The taxic approach. *Evol. Biol.*, **27**, 1–32.

Rieppel, O.C. (1993b) Studies on skeletal formation in reptiles: Patterns of ossification in the skeleton of *Chelydra serpentina* (Reptilia, Testudines). *J. Zool. Lond.*, **231**, 487–509.

Rieppel, O.C. (1994) Studies on skeletal formation in reptiles: Patterns of ossification in the limb skeleton of *Gehyra oceanica* (Lesson) and *Lepidodactylus lugubris* (Dumeril & Bibron). *Ann. Sci. Nat. Zool. Biol. Anim.*, **15**, 83–91.

Rieppel, O.C. and de Braga, M. (1996) Turtles as diapsid reptiles. *Nature*, **384**, 453–5.

Rijli, F.M., Mark, M., Lakkaraju, S. *et al.* (1993) A homeotic transformation is generated in the rostral branchial region of the head by disruption of *Hoxa-2*, which acts as a selector gene. *Cell*, **75**, 1333–49.

Rijli, F.M., and Chambon, P. (1997) Genetic interactions of *Hox* genes in limb development: learning from compound mutants. *Curr. Opin. Genet. Devel.*, **7**, 481–7.

Rinard, R.G. (1981) The problem of the organic individual: Ernst Haeckel and the development of the biogenetic law. *J. Hist. Biol*, **14**, 249–75.

Riska, B., Rutledge, J.J. and Atchley, W.R. (1985) Covariance between direct and maternal genetic effects in mice, with a model of persistent environmental influences. *Genet. Res.*, **45**, 287–97.

Rivera-Pérez, J.A., Mallo, M., Gendron-Maguire, M. *et al.* (1995) *Goosecoid* is not an essential component of the mouse gastrula organizer but is required for craniofacial and rib development. *Development*, **121**, 3005–12.

Robinson, G. W. and Hennighausen, L. (1997) Inhibins and activins regulate mammary epithelial cell differentiation through mesenchymal-epithelial interactions. *Development*, **124**, 2701–8.

Rocek, Z. (1996) Skull of the neotenic salamandrid amphibian *Triturus alpestris* and abbreviated development in the Tertiary Salamandridae. *J. Morph.*, **230**, 187–97.

Rodriguez-Esteban, C., Schwabe, K.W.R., De la Pena, J. *et al.* (1997) *Radical fringe* positions the apical ectodermal ridge at the dorsoventral boundary of the vertebrate limb. *Nature*, **386**, 360–6.

Roe, S.A. (1979) Rationalism and embryology: Caspar Friedrich Wolff's theory of epigenesis. *J. Hist. Biol.*, **12**, 1–43.

Roe, S.A. (1981) *Matter, Life and Generation. 18th Century Embryology and the Haller-Wolff Debate.* Cambridge University Press, Cambridge.

Roff, D.A. (1992) *The Evolution of Life Histories: Theory and Analysis*, Chapman & Hall, New York.

Roff, D.A. (1996) The evolution of threshold traits in animals. *Quart. Rev. Biol.*, **71**, 3–35.

Roger, J. (1997) *Buffon*, translated by S. L. Bonnefoi, Cornell University Press, Ithaca, NY.

Rohde, K. (1996) Robust phylogenies and adaptive radiations: A critical examinations of methods used to identify key innovations. *Amer. Nat.*, **148**, 481–500.

Roll-Hansen, N. (1984) E.S. Russell and J.H. Woodger: The failure of two twentieth-century opponents of mechanistic biology. *J. Hist. Biol.*, **17**, 399–428.

Rollhäuser-ter-Horst, J. (1980) Neural crest replaced by gastrula ectoderm in amphibia. Effect on neurulation, CNS, gills and limbs. *Anat. Embryol.*, **160**, 203–12.

Rollo, C.D. (1994) *Phenotypes: Their Epigenetics, Ecology and Evolution*, Chapman & Hall, London.

Romer, A.S. (1949) Time series and trends in animal evolution, in *Genetics, Paleontology and Evolution* (eds G.L. Jepsen, E. Mayr and G.G. Simpson), Princeton University Press, Princeton, NJ., pp. 103–20.

Romer, A.S. (1960) *The Vertebrate Body* 2nd edn, W.B. Saunders Co., Philadelphia, PA.

Ros, M.A., Sefton, M., and Nieto, M.A. (1997) *Slug,* a zinc finger gene previously implicated in the early patterning of the mesoderm and the neural crest, is also involved in chick limb development. *Development*, **124**, 1821–9.

Rose, M.R. and Lauder, G.V. (1996a) Post-Spandrel adaptationism, in *Adaptation* (eds M.R. Rose and G.V. Lauder), Academic Press, San Diego, CA, pp. 1–8.

Rose, M.R. and Lauder, G.V. (eds) (1996b) *Adaptation*, Academic Press, San Diego, CA.

Rosenquist, G.C. (1981) Epiblast origin and early migration of neural crest cells in the chick embryo. *Devel. Biol.*, **87**, 201–11.

Roth, G., Dicke, U. and Nishikawa, K. (1992) How do ontogeny, morphology, and physiology of sensory systems constrain and direct the evolution of amphibians? *Amer. Nat.*, **139**, Suppl., S105–S124.

Roth, G. and Wake, D.B. (1985) Trends in the functional morphology and sensorimotor control of feeding behavior in salamanders: An example of internal dynamics in evolution. *Acta Biotheor.*, **34**, 175–92.

Roth, S., Müller, K., Fischer, D.-C. and Dannhauer, K.-H. (1997) Specific properties of the extracellular chondroitin sulphate proteoglycans in the mandibular condylar growth centre in pigs. *Archs Oral Biol.*, **42**, 63–76.

Roth, V.L. (1984) On homology. *Biol. J. Linn. Soc.,* **22**, 13–29.

Roth, V.L. (1988) The biological basis of homology, in *Ontogeny and Systematics* (ed. C.J. Humphries), Columbia University Press, New York, pp. 1–26.

Roth, V.L. (1991) Homology and hierarchies: Problems solved and unresolved. *J. Evol. Biol.*, **4**, 167–94.

Roth, V.L. (1992) Inferences from allometry and fossils: Dwarfing of elephants on islands, in *Oxford Surveys in Evolutionary Biology* (eds D. Futuyma and J. Antonovics), Oxford University Press, Oxford, pp. 259–88.

Roth, V.L. and Klein, M.S. (1986) Maternal effects on body size of large insular *Peromyscus maniculatus*: Evidence from embryo transfer experiments. *J. Mammal.*, **67**, 37–45.

Rothe, M., Pehl, M., Taubert, H. and Jäckle, H. (1992) Loss of gene function through rapid mitotic cycles in the *Drosophila* embryo. *Nature*, **359**, 156–9.

Roughgarden, J. (1979) *Theory of Population Genetics and Evolutionary Ecology, An Introduction.* Macmillan Co., Inc., New York.

Rougier, G.W., de la Fuente, M.S. and Arcucci, A.B. (1995) Late Triassic turtles from South America. *Science*, **268**, 855–8.

Roux, W. (1905) *Die Entwicklungsmechanik.* Engelmann, Leipzig.

Rowe, T. (1996a) Coevolution of the mammalian middle ear and neocortex. *Science*, **273**, 651–4.

Rowe, T. (1996b) Brain heterochrony and origin of the mammalian middle ear, in *New Perspectives on the History of Life* (eds M.T. Ghiselin and G. Pinna), Memoirs Cal. Acad. Sci., number 20,

California Academy of Science, San Francisco, CA, pp. 71–95.

Ruberte, E., Wood, H.B. and Morriss-Kay, G.M. (1997) Prorhombomeric subdivision of the mammalian embryonic hindbrain: Is it functionally meaningful? *Int. J. Devel. Biol.*, **41**, 213–22.

Ruch, J.V. (ed.) (1995) Odontogenesis. *Int. J. Devel. Biol.*, **39**, 1–297.

Ruddle, F.H., Bentley, K.L., Murtha, M.T. and Risch, N. (1994) Gene loss and gain in the evolution of the vertebrates. *Development* 1994 Suppl., 155–61.

Rudwick, M.J.S. (1985) *The Meaning of Fossils: Episodes in the History of Palaeontology*, The University of Chicago Press, Chicago, IL.

Rugh, R. (1965) *Experimental Embryology. Techniques and Procedures*, Burgess Publishing Co., Minneapolis, Minn.

Ruiz i Altaba, A. and Jessell, T. (1991) Retinoic acid modifies mesodermal patterning in early *Xenopus* embryos. *Genes & Devel.*, **5**, 175–87.

Runnegar, B. (1980) Hyolitha: Status of the phylum. *Lethaia*, **13**, 21–5.

Runnegar, B., Pojeta, J., Morris, N.J. *et al.* (1975) Biology of the Hyolitha. *Lethaia*, **8**, 181–91.

Rupke, N.A. (1994) *Richard Owen: Victorian Naturalist*, Yale University Press, New Haven, CT.

Ruse, M. (1975) Woodger on genetics, a critical evaluation. *Acta Biotheoret.*, **24**, 1–13.

Ruse, M. (1979) *The Darwinian Revolution. Science Red in Tooth and Claw.* The University of Chicago Press, Chicago, IL.

Ruse, M. (1996) *Monad to Man: The Concept of Progress in Evolutionary Biology*, Harvard University Press, Cambridge, MA.

Russell, E.S. (1916) *Form and Function. A Contribution to the History of Animal Morphology.* John Murray, London. (Reprinted, 1972 by Gregg International Publishers Ltd., Westmead, England; 1982 by The University of Chicago Press, Chicago, IL, with a new introduction by G.V. Lauder).

Russell, E.S. (1930) *The Interpretation of Development and Heredity. A Study in Biological Method.* Clarendon Press, Oxford.

Russo, V.E.A., Martienssen, R.A. and Riggs, A.D. (eds) (1996) *Epigenetic Mechanisms of Gene Regulation.* Monograph 22. Cold Spring Harbor Laboratory Press, New York.

Rutledge, J.J., Eisen, E.J. and Legates, J.E. (1974) Correlated response in skeletal traits and replicate variation in selected lines of mice. *Theor. & Appl. Genet.*, **45**, 26–31.

Rybczynski, W. (1986) *Home. A Short History of an Idea*, Penguin Books, Harmondsworth, England.

Sachs, T. (1988) Epigenetic selection: An alternative mechanism of pattern formation. *J. Theor. Biol.*, **134**, 547–59.

Sage, R.D., Atchley, W.R. and Capanna, E. (1993) House mice as models in systematic biology. *Syst. Biol.*, **42**, 523–61.

Sage, R.D., Loiselle, P.V., Basasibwaki, P. and Wilson, A.C. (1984) Molecular versus morphological change among cichlid fishes of Lake Victoria, in *Evolution of Fish Species Flocks* (eds A.A. Eichelle and I. Kornfield), University of Maine at Orono Press, Orono, ME, pp. 185–202.

Saha, M.S., Spann, C.L. and Grainger, R.M. (1989) Embryonic lens induction: More than meets the optic vesicle. *Cell Differ. Devel.*, **28**, 153–72.

Salser, S.J. and Kenyon, C. (1994) Patterning of *C. elegans*: Homeotic cluster genes, cell fates and cell migrations. *Trends Genet.*, **10**, 159–64.

Salthe, S.N. (1993) *Development and Evolution: Complexity and Change in Biology*, The MIT Press, Cambridge, MA.

Salzgeber, B. and Guénet, J.-L. (1984) Studies on "Repeated Epilation" mouse mutant embryos: II. Development of limb, tail, and skin defects, *J. Craniofac. Genet. Devel. Biol.*, **4**, 95–114.

Sander, K. (1976) Specification of the basic body pattern in insect embryogenesis. *Adv. Insect Physiol.*, **12**, 125–238.

Sander, K. (1983) The evolution of patterning mechanisms: Gleanings from insect embryogenesis and spermatogenesis, in *Development and Evolution* (eds B.C. Goodwin, N. Holder and C.C. Wylie), The Sixth Symposium of the British Society for Developmental Biology, Cambridge University Press, Cambridge, pp. 137–60.

Sander, K. (1994) The evolution of insect patterning mechanisms: A survey of progress and problems in comparative molecular embryology. *Development*, 1994 Suppl., 187–91.

Sanders, E.J. (1988) The roles of epithelial-mesenchymal cell interactions in developmental processes. *Biochem. Cell Biol.*, **66**, 530–40.

Sanders, E.J. (1989) *The Cell Surface in Embryogenesis and Carcinogenesis.* The Telford Press, Caldwell, NJ.

Sanderson, M.J. and Hufford, L. (eds) (1996) *Homoplasy: The Recurrence of Similarity in Evolution*, Academic Press, San Diego, CA.

Sanford, L.P., Ormsby, I., Gittenberger-de Groot, A.C. *et al* (1997) TGFß2 knockout mice have multiple developmental defects that are non-overlapping with other TGFß knockout phenotypes. *Development*, **124**, 2659–70.

Sansom, I.J. (1996) *Pseudoneotodus* – a histological study of an Ordovician to Devonian vertebrate lineage. *Zool. J. Lin. Soc.*, **118**, 47–57.

Sansom, I.J., Smith, M.P., Armstrong, H.A. and Smith, M.M. (1992) Presence of the earliest vertebrate hard tissues in conodonts. *Science*, **256**, 1308–11.

Sansom, I.J., Smith, M.P. and Smith, M.M. (1994) Dentine in conodonts. *Nature*, **368**, 591.

Sasai, Y. and De Robertis, E.M. (1997) Ectodermal patterning in vertebrate embryos. *Devel. Biol.*, **182**, 5–20.

Sasai, Y., Lu, B., Steinbesser, H. and de Robertis, E.M. (1995) Regulation of neural induction by the Chd and Bmp-4 antagonistic patterning signals in *Xenopus*. *Nature*, **376**, 333–6.

Sasaki, Y., Nomaka, K. and Nakata, M. (1994) The effect of embryo transfer on the intrauterine growth of the mandible in mouse fetuses. *J. Craniofac. Genet. Devel. Biol.*, **14**, 111–117, 118–23.

Sasaki, Y., Nomaka, K. and Nakata, M. (1995) The strain effect of dam on intrauterine incisal growth in mouse fetuses. *J. Craniofac. Genet. Devel. Biol.*, **15**, 140–5.

Sato, S.M. and Sargent, T.D. (1990) Molecular approach to dorso-anterior development in *Xenopus laevis*. *Devel. Biol.*, **137**, 135–41.

Satoh, N. (1982) Timing mechanisms in early embryonic development. *Differentiation*, **22**, 156–63.

Satoh, N. (1994) *Developmental Biology of Ascidians*, Cambridge University Press, Cambridge.

Satoh, N. and Jeffery, W.R. (1995) Chasing tails in ascidians: Developmental insights into the origin and evolution of chordates. *Trends Genet.*, **11**, 354–9.

Satokata, I. and Mass, R. (1994) *Msx*-1 deficient mice exhibit cleft palate and abnormalities of craniofacial and tooth development. *Nature Genet.*, **6**, 348–56.

Sattler, R. (1984) Homology – a continuing challenge. *Syst. Bot.*, **9**, 382–94.

Sattler, R. (1992) Process morphology: Structural dynamics in development and evolution. *Can. J. Bot.*, **70**, 708–14.

Sattler, R. (1994) Homology, homeosis, and process morphology in plants, in *Homology: The Hierarchical Basis of Comparative Biology* (ed. B.K. Hall), Academic Press, San Diego, CA, pp. 423–75.

Sattler, R. and Jeune, B. (1992) Multivariate analysis confirms the continuum view of plant form. *Ann. Bot.*, **69**, 249–62.

Sawin, P.B. and Edmonds, H.W. (1949) Morphological studies of the rabbit. VII. Aortic arch variations in relation to regionally specific growth differences. *Anat. Rec.*, **96**, 183–200.

Saxén, L. and Toivonen, S. (1962) *Primary Embryonic Induction*, Academic Press, London.

Scadding, S.R. (1991) Skeletal patterns in the autopodium of native and regenerated limbs of the larval axolotl, *Ambystoma mexicanum*. *Can. J. Zool.*, **69**, 1–6.

Schaefer, S.A. and Lauder, G.V. (1996) Testing historical hypotheses of morphological change: Biomechanical decoupling in loricarioid catfishes. *Evolution*, **50**, 1661–75.

Schaeffer, B. (1948) The origin of mammalian ordinal characters. *Evolution*, **2**, 164–75.

Schaeffer, B. and Thomson, K.S. (1980) Reflections on agnathan-gnathosome relations, in *Aspects of Vertebrate History. Essays in Honor of Edwin Harris Colbert* (ed. L.L. Jacobs), Museum of Northern Arizona Press, Flagstaff, AZ, pp. 19–33.

Scharloo, W. (1987) Constraints in selection response, in *Genetic Constraints in Adaptive Evolution* (ed. V. Loeschcke), Springer-Verlag, Berlin, pp. 125–49.

Scharloo, W. (1990) The effect of developmental constraints on selection response, in *Organizational Constraints on the Dynamics of Evolution* (eds J. Maynard Smith and E. Vrba), Manchester University Press, Manchester, pp. 197–210.

Scharloo, W. (1991) Canalization: Genetic and developmental aspects. *Annu. Rev. Ecol. Syst.*, **22**, 65–93.

Scherson, T., Serbedzija, G., Fraser, S. and Bronner-Fraser, M. (1993) Regulative capacity of the cranial neural tube to form neural crest. *Development*, **118**, 1049–61.

Schierenberg, E. (1987) Reversal of cellular polarity and early cell-cell interactions in the embryo of *Caenorhabditis elegans*. *Devel. Biol.*, **122**, 452–63.

Schierwater, B., Murtha, M., Dick, M. *et al.* (1991) Homeoboxes in cnidarians. *J. Exp. Zool.*, **260**, 413–16.

Schilling, T.F. (1997) Genetic analysis of craniofacial development in the vertebrate embryo. *BioEssays*, **19**, 459–68.

Schilling, T.F., Walker, C. and Kimmel, C.B. (1996a) The *chinless* mutation and neural crest cell interactions in zebrafish jaw development. *Development*, **122**, 1417–26.

Schilling, T.F., Piotrowski, T., Grandel, H. *et al.*

(1996b) Jaw and branchial arch mutants in zebrafish 1: Branchial arches. *Development*, **123**, 329–44.

Schindewolf, O.H. (1969) Über den "Typus" in der morpholigischen und phylogenetischen Biologie. *Abh. Akad. Wiss. u. Lit., Mainz, Math.-Nat. Kl.*, no. 4, 58–131.

Schlichting, C.D. and Pigliucci, M. (1993) Evolution of phenotypic plasticity via regulatory genes. *Amer. Nat.*, **142**, 366–70.

Schliewen, U.K., Tautz, D. and Pääbo, S. (1994) Sympatric speciation suggested by monophyly of crater lake cichlids. *Nature*, **368**, 629–32.

Schlosser, G. and Roth, G. (1997) Evolution of nerve development in frogs. 2. Modified development of the peripheral nervous system in the direct-developing frog *Eleutherodactylous coqui* (Leptodactyliidae). *Brain Behav. Evol.*, **50**, 94–128.

Schmalhausen, I.I. (1917) On the extremities of *Ranidens sibiricus* Kessl. *Rev. Zool. Russe*, **2**, 129–35.

Schmalhausen, I.I. (1938) *The Integrating Factors of Evolution.* Nature (Priroda), Leningrad.

Schmalhausen, I.I. (1942) *The Organism as a Whole in Development and Evolution* (2nd edn), Moscow.

Schmalhausen, I.I. (1949) *Factors of Evolution. The Theory of Stabilizing Selection.* Translated by I. Dordick, edited by T. Dobzhansky, Blakiston, Philadelphia, PA. (Reprinted, University of Chicago Press, Chicago, IL, 1986.)

Schoenwolf, G.C. (1977) Tail (end) bud contributions to the posterior region of the chick embryo, *J. Exp Zool.*, **201**, 227–46.

Schoenwolf, G.C., Chandler, N.B. and Smith, J.L. (1985) Analysis of the origins and early fates of neural crest cells in caudal regions of avian embryos. *Devel. Biol.*, **110**, 467–79.

Schoenwolf, G.C. and Nichols, D.H. (1984) Histological and ultrastructural studies on the origin of caudal neural crest cells in mouse embryos. *J. Comp. Neurol.*, **222**, 496–505.

Schram, F.R. and Emerson, M.J. (1991) Arthropod pattern theory: A new approach to arthropod phylogeny. *Mem. Queensland Mus.*, **31**, 1–18.

Schughart, K., Kappen, C. and Ruddle, F.H. (1989) Duplication of large genomic regions during the evolution of vertebrate homeobox genes. *Proc. Natl Acad. Sci. USA*, **86**, 7067–71.

Schumacher, A. and Magnuson, T. (1997) Murine *Polycomb-* and *trithorax*-group genes regulate homeotic pathways and beyond. *Trends Genet.*, **13**, 167–70.

Schwartz, J.S. (1990) Darwin, Wallace, and Huxley, and *Vestiges of the Natural History of Creation. J. Hist. Biol.*, **23**, 127–153.

Schwenk, K. (1994/95) A utilitarian approach to evolutionary constraint. *Zoology*, **98**, 251–62.

Scott, M.P. (1992) Vertebrate homeobox gene nomenclature. *Cell*, **71**, 551–3.

Scriven, P.N. and Bauchau, V. (1992) The effect of hybridization on mandible morphology in an island population of the house mouse. *J. Zool. Lond.*, **226**, 573–83.

Searls, R.L. and Zwilling, E. (1964) Regeneration of the apical ectodermal ridge of the chick limb bud. *Devel. Biol.*, **9**, 38–55.

Sedgwick, A. (1910) Embryology, in *Encyclopædia Britannica* 11th edn, Volume 9, pp. 314–29, Cambridge.

Seger, J. and Stubblefield, J.W. (1996) Optimization and adaptation, in *Adaptation* (eds M.R. Rose and G.V. Lauder), Academic Press, San Diego, CA, pp. 93–123.

Seidel, F. (1960) Körpergrundgestalt und Keimstruktur. Eine Erörterung über die Grundlagen der vergleichenden und experimentellen Embryologie und deren Gültigkeit bei phylogenetischen Überlegungen. *Zool. Anz.*, **164**, 245–305.

Seilacher, A. (1989) Vendozoa: Organismic construction in the Proterozoic biosphere. *Lethaia*, **22**, 229–39.

Seilacher, A. (1992) Vendobionta and Psammocorallia: Lost constructions of Precambrian evolution. *J. Geol. Soc. Lond.*, **149**, 607–13.

Seleiro, E.A.P., Connolly, D.J. and Cooke, J. (1996) Early developmental expression and experimental axis determination by the chicken *Vg1* gene. *Curr. Biol.*, **6**, 1476–86.

Selleck, M.A.J. and Bronner-Fraser, M. (1995) Origins of the avian neural crest: The role of neural plate-epidermal interactions. *Development*, **121**, 525–38.

Semper, C. (1875) Die Stammesverwandschaft der Wirbelthiere und Wirbellosen. *Arb. Zool.-Zool. Inst. Würzburg*, **2**, 25–76.

Semper, C. (1876–7) Die Verwandschaftsbeziehungen der gegliederten Thiere. *Arb. Zool.-Zool. Inst. Würzburg*, **3**, 115–404.

Serbedzija, G.N. and McMahon, A.P. (1997) Analysis of neural crest cell migration in Splotch mice using a neural crest-specific LacZ reporter. *Devel. Biol.*, **185**, 139–47.

Sereno, P.C. and Chenggang, R. (1992) Early evolution of avian flight and perching: New evidence from the Lower Cretaceous of China. *Science*, **255**, 845–8.

Serres, E.R.A. (1830) Anatomie transcendante – quartrième mêmoire: loi de symétrie et de conjugaison du système sanguin. *Ann. Sci. Nat.*, **21**, 5–49.

Serres, E.R.A. (1860) Principes d'embryogénie, de zoogénie et de teratogénie. *Mém. Acad. Sci. Paris*, **25**, 1–943.

Servetnick, M. and Grainger, R.M. (1991a) Homeogenetic neural induction in *Xenopus*. *Devel. Biol.*, **147**, 73–82.

Servetnick, M. and Grainger, R.M. (1991b) Changes in neural and lens competence in *Xenopus* ectoderm: Evidence for an autonomous developmental timer. *Development*, **112**, 177–88.

Sessions, S.K. (1992) Developmental evolution: It's all in the timing. *Quart. Rev. Biol.*, **67**, 498–501.

Sessions, S.K. and Larson, A. (1987) Developmental correlates of genome size in plethodontid salamanders and their implications for genome evolution. *Evolution*, **41**, 1239–51.

Seufert, D.W. and Hall, B.K. (1990) Tissue interactions involving cranial neural crest in cartilage formation in *Xenopus laevis* (Daudin). *Cell Differ. Devel.*, **32**, 153–66.

Severtzov, A.N. (1927) Über die Beziehungen zwischen der Ontogenese und der Phylogenese der Thiere. *Z. Naturwiss. Jena*, **56**, 51–180.

Severtzov, A.N. (1931) *Morphologische Gesetzmässigkeiten der Evolution*. Gustav Fischer, Jena.

Shaffer, H.B. (1993) Phylogenetics of model organisms: The laboratory axolotl, *Ambystoma mexicanum*. *Syst. Biol.*, **42**, 508–22.

Shaffer, H.B., Clark, J.M. and Kraus, F. (1991) When molecules and morphology clash: A phylogenetic analysis of the North American ambystomatid salamanders (Caudata: Ambytomatidae). *Syst. Zool.*, **40**, 284–303.

Shaffer, H.B. and Voss, S.R. (1996) Phylogenetic and mechanistic analysis of a developmentally integrated character complex: Alternate life history modes in Ambystomatid salamanders. *Amer. Zool.*, **36**, 24–35.

Sham, M.H., Hunt, P., Nonchev, S. *et al.* (1992) Analysis of the murine Hox-2.7 gene: Conserved alternate transcripts with differential distribution in the nervous system and the potential for shared regulatory elements. *EMBO J.*, **11**, 1825–36.

Shardo, J.D. (1995) Comparative embryology of teleostean fishes. I. Development and staging of the American Shad, *Alosa sapidissima* (Wilson, 1811). *J. Morph.*, **225**, 125–67.

Sharman, A.C. and Holland, P.W.H. (1996) Conservation, duplication, and divergence of developmental genes during chordate evolution. *Neth. J. Zool.*, **46**, 47–67.

Sharpe, C.R. (1990) Regional neural induction in *Xenopus laevis*. *BioEssays*, **12**, 591–6.

Sharpe, C.R., Pluck, A. and Gurdon, J.B. (1989) XIF3, a *Xenopus* peripherin gene requires an inductive signal for enhanced expression in anterior neural tissue. *Development*, **107**, 701–14.

Shawlot, W. and Behringer, R.R. (1995) Requirement for *Lim1* in head-organizer function. *Nature*, **374**, 425–30.

Shea, B.T. (1983a) Allometry and heterochrony in the African apes. *Amer. J. Phys. Anthrop.*, **62**, 275–89.

Shea, B.T. (1983b) Paedomorphosis and neoteny in the pygmy chimpanzee. *Science*, **222**, 521–2.

Shea, B.T. (1985) Ontogenetic allometry and scaling: A discussion based on the growth and form of the skull in African apes, in *Size and Scaling in Primate Biology* (ed. W.L. Jungers), Plenum Press, New York, pp. 175–205.

Shea, B.T. (1989) Heterochrony in human evolution: The case for neoteny reconsidered. *Yearbook Phys. Anthropol.*, **32**, 69–101.

Shea, B.T. (1992) Developmental perspective on size change and allometry in evolution. *Evol. Anthrop.*, **1**, 125–33.

Shea, B.T. (1993) Bone growth and primate evolution, in *Bone, Volume 7. Bone Growth-B* (ed. B.K. Hall), CRC Press, Boca Raton, FL, 133–58.

Shea, B.T. and Bailey, R.C. (1996) Allometry and adaptation of body proportions and stature in African pygmies. *Amer. J. Phys. Anthropol.*, **100**, 311–40.

Shea, B.T., Hammer, R.E., Brinster, R.L. and Ravosa, M.R. (1990) Relative growth of the skull and postcranium in giant transgenic mice. *Genet. Rec. Camb.*, **56**, 21–34.

Shear, W.A. (1991) The early development of terrestrial ecosystems. *Nature*, **351**, 283–9.

Shimamura, M., Yasue, H., Ohshima, K. *et al.* (1997) Molecular evidence from retroposons that whales form a clade within even-toed ungulates. *Nature*, **388**, 666–70.

Shimeld, S.M. (1996) Retinoic acid, *Hox* genes and the anterior-posterior axis in chordates. *BioEssays*, **18**, 613–16.

Shimeld, S.M. (1997) Characterization of amphioxus HNF-3 genes: Conserved expression in the notochord and floor plate. *Devel. Biol.*, **183**, 74–85.

Shine, R. (1985) The evolution of viviparity in reptiles: An ecological analysis, in *Biology of the Reptilia*, Volume 15 (eds C. Gans and F. Billett), John Wiley and Sons, New York, pp. 605–94.

Shipman, P. (1997) Birds do it . . . did dinosaurs? *New Scientist*, **153**, 27–31.

Shishkin, M.A. (1968) Morphogenetic factors and the irreversibility of evolution. *Paleontol. J.*, **22**, 293–9.

Shu, D.-G., Conway Morris, S. and Zhang, X.-L. (1996) A *Pikaia*-like chordate from the Lower Cambrian of China. *Nature*, **384**, 157–8.

Shu, D.-G., Zhang, X.-L. and Chen, L. (1996) Reinterpretation of *Yunnanozoon* as the earliest known hemichordate. *Nature*, **380**, 428–30.

Shubin, N.H. (1991) The implications of "the Bauplan" for development and evolution of the tetrapod limb, in *Developmental Patterns of the Vertebrate Limb* (eds J.R. Hinchliffe, J.M. Hurle and D. Summerbell), Plenum Press, New York, pp. 411–21.

Shubin, N.H. (1994a) History, ontogeny, and evolution of the archetype, in *Homology, The Hierarchical Basis of Comparative Biology* (ed. B.K. Hall), Academic Press, San Diego, CA, pp. 250–71.

Shubin, N.H. (1994b) The phylogeny of development and the origin of homology, in *Interpreting the Hierarchy of Nature* (eds L. Grande and O. Rieppel), Academic Press, San Diego, CA, pp. 201–25.

Shubin, N.H. (1995) The evolution of paired fins and the origin of tetrapod limbs: Phylogenetic and transformational approaches. *Evol. Biol.*, **28**, 39–86.

Shubin, N.H. and Alberch, P. (1986) A morphogenetic approach to the origin and basic organization of the tetrapod limb. *Evol. Biol.*, **20**, 319–87.

Shubin, N., Tabin, C., and Carroll, S. (1997) Fossils, genes and the evolution of animal limbs. *Nature*, **388**, 639–48.

Shubin, N.H. and Wake, D.B. (1996) Phylogeny, variation, and morphological integration. *Amer. Zool.*, **36**, 51–60.

Shubin, N.H., Wake, D.B. and Crawford, A.J. (1995) Morphological variation in the limbs of *Taricha granulosa* (Caudata: Salamandridae): Evolutionary and phylogenetic implications. *Evolution*, **49**, 874–84.

Sibley, C.G. and Ahlquist, J.E. (1990) *Phylogeny and Classification of Birds: A Study in Molecular Evolution*, Yale University Press, New Haven, CT.

Sidow, A. (1992) Diversification of the *Wnt* gene family on the ancestral lineage of vertebrates. *Proc. Natl Acad. Sci. USA*, **89**, 5098–5102.

Simms, M.J. (1988) The role of heterochrony in the evolution of post-Palaeozoic crinoids, in *Echinoderm Biology* (eds R.D. Burke, P.V. Mladenov, P. Lambert and R.L. Parsley), A.A. Balkema, Rotterdam, Brookfield, pp. 97–102.

Simon, H.A. (1962) The architecture of complexity. *Proc. Amer. Phil. Soc.*, **106**, 467–82.

Simon, J. (1995) Locking in stable states of gene expression: Transcriptional control during *Drosophila* development. *Curr. Opin. Cell. Biol.*, **7**, 376–385.

Simonetta, A.M. and Conway Morris, S. (1991) *The Early Evolution of Metazoa and the Significance of Problematic Taxa*. Proceedings of an International Symposium held at the University of Camerino, 27–31 March, 1989. Cambridge University Press, Cambridge.

Simpson, G.G. (1944) *Tempo and Mode in Evolution*. Columbia University Press, New York.

Simpson, G.G. (1950) *The Meaning of Evolution. A Study of the History of Life and of its significance for Man*. Oxford University Press, London.

Simpson, G.G. (1953) *The Major Features of Evolution*. Columbia University Press, New York.

Simpson, G.G. (1959) The nature and origin of supra specific taxa. *Cold Spring Harb. Symp. Quant. Biol.*, **24**, 255–71.

Simpson, G.G. (1961) *Principles of Animal Taxonomy*. Columbia University Press, New York.

Sinervo, B. (1990) The evolution of maternal investment in lizards: An experimental and comparative analysis of egg size and its effects on offspring performance. *Evolution*, **44**, 279–94.

Sinervo, B. (1993) The effect of offspring size on physiology and life history. Manipulation of size using allometric engineering. *BioScience*, **43**, 210–18.

Sinervo, B. and Basolo, A.L. (1996) Testing adaptation using phenotypic manipulations, in *Adaptation* (eds M.R. Rose and G.V. Lauder), Academic Press, San Diego, CA, pp. 149–85.

Sinervo, B. and Huey, R.B. (1990) Allometric engineering: An experimental test of the causes of interpopulational differences in performance. *Science*, **248**, 1106–9.

Sinervo, B. and Licht, P. (1991) Hormonal and physiological control of clutch size, egg size, and egg shape in side-blotched lizards (*Uta stansburiana*): Constraints on the evolution of lizard life histories. *J. Exp. Zool.*, **257**, 252–64.

Sinervo, B. and McEdward, L.R. (1988) Developmental consequences of an evolutionary change in egg size: An experimental test. *Evolution*, **42**, 885–99.

Singer, C. (1959) *A History of Biology*. 3rd edn, Abelard- Schuman, London and New York.

Singer, C. (1959) *A History of Biology to about the year 1900. A General Introduction to the Study of Living Things* 3rd and revised edn, Abelard-Schuman, London and New York.

Skelton, P. (ed.) (1994) *Evolution: A Biological and Palaeontological Approach*, Addison-Wesley, Wokingham, UK.

Slack, J.M.W. (1985) *From Egg to Embryo. Determinative Events in Early Development*. Cambridge University Press, Cambridge.

Slack, J.M.W., Dale, L. and Smith, J.C. (1984) Analysis of embryonic induction by using cell lineage markers. *Phil. Trans. R. Soc. B*, **307**, 331–6.

Slack, J.M.W., Holland, P.W.H. and Graham, C.F. (1993) The zootype and the phylotypic stage, *Nature*, **361**, 490–2.

Slack, J.M.W. and Isaacs, H.V. (1989) Presence of basic fibroblast growth factor in the early *Xenopus* embryo. *Development*, **105**, 147–53.

Slatkin, M. (1987) Quantitative genetics of heterochrony. *Evolution*, **41**, 799–811.

Slavkin, H.C., Sasano, Y., Kikunaga, S. *et al.* (1990) Cartilage, bone and tooth induction during early embryonic mouse mandibular morphogenesis using serumless, chemically-defined medium. *Connect. Tiss. Res.*, **24**, 41–52.

Sloan, P.R. (ed.) (1992) *Richard Owen. The Hunterian Lectures in Comparative Anatomy May–June, 1837*, With an Introductory Essay and Commentary, The University of Chicago Press, Chicago, IL.

Sluys, R. (1996) The notion of homology in current comparative biology. *J. Zool. Syst. Evol. Res.*, **34**, 145–52.

Sluys, R., Hauser, J. and Wirth, Q.J. (1997) Deviation from the groundplan: A unique new species of freshwater planarian from South Brazil (Platyhelminthes, Tricladida, Paludicola). *J. Zool. Lond.*, **241**, 593–601.

Smirnov, S.V. (1992) The influence of variation in larval period on adult cranial diversity in *Pelobates fuscus* (Anura: Pelobatidae). *J. Zool. Lond.*, **226**, 601–12.

Smirnov, S.V. (1994) Postmaturation skull development in *Xenopus laevis* (Anura, Pipidae): Late-appearing bones and their bearing on the pipid ancestral morphology. *Russian J. Herpetol.*, **1**, 21–9.

Smirnov, S.V. (1995) Extra bones in the *Pelobates* skull as evidence of the paedomorphic origin of the Anura. *Zh. Obs. Biol.*, **56**, 317–28.

Smith, A.B. (1992) Echinoderm phylogeny: Morphology and molecules approach accord. *Trends Ecol. Evol.*, **7**, 224–9.

Smith, A.B., Littlewood, D.T.J. and Wray, G.A. (1995) Comparing patterns of evolution: larval and adult life history stages and ribosomal RNA of post-Palaeozoic echinoids. *Phil. Trans. R. Soc. Lond. B*, **349**, 11–18.

Smith, J.C. (1987) A mesoderm-inducing factor is produced by a *Xenopus* cell line. *Development*, **99**, 3–14.

Smith, J.C. (1989) Induction and early embryonic development. *Curr. Opin. Cell Biol.*, **1**, 1061–70.

Smith, J.C., Price, B.M.J., Van Nimmen, K. and Huylebroeck, D. (1990) Identification of a potent *Xenopus* mesoderm-inducing factor as a homologue of activin A. *Nature*, **345**, 729–31.

Smith, K.C. (1992a) Neo-rationalism versus neo-Darwinism: Integrating development and evolution. *Biol. Phil.*, **7**, 431–51.

Smith, K.C. (1992b) The new problem of genetics: A response to Gifford. *Biol. Philos.*, **7**, 331–48.

Smith, K.K. (1994) Development of craniofacial musculature in *Monodelphis domestica* (Marsupialia, Didelphidae). *J. Morph.*, **222**, 149–73.

Smith, K.K. and van Nievelt, A.F.H. (1997) Comparative rates of development in *Monodelphis* and *Didelphis*. *Science*, **275**, 683–4.

Smith, M.M. and Hall, B.K. (1990) Developmental and Evolutionary origins of vertebrate skeletogenic and odontogenic tissues. *Biol. Rev.*, **65**, 277–374.

Smith, M.M. and Hall, B.K. (1993) A developmental model for evolution of the vertebrate exoskeleton and teeth: The role of cranial and trunk neural crest. *Evol. Biol.*, **27**, 387–448.

Smith, M.M., Hickman, A., Amanze, D. *et al.* (1994) Trunk neural crest origin of caudal fin mesenchyme in the zebrafish *Brachydanio rerio*. *Proc. R. Soc. Lond. B*, **256**, 137–45.

Smith, S.C., Graveson, A.C. and Hall, B.K. (1994) Evidence for a developmental and evolutionary link between placodal ectoderm and neural crest. *J. Exp Zool.*, **270**, 292–301.

Smith, T.B. and Skúlason, S. (1996) Evolutionary significance of resource polymorphism in fishes, amphibians, and birds. *Annu. Rev. Ecol. Syst.*, **27**, 111–33.

Smocovitis, V.B. (1996) *Unifying Biology: The Evolutionary Synthesis and Evolutionary Biology*, Princeton University Press, Princeton.

Smothers, J.F., von Dohlen, C.D., Smith, L.H. and Spall, R.D. (1994) Molecular evidence that the myxozoan protists are metazoans. *Science*, **265**, 1719–21.

Snow, M.H.L. and Tam, P.P.L. (1979) Is compensatory growth a complicating factor in mouse teratology? *Nature*, **279**, 555–7.

Snow, N. (1996) The phylogenetic utility of lemmatal micromorphology in *Leptochloa* S.L. and related genera in subtribe Eleusininae (Poaceae, Chloridoideae, Eragrostideae). *Ann. Missouri Bot. Gard.*, **83**, 504–29.

Sokol, S., Christian, J.L., Moon, R.T. and Melton, D.A. (1991) Injected Wnt RNA induces a complete body axis in *Xenopus* embryos. *Cell*, **67**, 741–52.

Sommer, R.J. (1997) Evolution and development – The nematode vulva as a case study. *BioEssays*, **19**, 225–31.

Sommer, R.J. and Tautz, D. (1991) Segmentation gene expression in the housefly *Musca domestica*. *Development*, **113**, 419–30.

Sommer, R.J. and Tautz, D. (1993) Involvement of an orthologue of the *Drosophila* pair-rule gene hairy in segment formation of the short-germ-band embryo of *Tribolium* (Coleoptera). *Nature*, **361**, 448–50.

Song, H-K. and Sawyer, R.H. (1996) Dorsal dermis of the scaleless (*sc*/*sc*) embryo directs normal feather pattern formation until day 8 of development. *Devel. Dyn.*, **205**, 82–91.

Sordino, P. and Duboule, D. (1996) A molecular approach to the evolution of vertebrate paired appendages. *Trends Ecol. Evol.*, **11**, 114–19.

Sordino, P., van der Hoeven, F. and Duboule, D. (1995) *Hox* gene expression in teleost fins and the origin of vertebrate digits. *Nature*, **375**, 678–81.

Soulé, M. (1967) Phenetics of natural populations. II. Asymmetry and evolution in a lizard. *Amer. Nat.*, **101**, 141–60.

Soulé, M. (1979) Heterozygosity and developmental stability: Another look. *Evolution*, **33**, 396–401.

Southward, E.C. (1963) Pogonophora. *Oceanogr. Mar. Biol. Ann. Rev.*, **1**, 405–28.

Spemann, H. (1915) Zur Geschichte und Kritik des Begriffs der Homologie, in *Die Kultur der Gegenwart* (ed. P. Hinneberg), Verl., Teubner, Leipzig, pp. 63–86.

Spemann, H. (1938) *Embryonic Development and Induction*. Yale University Press, New Haven, CT.

(Reprinted 1962, Hafner Publishing Co., New York; 1988, Garland Publishing Inc., New York.)

Spemann, H. and Mangold, H. (1924) Über Induktion von Embryonalanlagen durch Implantation artfremder Organisatoren. *Wilhelm Roux Arch. EntwMech. Org.*, **100**, 599–628.

Spencer, H. (1852) The Development Hypothesis. *Leader*, 20 March, p. 1.

Spencer, H. (1857) Progress: Its law and cause. *Westminster Review*, **67**, 445–85.

Spencer, H. (1886) *The Principles of Biology. 2 Volumes*, D. Appleton, New York.

Stanley, S.M. (1973) Effects of competition on rates of evolution, with special reference to bivalve molluscs and mammals. *Syst. Zool.*, **22**, 486–506.

Stanley, S.M. (1979) *Macroevolution: Pattern and Process*. W. H. Freeman and Co., San Francisco, CA.

Stanley, S.M. (1981) *The New Evolutionary Timetable. Fossils, Genes, and the Origin of Species*. Basic Books, New York.

Stanley, S.M. (1982) Gastropod torsion: Predation and the opercular imperative. *N. Jb. Geol. Paläont. Abh.*, **164**, 95–106.

Starck, J.M. (1993) Evolution of avian ontogenies. *Curr. Ornithol.*, **10**, 275–366.

Starck, J.M. (1996) Comparative morphology and cytokinetics of skeletal growth in hatchlings of altricial and precocial birds. *Zool. Anz.*, **235**, 53–75.

Steadman, P. (1979) *The Evolution of Designs. Biological Analogy in Architecture and the Applied Arts*, Cambridge University Press, Cambridge.

Stearns, S.C. (1976) Life-history tactics: A review of the ideas. *Quart Rev Biol*, **51**, 3–47.

Stearns, S.C. (1982) The role of development in the evolution of life histories, in *Evolution and Development* (ed. J.T. Bonner), Springer-Verlag, Berlin, pp. 237–58.

Stearns, S.C. (1986) Natural selection and fitness, adaptation and constraint, in *Patterns and Processes in the History of Life* (eds D.M. Raup and D. Jablonski), Springer-Verlag, Berlin, pp. 23–44.

Stearns, S.C. (1989) The evolutionary significance of phenotypic plasticity. *Bioscience*, **37**, 436–45.

Stearns, S.C. (1992) *The Evolution of Life Histories*, Oxford University Press, Oxford.

Stebbins, G.L.Jr. (1950) *Variation and Evolution in Plants*, Columbia University Press, New York.

Stebbins, G.L.Jr. (1968) Integration of development and evolutionary progress, in *Population Biology and Evolution* (ed. R.C. Lewontin), Syracuse University Press, Syracuse, NY, pp. 17–36.

Stebbins, G.L.Jr. (1973) Adaptive radiation and the origin of form in the earliest multicellular organisms. *Syst. Zool.*, **22**, 478–85.

Stein, S., Fritsch, R., Lemaire, L. and Kessel, M. (1996) Checklist: vertebrate homeobox genes. *Mech. Devel.*, **55**, 91–108.

Steinberg, M.S. and Poole, T.J. (1982) Cellular adhesive differentials as determinants of morphogenetic movements and organ segregation, in *Developmental Order: Its Origin and Regulation* (eds S. Subtelny and P.B. Green), Alan R. Liss, New York, pp. 351–78.

Stemberger, R.S. and Gilbert, J.J. (1987) Multiple-species induction of morphological defenses in the rotifer *Keratella testudo*. *Ecology*, **68**, 370–8.

Stennard, F., Ryan, K. and Gurdon, J.B. (1997) Markers of vertebrate mesoderm induction. *Curr. Opin. Genet. Devel.*, **7**, 620–7.

Stern, C. (1958) Selection for subthreshold differences and the origin of pseudoexogenous adaptations. *Amer. Nat.*, **92**, 313–16.

Stern, C. (1959) Variation and hereditary transmission. *Proc. Amer. Phil. Soc.*, **103**, 183–9.

Stern, C.D. and Holland, P.W.H. (1993) *Essential Developmental Biology. A Practical Approach*, Oxford University Press, Oxford.

Sterrer, W. (1972) Systematics and evolution within the Gnathostomulida. *Syst. Zool.*, **21**, 151–73.

Stevens, P.S. (1984) Homology and phylogeny: Morphology and systematics. *Syst. Bot.*, **9**, 395–409.

Stiassny, M.L.J. (1992) Atavisms, phylogenetic character reversals, and the origin of evolutionary novelties. *Neth. J. Zool.*, **42**, 260–76.

Stiassny, M.L.J. and Jensen, J. (1987) Labroid interrelationships revisited: Morphological complexity, key innovations, and the study of comparative diversity. *Bull. Mus. Comp. Zool. Harv. Univ.*, **151**, 269–319.

Stirling, G. and McQueen, D.J. (1987) The cyclomorphic response of *Daphnia galeata mendatae*: Polymorphism or phenotypic diversity. *J. Plankton Res.*, **9**, 1093–1112.

Stockard, C.R. (1930) The presence of a factorial basis for characters lost in evolution. The atavistic reappearance of digits in mammals. *Amer. J. Anat.*, **45**, 345–78.

Stocum, D. (1995) *Wound repair, Regeneration and Artificial Tissues*, R. G. Landes Co., New York.

Stokes, M.D. and Holland, N.D. (1995) Embryos and larvae of a lancelet, *Branchiostoma floridae*, from hatching through metamorphosis: Growth in the laboratory and external morphology. *Acta Zool.*, **76**, 105–20.

Stoleson, S.H. and Beissinger, S.R. (1995) Hatching asynchrony and the onset of incubation in birds, revisited. When is the critical period? *Curr. Ornithol.*, **12**, 191–270.

Storm, E.E. and Kingsley, D.M. (1996) Joint patterning defects caused by single and double mutations in members of the bone morphogenetic (BMP) family. *Development*, **122**, 3969–79.

Strathmann, R.R. (1978) The evolution of loss of feeding larval stages of marine invertebrates. *Evolution*, **32**, 894–906.

Strathmann, R.R. (1985) Feeding and nonfeeding larval development and life-history evolution in marine invertebrates. *Annu. Rev. Ecol. Syst.*, **16**, 339–61.

Strathmann, R.R. (1988) Larvae, phylogeny, and von Baer's law, in *Echinoderm Phylogeny and Evolutionary Biology* (eds C.R.C. Paul and A.B. Smith), Clarendon Press, Oxford, pp. 53–68.

Strathmann, R.R. (1993) Hypotheses on the origins of marine larvae. *Annu. Rev. Ecol. Syst.*, **24**, 89–117.

Strathmann, R.R., Fenaux, L. and Strathmann, M.F. (1992) Heterochronic developmental plasticity in larval sea urchins and its implications for evolution of nonfeeding larvae. *Evolution*, **46**, 972–86.

Strauss, R.E. (1987) On allometry and relative growth in evolutionary studies. *Syst. Zool.*, **36**, 72–5.

Strauss, R.E. (1990) Heterochronic variation in the developmental timing of cranial ossifications in poeciliid fishes (Cyprinodontiformes). *Evolution*, **44**, 1558–67.

Strauss, R.E. and Altig, R. (1992) Ontogenetic body form changes in three ecological morphotypes of anuran tadpoles. *Growth Devel. Aging*, **56**, 3–16.

Streicher, J. (1991) Plasticity in skeletal development: Knee-joint morphology in fibula-deficient chick embryos, in *Developmental Patterns of the Vertebrate Limb* (eds J.R. Hinchliffe, J.M. Hurle, and D. Summerbell), Plenum Press, New York, pp. 407–9.

Streicher, J. and Müller, G.B. (1992) Natural and experimental reduction of the avian fibula: Developmental thresholds and evolutionary constraint. *J. Morph.*, **214**, 269–85.

Striedter, G.F. (1997) The telencephalon in tetrapods in evolution. *Brain Behav. Evol.*, **49**, 179–213.

Striedter, G.F. and Northcutt, R.G. (1991) Biological hierarchies and the concept of homology. *Brain Behav. Evol.*, **38**, 177–89.

Sturmbauer, C. and Meyer, A. (1992) genetic divergence, speciation and morphological stasis in a lineage of African cichlid fishes. *Nature*, **358**, 578–81.

Sturtevant, A.H. (1923) Inheritance of direction of coiling in *Limnaea*. *Science*, **58**, 269–70.

Sun, H., Rodin, A., Zhou, Y. *et al* (1997) Evolution of paired domains: Isolation and sequencing of jellyfish and hydra *PAX* genes related to *Pax-5* and *Pax-6*. *Proc. Natl Acad. Sci. USA*, **94**, 5156–61.

Sundin, O.H. and Eichele, G. (1990) A homeodomain protein reveals the metameric nature of the developing chick hindbrain. *Genes & Devel.*, **4**, 1267–76.

Surani, M.A., Kothary, R., Allen, N.D. *et al.* (1990) Genome imprinting and development in the mouse. *Development*, 1990, Suppl., 89–98.

Suzuki, A., Kaneko, E., Ueno, N *et al.* (1997a) Regulation of epidermal induction by BMP2 and BMP7 signaling. *Devel. Biol.*, **189**, 112–22.

Suzuki, A., Ueno, N., and Hemmati-Brivanlou, A. (1997b) *Xenopus msx1* mediates epidermal induction and neural inhibition by BMP4. *Development*, **124**, 3037–44.

Swain, D.P. (1992a) The functional basis of natural selection for vertebral traits of larvae in the stickleback *Gasterosteus aculeatus*. *Evolution*, **46**, 987–97.

Swain, D.P. (1992b) The functional basis of natural selection for vertebral traits of larvae in the stickleback *Gasterosteus aculeatus*: Reversal in the direction of selection at different larval sizes. *Evolution*, **46**, 998–1013

Swalla, B.J. (1992) The role of maternal factors in ascidian muscle development. *Sem. Devel. Biol.*, **3**, 287–95.

Swalla, B.J. (1996) Strategies for cloning developmental genes using closely related species, in *Molecular Zoology: Advances, Strategies and Protocols* (eds J.D. Ferraris and S.R. Palumbi), Wiley-Liss Inc., New York, pp. 197–208, 549–54.

Swalla, B.J. and Jeffery, W.R. (1990) Interspecific hybridization between an anural and urodele ascidian: Differential expression of urodele features suggests multiple mechanisms control anural development. *Devel. Biol.*, **142**, 319–34.

Swalla, B.J. and Jeffery, W.R. (1995) A maternal RNA localized in the yellow crescent is segregated to the larval muscle cells during ascidian development. *Devel. Biol.*, **170**, 353–64.

Swalla, B.J. and Jeffery, W.R. (1996) Requirement of the *Manx* gene for expression of chordate features in a tailless ascidian larva. *Science*, **274**, 1205–8.

Swalla, B.J., Makabe, K.W., Satoh, N. and Jeffery, W.R. (1993) Novel genes expressed differentially in ascidians with alternate modes of development. *Development*, **119**, 307–18.

Swalla, B.J., White, M.E., Zhou, J. and Jeffery, W.R. (1994) Heterochronic expression of an adult muscle actin gene during ascidian larval development. *Devel. Genet.*, **15**, 51–63.

Symonds, J.A. (1871) *Miscellanies by John Addington Symonds*, M.D., Macmillan, London.

Szathmáry, E. and Maynard Smith, J. (1995) The major evolutionary transition. *Nature*, **374**, 227–32.

Tabin, C.J. and Laufer, E. (1993) *Hox* genes and serial homology. *Nature*, **361**, 692–3.

Tague, R.G. (1997) Variability of a vestigial structure: First metacarpal in *Colobus guereza* and *Ateles geoffroyi*. *Evolution*, **51**, 595–605.

Taira, M., Saint-Jeannet, J.-P. and Dawid, I.B. (1997) Role of the *Xlim-1* and *Xbra* genes in anteroposterior patterning of neural tissue by the head and trunk organizer. *Proc. Natl Acad. Sci. USA*, **94**, 895–900.

Tait, J. (1928) Homology, analogy and plasis. *Quart. Rev. Biol.*, **3**, 151–73.

Takahashi, H., Ishida, K., Makabe, K.W. and Satoh, N. (1997) Isolation of cDNA clones for genes that are expressed in the tail region of the ascidian tailbud embryo. *Int. J. Devel. Biol.*, **41**, 691–9.

Takahashi, Y. and Le Douarin, N.M. (1990) cDNA cloning of a quail homeobox gene and its expression in neural crest-derived mesenchyme and lateral plate mesoderm. *Proc. Natl Acad. Sci. USA*, **87**, 7482–86.

Takeichi, M. (1987) Cadherins: A molecular family essential for selective cell-cell adhesion and animal morphogenesis. *Trends Genet.*, **3**, 213–17.

Tam, P.P.L. and Quinlan, G.A. (1996) Mapping vertebrate embryos. *Curr. Biol.*, **6**, 104–6.

Tam, P.P.L., and Selwood, L. (1996) Development of lineages of primary germ layers, extra-embryonic membranes and fetus. *Reprod. Fertil. Devel.*, **8**, 803–5.

Tanaka, M., Tamura, K., Noji, S. *et al.* (1997) Induction of additional limb at the dorsal-ventral boundary of a chick embryo. *Devel. Biol.*, **182**, 191–203.

Tarkowski, A.K. (1959) Experiments on the development of isolated blastomeres of mouse eggs. *Nature*, **184**, 1286–7.

Taylor, P.J. (1987) Historical versus selectionist explanations in evolutionary biology. *Cladistics*, **3**, 1–13.

Teather, K. (1996) Patterns of growth and asymmetry in nestling tree swallows. *J. Avian Biol.*, **27**, 302–10.

Ten Cate, G. (1953) *The Intrinsic Development of Amphibian Embryos*. North-Holland, Amsterdam.

Terazawa, K., and Satoh, N. (1995) Spatial expression of the amphioxus homologue of *Brachyury* (*T*) gene during early embryogenesis of *Branchiostoma belcheri*. *Devel. Growth Differ.*, **37**, 395–401.

Thacher, J.K. (1877) Median and paired fins, a contribution to the history of vertebrate limbs. *Trans. Connecticut Acad. Sci.*, **3**, 281–308, Plates XLIX–LX.

Theiler, K. (1972) *The House Mouse. Development and Normal Stages from Fertilization to 4 Weeks of Age*. Springer-Verlag, Berlin.

Thesleff, I. (1995) Homeobox genes and growth factors in regulation of craniofacial and tooth morphogenesis. *Acta Odontol. Scand.*, **53**, 129–34.

Thesleff, I. and Sahlberg, C. (1996) Growth factors as inductive signals regulating tooth morphogenesis. *Sem. Cell Devel. Biol.*, **7**, 185–93.

Thesleff, I., Vaahtokari, A. and Partanen, A-M. (1995a) Regulation of organogenesis. Common molecular mechanisms regulating the development of teeth and other organs. *Int. J. Devel. Biol.*, **39**, 35–50.

Thesleff, I., Vaahtokari, A., Kettungen, P. and Äberg, T. (1995b) Epithelial-mesenchymal signalling during tooth development. *Conn. Tiss. Res.*, **32**, 9–15.

Thewissen, J.G.M. and Babcock, S.K. (1992) The origin of flight in bats: To go where no mammal has gone before. *BioScience*, **42**, 340–45.

Thewissen, J.G.M. and Fish, F.E. (1997) Locomotor evolution in the earliest cetaceans: functional model, modern analogues, and paleontological evidence. *Paleobiology*, **23**, 482–90.

Thewissen, J.G.M. and Hussain, S.T. (1993) Origin of underwater hearing in whales. *Nature*, **361**, 444–5.

Thewissen, J.G.M., Hussain, S.T. and Arif, M. (1994) Fossil evidence for the origin of aquatic locomotion in Archaeocete whales. *Science*, **263**, 210–12.

Thom, R. (1989) An inventory of Waddingtonian concepts, in *Theoretical Biology* (eds B. Goodwin and P. Saunders), Edinburgh University Press, Edinburgh, pp. 1–7.

Thomas, A.L.R. (1997) The breath of life – did increased oxygen levels trigger the Cambrian explosion? *Trends Ecol. Evol.*, **12**, 44–5.

Thomas, R.D.K. and Reif, W.-E. (1991) Design elements employed in the construction of animal skeletons, in *Constructional Morphology and Evolution* (eds N. Schmidt-Kittler and K. Vogel), Springer-Verlag, Berlin, pp. 283–94.

Thompson, D'A.W. (1942) *Growth and Form* 2nd edn, Macmillan & Co., New York.

Thompson, J.D. (1991) Phenotypic plasticity as a component of evolutionary change. *Trends Ecol. Evol.*, **6**, 246–9.

Thompson, J.N. and Thoday, J.M. (1975) Genetic assimilation of part of a mutant phenotype. *Genet. Res.*, **26**, 149–62.

Thomsen, G., Woolf, T., Whitman, M. *et al.* (1990) Activins are expressed early in *Xenopus* embryogenesis and can induce axial mesoderm and anterior structures. *Cell*, **63**, 485–93.

Thomson, K.S. (1966) The evolution of the tetrapod middle ear in the rhipidistian-tetrapod transition. *Amer. Zool.*, **6**, 379–97.

Thomson K.S. (1975) On the biology of cosmine. *Bull. Peabody Mus. Yale Univ.*, **40**, 1–50.

Thomson, K.S. (1986) Essay review: The relationship between development and evolution. *Oxford Surv. Evol. Biol.*, **2**, 220–33.

Thomson, K.S. (1987) The neural crest and the morphogenesis and evolution of the dermal skeleton in vertebrates, in *Developmental and Evolutionary Aspects of the Neural Crest* (ed. P.F.A. Maderson), John Wiley & Sons, New York, pp. 301–38.

Thomson, K.S. (1988) *Morphogenesis and Evolution*. Oxford University Press, Oxford.

Thomson, K.S. (1991a) *Living Fossil. The Story of the Coelacanth*, W. W. Norton & Co., New York.

Thomson, K.S. (1991b) Where did tetrapods come from? *Amer. Sci.*, **79**, 488–90.

Thomson, K.S. (1991c) Parallelism and convergence in the horse limb: The internal-external dichotomy, in *New Perspectives on Evolution* (eds L. Warren and H. Koprowski), Wiley-Liss, New York, pp. 101–22.

Thomson, K.S. (1993) Segmentation, the adult skull and the problem of homology, in *The Vertebrate Skull. Volume 1. Development* (eds J. Hanken and B.K. Hall), The University of Chicago Press, Chicago, IL, pp. 36–68.

Thomson, K.S. (1997) Natural theology. *Amer. Sci.*, **85**, 219–21.

Thorogood, P.V. (1991) The development of the teleost fin and implications for our understanding of tetrapod limb evolution, in *Developmental Patterning of the Vertebrate Limb* (eds J.R.

Hinchliffe, J.M. Hurle and D. Summerbell), NATO ASI Series A: Life Sciences, Plenum Press, New York, Vol. 205, pp. 347–54,

Thorogood, P.V. (1993) Differentiation and morphogenesis of cranial skeletal tissues, in *The Skull, Volume 1. Development* (eds J. Hanken and B.K. Hall), The University of Chicago Press, Chicago, IL, pp. 112–52.

Thorogood, P.V. and Smith, L. (1984) Neural crest cells: The role of extracellular matrix in their differentiation and migration, in *Matrices and Cell Differentiation* (eds R.B. Kemp and J.R. Hinchliffe), Alan R. Liss, Inc., New York, pp. 171–85.

Thorson, G. (1950) Reproduction and larval ecology of marine bottom invertebrates. *Biol. Rev.*, **25**, 1–45.

Tickle, C. (1996) Vertebrate limb development. *Sem. Cell Devel. Biol.*, **7**, 137–43.

Tiedermann, H., Grunz, H., Loppnow-Blinde, B. and Tiedermann, H. (1994) Basic fibroblast growth factor can induce exclusively neural tissue in *Triturus* ectoderm explants. *Roux's Arch. Devel. Biol.*, **203**, 304–9.

Tigano, C. and Parenti, L.R. (1988) Homology of the median ethmoid ossification of *Aphanius fasciatus* and other Atherinomorph fishes. *Copeia*, **1988**, 866–70.

Tinsley, R.C. and Kobel, H.R. (eds) (1996) The Biology of *Xenopus*. Symposium Number 68 of the Zoological Society of London, Clarendon Press, Oxford.

Tintant, H. and Devillers, C. (1995) Atavism in present and past – its function in evolution. *Bull. Soc. Zool. Fr.*, **120**, 327–34.

Tran, S. and Hall, B.K. (1989) Growth of the clavicle and development of clavicular secondary cartilage in the embryonic mouse. *Acta Anat.*, **135**, 200–7.

Trasler, J.M., Trasler, D.G., Bestor, T.H. *et al.* (1996) DNA methyltransferase in normal and $Dnmt^n$/$Dnmt^n$ mouse embryos. *Devel. Dyn.*, **206**, 239–47.

Tredern, L.S. (1808) *Dissertatio inauguralis Medica Sistens Ovi Avium Historiæ et Incubationis Prodromum*, Etzdorf, Jena.

Trewavas, E., Green, J. and Corbet, S.A. (1972) Ecological studies on crater lakes in West Cameroon. Fishes of Barombi Mbo. *J. Zool. Lond.*, **167**, 41–95.

Trueb, L. (1985) A summary of the osteocranial development in anurans with notes on the sequence of cranial ossification in *Rhinophrynus dorsalis* (Anura: Pipoidea: Rhinophrynidae). *S. Afr. J. Sci.*, **81**, 181–5.

Trueb, L. (1996) Historical constraints and morphological novelties in the evolution of the skeletal system of pipid frogs (Anura: Pipidae), in *The Biology of* Xenopus (eds R.C. Tinsley and H.R. Kobel), Symposium Number 68 of the Zoological Society of London, Clarendon Press, Oxford, pp. 349–77.

Trueb, L. and Hanken, J. (1992) Skeletal development in *Xenopus laevis* (Anura: Pipidae). *J. Morph.*, **214**, 1–41.

Tsonis, P.A. (1996) *Limb Regeneration*, Cambridge University Press, Cambridge.

Tung, T.C. (1934) L'organisation de l'oeuf fécondé d'*Ascidiella scabra* au début de la segmentation. *C. r. Séance. Soc. Biol.*, **115**, 1375–8.

Tung, T.C., Wu, S.C. and Tung, T.T.F. (1962) Experimental studies on the neural induction in *Amphioxus*. *Scient. Sinica*, **11**, 805–20.

Tyler, S. (1988) The role of function in determination of homology and convergence – examples from invertebrate adhesive organs. *Fortsch. Zool.*, **36**, 331–47.

Urbanek, P., Fetka, I., Meisler, M.H. and Busslinger, M. (1997) Cooperation of *Pax2* and *Pax5* in midbrain and cerebellum development. *Proc. Natl Acad. Sci. USA*, **94**, 5703–8.

Vaahtokari, A., Äberg, T., Jernvall, J. *et al.* (1996a) The enamel knot as a signalling center in the developing mouse tooth. *Mech. Devel.*, **54**, 39–43.

Vaahtokari, A., Äberg, T. and Thesleff, I. (1996b) Apoptosis in the developing tooth: Association with an embryonic signalling center and suppression by EGF and FGF-4. *Development*, **122**, 121–9.

Vacelet, J. and Boury-Esnault, N. (1995) Carnivorous sponges. *Nature*, **373**, 333–5.

Vaglia, J.L., Babcock, S.K., and Harris, R.N. (1997) Tail development and regeneration throughout the life cycle of the four-toed salamander *Hemidactylium scutatum*. *J. Morph.*, **233**, 15–29.

Vainio, S., Karavanova, I., Jowett, A. and Thesleff, I. (1993) Identification of BMP-4 as a signal mediating secondary induction between epithelial and mesenchymal tissues during early tooth development. *Cell*, **75**, 45–58.

Valentine, J.W. (1985) The evolution of complex animals, in *What Darwin Began* (ed. L.R. Godfrey), Allyn and Bacon, Boston, MA, pp. 258–73.

Valentine, J.W. (1986) Fossil record of the origin of Baupläne and its implications, in *Patterns and Processes in the History of Life* (eds D.M. Raup and D. Jablonski), Dahlem Conference 1986, Springer-Verlag, Berlin, pp. 209–22.

Valentine, J.W. (1992a) *Dickinsonia* as a polypoid organism. *Paleobiology*, **18**, 378–82.

Valentine, J.W. (1992b) The macroevolution of phyla, in *Origin and Early Evolution of the Metazoa* (eds P.W. Signor and J.H. Lipps), Plenum Press New York, pp. 525–53.

Valentine, J.W. (1997) Cleavage patterns and the topology of the metazoan tree of life. *Proc. Natl Acad. Sci. USA*, **94**, 8001–5.

Valentine, J.W., Awramik, S.M., Signor, P.W. and Sadler, P.M. (1991) The biological explosion at the Precambrian-Cambrian boundary. *Evol. Biol.*, **25**, 279–356.

Valentine, J.W., Collins, A.G. and Meyer, C.P. (1994) Morphological complexity increase in metazoans. *Paleobiology*, **20**, 131–42.

Valentine, J.W., Erwin, D.H. and Jablonski, D. (1996) Developmental evolution of metazoan body plans: The fossil evidence. *Devel. Biol.*, **173**, 373–81.

Valentine, J.W. and May, C.L. (1996) Hierarchies in biology and paleontology. *Paleobiology*, **22**, 23–33.

Van Beneden, E. (1874) De la distinction originelle du testicule et de l'ovaire; caractère sexuel des feuillets primordiaux de l'embryon; hermaphrodisme morphologique de toute individualité animale: essai d'une théorie de la fécondation. *Bruxelles Acad. Sci. Bull.* **37**, 530–95.

Van Beneden, E. (1875) La maturation de l'oeuf, la fécondation et les premières de développement embryonnaire des Mammifères d'après les recherches faites chez le lapin. *Bruxelles Acad. Sci. Bull.* **40**, 686–736.

Van Beneden, E. (1876) Contribution à l'histoire de la vésicule germinative et du premier noyau embryonnaire. *Bruxelles Acad. Sci. Bull.*, **41**, 38–58; *Quart. J. Microsc. Sci.*, **16**, 153–82.

van den Eijnden-Van Raaij, A.J.M., van Zoelent, E.J.J., van Nimmen, K. *et al.* (1990) Activin-like factor from a *Xenopus laevis* cell line responsible for mesoderm induction. *Nature*, **345**, 732–4.

Van der Hammen, L. (1988) *Unfoldment and Manifestation*. SPB Academic, The Hague.

Van der Hoeven, F., Sordino, P., Fraudeau, N. *et al.* (1996) Teleost *HoxD* and *HoxA* genes: Comparison with tetrapods and functional evolution of the *HoxD* complex. *Mech. Devel.*, **54**, 9–21.

Van de Water, T.R. and Galinovic-Schwartz, V. (1986) Dysmorphogenesis of the inner ear: Disruption of extracellular matrix (ECM) formation by an L-proline analog in otic explants. *J. Craniofac. Genet. Devel. Biol.*, **6**, 113–30.

Van Valen, L. (1962) A study of fluctuating asymmetry. *Evolution* **16**, 125–42.

Van Valen, L. (1973) Festschrift. *Science*, **180**, 488.

Van Valen, L. (1982) Homology and causes. *J. Morph.*, **173**, 305–12.

Vargesson, N., Clarke, J.D.W., Vincent, K. *et al.* (1997) Cell fate in the chick limb bud and relationship to gene expression. *Development*, **124**, 1909–18.

Verhulst, J. (1996) Atavisms in *Homo sapiens* – A Bolkian heterodoxy revisited. *Acta Biotheor.*, **44**, 59–73.

Vermeij, G.J. (1996) Animal origins. *Science*, **274**, 525–36.

Verraes, W. (1981) Theoretical discussion on some functional-morphological terms and some general reflections on explanations in biology. *Acta Biotheor.*, **30**, 255–73.

Verraes, W. (1989) A theoretical reflection on some crucial concepts in functional morphology. *Acta Morphol. Neerl.-Scand.*, **27**, 75–81.

Via, S. and Lande, R. (1985) Genotype-environment interaction and the evolution of phenotypic plasticity. *Evolution*, **39**, 505–22.

Via, S., Gomulkiewicz, R., de Jong, G. *et al.* (1995) Adaptive phenotypic plasticity: Consensus and controversy. *Trends Ecol. Evol.*, **10**, 212–17.

Vieille-Grosjean, I., Hunt, P, Gulisano, M. *et al.* (1997) Branchial HOX gene expression and human craniofacial development. *Devel. Biol.*, **183**, 49–60.

Vogl, C., Atchley, W.R. and Xu, S. (1994) The ontogeny of morphological differences in the mandible in two inbred strains of mice. *J. Craniofac. Genet. Devel. Biol.*, **14**, 97–110.

Vogl, C., and Rienesl, J. (1991) Testing for developmental constraints: carpal fusions in urodeles. *Evolution*, **45**, 1516–19.

Vollbrecht, E., Veit, B., Sinha, N. and Hake, S. (1991) The developmental gene *Knotted*-1 is a member of a maize homeobox gene family. *Nature*, **350**, 241–3.

Vorobyeva, E. and Hinchliffe, J.R. (1996) From fins to limbs: Developmental perspectives on paleontological and morphological evidence. *Evol. Biol.*, **29**, 263–311.

Vortkamp, A. (1997) Skeletal morphogenesis: Defining the skeletal elements. *Curr. Biol.*, **7**, R104–R107.

Vortkamp, A., Lee, K., Lanske, B. *et al* (1996) Regulation of rate of cartilage differentiation by Indian hedgehog and PTH-related protein. *Science*, **273**, 613–22.

Vrba, E.S. (1983a) Macroevolutionary trends: New perspectives on the role of adaptation and incidental effect. *Science*, **221**, 387–89.

Vrba, E.S. (1983b) Evolutionary pattern and process in the sister-group Alcelaphini-Aepycerotini (Mammalia: Bovidae), in *Living Fossils* (eds N. Eldredge and S.M. Stanley) Springer-Verlag, New York.

Vrba, E.S. (1984) Patterns in the fossil record and evolutionary processes, in *Beyond Neo-Darwinism. An introduction to the new evolutionary paradigm* (eds M.-W. Ho. and P.T. Saunders), Academic Press, London, pp. 115–42.

Vrba, E.S., Denton, G.H., Partridge, T.C. and Burckle, L.H. (1996) *Paleoclimate and Evolution, with Emphasis on Human Origins*, Yale University Press, New Haven, CT.

Vrba, E.S. and Eldredge, N. (1984) Individuals, hierarchies, and processes: Towards a more complete evolutionary theory. *Paleobiology*, **10**, 146–71.

Wada, H., Holland, P.W.H. and Satoh, N. (1996) Origin of patterning in neural tubes. *Nature*, **384**, 123.

Wada, H. and Satoh, N. (1994) Details of the evolutionary history from invertebrates, as deduced from the sequences of 18s rDNA. *Proc. Natl Acad. Sci. USA*, **91**, 1801–4.

Waddington, C.H. (1938) Regulation of amphibian gastrulae with added ectoderm. *J. Exp. Zool.*, **15**, 377–81.

Waddington, C.H. (1940) *Organizers and Genes*. Cambridge University Press, Cambridge.

Waddington, C.H. (1942) Canalization of development and the inheritance of acquired characters. *Nature*, **150**, 563–5.

Waddington, C.H. (1943) Polygenes and oligogenes. *Nature*, **151**, 394.

Waddington, C.H. (1953a) Genetic assimilation of an acquired character. *Evolution*, **7**, 118–26.

Waddington, C.H. (1953b) The "Baldwin effect", "genetic assimilation" and "homeostasis". *Evolution*, **7**, 386–7.

Waddington, C.H. (1956) Genetic assimilation of the *bithorax* phenotype. *Evolution*, **10**, 1–13.

Waddington, C.H. (1957a) *The Strategy of the Genes. A Discussion of some Aspects of Theoretical Biology.* George Allen & Unwin, London.

Waddington, C.H. (1957b) The Genetic basis of the assimilated *bithorax* stock. *J. Genetics*, **55**, 240–5.

Waddington, C.H. (1959) Canalization of development and genetic assimilation of acquired characters. *Nature*, **183**, 1654–5.

Waddington, C.H. (1961) Genetic assimilation. *Adv. Genetics*, **10**, 257–93.

Waddington, C.H. (1962a) *New Patterns in Genetics and Development.* Columbia University Press, New York.

Waddington, C.H. (1962b) *The Nature of Life.* Athenaeum, New York.

Waddington, C.H. (1975) *The Evolution of an Evolutionist.* Cornell University Press, Ithaca, NY.

Waddington, C.H., Graber, H. and Woolf, B. (1957) Iso-alleles and the response to selection. *J. Genetics*, **55**, 246–50.

Waddington, C.H. and Robertson, E. (1966) Selection for developmental canalisation. *Genet. Res. Camb.*, **7**, 303–12.

Waggoner, B.M. (1996) Phylogenetic hypotheses of the relationship of arthropods to Precambrian and Cambrian problematic fossil taxa. *Syst. Biol.*, **45**, 190–222.

Wagner, G.P. (1986) The systems approach: An interface between development and population genetic aspects of evolution, in *Patterns and Processes in the History of Life* (eds D.M. Raup and D. Jablonski), Dahlem Conference 1986, Springer-Verlag, Berlin, pp. 149–65.

Wagner, G.P. (1989a) The origin of morphological characters and the biological basis of homology. *Evolution*, **43**, 1157–71.

Wagner, G.P. (1989b) The biological homology concept. *Annu. Rev. Ecol. Syst.*, **20**, 51–69.

Wagner, G.P. (1995) The biological basis of homologues: A building block hypothesis. *N. Jb. Geol. Paläont. Abh.*, **195**, 279–88.

Wagner, G.P. (1996) Homologues, natural kinds and the evolution of modularity. *Amer. Zool.*, **36**, 36–43.

Wagner, G.P. and Altenberg, L. (1996) Complex adaptations and the evolution of evolvability. *Evolution*, **50**, 967–76.

Wagner, G.P., Booth, G. and Bagheri-Chaichian, H. (1997) A population genetic theory of canalization. *Evolution*, **51**, 329–47.

Wagner, G.P. and Misof, B.Y. (1993) How can a character be developmentally constrained despite variation in developmental pathway? *J. Evol. Biol.*, **6**, 449–55.

Wainwright, S.A. (1988) *Axis and Circumference. The Cylindrical Shape of Plants and Animals.* Harvard University Press, Cambridge, MA.

Wainwright, S.A., Briggs, W.D., Currey, J.D. and Gosline, J.M. (1976) *Mechanical Design in Organisms.* Edward Arnold, London.

Wake, D.B. (1980) Evidence of heterochronic evolution: A nasal bone in the Olympic Salamander, *Rhyacotriton olympicus. J. Herpet.*, **14**, 292–5.

Wake, D.B. (1986) Foreword, in *Factors of Evolution: The Theory of Stabilizing Selection* by I.I. Schmalhausen, The University of Chicago Press, Chicago, IL, pp. v–xii.

Wake, D.B. (1989) Phylogenetic implications of ontogenetic data, in *Ontogenèse et Evolution* (eds B. David, J.L. Dommergues, J. Chaline and B. Laurin), Geobios, mém. spécial no 12, pp. 369–78.

Wake, D.B. (1991) Homoplasy: The result of natural selection, or evidence of design limitations? *Amer. Nat.*, **138**, 543–67.

Wake, D.B. (1996a) Schmalhausen's evolutionary morphology and its value in formulating research strategies. *Mem. Soc. Ital. Sci. Nat. Mus. Civ. Storia Nat. Milano*, **28**, 129–32.

Wake, D.B. (1996b) Introduction, in *Homoplasy: The Recurrence of Similarity in Evolution* (eds M.J. Sanderson and L. Hufford), Academic Press, San Diego, CA, pp. xvii–xxv.

Wake, D.B. (1996c) Evolutionary developmental biology – Prospects for an evolutionary synthesis at the developmental level. *Mem. Cal. Acad. Sciences*, **20**, 97–107.

Wake, D.B. (1996d) Incipient species formation in salamanders of the Ensatina complex. *Proc. Natl Acad. Sci. USA.*, **94**, 7761–7.

Wake, D.B. and Hanken, J. (1996) Direct development in the lungless salamanders: What are the consequences for developmental biology, evolution and phylogenesis. *Int. J. Devel. Biol.*, **40**, 859–69.

Wake, D.B. and Larson, A. (1987) Multidimensional analysis of an evolving lineage. *Science*, **238**, 42–8.

Wake, D.B. and Roth, G. (1989) *Complex Organismal Functions: Integration and Evolution in Vertebrates.* Report of the Dahlem Workshop on Complex Organismal Functions: Integration and Evolution in Vertebrate, . Berlin 1988, August 28–September 2, John Wiley & Sons, Chichester.

Wake, M.H. (ed.) (1979) *Hyman's Comparative Vertebrate Anatomy* 3rd edn, The University of Chicago Press, Chicago, IL.

Wake, M.H. (1986) The morphology of *Idiocranium russeli* (Amphibia: Gymnophiona), with comments on miniaturization through heterochrony. *J. Morph.*, **189**, 1–16.

Wake, M.H. (1989) Phylogenesis of direct development and viviparity in vertebrates, in *Complex Organismal Functions: Integration and Evolution in Vertebrates* (eds D.B. Wake and G. Roth), Wiley, New York, pp. 235–50.

Wake, M.H. (1990) The evolution of integration of biological systems: An evolutionary perspective through studies of cells, tissues, and organs. *Amer. Zool.*, **30**, 897–906.

Wake, M.H. (1992a) Evolutionary scenarios, homology and convergence of structural specializations for vertebrate viviparity. *Amer. Zool.*, **32**, 256–63.

Wake, M.H. (1992b) Morphology, the study of form and function, in modern evolutionary biology, in *Oxford Surveys in Evolutionary Biology* (eds D. Futuyma and J. Antonovics), Oxford University Press, New York, pp. 289–346.

Walcott, C.D. (1892) Preliminary note on the discovery of a vertebrate fauna in Silurian (Ordovician) strata. *Bull. Geol. Soc. Amer.*, **3**, 153–72.

Walcott, C.D. (1911a) Middle Cambrian holothurians and medusae. *Smithson. Misc. Collect.*, **57**, 41–68.

Walcott, C.D. (1911b) Middle Cambrian annelids. *Smithson. Misc. Collect.*, **57**, 109–44.

Walcott, C.D. (1931) Addenda to descriptions of Burgess Shale fossils (with explanatory notes by Charles E. Resser). *Smithson. Misc. Collect.*, **85**, 1–46.

Walker, J.A. (1997) Ecological morphology of lacustrine threespine stickleback *Gasterosteus aculeatus* L. (Gasterosteidae) body shape. *Biol. J. Linn. Soc.*, **61**, 3–50.

Walker, M.H. (1995) Relatively recent evolution of an unusual pattern of early embryonic development (long germ band?) in a South African onychophoran, *Opisthopatus cinctipes* Purcell (Onychophora: Peripatopsidae). *Zool. J. Linn. Soc.*, **114**, 61–75.

Wallis, G.A. (1996) Bone growth: coordinating chondrocyte differentiation. *Curr. Biol.*, **5**, 1577–80.

Walsh, J.B. (1995) How often do duplicated genes evolve a new function? *Genetics*, **139**, 421–8.

Walsh, J.B. (1996) The emperor's new genes. *Evolution*, **50**, 2115–18.

Warren, R.W. and Carroll, S.B. (1995) Homeotic genes and diversification of the insect body plan. *Curr. Biol.*, **5**, 459–65.

Warren, R.W., Nagy, L., Selegue, J. *et al.* (1994) Evolution of homeotic gene regulation and function in flies and butterflies. *Nature*, **372**, 458–61.

Watanabe, Y. and Le Douarin, N.M. (1996) A role for BMP-4 in the development of subcutaneous cartilage. *Mech. Devel.*, **57**, 69–78.

Waterman, A.J., Frye. B.E., Johansen, K. *et al.* (1971) *Chordate Structure and Function*, The Macmillan Co., New York.

Wayne, R.K. (1986a) Limb morphology of domestic and wild canids: The influence of development on morphologic change. *J. Morph*, **187**, 301–19.

Wayne, R.K. (1986b) Cranial morphology of domestic and wild canids: The influence of development on morphological change. *Evolution*, **40**, 243–61.

Wayne, R.K. and Ruff, C.B. (1993) Domestication and bone growth, in *Bone. Volume 7: Bone Growth-B* (ed. B.K. Hall), CRC Press, Boca Raton, FL, pp. 105–32.

Webb, J.F. (1990) Ontogeny and phylogeny of the trunk lateral line system in cichlid fishes. *J. Zool. Lond.*, **221**, 405–18.

Webb, J.F. and Noden, D.M. (1993) Ectodermal placodes: Contributions to the development of the vertebrate head. *Amer. Zool*, **33**, 434–47.

Webster, G. (1984) The relations of natural forms, in *Beyond Neo-Darwinism. An introduction to the new evolutionary paradigm* (eds M.-W. Ho. and P.T. Saunders), Academic Press, London, pp. 193–217.

Webster, G. and Goodwin, B.C. (1981) History and structure in biology. *Persp. Biol. Med.*, **25**, 39–62.

Webster, G. and Goodwin, B.C. (1982) The origin of species: A structuralist approach. *J. Soc. Biol. Struct.*, **5**, 15–47.

Weigel, D. and Meyerowitz, E.M. (1994) The ABCs of floral homeotic genes. *Cell*, **78**, 203–9.

Weisblat, D.A., Wedeen, C.J. and Kostriken, R.G. (1994) Evolution of developmental mechanisms: Spatial and temporal modes of rostrocaudal patterning, *Curr. Top. Devel. Biol*, **29**, 101–34.

Weismann, A. (1883a) *Die Entstehung des Sexualzellen bei den Hydromedusen. Zugleich ein Beitrag zur Kenntniss des Baues und der Lebenserscheinungen dieser Gruppe*. Gustav Fischer, Jena.

Weismann, A. (1883b) *Über die Vererbung*, Gustav Fischer, Jena.

Weismann, A. (1885) *Die Kontinuität des Keimplasmas als Grundlage einer Theorie der Vererbung*. Gustav Fischer, Jena. (Published in 1893 as *The Germ Plasm: A Theory of Heredity*, translated by W. Newton Parker and H. Ronnfeld, Walter Scott Ltd., London.

Weismann, A. (1889) *Essays upon Heredity and Kindred Biological Problems*, Clarendon Press, Oxford. (Reprinted with an introduction by J.A. Mazzeus, 1977, Dabor Science Publications, Oceanside, NY.)

Weiss, K.M., Bollekens, J., Ruddle, F.H. and Takashita, K. (1994) *Distalless* and other homeobox genes in the development of the dentition. *J. Exp. Zool.*, **270**, 273–84.

Weiss, K.M., Ruddle, F.H. and Bollekens, J. (1995) Dlx and other homeobox genes in the morphological development of the dentition. *Conn. Tiss. Res.*, **32**, 35–40.

West-Eberhard, M.J. (1986) Alternative adaptations, speciation, and phylogeny. *Proc. Natl Acad. Sci. USA*, **83**, 1388–92.

West-Eberhard, M.J. (1992) Adaptation: Current Usages, in *Keywords in Evolutionary Biology* (eds E. Fox Keller and E.A. Lloyd), Harvard University Press, Cambridge, MA, pp. 13–18.

Westerfield, M. (1995) *The Zebrafish Book: A Guide for the Laboratory use of Zebrafish* (Danio rerio), University of Oregon Press, Eugene, OR.

Weston, J.A. (1963) A radioautographic analysis of the migration and localization of trunk neural crest cells in the chick. *Devel. Biol.*, **6**, 279–310.

Wheeler, D.E. and Nijhout, H.F. (1983) Soldier determination in ants: New role for juvenile hormone. *Science*, **213**, 361–3.

Wheeler, D.E. and Nijhout, H.F. (1986) Soldier determination in the ant *Pheidole bicarinata*: Inhibition by adult soldiers. *J. Insect Physiol.*, **30**, 127–35.

Whiteman, H.H. (1994) Evolution of facultative paedomorphosis in salamanders. *Quart. Rev. Biol.*, **69**, 205–21.

Whitlock, M. (1996) The heritability of fluctuating asymmetry and the genetic control of developmental stability. *Proc. R. Soc. Lond. B*, **263**, 849–54.

Whitman, C.O. (1894) Evolution and epigenesis. *Biol. Lect. Mar. Biol. Lab. Woods Hole*, **1894**, 205–24.

Whitman, C.O. (1919) Posthumous works of Charles Otis Whitman (ed. H.A. Carr): I. Orthogenetic evolution in Pigeons, Carnegie Institute, Washington, DC.

Whitrow, G.J. (1990) *The Natural Philosophy of Time* 2nd edn, Clarendon Press, Oxford.

Whittaker, R.H. (1959) On the broad classification of organisms. *Quart. Rev. Biol.*, **34**, 210–26.

Whittington, H.B. (1985) *The Burgess Shale*. Yale University Press, New Haven, CT.

Whyte, L.L. (1965) *Internal Factors in Evolution*. Tavistock Publications, London.

Wilkens, H. (1988) Evolution and genetics of epigean and cave *Astyanax fasciatus* (Characidae, Pisces). *Evol. Biol.*, **23**, 271–367.

Wilkins, A.S. (1997) Canalization: A molecular genetic perspective. *BioEssays*, **19**, 257–62.

Wilkinson, D.G. (1990) Segmental gene expression in the developing mouse hindbrain. *Sem. Devel. Biol.*, **1**, 127–34.

Wilkinson, D.G. and Krumlauf, R. (1990) Molecular approaches to the segmentation of the hindbrain. *Trends Neurosci.*, **13**, 335–9.

Willey, A. (1894) *Amphioxus and the Ancestry of the Vertebrates*. Macmillan, New York and London.

Williams, D.M., Scotland, R.W. and Blackmore, S. (1990) Is there a direct ontogenetic criterion in systematics? *Biol. J. Linn. Soc.*, **39**, 99–108.

Williams, G.C. (1992) *Natural Selection: Domains, Levels and Challenges*. Oxford University Press, Oxford.

Williams, N.A. and Holland, P.W.H. (1996) Old head on young shoulders. *Nature*, **383**, 490.

Williams, T.A. and Müller, G.B. (1996) Limb development in a primitive crustacean, *Triops longicaudatus*: Subdivision of the early limb bud gives rise to multibranched limbs. *Devel. Genes Evol.*, **206**, 161–8.

Williams, T.A. and Nagy, L.M. (1995) Brine shrimp add salt to the stew. *Curr. Biol.*, **5**, 1330–3.

Willmer, E.N. (1960) *Cytology and Evolution*. Academic Press, New York.

Willmer, P.G. (1990) *Invertebrate Relationships: Patterns in animal evolution*. Cambridge University Press, Cambridge.

Willmer, P.G. and Holland, P.W.H. (1991) Modern approaches to metazoan relationships. *J. Zool. Lond.*, **224**, 689–94.

Wilson, E.B. (1891) Some problems in annelid morphology. *Biol. Lect. MBL, Woods Hole, 1890*, 53–78.

Wilson, E.B. (1892) The cell lineage of *Nereis*. A contribution to the cytology of the annelid body. *J. Morph.*, **6**, 361–480.

Wilson, E.B. (1894) The embryological criterion of homology. *Biol. Lect. MBL, Woods Hole, 1894*, 101–24.

Wilson, E.B. (1925) *The Cell in Development and Heredity*. 3rd edn, Macmillan, New York.

Wilson, E.O. (1971) *The Insect Societies*, Harvard University Press, Cambridge, MA.

Wilson, E.O. (1981) Epigenesis and the evolution of social systems. *J. Hered.*, **72**, 70–7.

Wilson, P.A. and Hemmati-Brivanlou, A. (1995) Induction of epidermis and inhibition of neural fate by Bmp-4. *Nature*, **376**, 331–3.

Wimsatt, W.C. (1986) Developmental constraints, generative entrenchment, and the innate-acquired distinction, in *Integrating Scientific Disciplines* (ed. W. Bechtel), Martinus Nijhoff, Dordrecht, Holland, pp. 185–208.

Winograd, J., Reilly, M.P., Roe, R. *et al* (1997) Perinatal lethality and multiple craniofacial malformations in MSX2 transgenic mice. *Human Molec. Genet.*, **6**, 369–79.

Withers, R.F.J. (1964) Morphological correspondence and the concept of homology, in *Form and Strategy in Science. Studies dedicated to Joseph Henry Woodger on the Occasion of his Seventieth Birthday* (eds J.R. Gregg and F.T.C. Harris), D. Reidel Publishing Co., Dordrecht-Holland, pp. 378–94.

Witte, F., Barel, C.D.N. and Hoogerhoud, R.J.C. (1990) Phenotypic plasticity of anatomical structures and its ecomorphological significance. *Neth. J. Zool.*, **40**, 278–98.

Woellwarth, C.V. (1961) Die rolle des Neuralleistenmaterials und der Temperatur bei der Determination der Augenlinse. *Embyologia*, **6**, 219–42.

Woese, C.R. (1981) Archaebacteria. *Sci. Amer.*, **244**, 92–122,

Woese, C.R., Kandler, O. and Whellis, M.L. (1990) Toward a natural system of organisms: Proposal for the domains Archaea, Bacteria and Eucarya. *Proc. Natl Acad. Sci. USA.*, **87**, 4576–9.

Wolff, C.F. (1759) *Theoria Generationis*. Halae ad Salam.

Wolff, C.F. (1764) *Theorie von der Generation in zwo Abhandlungen*. Friedrich Wilhelm Birnstiel, Berlin.

Wolff, C.F. (1767) *De formatione intestinorum praecipue*. Novi Commentarii Academine Scientarum Imperalis Petropolitane.

Wolpert, L. (1981) Positional information and pattern formation. *Phil. Trans. R. Soc. Lond.*, **295**, 441–50.

Wolpert, L. (1990) The evolution of development. *Biol. J. Linn. Soc.*, **39**, 109–24.

Wolpert, L. (1992) Gastrulation and the evolution of development. *Development*, 1992, Suppl., 7–13.

Wolpert, L. (1994) The evolutionary origin of development: Cycles, patterning, privilege and continuity. *Development* 1994, Suppl., 79–84.

Woltereck, R. (1909) Weiterer experimentelle Untersuchungen über Artveränderung, Speziell über das Wessen Quantitätiver Arunterschiede bei *Daphniden*. *Vers. Deutsch Zool. Gesell.*, **19**, 110–72.

Wong, M. and Carter, D.R. (1990) A theoretical model of endochondral ossification and bone architectural construction in long bone ontogeny. *Anat. Embryol.*, **181**, 523–32.

Wonsettler, A.L., and Webb, J.F. (1997) Morphology and development of the multiple lateral line canals on the trunk in two species of *Hexagrammos* (Scorpaeniformes, Hexagrammidae). *J. Morph.*, **233**, 195–214.

Woo, K. and Fraser, S.E. (1997) Specification of the zebrafish nervous system by nonaxial signals. *Science*, **277**, 254–7.

Wood, S.W. (1995) The first use of the terms "homology" and "analogy" in the writings of Richard Owen. *Archs Nat. Hist.*, **22**, 255–9.

Wood, S.W. (1997) Richard Owen and the politics of homology. *Archs Nat. Hist.*, (in press).

Woodger, J.H. (1929) *Biological Principles. A Critical Study.* K. Paul, Trench, Trubner & Co., London. (Reprinted in 1967 with a new introduction by Woodger, by Routledge & K. Paul, London and Humanities Press, New York.)

Woodger, J.H. (1937) *The Axiomatic Method in Biology.* Cambridge University Press, Cambridge.

Woodger, J.H. (1945) On biological transformations, in *Essays on Growth and Form Presented to D'Arcy Wentworth Thompson* (eds W.E. Le Gros Clark and P.B. Medawar), Cambridge University Press, Cambridge, pp. 95–120.

Woodger, J.H. (1952) *Biology as Language, an introduction to the methodology of the biological sciences, including medicine.* The Tanner Lectures 1949–50. Cambridge University Press, Cambridge.

Wray, G.A. (1992) The evolution of larval morphology during the post-Paleozoic radiation of echinoids. *Paleobiology*, **18**, 258–87.

Wray, G.A. (1994) The evolution of cell lineage in echinoderms. *Amer. Zool.*, **34**, 353–63.

Wray, G.A. and Bely, A.E. (1994) The evolution of echinoderm development is driven by several distinct factors. *Development*, 1994 Suppl., 97–106.

Wray, G.A., Levinton, J.S. and Shapiro, L.H. (1996) Molecular evidence for deep Precambrian divergences among metazoan phyla. *Science*, **274**, 568–73.

Wray, G.A. and McClay, D.R. (1988) The origin of spicule-forming cells in a 'primitive' sea urchin (*Eucidaris tribuloides*) which appears to lack primary mesenchymal cells. *Development*, **103**, 305–15.

Wray, G.A. and Raff, R.A. (1989) Evolutionary modification of cell lineage in the direct-developing sea urchin, *Heliocidaris erythrogramma. Devel. Biol.*, **132**, 458–70.

Wray, G.A. and Raff, R.A. (1990a) Novel origins of lineage founder cells in the direct-developing sea urchin *Heliocidaris erythrogramma. Devel. Biol.*, **141**, 41–54.

Wray, G.A. and Raff, R.A. (1990b) Pattern and process heterochronies in the early development of the sea urchin. *Sem. Devel. Biol.*, **1**, 249–52.

Wray, G.A. and Raff, R.A. (1991) Rapid evolution of gastrulation mechanisms in a sea urchin with lecithotrophic larvae. *Evolution*, **45**, 1741–50.

Wright, S. (1934) The method of path coefficients. *J. Agric. Res.*, **20**, 557–85.

Wright, S. (1967) Comments on the preliminary working papers of Eden and Waddington, in *Mathematical Challenges to the neo-Darwinian Interpretation of Evolution* (eds P.S. Moorhead and M.M. Kaplan), Wistar Institute Symposium Monograph No. 5, Wistar Institute Press, Philadelphia, PA, pp. 117–20.

Wright, S. (1968) *Evolution and the Genetics of Populations.* Volume 1, The University of Chicago Press, Chicago, IL.

Xu, R.H., Jaebong, K., Taira, M. *et al.* (1995) A dominant negative bone morphogenetic protein 4 receptor causes neuralization of *Xenopus* ectoderm. *Biochem. Biophys. Res. Comm.*, **212**, 212–19.

Xue, Z.-G., Gehring, W.J. and Le Douarin, N.M. (1991) *Quox-1*, a quail homeobox gene expressed in the embryonic central nervous system, including the forebrain. *Proc. Natl Acad. Sci. USA*, **88**, 2427–31.

Yablokov, A.V. (1974) *Variability of Mammals*, Amerind Publishing Co., New Delhi.

Yamada, T. (1990) Regulations in the induction of the organized neural system in amphibian embryos. *Development*, **110**, 653–9.

Yamamoto, T. (1967) Medaka, in *Methods in Developmental Biology* (eds F.H. Wilt and N.K. Wessells), Thomas Y. Crowell Co., New York, pp. 101–11.

Yi, Q. (1977) *Hyolitha* and some problematica from the lower Cambrian Meishucun stage in Central and S. W. China. *Acta Palaeont. Sinica*, **16**, 255–75.

Yntema, C.L. (1955) Ear and nose, in *Analysis of Development* (eds B.H. Willier, P.A. Weiss, and V. Hamburger), W.B. Saunders, Philadelphia, PA, pp. 415–28.

Yochelson, E.L. (1977) Agmata, a proposed extinct phylum of early Cambrian age. *J. Paleontol.*, **51**, 437–54.

Yokouchi, Y., Nakazato, S., Yamamoto, M. *et al.* (1995) Misexpression of *Hoxa*-13 induces cartilage homeotic transformation and changes cell adhesiveness in chick limb buds. *Genes Devel.*, **9**, 2509–22.

Yokouchi, Y., Sakiyama, J.-i., Kameda, T. *et al.* (1996) BMP-2/-4 mediate programmed cell death in chick limb buds. *Development*, **122**, 3725–34.

Young, C.M., Vázquez, E., Metaxas, A. and Tyler, P.A. (1996) Embryology of vestimentiferan tube worms from deep-sea methane/sulphide seeps. *Nature*, **381**, 514–16.

Young, G.C., Karatajute-Talimaa, V.N. and Smith, M.M. (1996) A possible Late Cambrian vertebrate from Australia. *Nature*, **383**, 810–12.

Young, J.Z., (1958) *The Life of vertebrates*, Clarendon Press, Oxford.

Yoxen, E. (1986) Form and strategy in biology: Reflections on the career of C. H. Waddington, in *A History of Embryology* (eds T.J. Horder, J.A. Witkowski and C.C. Wylie), Cambridge University Press, Cambridge, pp. 309–29.

Zákány, J. and Duboule, D. (1996) Synpolydactyly in mice with a targeted deficiency in the *HoxD* complex. *Nature*, **384**, 69–71.

Zardoya, R., Abouheif, E. and Meyer, A. (1996a) Evolutionary analyses of *hedgehog* and *Hoxd-10* genes in fish species closely related to the zebrafish. *Proc. Natl Acad. Sci. USA*, **93**, 13036–41.

Zardoya, R., Abouheif, E. and Meyer, A. (1996b) Evolution and orthology of hedgehog genes. *Trends Genet.*, **12**, 496–7

Zardoya, R. and Meyer, A. (1996a) Evolutionary relationships of the coelacanth, lungfishes, and tetrapods based on the 28S ribosomal RNA gene. *Proc. Natl Acad. Sci. USA*, **93**, 5449–54.

Zardoya, R. and Meyer, A. (1996b) The complete nucleotide sequence of the mitochondrial genome of the lungfish (*Protopterus dolloi*) supports its phylogenetic position as a close relative of land vertebrates. *Genetics*, **142**, 1249–63.

Zelditch, M.L., Bookstein, F.L. and Lundrigan, B.L. (1992) Ontogeny of integrated skull growth in the cotton rat, *Sigmodon fulviventer. Evolution*, **46**, 1164–80.

Zelditch, M.L., Bookstein, F.L. and Lundrigan, B.L. (1993) The ontogenetic complexity of developmental constraints. *J. Evol. Biol.*, **6**, 621–41.

Zelditch, M.L. and Fink, W.L. (1995) Allometry and developmental integration of body growth in a piranha, *Pygocentrus nattereri* (Teleostei: Ostariophysi). *J. Morph.*, **223**, 341–55.

Zelditch, M.L. and Fink, W.L. (1996) Heterochrony and heterotopy: Stability and innovation in the evolution of form. *Paleobiology*, **22**, 241–54.

Zeller, R., and Duboule, D. (1997) Dorsoventral limb polarity and origin of the ridge: on the fringe of independence. *BioEssays*, **19**, 541–6.

Zhang, H. and Bradley, A. (1996) Mice deficient for BMP2 are nonviable and have defects in amnion/chorion and cardiac development. *Development*, **122**, 2977–86.

Zhang, X.-g. and Pratt, B.R. (1994) Middle Cambrian arthropod embryos with blastomeres. *Science*, **266**, 637–9.

Zouros, E., Ball, A.O., Saavedra, C. and Freeman, K.R. (1994) An unusual type of mitochondrial DNA inheritance in the blue mussel *Mytilus. Proc. Natl Acad. Sci. USA.*, **91**, 7463–7.

Zuckerkandl, E. (1994) Molecular pathways to parallel evolution: I. Gene nexuses and their morphological correlates. *J. Mol. Evol.*, **39**, 661–78.

Zusi, R.L. and Jehl, J.R.Jr. (1970) The systematic relationships of *Aechmorhynchus, Prosobonia* and *Phegornis* (Charadriformes: Charadrii). *Auk*, **87**, 760–80.

Zweers, G. (1991) Path and space for evolution of feeding mechanisms in birds, in *The Unity of Evolutionary Biology* (ed. E.C. Dudley), Dioscorides Press, Portland, OR, pp. 530–47.

Zweers, G.A. (1992) Behavioural mechanisms of avian drinking. *Neth. J. Zool.*, **42**, 60–84.

Zweers, G.A. and Gerritsen, A.F.C. (1997) Transitions from pecking to probing mechanisms in waders. *Neth. J. Zool.*, **47**, 161–208.

Zweers, G.A. and Vanden Berge, J.C. (1997) Evolutionary transitions in the trophic system of the wader-waterfowl complex. *Neth. J. Zool.*, **47**, 255–87.

INDEX

References to figures are in bold; references to tables are in italics. Species and gene names are also italicized.